Convective Heat Transfer
in Porous Media

Energy Systems: From Design to Management

Series Editor
Vincenzo Bianco
Università di Genova, Italy

Analysis of Energy Systems: Management, Planning and Policy
Vincenzo Bianco

Solar Cooling Technologies
Sotirios Karellas, Tryfon C Roumpedakis, Nikolaos Tzouganatos, Konstantinos Braimakis

Biomass in Small-Scale Energy Applications: Theory and Practice
Mateusz Szubel, Mariusz Filipowicz

Convective Heat Transfer in Porous Media
Edited by Yasser Mahmoudi, Kamel Hooman, Kambiz Vafai

For more information about this series, please visit:
https://www.crcpress.com/Energy-Systems/book-series/CRCENESYSDESMAN

Convective Heat Transfer in Porous Media

Edited by
Yasser Mahmoudi
Kamel Hooman
Kambiz Vafai

CRC Press is an imprint of the
Taylor & Francis Group, an **informa** business

CRC Press
Taylor & Francis Group
6000 Broken Sound Parkway NW, Suite 300
Boca Raton, FL 33487-2742

© 2020 by Taylor & Francis Group, LLC
CRC Press is an imprint of Taylor & Francis Group, an Informa business

No claim to original U.S. Government works

Printed on acid-free paper

International Standard Book Number-13: 978-0-367-03080-3 (Hardback)

This book contains information obtained from authentic and highly regarded sources. Reasonable efforts have been made to publish reliable data and information, but the author and publisher cannot assume responsibility for the validity of all materials or the consequences of their use. The authors and publishers have attempted to trace the copyright holders of all material reproduced in this publication and apologize to copyright holders if permission to publish in this form has not been obtained. If any copyright material has not been acknowledged please write and let us know so we may rectify in any future reprint.

Except as permitted under U.S. Copyright Law, no part of this book may be reprinted, reproduced, transmitted, or utilized in any form by any electronic, mechanical, or other means, now known or hereafter invented, including photocopying, micro-filming, and recording, or in any information storage or retrieval system, without written permission from the publishers.

For permission to photocopy or use material electronically from this work, please access www.copyright.com (http://www.copyright.com/) or contact the Copyright Clearance Center, Inc. (CCC), 222 Rosewood Drive, Danvers, MA 01923, 978-750-8400. CCC is a not-for-profit organization that provides licenses and registration for a variety of users. For organizations that have been granted a photocopy license by the CCC, a separate system of payment has been arranged.

Trademark Notice: Product or corporate names may be trademarks or registered trademarks, and are used only for identification and explanation without intent to infringe.

Visit the Taylor & Francis Web site at
http://www.taylorandfrancis.com

and the CRC Press Web site at
http://www.crcpress.com

Contents

Preface vii
Editors xi
Contributors xiii

Section I Fundamentals of Convection in Porous Media

1. **Introduction to Fluid Flow and Heat Transfer in Porous Media** 3
   Meysam Nazari, Yasser Mahmoudi, and Kamel Hooman

2. **Natural Convection in Porous Media** 19
   Kamel Hooman and Donald A. Nield

3. **Forced Convection in Porous Media** 37
   Pourya Forooghi and Benjamin Dietrich

Section II Advanced Topics of Convection in Porous Media

4. **Convective Heat Transfer of Nanofluids in Porous Media** 55
   Bernardo Buonomo, Davide Ercole, Yasser Mahmoudi, Oronzio Manca, and Sergio Nardini

5. **Pore-Network Simulation of Drying of Heterogeneous and Stratified Porous Media** 87
   Hassan Dashtian, Nima Shokri, and Muhammad Sahimi

6. **Wicking of Liquids under Non-Isothermal and Reactive Conditions: Some Industrial Applications** 103
   Mohammad Amin Faghihi Zarandi, and Krishna M. Pillai

7. **Thermal Effect on Capillary Imbition in Porous Media** 125
   Jianchao Cai, Wei Wei, and Yasser Mahmoudi

8. **Convection in Bi-Disperse Porous Media** 137
   Arunn Narasimhan

9. **Pore Scale Analysis in Forced Convection Heat Transfer in Porous Media** 153
   Hasan Celik, Moghtada Mobedi, and Akira Nakayama

10. **Lattice Boltzmann Method for Modeling Convective Heat Transfer in Porous Media** 173
    Gholamreza Imani and Kamel Hooman

v

Section III   Advanced Engineering Applications of Convection in Porous Media

11. Modeling Thermohydraulic Process in Enhanced Geothermal System Based on Two-Equation Thermal Model for Porous Media 203
Wenbo Huang, Wenjiong Cao, Guoling Wei, Yunlong Jin, and Fangming Jiang

12. Mixed Convection and Radiation Heat Transfer in Porous Media for Solar Thermal Applications 227
Simone Silvestri and Dirk Roekaerts

13. Transpiration Cooling Using Porous Material for Hypersonic Applications 263
Adriano Cerminara, Ralf Deiterding, and Neil D. Sandham

14. Thermal Management and Heat Transfer Enhancement Using Porous Materials 287
Cong Qi, Kuo Huang, Jiaan Liu, Guohua Wang, and Yuying Yan

15. Metal Foam Heat Exchangers 309
Simone Mancin

16. Heat and Fluid Flow in Porous Media for Polymer-Electrolyte Fuel Cells 341
Prodip Kumar Das and Deepashree Thumbarathy

17. Combustion in Porous Media for Porous Burner Application 361
Muhammad Abdul Mujeebu

Index 377

Preface

Fundamental and applied research in flow and convective heat transfer in porous media has received increased attention among researchers in academia and industry during the past several decades. This is due to the importance of this research field in a wide range of engineering applications, which either involves a porous material or can be modeled as porous media. These include geographical application (i.e., enhanced geothermal system and carbon storage), biological systems, solar thermal systems, metal foam heat exchangers, porous burners, transpiration cooling for aerospace systems, thermal management for electronic devices, and polymer electrolyte fuel cells (PEFCs). Other examples of applications include drying technology, catalytic reactors, tissue replacement, drug delivery, advanced medical imaging, and porous scaffolds for tissue engineering.

This wide range of ironic applications encouraged us to work and research in this field for years, through which we learned loads of information about convective heat transfer in porous materials. After thorough research in this field, we found that there is an abundance of mathematical, numerical, and experimental methods and approaches performed in the field of convection in porous media, and they have been included by the existing books and publications in this matter. Nonetheless, certain new fundamental findings in convection in porous media (e.g., heat flux bifurcation in porous media), advanced engineering applications (e.g., fuel cells), and new numerical approaches (e.g., lattice Boltzmann method) have not been yet included in the existing books. So that, this book tries to present and discuss these new aspects of the convective heat transfer in porous media with the most concentration on the practical approaches and their advanced applications. Despite the fact that we have made a thorough effort to cover the most significant newfound and ironic methods, challenges, and applications of convection in porous materials, some aspects might have been missed by authors. We hope that the book provides readers (students, professors, scientists, and engineers) with practical approaches and applications and the most fruitful information in the field of convective heat transfer in porous materials.

Overall, to be well-organized, the proposed book is supposed to be comprised of 3 sections and 17 chapters. Section I is devoted to the fundamentals of convection (natural and forced) in porous media. Section II is allocated to advanced topics of convection in porous media, in which advances in wicking and drying in porous media, convection in bi-disperse porous media, pore-scale analysis, and the lattice Boltzmann method will be discussed. Section III is devoted to the most recent and interesting applications of convection in porous media. Hence, in this section, the newfound industrial applications are presented.

Chapter 1 is devoted to an introduction to fundamentals in porous media. Porous media will be introduced, and different types of porous media (e.g., foams and pebble beds) and their applications will be discussed. The chapter further covers main approaches on modeling fluid flow and heat transfer in porous media, including local thermal equilibrium (LTE) and local thermal non-equilibrium approaches. The chapter then covers the challenges in utilizing LTNE models when solving the interface between a porous medium and a fluid layer or a porous medium and an impermeable wall subjected to a heat flux. Then, some important physics of convection in porous media known as heat flux splitting and bifurcation in porous media are presented and discussed.

Chapter 2 presents free convection of incompressible Newtonian fluids in porous media. This chapter further presents scale analysis, intersection of asymptotes, and thermal and pore resistor networks as approaches to solve complex problems to a good accuracy at the pen-and-paper level. Practical problems associated with internal heat generation with and without moisture transfer, natural draft dry cooling towers, and free convection in porous-matrix heat exchangers will be investigated using the above techniques applied to either or both of the micro- and macrodomains of a porous medium.

Chapter 3, known as forced convection in porous media, provides the reader with an overview of the problem of forced convection in porous media focusing on heat transfer in ducts fully or partially filled with porous media. The importance of accurate determination of effective thermal conductivity and

vii

solid-to-fluid heat transfer coefficient, in the local thermal equilibrium (LTE) and local thermal non-equilibrium (LTNE) frameworks, respectively, are highlighted, and recent experimental results on their calculation in ceramic open-celled foams are presented.

Chapter 4 discusses convective heat transfer of nanofluids in porous media. It provides a review of the recent researches on natural, forced, and mixed convection of nanofluids in porous media. The equations governing the flow and heat transfer of nanofluids in porous media are provided. The chapter further discusses the effect of the system's pertinent parameters, such as permeability and thermal conductivity on the enhancement of heat transfer in confined areas using nanofluids combined with porous materials.

Chapter 5 is devoted to drying in porous materials and covers fundamentals, theory, and advanced models to describe drying of porous media. In particular, the chapter presents pore-network simulation of drying of a porous medium. Using large 3D pore networks, the effects of the correlations between the sizes of the pores, as well as anisotropy induced by stratification on drying of porous media are discussed in this chapter.

Chapter 6 covers wicking of liquids under non-isothermal and reactive conditions. The chapter presents four different applications of wicking of liquids into porous media that are coupled with heat transfer, sometimes under reactive conditions. In all the applications, the effect of heat transfer on the imbibition process is analyzed. The emphasis is on presenting the modeling approaches in terms of important governing equations along with the necessary constitutive relations and equations of state. This is done in order to highlight the commonality of employing similar/identical two-phase flow physics under reactive, non-isothermal conditions in these diverse models.

Chapter 7 presents the thermal effect on spontaneous imbibition of wetting liquid in porous media through theoretical, numerical, and experimental methods. First, for imbibition with no thermal effect, the basic equations for capillary rise in a single capillary and spontaneous imbibition in porous media are introduced, and the influence factors on imbibition are given from the fractal theory. Second, the temperature effect on liquid viscosity and contact angle is discussed from experimental result and empirical corrections. Last, the imbibition properties in water-wet, oil-wet, and mixed-wet reservoirs and in cellulosic materials are presented.

Chapter 8 provides a discussion on the research literature on convection in bi-disperse porous media as well as convection research in tri-disperse porous media. The chapter details forced and free convection in bi-disperse porous media under local thermal equilibrium and local thermal non-equilibrium conditions. The chapter further discusses conjugate and mixed-forced convection and change of phase on the thermal characteristics of the bi-disperse porous media.

Chapter 9 introduces pore-scale analysis of flow and heat transfer in both periodic and stochastic porous media. It first discusses pros and cons of the volume-averaged approach in studying fluid flow and heat transfer in porous media. Then it details recent developments in the field of image processing and X-ray microcomputed tomography for reproducing three-dimensional structures of realistic porous media. Then and pore-scale analysis of realistic porous media systems will be presented, which enables researchers to observe all details of flow and heat, even in the smallest pore of in porous media.

Chapter 10 discusses the lattice Boltzmann method (LBM) as an alternative numerical tool for pore and macroscale convection heat transfer modeling in porous media. First, the derivation of the standard LBM from the Boltzmann transport equation (BTE) is briefly presented. Next, the double-distribution function (DDF) thermal LBM based on the kinetic theory is discussed. After reviewing the LBM no-slip and conjugate heat transfer schemes, a brief survey of the applications of LBM in pore-scale fluid flow and heat transfer simulations in porous media is presented. At the end, LBM macroscale models for convection heat transfer in porous media based on both local thermal equilibrium (LTE) and local thermal non-equilibrium (LTNE) assumptions are discussed and some applications of the models are reviewed.

Chapter 11 presents an understanding on the subsurface thermohydraulic process in enhanced geothermal systems (EGS). EGS reservoir structure and modeling issues are discussed. The chapter then reports on a three-dimensional transient model of the subsurface thermohydraulic process in EGS. The model treats the geothermal reservoir as an equivalent porous medium while it considers local thermal non-equilibrium between solid rock matrix and fluid flowing in the fractures and employs two energy conservation equations to describe heat transfer in the rock matrix and in the fractures, respectively. Since large changes at temperature and pressure exist during EGS heat extraction process, a module modeling the pressure- and temperature-dependent thermophysical properties of working fluid is introduced into this model.

Preface

Chapter 12 provides the relevant tools and guidelines to simulate mixed convection and radiation in porous media, within the framework of solar thermal absorbers. First, a general introduction of modeling strategies for mixed convection and radiation is presented. The main challenges will be addressed and the relevant solution methodologies are reported. Second, different approaches for characterizing the radiative properties of a porous medium are discussed, and the most common radiative transfer solution methods are illustrated. Different numerical results regarding the effect of radiation on the heat transfer in a porous medium are reported, and finally, the practical case of volumetric solar absorbers based on porous media is briefly addressed.

Chapter 13 will explore the principal features and main parameters of the mechanism of blowing through a porous surface in a hypersonic flow for wall-cooling applications. The cooling techniques for the thermal protection system of hypersonic vehicles, namely, active cooling and passive cooling, are introduced. Then, an in-depth discussion on the principle of active cooling systems including film and transpiration cooling (using porous materials) techniques are provided. Results from direct numerical simulations are presented for different configurations, including blowing through discrete slots and through a layer of regular distributed porosity, and discussed with emphasis on their use for the validation of simplified theoretical models.

Chapter 14 outlines aspects of thermal management and heat transfer enhancement using porous materials. The chapter first introduces the background of using porous materials for thermal management application. Different kinds of preparation for porous materials are introduced, namely, the sintering method, deposition method, casting method, and foaming method. The chapter further discusses two methods for reconstruction of porous materials for numerical studies: experimental reconstruction (CT scanning method and sequence slice grouping method) and numerical reconstruction (simulated annealing method, sequential indicator method, Gaussian field method, multiple-point statistics, and process simulation method). Finally, the chapter presents current numerical and experimental research progresses of using porous materials for applications to heat transfer enhancement in heat exchangers and solar power system, and thermal management in lithium ion batteries.

Chapter 15 presents a comprehensive description of the application of metal foams in heat exchangers. It critically reviews the literature to highlight the most interesting applications, by subdividing the topic in two main categories: gas (air) heat exchangers and liquid (including two-phase) heat exchangers. The chapter describes some interesting applications of metal foam for HVAC&R heat exchangers, air-cooled condensers, exhaust gas recirculation (EGR) systems for vehicles applications, waste-energy recovery, and heat rejection from fuel cells.

Chapter 16 summarizes, in brief, the transport processes involved in polymer-electrolyte fuel cell (PEFC) porous media, both gas-diffusion electrode (GDE) and catalyst layer (CL). It provides insights to understand the structures of various components of PEFCs and the associated phases. The governing conservation equations provide understanding of the transport conditions of gas and liquid phases inside the GDEs and CLs. The chapter highlights the vast complexities of transport within the polymer electrolyte fuel cells and the governing equations, and transport properties for PEFCs are influenced by the physical structure of the cell and the co-existing phases.

Chapter 17 discusses combustion in porous materials for application in porous burners. It provides a brief outline of the fundamentals of porous media combustion (PMC) and then discusses flame stabilization in porous media burners, PMC with liquid fuels, and reverse combustion in porous media. Next, it presents development of PMC-based burners for various applications in micro- and macroscales for household and industrial applications. Finally, the chapter introduces the catalytic porous media burners and their industrial applications.

Yasser Mahmoudi
Queen's University Belfast

Kamel Hooman
The University of Queensland

Kambiz Vafai
University of California, Riverside

Editors

Yasser Mahmoudi received his BS degree from Shiraz University and his MSc and PhD degrees from Tarbiat Modares University. He was a postdoctoral researcher at Delft University of Technology and the University of Cambridge. He is now a lecturer of Mechanical Engineering at Queen's University Belfast and a Fellow of Higher Education Academy. He is an associate editor of *Special Topics & Reviews in Porous Media—An International Journal* and was awarded the Outstanding Reviewer recognition by *International Journal of Heat and Mass Transfer, International Journal of Thermal Sciences*, and *Applied Thermal Engineering*. Yasser has published over 35 high-quality journal papers and over 40 peer-reviewed conference papers with Google h-index of 17 and i10-index of 23. He has been a principal investigator or co-investigator for a number of research programs, funded by research councils and industry. His research areas of interests are flow in porous media, convection heat transfer, thermal energy storage, and renewable energy.

Kamel Hooman is the director of the Renewable Energy Conversion Centre of Excellence, with over 1 million average annual budget, working closely with the industry in the field of energy. An author of over 150 archival journal articles, 5 book chapters, and over 50 conference papers, he has given numerous national and international invited lectures, keynote addresses, and presentations. He has been awarded fellowships from Emerald, Australian Research Council, National Science Foundation China, Australian Academy of Sciences, and Chinese Academy of Sciences with visiting professor/researcher positions at University of Padova, Krakow Institute of Technology, Ecole Centrale Paris, University of Malaya, Karlsruhe Institute of Technology, and Shandong University. He is the associate editor for the *International Journal of Heat and Mass Transfer, Heat Transfer Engineering, Journal of Porous Media and Special Topics*, and *Reviews in Porous Media—an International Journal* and serves on the editorial advisory board of the *International Journal of Exergy, Energies*, and *Thermal Science and Engineering Progress*. He has been the organizer and chair of the International Conference on Cooling Tower and Heat Exchanger sponsored by IAHR. He has supervised 11 doctoral students, and has directed over 10 post docs. He has an h-index of 37 with an i10-index of 122. He has carried out various sponsored research projects through companies, governmental funding agencies, and national labs. He has also consulted for various companies and governments, in Australia, and overseas.

Kambiz Vafai received his BS degree from the University of Minnesota (with the highest honors), Minneapolis, and his MS and PhD degrees from the University of California, Berkeley. He is the distinguished professor of Mechanical Engineering at University of California, Riverside (UCR), where he started as the presidential chair in the Department of Mechanical Engineering. Author of over 350 archival journal articles, book chapters, books (Ed.), and symposium volumes (Ed.), he has given numerous national and international invited lectures, keynote addresses, and presentations. He is a Fellow of American Association for Advancement of Science, American Society of Mechanical Engineers, World Innovation Foundation, and an Associate Fellow of the American Institute of Aeronautics and Astronautics. He is the editor-in-chief of the *Journal of Porous Media and Special Topics* and *Reviews in Porous Media—an International Journal*, an editor for the *International Journal of Heat and Mass Transfer*, and serves on the editorial advisory board of *International Communications in Heat and Mass Transfer, Numerical Heat Transfer, International Journal of Numerical Methods for Heat and Fluid Flow, International Journal of Heat and Fluid Flow*, and *Experimental Heat Transfer*. He is the editor of the all three editions of the *Handbook of Porous Media*, which became a best seller. Kambiz has been the director/chair of the First to Sixth International Conferences on Porous Media, all sponsored by ECI and NSF. He has supervised 65 doctoral and masters students and has directed over 66 post docs and visiting scholars. He has worked on a multitude of fundamental research investigations, a number of which have addressed some highly pertinent concepts presented for the first time. He is among the very few engineering scientists who have been within the prestigious ISI highly cited category with over 13,600 ISI citations covering a wide spectrum of disciplines and journals, an ISI h-index of 56, over 25,870 Google Scholar Citations, and a Google h-index of 74 and i10-index of 263. He has carried out various sponsored research projects through companies, governmental funding agencies, and national labs. He has also consulted for various companies and national labs and has been granted 13 US patents. He was the recipient of the ASME Classic Paper Award in 1999 and has received the 2006 ASME Heat Transfer Memorial Award, which are among the most selective awards in the field of heat transfer. He was given the International Society of Porous Media (InterPore) Highest Award in 2011, and he was also the recipient of the 75th Anniversary Medal of ASME Heat Transfer Division.

Contributors

Bernardo Buonomo
Dipartimento di Ingegneria
Università degli Studi della Campania
 "Luigi Vanvitelli"
Caserta, Italia

Jianchao Cai
State Key Laboratory of Petroleum Resources and
 Prospecting
China University of Petroleum
Beijing, China

Wenjiong Cao
CAS Key Laboratory of Renewable Energy
Guangzhou Institute of Energy Conversion
Chinese Academy of Sciences (CAS)
Guangzhou, China

Hasan Celik
Mechanical Engineering Department
Izmir University of Economics
Izmir, Turkey

Adriano Cerminara
Aerodynamics and Flight Mechanics
 Research Group
University of Southampton
Southampton, United Kingdom

Prodip Kumar Das
School of Engineering
Newcastle University
Newcastle, United Kingdom

Hassan Dashtian
Mork Family Department of Chemical
 Engineering and Materials Science
University of Southern California
Los Angeles, California

Ralf Deiterding
Aerodynamics and Flight Mechanics Research
 Group
University of Southampton
Southampton, United Kingdom

Benjamin Dietrich
Institute of Thermal Process Engineering
Karlsruhe Institute of Technology (KIT)
Karlsruhe, Germany

Davide Ercole
Dipartimento di Ingegneria
Università degli Studi della Campania
 "Luigi Vanvitelli"
Caserta, Italia

Pourya Forooghi
Institute of Fluid Mechanics
Karlsruhe Institute of Technology (KIT)
Karlsruhe, Germany

Kamel Hooman
School of Mechanical and Mining Engineering
The University of Queensland
Brisbane, Queensland, Australia

Kuo Huang
Faculty of Engineering
University of Nottingham
Nottingham, United Kingdom

Wenbo Huang
CAS Key Laboratory of Renewable Energy
Guangzhou Institute of Energy Conversion
Chinese Academy of Sciences (CAS)
Guangzhou, China

Gholamreza Imani
Department of Mechanical Engineering
Persian Gulf University
Bushehr, Iran

Fangming Jiang
CAS Key Laboratory of Renewable Energy
Guangzhou Institute of Energy Conversion
Chinese Academy of Sciences (CAS)
Guangzhou, China

Yunlong Jin
Guangdong Hydrogeology Battalion
Guangzhou, China

Jiaan Liu
Key Laboratory of Automobile Materials
  (Ministry of Education)
College of Materials Science and Engineering
Jilin University
Changchun, China

Yasser Mahmoudi
School of Mechanical and Aerospace Engineering
Queen's University Belfast
Belfast, United Kingdom

Oronzio Manca
Dipartimento di Ingegneria
Università degli Studi della Campania
  "Luigi Vanvitelli"
Caserta, Italia

Simone Mancin
Department of Management and Engineering
University of Padova
Vicenza, Italy

Moghtada Mobedi
Faculty of Engineering
Shizuoka University
Shizuoka, Japan

Muhammad Abdul Mujeebu
Department of Building Engineering
College of Architecture and Planning
Imam Abdulrahman Bin Faisal University
Dammam, Kingdom of Saudi Arabia

Akira Nakayama
Department of Mechanical Engineering
Shizuoka University
Shizuoka, Japan

Arunn Narasimhan
Department of Mechanical Engineering
Indian Institute of Technology Madras
Chennai, India

Sergio Nardini
Dipartimento di Ingegneria
Università degli Studi della Campania
  "Luigi Vanvitelli"
Caserta, Italia

Meysam Nazari
Department of Forest Biomaterials and
  Technology
Swedish University of Agricultural Sciences
Uppsala, Sweden

Donald A. Nield
Department of Engineering Science
University of Auckland
Auckland, New Zealand

Krishna M. Pillai
Department of Mechanical Engineering
University of Wisconsin–Milwaukee
Milwaukee, Wisconsin

Cong Qi
Faculty of Engineering
University of Nottingham
Nottingham, United Kingdom

Dirk Roekaerts
Department of Process & Energy Delft University
  of Technology
Delft, The Netherlands

and

Department of Mechanical Engineering
Eindhoven University of Technology
Eindhoven, The Netherlands

Muhammad Sahimi
Mork Family Department of Chemical
  Engineering and Materials Science
University of Southern California
California, Los Angeles

Neil D. Sandham
Aerodynamics and Flight Mechanics
  Research Group
University of Southampton
Southampton, United Kingdom

Nima Shokri
School of Chemical Engineering and
  Analytical Science
The University of Manchester
Manchester, United Kingdom

Simone Silvestri
Department of Process & Energy
Delft University of Technology
Delft, The Netherlands

Deepashree Thumbarathy
School of Engineering
Newcastle University
Newcastle, United Kingdom

Guohua Wang
Faculty of Engineering
University of Nottingham
Nottingham, United Kingdom

Guoling Wei
Guangdong Hydrogeology Battalion
Guangzhou, China

Wei Wei
Institute of Rock and Soil Mechanics
Chinese Academy of Science
Wuhan, China

Yuying Yan
Faculty of Engineering
University of Nottingham
Nottingham, United Kingdom

Mohammad Amin Faghihi Zarandi
Department of Mechanical Engineering
University of Wisconsin–Milwaukee
Milwaukee, Wisconsin

Section I

Fundamentals of Convection in Porous Media

1

Introduction to Fluid Flow and Heat Transfer in Porous Media

Meysam Nazari, Yasser Mahmoudi, and Kamel Hooman

CONTENTS

1.1 Introduction 3
1.2 Porosity and Permeability 6
1.3 Continuity Equation in Porous Media 6
1.4 Momentum Equation in Porous Media 6
1.5 Modeling the Porous-Fluid Interface 10
1.6 Heat Transfer in Porous Media 11
    1.6.1 Challenges in Modeling Based on LTNE 13
    1.6.2 Radiation in Porous Materials 14
1.7 Conclusions 15
Nomenclature 15
References 16

1.1 Introduction

Porous materials (media) are solids, which are permeated by a network of pores. The interconnection of the pores allows the flow of one or more fluids through the material. Porous materials exist widely in nature and made industrially. Sandstone, limestone, rye bread, wood, and the human lung could be exemplified as natural porous media (Nield and Bejan 2006; Kampman et al. 2014; Suchanek and Olejniczak 2015). Figures 1.1 and 1.2 show two natural porous media: sandstone and wood. Man-made porous media include metallic and nonmetallic foams, porous ceramics, and packed bed (Nield and Bejan 2006), which are widely used in industrial applications. Metallic foams are mostly used in heat exchangers (Shikh Anuar et al. 2018; T'Joen et al. 2010; Chumpia and Hooman 2014; Odabaee and Hooman). Packed beds have been also utilized for thermal storage applications (Klein et al. 2014; Oró et al. 2013; Anderson et al. 2015). Figures 1.3 and 1.4, respectively, illustrate aluminum foam utilized in heat exchangers and packed bed in energy storage applications.

On the pore scale, the flow quantities (velocity, pressure, temperature, etc.) will be clearly irregular. In addition, local on-site measurements of fluid dynamics at the pore level are very difficult due to the small and complex flow passage geometry within the pores. Therefore, theoretical studies based on solving Navier–Stokes are widely used to analyze fluid and heat transport in porous media. There are two main approaches for the theoretical modeling of fluid and heat transport in porous media: the pore-scale approach and the continuum approach (Nield and Bejan 2006). In the theoretical studies, the complex microscopic transport phenomena at the pore level in the porous region is important, as it directly accounts for the fundamental physical processes that affect the fluid behavior at the pore scale. However, the complexity of the cellular morphology in any typical porous medium (such as sand, packed bed, or metallic foam) precludes a detailed microscopic investigation of the transport phenomena at the pore

3

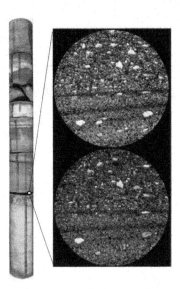

FIGURE 1.1 Core photograph of the Entrada Sandstone and siltstone. (From Kampman, N. et al., *Chem. Geol.*, 369, 22–50, 2014.)

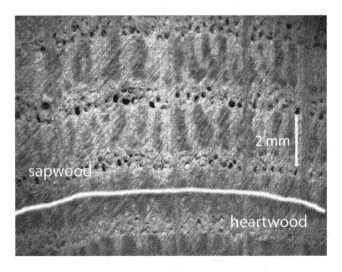

FIGURE 1.2 The microscopic image of the open vessels in the oak sapwood. (From Suchanek, M., et al., *Int. J. Multiphase Flow*, 72, 83–87, 2015.)

level. Such complexity makes it extremely tedious to create a mesh for computational studies, and it is extremely computationally expensive to solve for transport phenomena in a porous medium for large-scale realistic porous media systems. Hence, such a microscopic approach is almost never adopted for solving flows in any reasonably sized porous domains. Therefore, for realistic applications of transport in porous media, the continuum (macroscopic) approach is widely adopted. In this approach, the general transport equations are spatial averaging, that is, integrated over a representative elementary volume (REV) (see Figure 1.5), which includes the fluid and the solid phases within a porous medium (Nield and Bejan 2006). As a result, fluid flow and heat transport at the pore scale are unresolved, and the aggregate effect of the pore-scale processes are taken into account through various effective constitutive parameters, such as porosity, permeability, specific surface area, and fluid-to-solid heat transfer coefficient (Nield and Bejan 2006). Though the loss of information is inevitable with this approach, it provides a low-cost analysis for studying transport in porous media systems.

Introduction to Fluid Flow and Heat Transfer in Porous Media

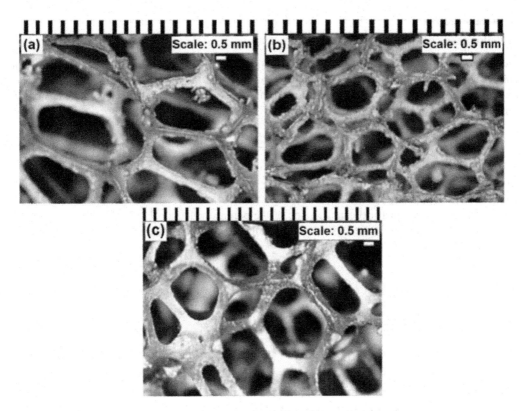

FIGURE 1.3 Aluminum open-cell metal foam microstructure: (a) 10 pores per inch (PPI); (b) 30 PPI; and (c) 5 PPI. (From Shikh Anuar, F. et al., *Exp. Therm. Fluid Sci.*, 99, 117–128, 2018.)

FIGURE 1.4 Packed bed used in thermal energy storage system. (From Klein, P. et al., *Energy Procedia*, 49, 840–849, 2014.)

1.2 Porosity and Permeability

Two characteristics of porous materials fundamentally describe them known as porosity and permeability. The porosity of a porous medium could be explained as the fraction of the total media occupied by void spaces (Nield and Bejan 2006). If total volume of the media and the solid volume are known, literally porosity could be defined as follow.

$$\varepsilon = 1 - \frac{V_s}{V_t}, \tag{1.1}$$

where ε is the porosity of the medium, V_s and V_t are respectively the volume of the solid phase and total volume of the medium.

Permeability is a geometrical characteristic of the porous media, as discussed by references (Nield and Bejan 2006; Vafai 2005), for particle beds and fibers, the permeability can be defined utilizing the Carman–Kozeny theory as follow.

$$K = \frac{\varepsilon^3 d_p^2}{180(1-\varepsilon)^2}, \tag{1.2}$$

where d_p is the average particle or fiber diameter (Nield and Bejan 2006).

It has been discussed that Equation (1.2) is well satisfied by the media made of spherical particles, and also there are certain modifications done on the relation to be satisfactory for some other cases (Nield and Bejan 2006).

1.3 Continuity Equation in Porous Media

Considering the concept of REV and taking a volume element that is larger than the pore volumes for the volume averaging, we can obtain an average fluid velocity (U) in the porous media. This velocity is related to another form of velocity $u = (u_x, u_y, u_z)$ in which the porosity of the media is taken into account by the Dupuit–Forchheimer relationship $u = \varepsilon U$ (Nield and Bejan 2006). Hence, the continuity equation in the porous media can be given as (Nield and Bejan 2006):

$$\varepsilon \frac{d\rho_f}{dt} + \nabla.(\rho_f u) = 0, \tag{1.3}$$

where ρ_f, u, and t are respectively fluid density, velocity, and time. The velocity coming from the Dupuit–Forchheimer relationship (u) has been named differently by different researchers, such as Darcy velocity, superficial velocity, filtration velocity, and seepage velocity (Nield and Bejan 2006).

1.4 Momentum Equation in Porous Media

Darcy (1856) experimentally studied the flow of water in a uniform bed made of particles. For a steady-state condition and low Reynolds number defined based on the particle diameter, he introduced the following relation for the momentum equation in the porous media (Vafai 2005):

$$\frac{dp}{dx} = -\frac{\mu u}{K}, \tag{1.4}$$

where p and μ are respectively pore pressure and viscosity of the fluid. K is the permeability of the media obtained using Equation (1.2). The Darcy relation is valid for a low Reynolds number and is a linear equation (Nield and Bejan 2006; Vafai 2005). Forchheimer (1901) added a nonlinear term to the Darcy relation representing the effect of drag and advection inertia terms as (Nield and Bejan 2006):

Introduction to Fluid Flow and Heat Transfer in Porous Media

$$\frac{dp}{dx} = -\frac{\mu u}{K} - \frac{F\rho_f |u| u}{\sqrt{K}},\qquad(1.5)$$

where ρ_f and F are respectively fluid density and Forchheimer coefficient, which is obtained using the following relation (Ergun 1952):

$$F = \frac{b}{\sqrt{180\varepsilon^3}},\qquad(1.6)$$

where b is a constant to parameterize the microscopic geometry of the porous media (Nield and Bejan 2006). Lage (1998) stated that the Forchheimer coefficient (F) is related to the geometry of a porous medium, which represents the effect of any solid surface interfering with the fluid flow and has been named form drag or form coefficient. Joseph et al. (1982) proposed the term $\frac{F\rho_f |u| u}{\sqrt{K}}$ in Equation (1.5) based on the original relation presented by Forchheimer (1901). Later, Ward (1964) suggested the coefficient (F) might be a universal constant of 0.55. Nield and Kuznetsov (2013) in a review paper stated that this coefficient is approximately constant for some particular materials, such as for foamed metals, to be 0.1. Nield and Kuznetsov (2013) further discussed that F depends on the geometry of the porous media, and it is dimensionless. Lage and Antohe (2000) argued that the permeability of a porous medium relates to the effective surface area of the solid phase, so that the F coefficient should depend on the form of a porous matrix. Lage and Antohe (2000) more argued that it is crucial to study the effect of the geometry of a porous medium on the form drag, because very little is clear in this matter. Hooman and Dukhan (2013) theoretically found that permeability and the form drag coefficient (F) are dependent on the porosity of the porous media. They (Hooman and Dukhan 2013) developed correlations for the permeability and the form drag coefficient for high porosity foams and experimentally validated their correlations. In another work, Hooman and Gurgenci (2010) showed that for a porous matrix made of finned tube bundles, the pressure drop and the form drag depend on porosity of the medium. They (Hooman and Gurgenci 2010) also proposed relations for pressure drop and form drag coefficient for porous media with porosity in the range of 0.63–0.78. They (Hooman and Gurgenci 2010) further found that both pressure drop and form drag coefficients decrease with the increase in porosity of the porous media.

An alternative to Darcy's equation is what is commonly known as Brinkman's equation, considering flow shear effect and with the inertial terms omitted as (Nield and Bejan 2006):

$$\frac{dp}{dx} = -\frac{\mu u}{K} - \tilde{\mu}\nabla^2 u.\qquad(1.7)$$

The relation has two terms respectively, including the Darcy term and analogous to the Laplacian term, which exists in the common Navier–Stokes equation (Nield and Bejan 2006). The coefficient $\tilde{\mu}$ is the effective viscosity, which Brinkman considered equal to the fluid viscosity μ, while in general that is not true (Nield and Bejan 2006).

In an analytical study, Hsu and Cheng (1990) proposed another term to be added to Equation (1.5) representing the effect of viscous boundary layer as:

$$\frac{dp}{dx} = -\frac{\mu u}{K} - \frac{F\rho_f |u| u}{\sqrt{K}} - \frac{H\sqrt{\rho_f \mu |u|} u}{K^{0.75}},\qquad(1.8)$$

where H is a dimensionless coefficient and is a function of porosity and microscopic solid geometry (Hsu and Cheng 1990; Vafai 2005). Irmay (1958) proposed an alternative relation for the Forchheimer relation in Equation (1.5) as follows:

$$\frac{\Delta P}{L} = \beta \frac{(1-\varepsilon)^2 \mu U}{\varepsilon^3 d^2} + \alpha \frac{(1-\varepsilon)\rho_f U^2}{\varepsilon^3 d},\qquad(1.9)$$

where β and α are shape factors and should be experimentally determined (Nield and Bejan 2006). Ergun (1952) determined $\beta = 150$ and $\alpha = 1.75$ and correlated a new relation for fluid flow in porous structures as:

$$\frac{\Delta P}{L} = 150\frac{(1-\varepsilon)^2 \mu U}{\varepsilon^3 d^2} + 1.75\frac{(1-\varepsilon)\rho_f U^2}{\varepsilon^3 d}. \tag{1.10}$$

In another attempt, Vafai et al. (2006) defined a relation for turbulent fluid flow in porous materials as:

$$\frac{\Delta P}{L} = 120\frac{(1-\varepsilon)^2 \mu u}{\varepsilon^3 d^2} + 2.3\frac{(1-\varepsilon)\rho_f u^2}{\varepsilon^3 d}. \tag{1.11}$$

Lee and Ogawa (1994) proposed following relation for fluid flow in porous media:

$$\frac{\Delta P}{L} = 12.5\frac{(1-\varepsilon)^2 \rho_f U^2}{2\varepsilon^3 d}\left(29.32 Re_d^{-1} + 1.56 Re_d^{-n} + 0.1\right) \tag{1.12}$$

where $Re_d = \frac{\rho_f U d}{\mu}$ is the Reynolds number based on the sphere's diameter and $n = 0.352 + 0.1\varepsilon + 0.275\varepsilon^2$.

Vafai et al. (2006) stated that macroscopically there are different regimes for flow in porous media, which are based on Reynolds number based on pore diameter $\left(Re_{dp} = \frac{u d_p}{\vartheta_f}\right)$ and are categorized as follows: (1) Darcy regime ($Re_{dp} < 1$), (2) Forchheimer regime ($1 \sim 10 < Re_{dp} < 100$), (3) transition flow regime ($100 < Re_{dp} < 300$), and (4) fully turbulent regime ($300 < Re_{dp}$). It is further discussed that at higher Re_{dp}, Forchheimer flow resistance and dispersion are more significant. And also it was presented that the free stream velocity under certain circumstances could be in the order of Darcy velocity. In addition to Equation (1.10), Vafai et al. (2006) also introduced another equation for fluid flow in porous media similar to the Ergun equation but utilizing the Darcy velocity as:

$$\frac{\Delta P}{L} = 150\frac{(1-\varepsilon)^2 \mu u}{\varepsilon^3 d^2} + 1.75\frac{(1-\varepsilon)\rho_f u^2}{\varepsilon^3 d}. \tag{1.13}$$

Nazari et al. (2017) compared their experimental data on pressure drop (Nazari et al. 2017) obtained for a packed bed (with particles diameter d) with the results predicted using Equations (1.10)–(1.13) as shown in Figure 1.6. For comparison, experimental data of Lee and Ogawa (1994) is also included in Figure 1.6. It is seen that for a high value of Re_d, the Ergun equation (1952) overpredicts the experimental data (Lee and Ogawa 1994). However, relations presented by Vafai et al. (2006) given in Equations (1.11) and (1.13) can well predict the pressure drop measured in the experiment.

Efforts to obtain a comprehensive equation for momentum in porous media were performed by researchers in the past. Wooding (1957), using an analogy with the Navier–Stokes equation, obtained the following equation for the momentum transfer in porous media:

$$\frac{\rho_f}{\varepsilon}\left[\frac{\partial u}{\partial t} + \frac{1}{\varepsilon}(u \cdot \nabla)u\right] = -\nabla P - \frac{\mu_f}{K}u. \tag{1.14}$$

The right hand of Equation (1.14) is the Darcy relation. However, Beck (1972) stated that the term $(u \cdot \nabla)u$ was not appropriate because it increased the order of the differential equation, and this was not consistent with the slip boundary condition. Nield and Bejan (2006) suggested to drop this term in numerical works and propped the following equation:

$$\frac{\rho_f}{\varepsilon}\frac{\partial u}{\partial t} = -\nabla P - \frac{\mu_f}{K}u. \tag{1.15}$$

The left-hand side of Equation (1.15) is an inertial term. As discussed in (Nield and Bejan 2006), this term was obtained using a manipulation and rearrangement of the partial derivative with respect to time with a

Introduction to Fluid Flow and Heat Transfer in Porous Media

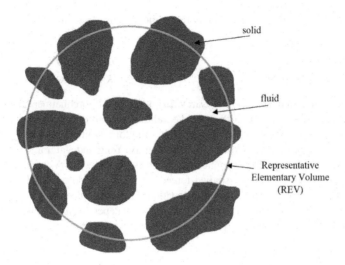

FIGURE 1.5 A representative elementary volume (REV) used for spatial averaging in porous media.

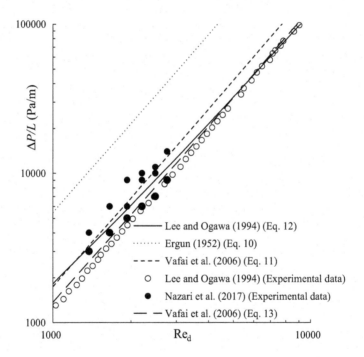

FIGURE 1.6 Pressure drop in packed bed versus Re_d number predicted using equation developed by Ergun. (From Ergun, S., *Chem. Eng. Prog.*, 48, 89–94, 1952.)

volume average, which in general this operation is not valid (Nield and Bejan 2006). Nield and Bejan (2006) pointed out that to tackle this inadequacy a factor needs to be added to relation Equation (1.15) in order to represent the geometry of the porous medium as:

$$\rho_f \cdot c_a \cdot \frac{\partial u}{\partial t} = -\nabla P - \frac{\mu_f}{K} u, \tag{1.16}$$

where c_a is a constant tensor to represent the geometry of the porous medium. Nield and Bejan (2006) called this coefficient the "acceleration coefficient tensor" of the porous medium.

With considering Darcy, Brinkman, and Forchheimer terms, Hsu and Cheng (1990) obtained a momentum equation as follows:

$$\frac{\rho_f}{\varepsilon}\left[\frac{\partial u}{\partial t}+\frac{1}{\varepsilon}\nabla\left(u\cdot u\right)\right]=-\nabla P+\frac{\mu_f}{\varepsilon}\nabla^2 u-\frac{\mu_f}{K}u-\frac{\rho_f\varepsilon F}{\sqrt{K}}|u|u. \tag{1.17}$$

In the right-hand side of Equation (1.17) the Darcy, Brinkman, and Forchheimer terms were considered. It was discussed by Nield and Bejan (2006) that the validity of considering Brinkman and Forchheimer terms together is not evident. As Brinkman's equation is valid for a high-porosity medium and considering the Forchheimer relation for this case raises some uncertainty. Yang and Vafai [(2011a) discussed that the inertia effect is significant in a number of applications, such as high-speed flows and high-porosity medium. In another study, Vafai and Kim (1989) studied forced convection in porous media utilizing Brinkman, Forchheimer, and Darcy's terms in the momentum equation. They concluded that for high-permeable porous media, the thickness of the momentum boundary layer not only depends on the Darcy number but also the inertia parameter, while for low-permeable porous media, the Darcy number is the only significant parameter.

1.5 Modeling the Porous-Fluid Interface

For modeling the interface between a clear fluid region and a fluid-saturated porous medium (porous-fluid interface), there are certain debates among researchers. Nield and Kuznetsov (2013) argued that when the Brinkman equation is considered for fluid flow in a porous medium, four matching conditions should be considered at the interface because the differential equations are of second order for both the clear and porous regions including tangential and normal velocities and stresses. For velocities, there is no problem with matching velocities between the clear and solid region, but it is different for stresses. For matching tangential stress over the pores portion in the interface, the velocity shear is continuous and consequently tangential stress is also continuous, while for the solid portion, the tangential shear stress is not continuous and is zero. However, for the near-clear fluid region, tangential shear stress is nonzero as discussed above. Therefore, when these adjacent regions are matched, some inaccuracies will arise (Nield and Kuznetsov 2013).

This problem is even more complicated when deploying the Darcy equation. In this situation, three matching conditions are required, including two continuity equations for the tangential fluid velocity and the normal velocity, and the third is the Beavers–Joseph (1967) boundary condition. Beavers–Joseph (1967) presented a dimensionless constant that is independent of the fluid viscosity, while it depends on the structure of the porous material. For normal stress in this situation, pressure is also needed to be taken into account, as most of the cases the continuity of stress reduces to continuity of pressure. Normal flow to a porous medium has rarely been studied, and most of the researchers have investigated tangential flow to the porous media. Beavers and Joseph (1967) showed that sharp gradients at the interface between the porous and fluid regions exist. Their work highlighted the existence of a slip velocity at the interface. Alazmi and Vafai (2001) classified various interface conditions into two main categories: slip and no-slip boundary conditions. They (Alazmi and Vafai 2001) also established five main categories for the hydrodynamic interface conditions. The different models mostly lead to comparable results except for few specific cases. To show the complexity of the problem, it is interesting to note that all these works were conducted for duct flows where there is no recirculation or wakes, which cannot be modeled as internal flows. Hooman (2014) and Anuar et al. (2018) experimentally analyzed interface problems for foamed bed porous media and presented a comparison between the thermohydraulic performances of heat exchangers composed of passages, which are fully or partially blocked by porous inserts. They (Anuar et al. 2018; Hooman 2014) experimentally compared pressure drop and flow visualization in channels fully or partially filled with metallic foams and discussed interface problems between the clear fluid region and porous block region. Anuar et al. (2018) discussed that a portion of the incoming flow passes through the porous media leaves the media and goes into the clear region section after a certain foam length. They (Anuar et al. 2018) further discussed that this phenomenon only occurs for foams with high PPI and high blockage ratio. Their results (Anuar et al. 2018) also showed that there is certain fluctuating flow velocity and changing in the flow direction affecting the total pressure drop and resistance coefficient.

Introduction to Fluid Flow and Heat Transfer in Porous Media 11

1.6 Heat Transfer in Porous Media

In this section we present equation that expresses the transport of energy in a porous medium. We assume that the medium is isotropic and, viscous dissipation and the work done by pressure changes are negligible. There are two approaches available in applying the volume-averaging technique for heat transfer investigations: one is averaging over a representative elementary volume containing both the fluid and solid phases, and the other is averaging separately over each of the phases, thus resulting in a separate energy equation for each individual phase in the porous medium. These two models are referred to as the local thermal equilibrium (LTE) model and local thermal non-equilibrium (LTNE) model. The LTE model is valid when the heat transfer between two phases is large and the local temperature difference between the two phases is negligible. Thus, in this case, only one equation is considered to show the energy equation of both phases in the porous media. In some applications (such as biological media (Mahjoob and Vafai 2009) and CO_2 storage (Jiang et al. 2013)), however, the temperature differences between the solid and fluid phases cannot be neglected, and thus LTNE needs to be utilized, which in two separate energy equations are considered for both solid and fluid phases. Considering the above for an incompressible and laminar flow, and neglecting the effect of natural convection, the governing equations for energy in a porous medium for solid and fluid phases under LTE condition (i.e., $T_f = T_s = T$) are written as:

$$\left(\rho c_p\right)_m \frac{\partial T}{\partial t} + \left(\rho c_p\right)_f u \nabla T = \nabla \cdot \left(k_m \nabla T\right) + \varepsilon q_m''', \tag{1.18}$$

where subscribe m implies the average of parameters as:

$$\left(\rho c_p\right)_m = \varepsilon \rho_f c_{p,f} + \left(1 - \varepsilon\right) \rho_s c_{p,s}, \tag{1.19}$$

$$k_m = \varepsilon k_f + \left(1 - \varepsilon\right) k_s, \tag{1.20}$$

$$q_m''' = \varepsilon q_f''' + \left(1 - \varepsilon\right) q_s''', \tag{1.21}$$

Subscribe f and s indicate the fluid and solid phases, respectively.

Under the LTNE condition, the two energy equations for the fluid and solid phases in the porous medium are written as:

For fluid phase

$$\varepsilon\left(\rho c_p\right)_f \frac{\partial T_f}{\partial t} + \left(\rho c_p\right)_f u \nabla T_f = \varepsilon \nabla \cdot \left(k_f \nabla T_f\right) + h_{sf} a_{sf} \left(T_s - T_f\right) + \varepsilon q_f''', \tag{1.22}$$

For solid phase

$$\left(1 - \varepsilon\right)\left(\rho c_p\right)_s \frac{\partial T_s}{\partial t} = \left(1 - \varepsilon\right) \nabla \cdot \left(k_s \nabla T_s\right) + h_{sf} a_{sf} \left(T_f - T_s\right) + \left(1 - \varepsilon\right) q_s''', \tag{1.23}$$

where h_{sf} is the internal heat transfer coefficient between the fluid and solid phases, and a_{sf} is the specific surface area (surface per unit volume). A critical aspect of using these equations lies in the determination of the appropriate value of h_{sf}. There are several correlations proposed for the fluid-to-solid Nusselt number $\left(\mathrm{Nu}_{sf} = \frac{h_{sf} d}{k}\right)$ as:

- Nie et al. (2011) proposed the following correlation for the convective heat transfer coefficient in a spherical packed bed based on an experimental investigation for the steady-state condition.

$$Nu_{sf} = 0.052 \frac{\left(1 - \varepsilon\right)^{0.14}}{\varepsilon} Re_d^{0.86} Pr^{1/3}. \tag{1.24}$$

- Wakao and Kaguei (1982) conducted experimental investigations and assembled both steady and unsteady data together. They proposed the following correlation for the interfacial convective heat transfer coefficient, which is valid only for packed bed with $\varepsilon \approx 0.4$.

$$Nu_{sf} = 2 + 1.1 Re_d^{0.6} Pr^{1/3}. \tag{1.25}$$

- Kuwahara et al. (2001) developed the following correlation from numerical experiments based on the two-dimensional model for square-shaped particles in a packed bed.

$$Nu_{sf} = \left(1 + \frac{4(1-\varepsilon)}{\varepsilon}\right) + \frac{1}{2}(1-\varepsilon)^{1/2} Re_d^{0.6} Pr^{1/3}. \tag{1.26}$$

- Whitaker (1972), using experimental data, recommended the following correlation for forced convection heat transfer coefficient in a packed bed.

$$Nu_{sf} = \left[0.5 Re_d^{-0.1}\left(\frac{(1-\varepsilon)^{0.5}}{\varepsilon}\right) + 0.2 Re_d^{1/15}\left(\frac{(1-\varepsilon)^{1/3}}{\varepsilon}\right)\right] Re_d^{0.6} Pr^{1/3}. \tag{1.27}$$

- Kays and London (1984) introduced the following correlation for the heat transfer coefficient in a packed bed.

$$Nu_{sf} = 0.26 \frac{(1-\varepsilon)^{0.3}}{\varepsilon} Re_d^{0.7} Pr^{1/3}. \tag{1.28}$$

- Nsofor and Adebiyi (2001) based on an experimental investigation correlate the following relation for the heat transfer coefficient in a packed bed made of cylindrical particles.

$$Nu_{sf} = 8.74 + 9.34 Re_d^{0.2} Pr^{1/3}. \tag{1.29}$$

- Incropera and DeWitt (1990) presented the following relation for heat transfer coefficient in a nonspherical packed bed.

$$Nu_{sf} = \frac{0.79}{\varepsilon} Re_d^{0.425} Pr^{1/3}. \tag{1.30}$$

- Bird et al. (1960) suggested the following correlation to use for the heat transfer coefficient in nonspherical packed bed.

$$Nu_{sf} = 0.534 Re_d^{0.59} Pr^{1/3}. \tag{1.31}$$

Figure 1.7 shows the average Nusselt number for the steady-state condition obtained in the experiment performed by Nazari et al. (2017) (with $d = 5.5$ mm and $\varepsilon = 0.47$), compared with the experimental data of Wakao and Kaguei (1982) and the correlations given in Equations (1.24)–(1.31). It is seen that the Nusselt number obtained based on the experiment done by Nazari et al. (2017) are in a good agreement with the results obtained using the correlation developed by Nie et al. (2011) [Equation (1.24)]. But the results of references (Nazari et al. 2017; Nie et al. 2011) show a discrepancy with other correlations. Correlations presented by Equations (1.26), (1.30), and (1.31) are valid for a nonspherical packed bed. In addition, Equations (1.25), (1.27), and (1.28) are valid for a spherical but not purely steady-state flow regime. The results predicted by Nie et al. (2011) and those obtained in the work of Nazari et al. (2017) are obtained for spherical particles under the steady-state condition. Kuwahara et al. (2001) commented that the correlation proposed by Wakao and Kaguei (1982), which was obtained by combining steady and unsteady data, needs to be corrected because their (Kuwahara et al. 2001) unsteady numerical solutions suggested that a separate equation should be established for correlating the unsteady data.

Introduction to Fluid Flow and Heat Transfer in Porous Media

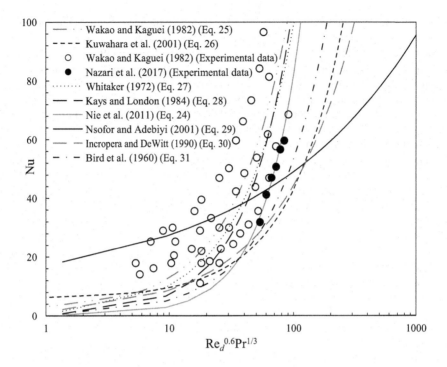

FIGURE 1.7 Average fluid-to-solid Nusselt number in the fully developed region obtained using the experimental data of Nazari et al. (2017) and Wakao and Kaguei (1982) in comparison with respect to the correlations given in refs. (From Nie, X. et al., *ASME J. Heat Transf.*, 133, 041601, 2011; Wakao, N. and Kaguei, S., *Heat and Mass Transfer in Packed Beds*, Vol. 1, pp. 243–295, Gorden and Breach, New York, 1982; Kuwahara, F. et al., *Int. J. Heat Mass Transf.*, 44, 1153–1159, 2001; Whitaker, S., *AIChE J.*, 18, 361–371, 1972; Kays, W.M. and London, A.L., *Compact Heat Exchangers*, 3rd ed., McGraw-Hill, New York, 1984; Nsofor, E.C. and Adebiyi, G.A., *Exp. Thermal Fluid Sci.*, 24, 1–9, 2001; Incropera, F.P. and DeWitt, D.P., *Introduction to Heat Transfer*, 2nd ed., John Wiley & Sons, New York, 1990; Bird, R.B. et al., *Transport Phenomena*, 1st ed., John Wiley & Sons, New York, 1960.)

Also, Nie et al. (2011) discusses that Nsofor and Adebiyi (2001) experimentally investigated cylindrical particles at very high temperature to correlate with Equation (1.29). In such high temperature, radiation and free-convection heat transfer must be contributed to develop a correlation. While these important effects were not included in the correlation developed by Nsofor and Adebiyi (2001) given by Equation (1.29), it is further discussed by Nie et al. (2011) that in the works by Whitaker (1972) and Kays and London (1984), the thermocouples used to measure the particles' surface temperature were not glued to the particles. Therefore, the temperature of the solid surface may be affected by the fluid temperature in references (Whitaker 1972; Kays and London 1984) and thus may not be accurately measured.

1.6.1 Challenges in Modeling Based on LTNE

The deployment of LTNE for analyzing porous media systems is being challenged when the boundary of the porous medium is subjected to a contact heat flux or is adjacent to a fluid layer (Yang and Vafai 2011; Vafai and Yang 2013). Here, there is some ambiguity about the distribution of interface flux between the two phases within the porous medium. Amiri et al. (1995) highlighted for the first time this complicity and presented two different approaches for boundary conditions under constant wall heat flux. They reported that the constant heat flux boundary condition could be viewed in two different ways (Amiri et al. 1995). The first is to assume that the total wall heat flux being divided between the two phases based on their effective conductivities and the corresponding temperature gradients. In the second approach, each of the two phases receives an equal amount of the total heat flux at the pipe wall (Amiri et al. 1995). The first and second approaches presented by Amiri et al. (1995) later led to the introduction of two interface models, respectively referred to as models A and B (Yang and Vafai 2010).

Alazmi and Vafai (2002) studied the effect of using different boundary conditions under constant wall heat flux and LTNE condition. Their study included six models based on the first approach (model (A) and two models based on the second approach (model B) of Amiri et al. (1995). They found that using different boundary conditions may lead to substantially different results, and certain physical properties, such as porosity, Darcy number, thermal conductivity, and inertia parameter, were found to have a significant effect on the heat transfer predictions using different models. Yang and Vafai (2011b) considered five forms of thermal boundary conditions at the interface between a porous medium and a fluid under the LTNE condition. They presented exact solutions for all of these models and further reported the restrictions in the validity of LTE in a partially filled system.

Furthermore, Yang and Vafai (2010) discussed that for the constant wall heat flux case and in the presence of internal heat generation in the solid phase, the direction of the temperature gradient for the fluid and solid phases for model B are different at the wall and leads to temperature bifurcation. In this case, a portion of the internal heat generation in the solid phase will transfer to the fluid phase through the thermal conduction at the wall, instead of through internal heat transfer exchange between the fluid and solid. This paves the way for the occurrence of the temperature gradient bifurcation at the wall. In the absence of internal heat generation, the temperature gradient directions for the fluid and solid phases at the wall are kept the same. They further discussed that in the case of a constant wall temperature boundary condition, the temperature gradient bifurcation phenomena for the fluid and solid phases at the wall occurs only over a given axial region. In another study, Yang and Vafai (2011c) investigated the bifurcation phenomena for the transient flow regime and introduced a region at which the phenomena occur. They stated that these phenomena only occur during the transient period and will disappear when the fluid reaches its steady state.

As pointed out in previous works, the problem of heat flux splitting at the interface between a porous medium and a clear fluid or a porous medium with an impermeable wall is still not resolved. This is primarily because the thermal characteristics at the interface between a porous medium and the fluid layer above is unknown, and the determination of the interface thermal boundary condition remains a scientific challenge. Moreover, the physics underlying the appearance of the bifurcation phenomenon in the porous medium is still unknown. These would offer a new and valuable direction to investigate heat transfer in porous media (e.g., Vafai and Yang 2013).

1.6.2 Radiation in Porous Materials

In case there is a high temperature difference between the fluid and solid phases of the porous media, a large temperature differential between these two phases occurs leading to increasing in the significance of the effect of radiative heat transfer between the phases (Mahmoudi 2014). Due to transparency of the fluid phase, radiation from the fluid phase could be ignored compared with that of the solid phase. Hence, radiative heat transfer is rationally considered from the solid phase, and the fluid phase is assumed to be nonradiative in comparison to solid radiation (Hossain et al. 2001; Jahanshahi Javaran et al. 2010; Keshtkar and GandjalikhanNassab 2009; Khosravy El-Hossaini et al. 2008; Maerefat et al. 2011; Mahmoudi 2014; Mendes et al. 2008). Therefore, a term of the radiative flux divergence ∇q_{rad} appears in the energy equation of the solid phase and Equation (1.23) is written as (Mahmoudi 2014):

$$(1-\varepsilon)(\rho c_p)_s \frac{\partial T_s}{\partial t} = (1-\varepsilon)\nabla \cdot (k_s \nabla T_s) + h_{sf} a_{sf} (T_f - T_s) + (1-\varepsilon) q_s''' - \nabla q_{rad}, \qquad (1.32)$$

If a cylindrical enclosure with absorbing-emitting-scattering medium is considered, radiation flux is calculated by using the radiative transfer equation (RTE) (Mahmoudi 2014). The heat source term ∇q_{rad} due to solid thermal radiation in Equation (1.32) is obtained from the radiative heat transfer equation. The general equation of radiative transfer for an absorbing, emitting, and anisotropically scattering medium along the direction of the vector s is written as (Mahmoudi 2014; Mendes et al. 2008; Modest 1993):

$$S.\nabla I = \gamma \left[(1-\omega)I_b - I + \frac{\omega}{4\pi} \int_{4\pi} I(S_i)\varphi(S_i,S)d\Omega_i \right], \qquad (1.33)$$

Introduction to Fluid Flow and Heat Transfer in Porous Media

$$\nabla q_{rad} = \gamma(1-\omega)\left(4\pi I_b - \int_{4\pi} I(S)d\Omega\right), \tag{1.34}$$

$$I = \frac{\sigma}{\pi}T^4. \tag{1.35}$$

In these equations, γ, ω, φ, and σ are respectively extinction coefficient, scattering albedo, phase function, and Stefan–Boltzmann constant. I is the radiation intensity, and the radiation intensity of the black body is expressed as $I_b = \frac{\sigma}{\pi}T_s^4$ (Mahmoudi 2014; Mendes et al. 2008; Modest 1993).

1.7 Conclusions

In this chapter, porous media is introduced, and certain significant and recent studies regarding heat transfer and fluid flow in porous media have been presented. Two main approaches for the theoretical modeling of fluid and heat transport in porous media, including the pore-scale approach and the continuum approach, are presented and discussed. Moreover, some new and interesting relations of fluid flow in porous media are compared with experimental data, founding that well-known Ergun equation (Ergun 1952) is not exact for a high Reynolds number (more than unity). However, those presented by Vafai et al. (2006) well predicted the pressure gradient in porous media for an extensive range of Reynolds numbers. Two theoretical approaches to model energy equation in porous media, including LTE and LTNE models, are discussed. According to the previous studies, the LTNE model is more accurate for modeling heat transfer in porous media. While, in conditions where there is large heat transfer between the two phases, the LTE model can be utilized. We further discussed the challenges in utilizing LTNE models when solving the interface between a porous medium and a fluid layer or a porous medium and an impermeable wall subjected to a heat flux. Then, some important physics of convection in porous media, known as heat flux splitting and bifurcation in porous media, are presented and discussed. These physics underlying the heat flux splitting and bifurcation are still unknown and would require attention by the porous media community to deploy pore-scale modeling in order to shed some light into the physics underlying these phenomena.

NOMENCLATURE

b microscopic geometry constant
c_a a constant tensor
c_p fluid specific heat [J/(kg K)]
d particle diameter
d_p pore or fiber diameter (m)
F Forchheimer constant
H dimensionless microscopic solid geometry coefficient
h heat transfer coefficient (W/m² K)
I radiation intensity (W/m²)
K permeability (m²)
k thermal conductivity (W/m K)
L length of the media (m)
Nu Nusselt number
Pr Prandtl number
P pressure (Pa)
q total heat flux (W/m²)
q''' volumetric internal heat generation (W/m³)

Re	Reynolds number
S	scattering direction vector
t	time (s)
T	temperature (K)
u	Darcy velocity (m/s)
U	average velocity (m/s)
V	volume (m^3)
x	longitudinal vector (m)
y	vertical vector (m)
z	longitudinal vector (m)

Greek Letters

ε	porosity
ρ	density (kg/m^3)
ω	scattering albedo
μ	viscosity (kg/m s)
α	shape factor
β	shape factor
γ	extinction coefficient
φ	phase change
σ	Stefan–Boltzmann constant
ϑ	dynamic viscosity (m^2/s)

Subscripts

b	black body
d	Darcy
f	fluid
p	pore or fiber
s	solid
rad	radiation
t	total

REFERENCES

Alazmi B., K. Vafai, Analysis of fluid flow and heat transfer interfacial conditions between a porous medium and a fluid layer, *Int. J. Heat Mass Trans.* 44 (2001) 1735–1749.

Alazmi B., K. Vafai, Constant wall heat flux boundary conditions in porous media under local thermal non-equilibrium conditions, *Int. J. Heat Mass Transf.* 45 (15) (2002) 3071–3087.

Amiri A., K. Vafai, T.M. Kuzay, Effects of boundary conditions on non-Darcian heat transfer through porous media and experimental comparisons, *Numer. Heat Transfer 27 Part A* 27 (1995) 651–664.

Anderson R., L. Bates, E. Johnson, J. F. Morris, Packed bed thermal energy storage: A simplified experimentally validated model, *J. Energy Storage* 4 (2015) 14–23.

Anuar F. S., I. Ashtiani Abdi, K. Hooman, Flow visualization study of partially filled channel with aluminium foam block, *Int. J. Heat Mass Trans.* 127 (2018) 1197–1211.

Beavers G. S., D. D. Joseph, boundary conditions at a naturally permeable wall, *J. Fluid Mech.* 30 (1967) 197–207.

Beck J. L., Convection in a box of porous material saturated with fluid, *Phys. Fluids* 15 (1972) 1377–1383.

Bird R.B., W.E. Stewart, E.N. Lightfoot, *Transport Phenomena*, 1st ed., John Wiley & Sons, New York, 1960.

Chumpia A., K. Hooman, Performance evaluation of single tubular aluminium foam heat exchangers, *Appl. Therm. Eng.* 66 (2014) 266–273.

Darcy H. P. G., Les Fontaines Publiques de la Ville de Dijon. Victor Dalmont, Paris (1856).

Ergun S., Fluid flow through packed columns, *Chem. Eng. Prog* 48 (2) (1952) 89–94.

Forchheimer P. Wasserbewegung durch Boden. Zeitschrift des Vereines Deutscher Ingenieuer, 45 edition, 1901.

Hooman K., Thermohydraulics of porous heat exchangers: Full or partial blockage? *Proceedings of the 5th International Conference on Porous Media and its Applications in Science and Engineering* ICPM5 June 22–27, 2014, Kona, Hawaii.

Hooman K., N. Dukhan, A Theoretical model with experimental verification to predict hydrodynamics of foams, *Transp Porous Med.* 100 (2013) 393–406.

Hooman K., H. Gurgenci, Porous medium modeling of air-cooled condensers, *Transp Porous Med.* 84 (2010) 257–273.

Hossain M. A., K. Khanafer, K. Vafai, The effect of radiation on free convection flow of fluid with variable viscosity from a porous vertical plate, *Int. J. Therm. Sci.* 40 (2001) 115–124.

Hsu C. T., P. Cheng, Thermal dispersion in a porous medium, *Int. J. Heat Mass Transfer* 33 (1990) 1587–1597.

Incropera F.P., D.P. DeWitt, *Introduction to Heat Transfer*, 2nd ed., John Wiley & Sons, New York, 1990.

Irmay S., On the theoretical derivation of Darcy and Forchheimer formulas, *Eos, Trans. AGU* 39 (1958) 702–707.

Jahanshahi Javaran E., S.A. Gandjalikhan Nassab, S. Jafari, Thermal analysis of a 2-D heat recovery system using porous media including lattice Boltzmann simulation of fluid flow, *Int. J. Therm. Sci.* 49 (2010) 1031–1041.

Jiang P.X., X.L. Ouyang, Z. Huang, R.N. Xu, Local thermal non-equilibrium model for heat transfer in porous media and its applications, in *The Thirteenth Tsinghua–Seoul National–Kyoto University Thermal Engineering Conference*, Tsinghua University, China, 2013.

Joseph D. D., D. A. Nield, G. Papanicolaou, Non-linear equation governing flow in a saturated porous medium, *Water Resour. Res.* 18 (4) (1982) 1049–1052.

Kampman N., M. Bickle, M. Wigley, B. Dubacq, Fluid flow and CO2–fluid–mineral interactions during CO2-storage in sedimentary basins, *Chem. Geol.* 369 (2014) 22–50.

Kays W.M., A.L. London, *Compact Heat Exchangers*, 3rd ed., McGrow-Hill, New York, 1984.

Keshtkar M. M., S.A. GandjalikhanNassab, Theoretical analysis of porous radiant burners under 2-D radiation field using discrete ordinates method, *J. Quant. Spectrosc. Radiat. Transf* 110 (2009) 1894–1907.

Khosravy El-Hossaini M., M. Maerefat, K. Mazaheri, Numerical investigation on the effects of pressure drop on thermal behavior of porous burners, *ASME J. Heat Transf.* 130 (2008), 32601.

Klein P., T. H. Roos, T. J. Sheer, Experimental investigation into a packed bed thermal storage solution for solar gas turbine systems, *Energy Procedia* 49 (2014) 840–849.

Kuwahara F., M. Shirota, A. Nakayama, A numerical study of interfacial convective heat transfer coefficient in two-energy equation model for convection in porous media, *Int. J. Heat Mass Transf.* 44 (6) (2001) 1153–1159.

Lage J. L., The fundamental theory of flow through permeable media from Darcy to turbulence. In: Ingham, D.B., Pop, I. (eds.) *Transport Phenomena in Porous Media*, Oxford, Pergamon, Turkey, (1998).

Lage J. L., B. V. Antohe, Darcy's experiments and the deviation to nonlinear flow regime, *J. Fluids Eng.-Trans. ASME* 122 (3), 619–625 (2000).

Lee J. S., K. Ogawa, Pressure drop through packed bed, *J. Chem. Eng. Jpn* 27 (5) (1994) 691–693.

Maerefat M., M. Khosravy, El-Hossaini, K. Mazaheri, Numerical modeling of two-dimensional cylindrical porous radiant burners with sidewall heat losses, *J. Porous Media* 14 (2011) 317–327.

Mahjoob S., K. Vafai, Analytical characterization and production of an isothermal surface for biological and electronic application, *ASME J. Heat Transf.* 131 (2009). 052604.

Mahmoudi Y., Effect of thermal radiation on temperature differential in a porous medium under local thermal non-equilibrium condition, *Int. J. Heat Mass Transf.* 76 (2014) 105–121.

Mendes M. A. A., J.M.C. Pereira, J.C.F. Pereira, A numerical study of the stability of one-dimensional laminar premixed flames in inert porous media, *Combust. Flame* 153 (2008) 525–539.

Modest M. F., *Radiative Heat Transfer*, McGraw-Hill, New York, 1993.

Nazari M., D. Jalali Vahid, R. Khoshbakhti Saray, Y. Mahmoudi, Experimental investigation of heat transfer and second law analysis in a pebble bed channel with internal heat generation, *Int. J. Heat Mass Transf.* 114 (2017) 688–702.

Nie X., R. Besant, R. Evitts, J. Bolster, A new technique to determine convection coefficients with flow through particle beds, *ASME J. Heat Transf.* 133 (4) (2011) 041601.

Nield D. A., A. Bejan, *Convection in Porous Media*, Springer, New York, 2006.

Nield D. A., A. V, Kuznetsov, An historical and topical note on convection in porous media, *J. Heat Transfer* 135 (6), 061201 (2013).

Nsofor E.C., G.A. Adebiyi, Measurements of the gas-particle convective heat transfer coefficient in a packed bed for high-temperature energy storage, *Exp. Thermal Fluid Sci.* 24 (1) (2001) 1–9.

Odabaee M., K. Hooman, Application of metal foams in air-cooled condensers for geothermal power plants: An optimization study, *Int. Commun. Heat Mass Transf.* 38 (7) (2011) 838–843.

Oró E., A. Castell, J. Chiu, V. Martin, L. F. Cabeza, Stratification analysis in packed bed thermal energy storage systems, *Appl. Energy* 109 (2013) 476–487.

Shikh Anuar F., I. Ashtiani Abdi, M. Odabaee, K. Hooman, Experimental study of fluid flow behavior and pressure drop in channels partially filled with metal foams, *Exp. Therm. Fluid Sci.* 99 (2018) 117–128.

Suchanek M., Z. Olejniczak, Visualization of fluid flow pathways in wood by low-field 1H and 3He contrast MRI, *Int. J. Multiphase Flow* 72 (2015) 83–87.

T'Joen C., P. De Jaeger, H. Huisseune, S. Van Herzeele, N. Vorst, M. De Paepe, Thermo-hydraulic study of a single row heat exchanger consisting of metal foam covered round tubes, *Int. J. Heat Mass Transf.* 53 (15–16) (2010) 3262–3274.

Vafai K., *Handbook of Porous Media* (2nd ed.), Taylor & Francis Group, Boca Raton, FL, 2005.

Vafai K., A. Bejan, W. J. Minkowycz, K. Khanafer, A Critical synthesis of pertinent models for turbulent transport through porous media, *Advances in Numerical Heat Transfer*, Vol. 2, Chapter 12, pp. 389–416, John Wiley & Sons, Hoboken, NJ, 2006.

Vafai K., S. J. Kim, Forced convection in a channel filled with a porous medium: An exact solution, *ASME J. Heat Transfer* 111 (1989) 1103–1106.

Vafai K., K. Yang, A note on local thermal non-equilibrium in porous media and heat flux bifurcation phenomenon in porous media, *J. Transp. Porous Media* 96 (2013) 169–172.

Wakao N., S. Kaguei, *Heat and Mass Transfer in Packed Beds*, Vol. 1, pp. 243–295, Gorden and Breach, New York, 1982.

Ward J. C., Turbulent flow in porous media, *ASCE J. Hydraul. Division*, 90, HY 5 (1964) 1–12.

Whitaker S., Forced convection heat transfer correlations for flow in pipes, past flat plates, single cylinders, single spheres, and for flow in packed beds and tube bundles, *AIChE J.* 18 (2) (1972) 361–371.

Wooding R. A., Steady state free thermal convection of liquid in a saturated permeable medium, *J. Fluid Mech.* 2 (1957) 273–285.

Yang K., K. Vafai, Analysis of temperature gradient bifurcation in porous media – an exact solution, *Int. J. Heat Mass Transf.* 53 (2010) 4316–4325.

Yang K., K. Vafai, Analysis of heat flux bifurcation inside porous media incorporating inertial and dispersion effects—An exact solution, *Int. J. Heat Mass Tran.* 54 (2011a) 5286–5297.

Yang K., K. Vafai, Restrictions on the validity of the thermal conditions at the porous–fluid interface: An exact solution, *ASME J. Heat Transf.* 133 (2011b) 112601-1–112601-12.

Yang K., K. Vafai, Transient aspects of heat flux bifurcation in porous media: An exact solution, *ASME J. Heat Transf* 133 (2011c) 052602-1–052602-12.

2

Natural Convection in Porous Media

Kamel Hooman and Donald A. Nield

CONTENTS

2.1 Introduction 19
2.2 Self-Heating of a Porous Medium 20
    2.2.1 Dry Porous Medium 21
    2.2.2 Wet Porous Medium 23
2.3 Natural Convection in a Porous-Saturated Box 24
2.4 Natural Convention in a Partly Porous Box 28
2.5 Natural Draft Dry Cooling Tower 31
2.6 Conclusion 33
Nomenclature 33
References 34

2.1 Introduction

We use the definition of Nield and Bejan (2017) for a porous medium as "a material consisting of a solid matrix with an interconnected void." This chapter is limited to cases where the solid matrix is rigid (no deformation) to keep the interconnectedness of the void (the pores), hence allowing the flow of a fluid through the material.

In this chapter, we use a multitude of approaches, including scale analysis, resistance/pore networks, and intersection of asymptotes to investigate the heat and fluid flow behavior at different scales depending on the "information" that may be of interest to the problem solver. According to Bejan (2004a, 2004b), the way flow through a porous medium is analyzed is a question of distance—the distance between the problem solver and the actual flow structure. Depending on the aspects of the flow features that may be of interest to the solver, the distance can be adjusted, as is the normal practice in engineering thermodynamics with the choice of control mass or control volumes. For a short distance, the observer sees only some flow channels, or a number of cavities, open or closed. Hence, the conventional thermofluids engineering knowledge is appropriate to give us the information at every point of the domain. With a longer distance, however, there are too many channels and cavities in the field of vision to allow for the (practical) application of a conventional approach. Hence, volume-averaging and global measurements are left as pragmatic options. These rely on the standard convection equations governing the flow and result in the macroscopic equations by averaging over volumes or areas containing many pores. The resultant governing equations are then "visually" different from conventional equations governing convective flows. For instance, the Darcy flow model assumes a linear relationship between the pressure drop across a porous medium and the flow rate. At the pore level, however, there are local changes in the flow direction, and mass flow rate might vary in each void. Such macroscopic models can be obtained through a number of different approaches including volume averaging. Lage (1998) presents a very interesting note on the historical development of the current models from Darcy's experiments in 1700s. It is beyond the scope of this chapter to go through different models. A detailed review of the literature on the existing models for mass, momentum, and heat transfer in porous media is presented in Nield and Bejan (2017).

Macroscopic models are preferred for most of the engineering applications due to their simplicity compared to microscopic modeling. These macroscopic models lend themselves well to numerical simulations and are easier to use to interpret experimental data. The former faces the challenge of computational requirement to perform a proper numerical simulation (creating the fine grids required to capture the boundary layers over interconnected solid surfaces at reasonably fine time-steps) while the latter has to rely on non-destructive measurement techniques to provide accurate pore-level information. Attempts are made to address these issues, but the challenge remains. For example, pore-scale modeling of the reservoir for an enhanced geothermal system is impossible mainly because the required grid size is smaller than the reservoir size by several orders of magnitude. Similarly, instrumentation required for experimental measurements of, say, local velocity, through a lab-scale-sized porous medium sample is prohibitively difficult while flow visualization techniques like particle image velocimetry (PIV) are only able to cover a small field of view, which again poses a significant restriction if a full-scale measurement is needed.

Having said these, most of the practical applications are concerned with global aspects of the heat and fluid flow, say, the total pressure drop across a porous medium or the total heat transfer to or from a plate attached to a porous medium. Hence, reasonably accurate but easy-to-solve (and use) governing equations are preferred. It is to be anticipated that some information would be lost through the averaging process. However, the problem-solver wants to be able to "choose" which part and level of details are to be dropped without causing vital errors. The obvious challenge is the trade-off between accuracy and simplicity. Hence, we try to present some approaches as simplifying tools to investigate complex engineering problems.

The remainder of this chapter focuses on some case studies with free convection as the common thread in all of them. We will consider a number of problems where shortcut procedures can be implemented to obtain easy and relatively accurate answers without the need for somewhat more involved solution procedures. Throughout this chapter, when dealing with volume-averaged equations, we add the gravitational term to the momentum equation—the Darcy equation or its appropriate extension. In governing equations, the gravitational force per unit volume of the fluid will appear. To be able to analyze thermal natural convection, the fluid density must be given as a function of the fluid temperature. Therefore, we need an equation of state to complement the equations of mass, momentum, and energy. The simplest equation of state is

$$\rho_f = \rho_0 \left[1 - \beta \left(T - T_0 \right) \right] \tag{2.1}$$

where ρ_0 is the fluid density at some reference temperature T_0, and β is the thermal expansion coefficient.

Throughout this chapter, we employ the Oberbeck–Boussinesq approximation, which assumes constant properties for the medium, except that the vital buoyancy term involving β is retained in the momentum equation. The approximation is valid provided that density changes remain small in comparison with the reference density throughout the flow region and provided that temperature variations are insufficient to cause the various properties of the medium (fluid and solid) to vary significantly from their mean values.

2.2 Self-Heating of a Porous Medium

Applications like grain storage, composting, food processing, and underground fire are associated with internal heat generation in a porous medium, see Baytas (2003), Celli et al. (2010), Merrikh et al. (2002, 2005a, 2005b), Bejan and Morega (1993), Ejlali et al. (2011), and Saulov et al. (2014). There are a number of purely theoretical investigations looking into such problems. However, our

Natural Convection in Porous Media

interest in the problem has been mainly driven by the hazard of coal spontaneous combustion faced by the coal and the power industry. Power plants consume thousands of tons of coal on a daily basis and thereby store it in piles next to the plant. Over time, internal chemical reactions (oxidation) generate heat and increase the pile temperature. If the increase in temperature continues, the coal will reach the thermal runaway temperature, and a catastrophic fire (given the pile size) will be inevitable. There are a number of cases, on top of lab experiments, that report and document this phenomenon. Hence, the pile temperature has to be monitored to comply with the safety requirements and to minimize the risk. Crosswinds can either expedite or slow down the process of coal stockpile spontaneous combustion. The industry relies on techniques like compaction and wetting of the pile, both of which are proven to worsen the scenario, at least in some cases, depending on the conditions (Ejlali 2012). Hence, the problem deserves more attention and a rigorous analysis. Ejlali (2012) conducted a thorough two-dimensional numerical simulation of the problem using the governing equations given below.

Mass continuity:

$$\frac{\partial u}{\partial x} + \frac{\partial v}{\partial y} = 0 \tag{2.2}$$

Momentum:

$$\frac{\partial u}{\partial t} + u\frac{\partial u}{\partial x} + v\frac{\partial u}{\partial y} = -\varepsilon\frac{\partial p}{\partial x} + v\left(\frac{\partial^2 u}{\partial x^2} + \frac{\partial^2 u}{\partial y^2}\right) - \varepsilon\left(\frac{vu}{K} + C_F\varepsilon u\sqrt{\frac{u^2+v^2}{K}}\right) \tag{2.3}$$

$$\frac{\partial v}{\partial t} + u\frac{\partial v}{\partial x} + v\frac{\partial v}{\partial y} = -\varepsilon\frac{\partial p}{\partial y} + v\left(\frac{\partial^2 v}{\partial x^2} + \frac{\partial^2 v}{\partial y^2}\right) - \varepsilon\left(\frac{vv}{K} + C_F\varepsilon v\sqrt{\frac{u^2+v^2}{K}}\right) + \varepsilon g\beta\left(T-T_0\right) \tag{2.4}$$

Energy:

$$\rho c_p \frac{\partial T}{\partial t} + \left(\rho c_p\right)_f\left(u\frac{\partial T}{\partial x} + v\frac{\partial T}{\partial y}\right) = k\left(\frac{\partial^2 T}{\partial x^2} + \frac{\partial^2 T}{\partial y^2}\right) + (1-\varepsilon)q + M\rho h_{vl} \tag{2.5}$$

Water liquid and vapor transfer:

$$\frac{\partial \omega_l}{\partial t} = D_{ml}\left(\frac{\partial^2 \omega_t}{\partial x^2} + \frac{\partial^2 \omega_t}{\partial y^2}\right) + D_{tl}\left(\frac{\partial^2 T}{\partial x^2} + \frac{\partial^2 T}{\partial y^2}\right) + M \tag{2.6}$$

$$\frac{\partial \omega_v}{\partial t} = D_{mv}\left(\frac{\partial^2 \omega_t}{\partial x^2} + \frac{\partial^2 \omega_t}{\partial y^2}\right) + D_{tv}\left(\frac{\partial^2 T}{\partial x^2} + \frac{\partial^2 T}{\partial y^2}\right) - M \tag{2.7}$$

Total moisture:

$$\omega_{tot} = \omega_l + \omega_v \tag{2.8}$$

We, however, rely on scale analysis in this chapter. We investigate two cases of dry and wet porous medium in the following subsections.

2.2.1 Dry Porous Medium

Coal stockpiles are usually extended in one direction (here in z), and shorter in the other two (x, y) allowing for a two-dimensional assumption to be made. The side angle is usually close to 45°, thereby the width and height of the pile are of the same order of magnitude (see Figure 2.1). We take the stockpile

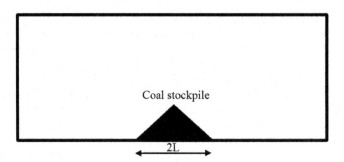

FIGURE 2.1 Side view of a coal stockpile with 45° side angle.

as our region of interest. We start with the steady-state case and assume that the heat generation rate remains constant (spatially and temporally); we will revisit an assumption in the next subsection. As dry porous medium is investigated here, only Equations (2.2) through (2.5) are needed, which are further simplified with $M = 0$ (M is the evaporation rate divided by the mass of the porous medium).

We consider two extremes for this problem to carry out scale analysis for the stockpile. One can think of an extreme when the permeability (the stockpile permeability not the coal permeability) is too low to allow the flow of air in the stockpile. This is a limiting case when buoyant forces are not strong enough to overcome viscous and form-drag forces. Hence, there is no fluid movement. In this limit, the mass continuity and momentum equations are already satisfied, and the energy equation reduces to a balance between generation and conduction, leading to:

$$\Delta T \sim \frac{(1-\varepsilon)qL^2}{4k} \tag{2.9}$$

On the other extreme, we assume another limiting case when the permeability is high enough to allow for a convection-dominant process. Hence, the energy equation reduces to a convection-generation balance leading to

$$\Delta T \sim \frac{(1-\varepsilon)qL}{2(\rho C_p)_f V} \tag{2.10}$$

A proper scale for the fluid velocity can be obtained by balancing the viscous drag with the driving force, here buoyancy, to observe that

$$V \sim \frac{K}{\mu}(\rho_f g \beta \Delta T) \tag{2.11}$$

Making use of the velocity scale, one gets a scale for the temperature as

$$\Delta T \sim \sqrt{\frac{\nu(1-\varepsilon)qL}{2(\rho C_p)_f K g \beta}} \tag{2.12}$$

Now, with the steady-state results in hand, one can go back to the unsteady governing equations to obtain the time taken to reach the steady state without the "need" to solve the unsteady problem. For the first case we considered here, that is, the extremely low permeability case, one balances the transient term with either the conduction or generation term (which are of the same order of magnitude under steady condition—Bejan's late regime) to obtain

$$\Delta t \sim \frac{L^2}{4\alpha} \tag{2.13}$$

Natural Convection in Porous Media

Similarly, for the other extreme, that is, the free convection-dominated problem, one balances the transient term with that of convection or generation, after the steady-state condition is reached, to obtain the time scale as

$$\Delta t \sim \frac{\rho c_p}{\left(\rho c_p\right)_f} \sqrt{\frac{\left(\rho c_p\right)_f vL}{2qKg\beta}} \qquad (2.14)$$

The above results were successfully compared with corresponding transient and steady-state results in Ejlali and Hooman (2011).

2.2.2 Wet Porous Medium

The main difference, here, compared with the dry porous medium, is the presence of another phase, namely water in the porous medium. Water is present in the ambient air (humidity) while it also fills the pores (of each wet coal particle) in its liquid phase. In essence, we are now dealing with a multiphase heat and mass-transfer problem where momentum transfer is also of interest. Furthermore, the chemical reactions are modeled through the addiction of an Arrhenius-type heat source term

$$q = \rho_f Q_1 A \exp\left(-\frac{E}{R^*T}\right) \qquad (2.15)$$

while mass transfer is best modeled through the use of Equations (2.6) through (2.9). Given the low porosity and permeability of the pile on top of its huge size, the heat generated in the pile, away from the coal–air interface, has to be either conducted through the solid and liquid phase or convected away by the air in the pile through free convection. In general, there will be three stages for the process. The first stage is mainly using the generated heat to increase the pile temperature until it reaches the saturation temperature. Second, once the saturation state is reached, the internally generated heat will be spent on evaporating the water content of the coal pile. Finally, when the pile is dry, then the heat of reaction will further increase the pile temperature to the thermal runaway temperature when a new chemical process kicks in. Once plotted on temperature–time chart, the first and last stage are best described by an increase of the stockpile temperature with time while the second stage, dominated by phase change, is an isothermal process. Mathematically, the first and the last stage are represented by the balance between the internal heat generation and the transient term, that is,

$$\rho c_p \frac{\Delta T}{t} \sim \left(1-\varepsilon\right)\rho_f Q_1 A \exp\left(-\frac{E}{R^*T}\right) \qquad (2.16)$$

with the only difference being in the coal property being wet during Stage 1 and dry during the final stage. Hence, the temperature scale is given by

$$\Delta T \sim \left[\left(1-\varepsilon\right)\frac{\rho_f}{\rho c_p}Q_1 A \exp\left(-\frac{E}{R^*T}\right)\right]t \qquad (2.17)$$

The second stage, as explained above, is associated with phase-change generation balance. That is, the temperature remains constant at the saturation or coal "dry" temperature, and during this stage coal dehumidification occurs. Hence, we have

$$\rho M h_{vl} \sim \left(1-\varepsilon\right)\rho_f Q_1 A \exp\left(-\frac{E}{R^*T}\right) \qquad (2.18)$$

The time taken for this process to finish is needed. Here, we have to rely on pore-scale information to find out how long does it take for the moisture to travel across the particle, say, to reach the particle surface where the moisture transfer between each particle and the convective flow can happen. Adding Equations (2.6) and (2.7) and making use of Equation (2.8), one balances the transient term with that of mass diffusion to get

$$t \sim \frac{d^2}{D_{ml} + D_{mv}} \qquad (2.19)$$

This time scale is the same as that taken for the stockpile to get to dry stage, that is, for the total moisture to transfer to the air, which is mathematically described as

$$t \sim \frac{\omega_t}{M} \qquad (2.20)$$

Now eliminating M between Equations (2.18) and (2.20) and using Equation (2.19), one has

$$T \sim \frac{E}{R^*} \frac{1}{Ln\left[\dfrac{\omega_t\left(D_{ml} + D_{mv}\right)\rho h_{vl}}{\left(1 - \varepsilon\right)\rho_f Q_1 A d^2}\right]} \qquad (2.21)$$

This is the scale for the temperature during the second stage, which is, as mentioned before, remaining constant throughout this stage. Following the completion of this stage, the pile temperature will increase beyond this temperature by a slope given in Equation (2.16). Hence, the three stages are now described in simple mathematical terms. Results are shown in Hooman and Maas (2014) to agree well with those of independent experimental and numerical data in the literature.

2.3 Natural Convection in a Porous-Saturated Box

While a classical porous media problem, it is a challenge to generate results from the volume-average-based models to match those of experiments. Beji and Goblin (1992) showed that the use of thermal dispersion and local thermal non-equilibrium (LTNE) models can improve the performance of volume-averaged models. An immediate question faced by the LTNE models is the use of a proper correlation for the solid–fluid heat transfer correlation. Most of the models rely on the correlations developed for forced convection problems where the velocity is prescribed or can be inferred based on inlet or a given velocity. For free convection problems, however, an iterative procedure has to be involved (though ignored by most authors) to get the correct velocity field, which can satisfy the coupled momentum and energy equation while furnishing the interfacial heat transfer coefficient in the LTNE term. The existing LTNE models need an average velocity that is realistic for a homogenous porous medium with uniform distribution of porosity and permeability across the medium. One can, to some extent, minimize the error through the use of tortuosity or dispersion models [Ozgumus et al. (2013) and Kuwahara and Nakayama (1999)]. However, with free convection this may not be as straight forward as desired. Pore-scale simulations conducted by Merrikh and Lage (2005) showed interesting flow features that are not "visible" to the solver when volume-averaged models are used. Similarly, Imani et al. (2012, 2013) questioned the validity of the heat flux splitting models at the solid–porous interface using Lattice Boltzmann method (LBM) to generate pore-scale results, which were then contrasted by the existing volume-averaged-based models. The clear flow-porous media interface proved to be as challenging to model as the wall-porous matrix, if not more. Sauret et al. (2014) and Anuar et al. (2018a, 2018b) have experimentally shown a non-uniform interface velocity as well as the formation of a boundary layer on the interface. Further to these, the flow departs the porous medium, halfway through or so along the way, leaving part of the downstream section at lower pressure giving rise to the formation of secondary flows and wake for a forced convection

Natural Convection in Porous Media

flow through and parallel to a porous block. Nield and Bejan (2017) argue that interface modeling for a free convection problem is less problematic than that of the force convection counterpart. The main point, however, is the limited nature of the information one can expect from the volume-averaged models. The thermal resistance approach comes across as a very powerful and pragmatic yet surprisingly less popular approach than more computationally involved ones. Tamayol and Hooman (2011) used the approach to generate a model to analyze porous media flows by modeling the solid phase as a network of resistors. On top of thermal resistors, Hooman and Dukhan (2013) have created a pore network to investigate local fluid flow resistances as well. Scale analysis was used to evaluate local flow resistances in the pores and around them for a regular and periodic porous medium, namely a metal foam. Besides, the intersection of asymptotes approach was undertaken to come up with an equation that simplifies the drag force calculation for a cylinder in cross flow and forms a theoretical basis for the Hazen–Dupuis–Darcy (a.k.a. Forchheimer) model. As seen, a high-resolution solution can be obtained using the resistance network approach, which mainly uses the microscale information as the input.

Here, we revisit the problem investigated by Hooman and Merrikh (2010) to demonstrate the application of the method to natural convention in a laterally heated box filled with conducting unconsolidated square blocks. The enclosure is replaced by a number of parallel and series resistances (Figure 2.2). The network of resistors starts from the heated wall. There is an immediate resistance, R_H, to the flow of heat toward the boundary layer adjacent to the wall before the blocks are heated. The blocks and the spaces in between them are replaced by a number of parallel resistances with their equivalent resistance, $R_{eq,p}$, being in series with the heated wall boundary layer resistance. The boundary layer at the cold vertical wall is also replaced by a thermal resistance, R_C. Hence, the laterally heated box is modeled as two convection resistances at the heated and the cold wall and an equivalent resistance taking into account the thermal resistance posed by the fluid within the voids and those of solid blocks. Under steady-state conditions, the same heat that is removed from the heated wall is transferred to the cold wall. The heat has to cross the fluid–solid cluster, which can be thought of as a porous medium. This allows for the development of an overall resistance, R_{tot}, to which the total heat is inversely proportional. That is

$$R_{tot} = \frac{T_H - T_C}{Q} \tag{2.22}$$

The overall resistance is the sum of the aforementioned resistances

$$R_{tot} = R_H + R_{eq,p} + R_C \tag{2.23}$$

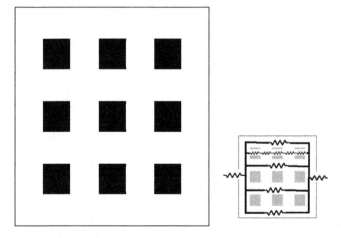

FIGURE 2.2 Natural convection in a laterally heated box with conducting identical blocks along with the constructed resistance network model.

One notes that while locally, along the wall, the convection resistances are different, their average value is the same. Hence, $R_C = R_H$ leading to

$$R_{tot} = 2R_H + R_{eq,p} \qquad (2.24)$$

where R_H is the average free convection heat transfer resistance of a heated wall, of height L and width W, given by

$$R_H = \frac{1}{hLW} = \frac{1}{k_f Nu_H W} \qquad (2.25)$$

Note that Nu_H can be obtained through the use of proper correlations in the literature but obtaining an expression for $R_{eq,p}$ is a bit more involved. This is mainly because one needs to account for block conduction resistances on top of conduction and convection resistances posed by the fluid flowing between the blocks. We subdivide the fluid–solid area to a number of horizontal layers—energy corridors from the heated to the cold wall. Some of these layers will be only fluid layers (horizontal channels); hence, the resistance will be convection resistance. Parallel to these fluid layers, there will be an adjacent horizontal layer, composed of parallel conducting solid blocks, along which the solid and fluid resistances are in series. The total resistance, of this solid–fluid assembly, will then be the equivalent of a number of parallel resistances.

The solid block resistance is expressed as

$$R_B = \frac{1}{k_s W} \qquad (2.26)$$

One assumes that the fluid in the horizontal channels between the blocks moves slowly enough to have the resistance dominated by that of fluid conduction like a stagnant layer, that is,

$$R_{SF} = \frac{1}{k_f W} \qquad (2.27)$$

With $\sqrt{N}$ blocks in each horizontal row, there will be $(\sqrt{N} - 1)$ spaces allocated to these stagnant fluid packets. Given that these resistances are in series, the equivalent thermal resistance of each horizontal layer is the summation of solid block and stagnant fluid resistances with their number as the weight functions, as follows

$$R_i = \sqrt{N} R_B + (\sqrt{N} - 1)R_{SF} \qquad (2.28)$$

The overall resistance of identical horizontal energy corridors, with R_i as their resistances, can be obtained as

$$R^{-1} = \sum_{i=1}^{\sqrt{N}} R_i^{-1} = \frac{\sqrt{N}}{\left(\sqrt{N} R_B + (\sqrt{N} - 1)R_{SF}\right)} \qquad (2.29)$$

The resistance of fluid channels is given by that of convection, which is manifested by the change in the enthalpy of the fluid in the channel, that is,

$$R_{conv} = \frac{1}{c_{p,f} \dot{m}} \qquad (2.30)$$

To get the average flow rate in each channel, one can rely on simple scale analysis results to correlate the local velocity with the Rayleigh number, see Hooman and Merrikh (2010). Skipping over the details presented in Hooman and Merrikh (2010), one gets to express the overall Nusselt number as a function

Natural Convection in Porous Media 27

of the number of blocks, solid-to-fluid thermal conductivity ratio, and the Rayleigh number (Nu_H is a function of Ra_f, too) as

$$Nu = \frac{Nu_H}{2 + \dfrac{Nu_H}{\dfrac{\sqrt{N}}{\sqrt{N}(1+\kappa)-1} + \dfrac{1}{2.5\sqrt{N}-1}\left(0.33Ra_f^{0.5} + \sqrt{N} - 2\right)}} \tag{2.31}$$

The preceding equation, which was obtained simply based on basics of fluid mechanics and heat transfer at the pore level using the thermal resistance network approach, was successfully validated against pore-scale numerical results of Merrikh and Lage (2005).

A possible extension to the preceding work is to build a hydrodynamic resistance network for the fluid flow as well to get a more accurate understanding of the local heat transfer coefficients and flow distributions, which are obtained here using simplifying assumptions. One notes that here the solver can "see" the extent of the boundary layer if it covers a conducting block (or more than one column might fall within the boundary layer). Hence, the equivalent resistance can be adjusted and the flow switch behavior reported in the Merrikh and Lage (2005) pore-scale simulation can be modeled properly. Interestingly, one can obtain a term that mainly manifests the solid–fluid interfacial heat transfer (which will then be shown to be proportional to $Ra_f^{0.5}$). This will be even easier through the use of correlations developed in Hooman and Merrikh (2010) based on the least squares model. The Nusselt number correlations developed in Hooman and Merrikh (2010) took into account effects of the Rayleigh number, the fluid-to-solid thermal conductivity ratio, and the block number density. The following correlation was found to work best across the range of parameters considered in that paper

$$Nu = 0.0155 Ra_f^{0.45} N^{-0.268} \kappa^{0.073} \tag{2.32}$$

The Nusselt number defined therein shows the heat transfer from the heated wall. It is the same as the heat transferred to the cooled wall. Note that solid blocks do not touch the walls. Hence, for steady-state conditions, the same heat crossing the walls have to be transferred through the porous matrix, here disconnected conducting blocks. Naturally, one can use the correlations for solid–fluid heat transfer for LTNE problems. Recasting in terms of an LTNE heat transfer coefficient, h_V, one has:

$$h_V = 0.0155 \frac{k_f}{\overline{V}Xa_{sf}} Ra_f^{0.45} \left(\frac{d}{L}\right)^{0.53} (1-\varepsilon)^{-0.268} \kappa^{0.073} \tag{2.33}$$

with $\overline{V}$ and X being the box volume and aspect ratio (width to height ratio), respectively.

One can round the Ra_f and d/L exponent to 0.5, at the expense of some accuracy, after some algebraic manipulation, to get

$$\frac{h_V L^2}{k_f} = C \frac{L^2}{\overline{V}Xa_{sf}} \left(Ra_f \frac{d}{L}\right)^{0.5} (1-\varepsilon)^{-0.25} \kappa^{0.073} \tag{2.34}$$

The preceding equation can be further simplified through the use of a permeability–porosity function, say, the Carman–Kozeny relationship, to replace the micro to macro length scale ratio (d/L) with a Darcy number. A Rayleigh–Darcy number can then be grouped if an effective-to-fluid thermal conductivity ratio is also incorporated as a multiplier. However, the use of Equation (2.34) seems to be more appropriate and is left as it is.

The preceding LTNE correlation has to be used with care. It has been developed based on two-dimensional microscale solutions for an unconsolidated matrix over the following range for the key parameters: $10^5 < Ra_f < 10^8$, $Pr_f = 1$, $\varepsilon = 0.64$, $9 < N < 144$, $0.1 < \kappa (= k_s/k_f) < 100$, and $5 < L/d < 20$. Despite these limitations, it has the obvious benefit of being independent from the pore velocity; therefore, it is much easier to implement in free convection problems.

2.4 Natural Convention in a Partly Porous Box

The Darcy–Bénard problem is a classical one in the literature with Elder's correlation predicting a linear Nu-Ra_K dependence. Higher Rayleigh values would mark the transition to form-drag-dominated flow, and the more generic model of Vafai-Tien or Hsu-Cheng [see Nield and Bejan (2017)] was applied to investigate different aspects of the flow and heat transfer in such bottom-heated enclosures. The problem lends itself well to modeling some practical applications including electronic cooling and hot sedimentary acquires. Recent experiments with metal foams in such settings showed excellent performance, and authors suggested the use of such foams in cooling electronics. Furthermore, Righetti et al. (2019) and Yang et al. (2010) reported the use of a metal foam layer to enhance pool boiling under both confined and unconfined conditions. In either case, the vertical channel was only partly filled by a porous layer. Results suggested a non-uniform trend for the Nusselt number versus the pores per inch (PPI). In some cases, a flat plate is found to work better than a foam-covered layer. While, as anticipated, the foam-covered plate outperformed the bare flat plate in most cases investigated therein. In order to understand that problem, it helps to start with single phase flow across a foam layer heated from below and insulated at vertical walls. We consider the vertical water channel in Righetti et al. (2019) partly filled with a thin foam layer. The heat is provided by an electric heater, transferred to a copper plate temperature of which is kept at T_s and then the aluminum foam (Figure 2.3). Sidewalls are assumed to be adiabatic, and heat is allowed to be transferred in the vertical direction. The heat will increase the water temperature, but the channel is long enough for the water surface temperature to be the same as that of air on top of it.

To use scale analysis, one relies on the momentum equation that suggests a balance between the form and viscous drag as resistances against the flow while the buoyancy term remains as the driving force described as

$$\Delta p = \rho_f \, g \, \beta \, \Delta T \, L. \qquad (2.35)$$

This has to balance the foam resistance and that of the fluid layer on top of the foam. The obvious assumption is that the former is the dominant one, and the latter is negligibly small. Hence, the force balance is that of buoyancy versus foam resistance being

$$\frac{\Delta p}{H} = \frac{\mu V}{K} + c_F \frac{\rho_f V^2}{K^{0.5}}. \qquad (2.36)$$

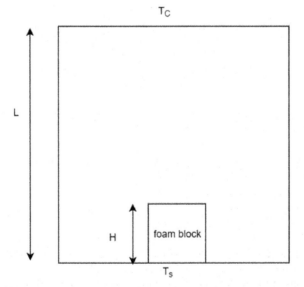

FIGURE 2.3 Foam block heated from below in a large reservoir with the tip kept at a constant temperature.

Natural Convection in Porous Media 29

For low velocity flow, the viscous term is dominant while at higher velocities the form-drag dominates as discussed in Hooman et al. (2019). In either case, comparing the preceding equations will lead to a proper scale for the fluid velocity. We consider both cases. For Darcy flow or negligible form-drag cases, one has

$$V \sim \frac{\rho_f \, g \, \beta \, \Delta T \, L K}{\mu H}, \tag{2.37}$$

which can be rearranged as

$$V \sim \frac{\alpha_f}{H} \frac{L}{H} Ra_f Da, \tag{2.38}$$

with Ra_f and Da defined as

$$Ra_f = \frac{g \, \beta \, \Delta T \, H^3}{\upsilon \alpha_f}, \tag{2.39}$$

and

$$Da = \frac{K}{H^2}. \tag{2.40}$$

Considering the other extreme, where the form drag is dominant, the force balance leads to

$$V \sim \frac{\alpha_f}{H} \left(\frac{L}{H} \right)^{0.5} \left(\frac{Bo_f}{C_F} \right)^{0.5} Da^{0.25}. \tag{2.41}$$

Here, a proper scale for velocity is obtained. For low-speed flows, the velocity is linearly proportional to the $Ra_f Da$ product, while at higher speeds associated with stronger convection patterns the velocity scales with $Bo^{0.5}$ ($Bo = Ra \, Pr$) and $Da^{0.25}$. It is interesting to note the presence of the Pr_f in form-drag-dominated free convection flow.

As we deal with a partly porous enclosure, an interesting question is to find out the required foam height to mark the transition from a viscous-drag-dominated to a form-drag-dominated flow. This can be done through intersecting the asymptotes, that is, by equating the velocity at different extremes. Once done, it will lead to the following transition criterion

$$\frac{H}{L} \sim C_F Gr_K, \tag{2.42}$$

where Gr_K is defined as

$$Gr_K = \frac{g \beta \Delta T K^{1.5}}{\upsilon^2}. \tag{2.43}$$

Note that the transition is controlled by the form-drag coefficient and the pore-Grashof number somewhat similar to the forced convection case where the pore-Reynolds number is used in the literature to mark the flow transition (to non-Darcy regime). The blockage, the foam-to-box height ratio, which is required for the flow behavior switch, is independent of the "external" length of the box, the size of the porous box, and the effective thermal conductivity. The porous structure, however, comes into play through the permeability and form-drag coefficient as one would expect.

The obtained velocity scale can be used to give us the total heat transfer from the base plate to the top plate (at the ambient air temperature), which is given by the change in the enthalpy of the fluid moving between the two horizontal plates, that is,

$$Q = \dot{m} c_p \Delta T \tag{2.44}$$

The Nusselt number defined as the ratio of the total heat transferred divided by that of pure conduction reads

$$Nu \sim \frac{UH}{\alpha}. \tag{2.45}$$

The Nusselt number for the viscous-drag- and form-drag-dominated cases are given, respectively, by

$$Nu \sim \frac{L}{H} \frac{\alpha_f}{\alpha} Ra_f Da, \tag{2.46}$$

and

$$Nu \sim \frac{\alpha_f}{\alpha} \left(\frac{L}{H}\right)^{0.5} \left(\frac{Bo}{C_F}\right)^{0.5} Da^{0.25}. \tag{2.47}$$

For the low-speed free convection flow in a bottom-heated box, the Nusselt number is linear in Ra. This is in agreement with the experimental data in the literature; see for instance Elder's correlation. On the other extreme, for high-speed or form-drag-dominated flow, the literature for both internal and external free convection suggests a parabolic Ra–Nu correlation. Figure 2.4 compares the preceding predictions with experimental data in Kathare et al. (2008).

Hooman et al. (2019) offers a thermal resistance network model for the same problem, which is not presented here. The simple approach reported therein leads to close results compared with independent experimental data reported in the literature and the preceding equations obtained based on scale analysis here.

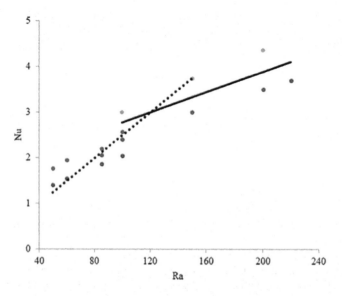

FIGURE 2.4 Our prediction for the Nusselt number versus the Rayleigh number compared with experimental data.

2.5 Natural Draft Dry Cooling Tower

All thermal power plants need to dump some heat to the ambient. This can be done through a cooling tower, which uses free convection to remove the heat from the power plant. Wet cooling towers rely on evaporative cooling thanks to direct contact between the hot water and cold air, but dry counterparts make use of air to cool the working fluid flowing through the heat exchangers with no direct contact. The hot working fluid from the turbine flows in heat exchangers, which are exposed to the air in a shell (Figure 2.5). The air at the heat exchanger is heated and gets lighter to move up while the fresh ambient air is sucked into the shell as a result of the created negative pressure. The overwhelming resistance in most cases is that of the air side, and the standard practice is to use area extension to compensate for it (Figure 2.6). Hence, densely finned tubes are used in (usually) huge cooling towers. Fins are as thin as 0.5 mm and are spaced as densely as 3 mm apart while the pipes are 5–10 m long and are bundled in multirows. Numerical simulation of such systems using direct approaches is practically impossible, as one needs to capture boundary layers in tiny fin–fin channels of 3 mm width for thousands of pipes in a heat exchanger. Hence, Hooman and Gurgenci (2010) came up with an alternative approach to model the heat exchangers as a porous medium. Bejan and Morega (1993) have reported a porous medium model for bare tube bundles and reported a permeability value for such bundles. Hooman and Gurgenci (2010), on the other hand, extended that work to the case of a finned tube bundle. They applied the existing f-Re correlations to different bundles (different fin size, fin number density, and pipe diameter) and mapped their findings to a porous medium model. Using the permeability function for bare tube bundles, the form-drag coefficient was then obtained as a function of bundle porosity. Hence, the complex bundle flow is reduced to a simple porous media flow,

FIGURE 2.5 The natural draft dry cooling tower built at UQ-Gatton campus.

FIGURE 2.6 Slightly tapered circular fins on a tube.

making it possible to run computational fluid dynamics (CFD) simulations of the cooling tower. In line with that, Hooman (2010) proposed a scale analysis approach to investigate free convection through a cooling tower treating the heat exchangers as a porous medium. The form-drag resistance is shown to be the dominant one for practical applications and was balanced with the driving force similar to what we have shown in the previous section. Table 2.1 shows a sample of results obtained from scale analysis versus those of CFD simulations of the tower with identical heat exchangers. Accuracy of the results based on simple scale analysis is further demonstrated when the predictions are compared to a full-scale tower (Figure 2.7).

TABLE 2.1
Bundle Velocity (in m/s) from Scale Analysis and 2D CFD Simulation Results

Scale Analysis	CFD	H (m)	Difference (%)
2.48	2.54	62.5	2.3
2.27	2.22	50	2.25
2.05	2.08	37.5	1.44
1.78	1.77	25	0.5

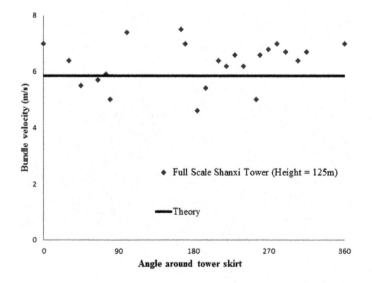

FIGURE 2.7 Theoretical model predictions for the air velocity versus field data for an operating cooling tower in China.

Natural Convection in Porous Media 33

A follow-up study, using the same approach for a case that the bundles are inclined (not normal to the air flow; A-frame or V-frame designs), showed a less successful comparison. Sakhaei (2014) used a heterogenic permeability and form-drag coefficient for the inclined bundle to note that the formulation would lead to a trend opposite to those in the experimental data. He concludes that for inclined bundles at higher velocities, the form-drag coefficient has to be a function of velocity; hence, a non-linear quadratic drag is to be formulated, see also Lage et al. (1997). Similarly, Auriault et al. (2007) indicated that the Forchheimer law does not generally survive upscaling the flow at the heterogeneity scale where the law is assumed to hold. This remains as an ongoing challenge to be addressed through further use of detailed numerical simulations and accurate pore-level experiments.

2.6 Conclusion

This chapter has presented a number of solution techniques applied to practical engineering problems with free convection as the mechanism behind the heat and fluid flow. In particular, scale analysis, intersection of asymptotes, and thermal and pore-resistor networks are presented as approaches to solve complex problems to a good accuracy at pen-and-paper level. Practical problems associated with internal heat generation with and without moisture transfer, natural draft dry cooling towers, and free convection in porous-matrix heat exchangers were investigated using the above techniques applied to either or both of the micro and macro domains of a porous medium.

NOMENCLATURE

A constant (1/s)
a_{sf} specific surface area (1/m)
Bo Boussinesq number
C constant
$\mathbf{C}_F$ form-drag coefficient
Cp specific heat at constant pressure (kJ/kg·K)
d block or particle size (m)
D mass transfer coefficient (m^2/s)
Da Darcy number
E activation energy (J/mol)
Gr Grashof number
h convection heat transfer coefficient (W/m^2 K)
h_V interfacial heat transfer coefficient (W/m^3 K)
h_{vl} latent heat
H height (m)
K permeability (m^2)
k thermal conductivity (W/mK)
L length (m)
M relative moisture evaporation rate (1/s)
$\dot{m}$ mass flow rate (kg/s)
Nu Nusselt number
P Pressure (*Pa*)
Pr Prandtl number
q volumetric heat (W/m^3)
Q heat transfer rate (W)
$Q1$ heat of reaction (J/kg)
R thermal resistance (K/W)
R^* Universal gas constant [J/ (mol·K)]
Ra Rayleigh number
t time (s)

T	temperature (K)
(u, v)	(x, y) velocity vector components (m/s)
V	velocity scale (m/s)
$\bar{V}$	volume (m^3)
W	width (m)
X	aspect ratio
x, y	Cartesian coordinates (m)

Greek Letters

α	thermal diffusivity (m^2/s)
β	thermal expansion coefficient (1/K)
ε	porosity
κ	solid–fluid thermal conductivity ratio
μ	dynamic viscosity (Pa·s)
ν	kinematic viscosity (m^2/s)
ω	moisture content
ρ	density (kg/m^2)

Subscripts

0	reference
B	block
C	cold
$Conv$	convection
eq	equivalent
f	fluid
H	heated
l	liquid
m	isothermal
p	porous
s	solid
t	non-isothermal
tot	total
v	vapor
V	volume

REFERENCES

Anuar, S. F., Ashtiani Abdi, I. and Hooman, K. 2018a. Flow visualization study of partially filled channel with aluminium foam block. *Int. J. Heat Mass Transfer* **127**, 1197–1211.

Anuar, S. F., Ashtiani Abdi, I., Odabaee, M. and Hooman, K. 2018b. Experimental study of fluid flow behaviour and pressure drop in channels partially filled with metal foams. *Exp. Thermal Fluid Sci.* **99**, 117–128.

Auriault, J. L., Geindreau, C. and Orgeas, L. 2007. Upscaling Forchheimer law. *Transp. Porous Media* **70**, 213–229.

Baytas, A. C. 2003. Thermal non-equilibrium natural convection in a square enclosure filled with a heat-generating solid phase, non-Darcy porous medium. *Int. J. Energy Res.* **27**, 975–988.

Bejan, A. 2004a. *Convection Heat Transfer*, 3rd ed., Wiley, New York.

Bejan, A. 2004b. Designed porous media: Maximal heat transfer density at decreasing length scales. *Int. J. Heat Mass Transf.* **47**, 3073–3083.

Bejan, A. and Morega, A. M. 1993. Optimal arrays of pin fins and plate fins in laminar forced convection. *ASME J. Heat Transf.* **115**, 75–81.

Beji, H. and Gobin, P. 1992. The effect of thermal dispersion on natural dispersion heat transfer in porous media. *Numer. Heat Transf. A* **23**, 487–500.

Celli, M., Rees, D. A. S. and Barletta, A. 2010. The effect of local thermal non-equilibrium on forced convection boundary layer flow from a heated surface in porous media. *Int. J. Heat Mass Transf.* **53**, 3533–3539.

Ejlali, A. 2012. Numerical modelling of a self-heating porous medium: Application to coal stockpiles, PhD Thesis, School of Mechanical and Mining Engineering, The University of Queensland.

Ejlali, A. and Hooman, K. 2011. Buoyancy effects on cooling a heat generating porous medium: Coal stockpile. *Transp. Porous Media* **88**, 235–248.

Ejlali, A., Mee, D. J., Hooman, K. and Beamish, B. 2011. Numerical modelling of the self-heating process of a wet porous medium. *Int. J. Heat Mass Transf.* **54**, 5200–5206.

Hooman, K. and Maas, U. 2014. Theoretical analysis of coal stockpile self-heating. *Fire Safety J.*, **67**, 107–112.

Hooman, K. 2010. Dry cooling towers as condensers for geothermal power plants. *Int. Comm. Heat Mass Transf.* **37**, 1215–1220.

Hooman, K. and Dukhan, N. 2013. A theoretical model with experimental verification to predict hydrodynamics of foams. *Transp. Porous Media* **100** (3), 393–406.

Hooman, K. and Gurgenci, H. 2010. Porous medium modeling of air-cooled condensers. *Transp. Porous Media* **84**, 257–273.

Hooman, K. and Merrikh, A. A. 2010. Thermal analysis of natural convection in an enclosure filled with disconnected conducting square solid blocks. *Transp. Porous Media* **85**, 641–651.

Hooman, K., Sadafi, H., Mancin, S., Righetti, G. and Xin, G. 2019. Theoretical analysis of free convection in a partially foam-filled enclosure. *Heat and Mass Transfer* **55**, 1937. doi:10.1007/s00231-018-2466-4.

Imani, G. R., Maerefat, M. and Hooman, K. 2012. Estimation of heat flux bifurcation at the heated boundary of a porous medium using a pore-scale numerical simulation. *Int. J. Therm. Sci.* **54**, 109–118.

Imani, G., Maerefat, M. and Hooman, K. 2013. Pore-scale numerical experiment on the effect of the pertinent parameters on heat flux splitting at the boundary of a porous medium. *Transp. Porous Media* **98**, 631–649.

Kathare, V., Davidson, J. H. and Kulacki, F. A. 2008 Natural convection in water-saturated metal foam. *Int. J. Heat Mass Transf.* **51**, 3794–3802.

Kuwahara, F. and Nakayama, A. 1999. Numerical determination of thermal dispersion coefficients using periodic porous structure. *ASME J. Heat Transf.* **121**, 160–163.

Lage, J. L. 1998. The fundamental theory of flow through permeable media: From Darcy to turbulence. *Transport Phenomena in Porous Media* (Eds. D.B. Ingham and I. Pop), Elsevier, Oxford, UK, pp. 1–30.

Lage, J. L., Antohe, B. V. and Nield, D. A. 1997. Two types of nonlinear pressure-drop versus flow rate relation observed for saturated porous media. *ASME J. Fluids Engng.* **119**, 701–706.

Merrikh, A. A. and Lage, J. L. 2005. From continuum to porous continuum: The visual resolution impact on modeling natural convection in heterogeneous media. In *Transport Phenomena in Porous Media III*, (Eds. D. B. Ingham and I. Pop), Elsevier, Oxford, UK, pp. 60–96.

Merrikh, A. A., Lage, J. L. and Mohamad, A. A. 2002. Comparison between pore-level and porous medium models for natural convection in a nonhomogeneous enclosure. *AMS Contemp. Math.* **295**, 387–396.

Merrikh, A. A., Lage, J. L. and Mohamad, A. A. 2005a. Natural convection in an enclosure with disconnected and conducting solid blocks. *Int. J. Heat Mass Transf.* **46**, 1361–1372.

Merrikh, A. A., Lage, J. L. and Mohamad, A. A. 2005b. Natural convection in non-homogeneous heat generating media: Comparison of continuum and porous-continuum models. *J. Porous Media* **8**, 149–163.

Nield, D. A. and Bejan, A. 2017. *Convection in Porous Media*, Springer, New York.

Ozgumus, T., Mobedi, M., Ozkol, U. and Nakayama, A. 2013. Thermal dispersion in porous media: A review on the experimental studies of packed beds. *Appl. Mech. Rev.* **65**, 031001.

Righetti, G., Doretti, L., Sadafi, H., Hooman, K. and Mancin, S. 2019. Water pool boiling across low pore density aluminum foams. *Heat Transfer Engineering*. doi:10.1080/01457632.2019.1640464.

Sakhaei, M. 2014. Inclined arrangement for heat exchanger bundles. PhD Thesis, School of Mechanical and Mining Engineering, The University of Queensland.

Saulov, D.N., Watanabe, S., Yin, J., Klimenko, D.A., Hooman, K., Feng, B., Cleary, M.J. and Klimenko, A.Y. 2014. Conditional methods in modeling CO2 capture from coal syngas. *Energies* **7** (4), 1899–1916. doi:10.3390/en7041899.

Sauret, E., Abdi, I.A. and Hooman, K. 2014. Fouling of waste heat recovery: Numerical and experimental results. In: Harun Chowdhury and Firoz Alam, Proceedings of the 19th Australasian Fluid Mechanics Conference. *19th Australasian Fluid Mechanics Conference*, Melbourne, VIC, Australia, (38.1–38.4). December 8–11, 2014.

Tamayol, A. and Hooman, K. 2011. Thermal assessment of forced convection metal foam heat exchangers. *ASME J. Heat Transf.* **133**, 1118011–1118017.

Yang Y., Ji X., and Xu J. 2010. Pool boiling heat transfer on copper foam covers with water as working fluid. *Int. J. Therm. Sci* **49**, 1227–1237. doi: 10.1016/j.ijthermalsci.2010.01.013.

3
Forced Convection in Porous Media

Pourya Forooghi and Benjamin Dietrich

CONTENTS

3.1 Introduction 37
3.2 Fundamental Principles 38
3.3 Effective Thermal Conductivity 40
3.4 Forced Convection in Ducts 44
    3.4.1 Fully Developed Flow 44
    3.4.2 Thermally Developing Flow 45
    3.4.3 Ducts Partially Filled with Porous Media 46
3.5 Local Thermal Non-Equilibrium 47
3.6 Conclusion 48
Nomenclature 48
References 49

3.1 Introduction

This chapter deals with forced convection, that is, heat transfer between a fluid flow, driven by an external source of power, and solid surfaces. The problem of forced convection in porous media has been studied in a variety of flow configurations ranging from canonical flows to the more complicated industrially relevant configurations, such as partially filled ducts (Jang and Chen, 1992; Huang and Vafai, 1994; Abu-Hijleh, 1997; Forooghi et al., 2011; Orihuela et al., 2018), porous layers wrapped around tubes (Odabaee et al., 2011; Rashidi et al., 2013; Chumpia and Hooman, 2015), jets impinging on porous blocks (Fu and Huang, 1996; Feng et al., 2014; Buonomo et al., 2016), and novel designs of foam-enhanced heat exchangers (Ejlali et al., 2009; Wang et al., 2014). The purpose of the present chapter is twofold. First, it outlines a brief overview of the topic, which also serves as a basis for the following chapters. Second, it discusses in more detail some key aspects of forced convection in solid open-celled foams, a class of porous materials with proven potential in several industrial applications. For the sake of brevity, we limit our scope to the flow in confined ducts and avoid discussing external flows as well as advanced effects, such as thermal dissipation, inhomogeneity of porous medium, and variable properties. Such effects are comprehensively reviewed by Nield and Bejan (2017).

Solid (metal or ceramic) open-celled foams have received increasing attention in the past few decades due to a combination of attractive features. Their high porosity—mostly larger than 0.75—leads to low pressure drop and lightness. At the same time, they offer attractive heat transfer properties due to large specific surface areas and continuity of the solid structure (Calmidi and Mahajan, 2000; Mahjoob and Vafai, 2008; Dietrich, 2010). In this chapter, we mainly focus on the recent results on ceramic open-celled foams regarding their pressure drop, effective thermal conductivity, and solid-to-fluid heat transfer coefficient. Ceramic open-celled foams are considered to be promising means of heat and mass-transfer enhancement particularly in applications such as solar receivers (Ávila-Marín, 2011), catalyst supports (Twigg and Richardson, 2002), and porous burners (Gauthier et al., 2008).

3.2 Fundamental Principles

We start with introducing the Forchheimer extension of the Darcy's law, that is,

$$-\nabla p = \frac{\mu_f}{K} \boldsymbol{u} + \frac{c_F \, \rho_f}{K^{1/2}} |\boldsymbol{u}| \boldsymbol{u} \tag{3.1}$$

as the governing equation for momentum transport in a porous medium (Joseph et al., 1982), where $\boldsymbol{u}$ denotes the superficial velocity. The permeability K and form-drag coefficient c_F are both geometric properties depending on the microscopic structure of the porous medium. According to Equation (3.1), the pressure gradient in the porous medium balances with contributions from viscous and form-drag terms, the first and second terms on the right-hand side also referred to as Darcy and Forchheimer terms, respectively. At low velocities, the quadratic Forchheimer term becomes insignificant and Equation (3.1) reduces to the Darcy's law. As will be discussed in Section 3.4 of this chapter, other extensions to the Darcy's law are also suggested in the literature, but we continue with Equation (3.1), as it serves our purpose for the time being.

It is important to note that by using a Darcy-type governing equation, we imply a macroscopic point of view toward the problem; that is, the flow phenomena occurring at the microscopic or "pore" scales are not captured. One can alternatively adopt a microscopic point of view and solve the complete set of Navier–Stokes equations while imposing the no-slip condition at the solid–fluid interface. This alternative approach necessitates considering the detailed geometry of the porous medium down to its smallest scales. While providing comprehensive information on the flow field, it is often impractical in non-academic problems due to its formidable computational cost. In fact, unless the pore-scale flow phenomena are of interest, one can satisfactorily use the macroscopic approach with the crucial condition that the two constants, K and C_F, are known a priori. It must be added that the Darcy and Forchheimer terms arise from the viscous and convection terms in the microscopic Navier–Stokes equations via volumetric averaging.

To continue the discussion on the issue of pressure drop in porous media, it is convenient to consider the one-dimensional flow in a duct with a constant cross section. In this flow, the macroscopic pressure gradient only exists in the axial direction, thereby Equation (3.1) reduces to the following algebraic relation

$$\frac{|\Delta p|}{\Delta L} = \frac{\mu_f}{K} u + \frac{c_F \, \rho_f}{K^{1/2}} u^2 \tag{3.2}$$

where Δp is the pressure change over the axial distance ΔL. For open-celled foams, the permeability and the form-drag coefficient can be obtained directly when using the definition of the friction factor (Equation 3.3) and the generally accepted ansatz $\xi = A / Re_{d_h} + B$, where A and B are fitting constants to experimental values (Dietrich et al., 2009).

$$\frac{|\Delta p|}{\Delta L} = \xi \, \rho_f \, u^2 / \psi^2 \, \frac{1}{d_h} \tag{3.3}$$

The pore hydraulic diameter d_h depends on the porosity and the specific solid–fluid surface area a_{sf}, which can be obtained using reconstruction techniques of pictures recorded by μCT (Dietrich et al., 2009; Dietrich, 2010).

Fitting A and B to more than 2,500 experimental values, the dimensionless correlation for the determination of pressure drop in open-celled foams stated in Equation (3.4) is obtained. The experimental data points used for the fitting procedure are obtained from experiments on a variety of ceramic and metal open-celled foams (a-Al_2O_3, cordierite, SiC, SiC-Al_2O_3, Al, Ni, Ni–Cr, Cu, FeCrAlY, Ti-6Al-4V, graphite) with different porosities ($0.7 < \psi < 0.98$), cell densities ($5 \ldots 100$ ppi) and on different

Forced Convection in Porous Media 39

fluids [air, water, glycerol, NaOH solution (1N)] flowing through the structure. The correlation is valid for $10^{-1} < Re_{d_h} < 10^5$ (Dietrich, 2012).

$$Hg = 110\ Re_{d_h} + 1.45\ Re_{d_h}^2 \tag{3.4}$$

Thus, the permeability and the form-drag constant are derived from a comparison of the coefficients in Equations (3.2) and (3.4):

$$\frac{1}{K} = \frac{110}{\psi d_h^2} \tag{3.5}$$

$$c_F = 0.14\psi^{-3/2} \tag{3.6}$$

At this point, it is necessary to discuss the validity of the Darcy–Forchheimer equation as given by Equation (3.1). There have been some fundamental investigations on the actual flow regimes prevailing in porous media (e.g., Scheidegger, 1958; Dybbs and Edwards, 1984). Using a "particle-diameter-based Reynolds number," the following three regimes have been distinguished: (1) laminar, steady-state at $Re < 150$; (2) laminar, transient at $150 < Re < 300$; and (3) turbulent, transient at $Re > 300$. This classification has also been transferred to open-celled foams. The pressure-drop contributions, which are quantified by the Darcy–Forchheimer equation in its form according to Equation (3.1), are assigned to regime (1). On the other hand, regimes (2) and (3) have been referred to as post-Forchheimer regimes (Della Torre et al., 2014), following a similar form but with different coefficients of the linear and quadratic terms. Since the definition of the characteristic length varies among different authors and flow phenomena inside consolidated porous media—like open-celled foams may differ from those observed for packings—direct transfer of the above Re limits is questionable. Meinicke et al. (2017) have conducted combined experimental and numerical investigations on this issue. By carrying out µPIV measurements to visualize the velocity field inside a glass foam structure and comparing these results to the predictions of a computational fluid dynamics (CFD) modeling technique, they were able to confirm the existence of the above-mentioned flow regimes in open-celled foams. According to this investigation, the existence of classical macroscopic turbulence in such porous media may be doubted. Demarcation of turbulent from laminar transient vortex shedding and the definition of flow regime boundaries remain an open issue calling for further research.

We continue the discussion with introducing the governing equations for heat transfer. Except in case of a deformable porous structure, it is enough to solve the momentum balance only for the fluid phase as that for the solid phase is trivial. The same does not hold for the energy balance as where both phases should be taken into account. To this end, one can simply use the assumption that the fluid and solid locally[*] possess the same temperature or, more generally, distinguish between the temperatures of the two phases. The latter approach is widely referred to as local thermal non-equilibrium (LTNE) and the former as local thermal equilibrium (LTE). Under LTNE, the thermal energy balance excluding all heat sources can be written as

$$\psi\left(\rho c_p\right)_f \frac{\partial \theta_f}{\partial t} + \left(\rho c_p\right)_f \boldsymbol{u} \cdot \nabla \theta_f = \psi \nabla \cdot \left(k_f \nabla \theta_f\right) + h_{sf}\, a_{sf}\left(\theta_s - \theta_f\right) \tag{3.7}$$

$$\left(1 - \psi\right)\left(\rho c\right)_s \frac{\partial \theta_s}{\partial t} = \left(1 - \psi\right)\nabla \cdot \left(k_s \nabla \theta_s\right) + h_{sf}\, a_{sf}\left(\theta_f - \theta_s\right) \tag{3.8}$$

Unlike in the momentum equation, here we do not drop the temporal terms in the energy equations to keep a more general form. Moreover, the macroscopic convection [the second term on the left-hand side

---

[*] The "locality" is a result of the macroscopic approach where the same space coordinate can be attributed to both fluid and solid phases.

of Equation (3.7)] and diffusion terms [the first terms on the right-hand sides of Equations (3.7) and (3.8)] are maintained, as they normally make significant contributions. Finally, due to local solid-to-fluid heat transfer, two additional terms with identical values but opposite signs appear on the right-hand sides of the equations (coupling terms). It should be noted that by using the same thermal conductivity in all directions we imply having an isotropic porous medium. As will be addressed later, this is not always the case in solid open-celled foams.

The coupling term in Equation (3.7) is the macroscopic representation of interfacial heat transfer in pore scale in the same way as the Darcy and Forchheimer terms represent the interfacial drag force. The solid-to-fluid heat transfer coefficient, h_{sf}, appearing in the coupling terms, must be determined a priori based on experiments or microscopic simulations. Section 3.5 of this chapter contains further details on the determination of h_{sf} in open-celled foams.

If the time scale of heat transfer between the two phases is adequately short, porous medium can be considered to be in LTE, meaning that $\theta = \theta_f = \theta_s$. All thermal properties are homogeneous in space, and the open-celled foams can be considered as quasi-continuous media. Combining Equations (3.7) and (3.8) yields

$$(\rho c)_{tot} \frac{\partial \theta}{\partial t} + (\rho c_p)_f \, \boldsymbol{u} \cdot \nabla \theta = \nabla \cdot (k_{eff} \nabla \theta) \tag{3.9}$$

In Equation (3.9), the overall heat capacity per unit total volume $(\rho c)_{tot}$ is the weighted sum of the intrinsic heat capacities, that is, $\varphi (\rho c_p)_f + (1 - \varphi)(\rho c)_s$. Ideally, the same type of expression should apply to the effective thermal conductivity k_{eff}. This, however, does not hold in general case, as k_{eff} is a function of the detailed geometry of the porous structure and may also need to contain the effect of radiation and dispersion (details in Section 3.3 of this chapter).

As will be discussed later, LTE is indeed a good approximation for a wide range of real-world problems. Sections 3.3 and 3.4 of the present chapter are based on the LTE assumption.

3.3 Effective Thermal Conductivity

We continue with discussing the steady and fully developed flow in a pipe fully filled with porous medium. This basic problem provides insight into the topic of this section and also serves as a departure point for the following section. In a hydrodynamically fully developed flow, only a constant pressure gradient in the axial direction exists. Replacing a constant dp/dx in Equation (3.1), a uniform axial velocity profile or a so-called slug flow, that is, $\boldsymbol{u} = (u_0, 0, 0)$, is obtained. The constant u_0 is determined algebraically based on the balance of the prescribed pressure gradient with Darcy and Forchheimer terms. Using a uniform axial velocity $(u(r) = u_0)$ and dropping the temporal term in Equation (3.9), one can analytically determine the Nusselt number at the wall

$$Nu_w = \frac{q_w}{\theta_w - \theta_b} \frac{D}{k_{eff}} \tag{3.10}$$

for a thermally fully developed flow (Bejan, 2013). Such a solution yields $Nu_w = 5.78$ for an isothermal wall (θ_w = constant) and $Nu_w = 8$ for an isoflux wall (θ_w = constant).[*] These values are larger than those of a laminar fully developed clear flow, that is, without porous medium, in a pipe ($Nu_w = 3.66$ for the isothermal and $Nu_w = 4.36$ for the isoflux case) but of the same order of magnitude. They are, however, smaller than the typical values in the turbulent regime. For example, the widely used Gnielinski correlation (Gnielinski, 1976) predicts a Nusselt number of 16.7 for a turbulent pipe flow at a Reynolds

---

[*] This result is based on the assumption that the same value of effective thermal conductivity is valid all the way down to the pipe wall, which implies that there is no extra thermal resistance due to poor contact between the porous structure and the solid wall. In reality, the "contact resistance" can be a critical issue. This issue is, however, out of the scope of the present chapter.

number as low as $Re_D = 5000$ and $Pr = 0.71$. It is important to note that the pressure-drop coefficient for a flow through porous medium is considerably larger than that of a clear fluid flow; therefore, much higher flow rates and Reynolds numbers can be reached using the same pumping power in the clear fluid.

Rearranging Equation (3.10),

$$h_w = \frac{q_w}{\theta_w - \theta_b} = \frac{k_{eff}}{D} Nu_w \qquad (3.11)$$

we obtain an expression for wall heat transfer coefficient h_w (not to be mistaken with the interfacial heat transfer coefficient). Based on Equation (3.11), both an increase in Nu_w and in k_{eff} can lead to an enhancement in the wall heat transfer coefficient. Considering the previous discussion on the comparison of wall Nusselt numbers with and without porous medium, it can be seen that the real potential of using porous media for heat transfer enhancement lies mainly in modifying the effective thermal conductivity rather than Nusselt number. One can utilize a simplified volume-weighted approach for estimation of k_{eff} to realize that even at small volume fractions, a highly conductive solid material can lead to a major augmentation in k_{eff}. The fact that the heat transfer coefficient in porous media is controlled to a high degree by the effective thermal conductivity highlights the need for its accurate determination.

The following discussion, especially the fitting parameters appearing in the relations, is focused on effective thermal conductivity of open-celled foams, whereas the general relations are valid for any kind of porous media. In general, the heat transfer in porous media can be controlled by heat conduction, thermal radiation, and dispersion effects due to convection. Thus, the effective thermal conductivity k_{eff} can be estimated with the help of the following additive approach:

$$k_{eff,i} = k_{eff,i,conduction} + k_{eff,i,radiation} + k_{eff,i,mix} \qquad i = x,r \qquad (3.12)$$

The effective thermal conductivity with stagnant fluid (i.e., without flow) can be divided into effects due to heat conduction $\left(k_{eff,i,conduction}\right)$ and due to thermal radiation $\left(k_{eff,i,radiation}\right)$. These effects can be determined experimentally using a guarded hot-plate test facility (e.g., Hsu and Howell, 1992; Zhao et al., 2004; Dietrich, 2010; Dietrich et al., 2010; Fischedick et al., 2015). The test setup consists of a heating and a cooling plate, a guard heater, two reference elements with well-known material properties, and the open-celled foam sample. The effective thermal conductivity of the sample can be calculated easily using Fourier's law and adapting the calculated temperature profiles to those determined experimentally. Since heat conduction and thermal radiation effects can only be determined together when applying this experimental technique, the contributions of the one or the other has to be differentiated when correlating the experimental values. This is possible because the heat conduction dominates at moderate temperature levels and the thermal radiation at high temperature levels.

In Figure 3.1 the effective thermal conductivity with stagnant fluid versus temperature for different ceramic open-celled foam types made of alumina, oxidic-bonded silicon carbide (OBSiC), and mullite is shown. In the range of ambient temperature up to 400°C, the experimental values show a decreasing trend with increasing temperature. This effect can be explained by the temperature dependence of the solid phase's thermal properties. It was found that the thermal conductivity of the ceramics decreases significantly with increasing temperature (Dietrich, 2010). Further elevation of the temperature leads to an increase of the effective thermal conductivity due to significant thermal radiation effects. A comparison of open-celled foams with different porosities and fixed cell density (10 ppi) shows that, at low temperatures, open-celled foams with high porosities have significant lower effective thermal conductivities compared to open-celled foams with low porosities. Due to the big difference between the thermal conductivity of air and the ceramic, the main part of heat is conducted by the solid structure. Open-celled foams with low porosities consist of more solid material than open-celled foams with high porosities leading to higher effective thermal conductivities. With increasing thermal radiation effects (increasing temperature) the difference decreases and the data points converge. Comparing open-celled foams with different cell densities and fixed porosity ($\psi = 0.85$), it can be seen that at low temperatures all open-celled foam types show the same effective thermal conductivities. With increasing temperature,

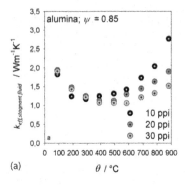 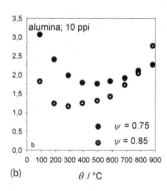 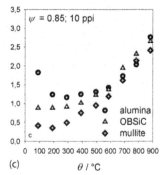

FIGURE 3.1 Dependence of the effective thermal conductivity due to heat conduction and thermal radiation on (a) the cell density; (b) the porosity; and (c) the material. The open-celled foams shown are (a) made of alumina with a porosity of 0.85 and cell densities in the range of 10–30 ppi; (b) made of alumina with a cell density of 10 ppi and porosities of 0.75–0.85; and (c) made of alumina, mullite, and OBSiC with a cell density of 10 ppi and a porosity of 0.85. (Adapted from Fischedick, T. et al., *Int. J. Therm. Sci.*, 96, 1–11, 2015; Fischedick, T. et al., *Int. J. Therm. Sci.*, 114, 98–113, 2017.)

the effective thermal conductivities of different open-celled foam types differ significantly due to radiation. As shown above, radiation is less relevant at low temperatures but not at high temperatures. Due to significant larger cell sizes, the free path length rises, and consequently the 10 ppi open-celled foam shows higher values compared to the 30 ppi sample. By comparing open-celled foams made of different materials, significant differences can only be noticed for moderate temperatures where conduction in the solid is the main heat transport mechanism. Here, alumina open-celled foams show the highest effective thermal conductivity, whereas mullite open-celled foams show the lowest one. This behavior goes along with the behavior of the pure ceramics where mullite shows the lowest solid thermal conductivity of the ceramics investigated (Dietrich, 2010; Fischedick et al., 2015, 2017).

In the case of heat conduction (moderate temperatures), the experimental trends can be correlated using an approach based on thermal resistances. Here, the two limiting cases (serial as well as parallel connection of the fluid and the solid thermal conductivity) are combined in a parallel way, as it can be seen in Equations (3.13)–(3.15) (Dietrich, 2010).

$$k_{eff,i,conduction} = b k_{serial} + (1-b) k_{parallel} \text{ with } b = 0.54 \tag{3.13}$$

$$k_{serial} = \frac{1}{\dfrac{\psi}{k_f(\theta)} + \dfrac{1-\psi}{k_s(\theta)}} \tag{3.14}$$

$$k_{parallel} = \psi k_f(\theta) + (1-\psi) k_s(\theta) \tag{3.15}$$

In the case of thermal radiation (high temperatures), complex models derived from the radiative transport equation (RTE) exist in literature. One of these approaches is the so-called Rosseland equation [see Equation (3.16)], which originally has been used for opaque systems (Rosseland, 1931). This equation is still widely used for the description of radiation effects in porous systems (Hsu and Howell, 1992; Hendricks and Howell, 1996; Baillis et al., 1999; Boomsma and Poulikakos, 2001; Zeghondy et al., 2006; and Zhao et al., 2004).

$$k_{eff,i,radiation} = \frac{16\sigma T}{3 E_R} \tag{3.16}$$

The Rosseland extinction coefficient E_R can be obtained from spectroscopic measurements (FTIR, Fourier-transform infrared) determining the spectral transmittance as a function of the sample length (for more details see Dietrich et al., 2014).

Forced Convection in Porous Media

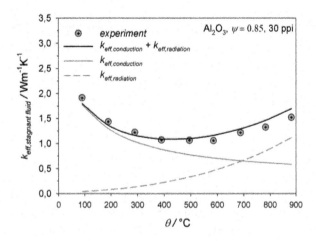

FIGURE 3.2 Correlation of the effective thermal conductivity in the case of stagnant fluid for alumina open-celled foams with a porosity of $\psi = 0.85$ and a cell density of 30 ppi including thermal conduction and thermal radiation effects. (Adapted from Fischedick, T. et al., *Int. J. Therm. Sci.*, 96, 1–11, 2015.)

Figure 3.2 shows the successful correlation of measured effective thermal conductivity values exemplarily for an alumina open-celled foam when heat conduction and thermal radiation are calculated based on the approaches previously described. As it can be seen, the decreasing trend at moderate temperatures can be captured by the theory of heat conduction if temperature-dependent material properties are used (gray line). Above 400°C, the effective thermal conductivity increases slightly with increasing temperature due to significant thermal radiation effects. These are described by the Rosseland equation, which is represented in Figure 3.2 by the dashed gray line. According to Equation (3.12), the expressions for heat conduction and heat radiation are summed to reflect the actual trend of experimental values leading to the black line in Figure 3.2 [the third term in Equation (3.12) is set to zero since a stagnant fluid is considered]. Remarkably, the line matches the experimental data points well (Dietrich, 2010; Fischedick et al., 2015).

If the open-celled foam is flowed through, dispersion effects must also be taken into account in addition to the heat transport mechanisms previously described [consequently, the term $k_{eff,i,mix}$ in Equation (3.12) can no more be set to zero]. In this case, the experimental determination of the effective thermal conductivity is carried out with the aid of a flow channel in which the temperature and volume flow rate of the fluid can be variably adjusted. The test section itself must have a heatable wall so that defined radial temperature gradients can be realized in the test samples. In the experiments, the temperature field in the axial and radial directions is determined after reaching steady-state conditions. The effective thermal conductivity can then be found by adjusting the temperature field calculated to the energy equation for homogeneous media by minimizing the error squares since the effective thermal conductivity sought is the only unknown parameter. In experiments, it is essential to ensure that flow distributer is installed in front of the actual test section in order to uniform the flow. It is also important to use sufficiently large samples in the test section to avoid running-in effects and, depending on the specimen used, to compensate for irregularities in the structure by averaging. An exemplary description of such an experimental setup and procedure can be found in Fischedick et al. (2017).

Experimental results clearly show that the effective thermal conductivity increases linearly with velocity due to increasing dissipative lateral mixing effects. Experiments carried out in the past on packed beds of spheres, cylinders, etc. have shown a similar trend. Since packed beds can in principle also be regarded as porous media, the findings obtained there support the finding for the open-celled foams in the experiments. Also, in the experiments carried out with through-flow above a temperature of 400°C, the effective thermal conductivity increases with increasing temperature due to thermal radiation. At lower temperatures, k_{eff} decreases as the temperature rises due to the decreasing thermal conductivity

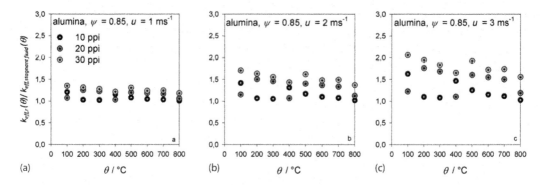

FIGURE 3.3 Dependence of the amount of the radial effective thermal conductivity due to forced convection on the cell density for alumina open-celled foams with a porosity of 0.85 and cell densities in the range of 10–30 ppi. The superficial air velocities are (a) $u = 1$ m/s; (b) $u = 2$ m/s; and (c) $u = 3$ m/s. (Adapted from Fischedick, T. et al., *Int. J. Therm. Sci.*, 114, 98–113, 2017.)

of the solid. This trend is consistent to the description above when talking about the effective thermal conductivity for open-celled foams with stagnant fluids. Thus, it can be concluded that the dominant part of the heat transport may be identified as the thermal conduction's contribution (Fischedick et al., 2017). This was also observed by Bianchi et al. (2012) for metal open-celled foams.

As presented in Figure 3.3 (exemplarily shown for an alumina open-celled foam with a porosity of 0.85), the amount of the radial effective thermal conductivity due to forced convection increases with increasing superficial air velocity because of increasing lateral mixing effects. When increasing the cell density, the amount of the struts per unit volume increases, which also leads to increasing lateral mixing effects and therefore to an increase of the effective thermal conductivity. Furthermore, in case of constant fluid velocity, the amount of the radial effective thermal conductivity due to forced convection decreases with increasing temperature because of decreasing fluid density (Fischedick et al., 2017).

The correlation in Equation (3.17) can be employed to include these findings in Equation (3.12). This correlation is based on a Péclet approach similar to that used for all types of packed beds. The characteristic length d_{mix} is derived from geometric considerations as described in detail by Fischedick et al. (2017).

$$k_{eff,r,mix} = CPe\, k_f \text{ with } Pe = \frac{u\, d_{mix}\left(\rho c_p\right)_f}{k_f} \quad (3.17)$$

The effective thermal conductivity in the axial direction can be determined using an experimental setup in which the open-celled foam is heated against the flow direction with an adiabatic wall. It has been shown that at flow velocities of about 1 m/s, the proportion of heat conduction in the axial direction is negligible compared to the proportion of the remaining heat transport mechanisms. At lower flow velocities, however, this must be taken into account and can be correlated using a Péclet approach as described in Dietrich (2010).

3.4 Forced Convection in Ducts

3.4.1 Fully Developed Flow

In the previous section, we started to discuss Nusselt number in pipes filled with porous medium for slug velocity profiles. In the following, we extend this discussion by considering non-slug velocity profiles. Such profiles occur when the Darcy–Brinkman equation

$$-\nabla p = \frac{\mu_f}{K}\boldsymbol{u} - \mu_{eff}\nabla^2\boldsymbol{u} \quad (3.18)$$

Forced Convection in Porous Media

is used as the momentum balance in porous medium. The Darcy–Brinkman equation is an alternative extension to Darcy's law, which is particularly believed to be an appropriate approach toward the porous media with high porosities (Nield et al., 2002). Equation (3.18) contains a "macroscopic" diffusion term, analogous to the diffusion term in Navier–Stokes equations. Here μ_{eff} is the effective viscosity, which is of the same order of magnitude as the intrinsic fluid viscosity μ_f, at least for highly porous material (Liu and Masliyah, 2005).

It can be easily shown that the relative magnitude of the Darcy and Brinkman terms in Equation (3.18) is $(\mu_f/\mu_{eff})Da^{-1}$, in which Da is the Darcy number. Square root of Da can be loosely interpreted as the microscopic to macroscopic scale ratio in the porous medium. At very low values of Da, Equation (3.18) tends to the Darcy's law and at very high values to the simplified Navier–Stokes equation. Figure 3.4 depicts the analytical solution of Equation (3.18) at different values of the Darcy number. It can be seen that at adequately small values of Da, the velocity profile tends to a slug profile. As Da grows, the profiles depart from the slug flow and gradually tend to the laminar parabolic profile. It should be noted that at $Da \sim 1$, the macroscopic–microscopic scale separation vanishes, and hence our "porous media" approach is no more valid. As a result, the larger values of the Darcy number are merely hypothetical and cannot be realized in reality.

Departure from a slug velocity profile translates to deviation from the Nusselt number values discussed in the previous section. Expectedly, as Da varies between zero and infinitely large values, the Nusselt number varies between that of the slug flow and a laminar pipe flow. The values of Nusselt number for these two limits and two other Darcy numbers are given in Table 3.1. Notably at $Da = 10^{-2}$, which is arguably the largest acceptable order of magnitude for Da, the Nusselt number is approximately halfway between its two limits. It implies that, for many real porous materials, the Brinkman term modifies the wall Nusselt number only moderately.

3.4.2 Thermally Developing Flow

In porous ducts, flow normally reaches a hydrodynamically fully developed state in a short distance. Using a simple scale analysis, based on the argument that the flow acceleration is of the same order of magnitude as the resistance, it can be shown that the hydrodynamic development length scales with $K u_b/v_f$ are often a small value (Nield and Bejan, 2017). Hence, in the following we focus on the problem of "thermal" development.

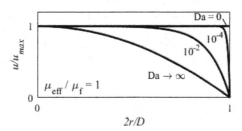

FIGURE 3.4 Velocity profiles in fully porous pipes with different Darcy numbers based on the Darcy–Brinkman model (based on the analytical solution in Kuznetsov et al., 2003). $Da \to \infty$ corresponds to the Hagen–Poisseuille solution in a laminar pipe.

TABLE 3.1

Values of Nu_w for Fully Developed Darcy–Brinkman Flow in Pipe at Different Darcy Numbers $(\mu_{eff}/\mu_f = 1)$; $Da \to \infty$ Corresponds to the Hagen–Poiseuille Solution in a Laminar Pipe

Da	Isothermal	Isoflux
0	5.78	8
10^{-5}	5.75	7.9
10^{-2}	4.90	5.99
$\to \infty$	3.66	4.36

We start the discussion with the slug flow, which corresponds to a Darcy–Forchheimer equation. Assuming that hydrodynamic development length is negligible, thermal development reduces to the solution of the classical Graetz problem for a slug flow, that is, the solution of the steady energy equation [Equation (3.9)] without the temporal term, with a uniform inlet temperature profile, that is, $\theta(x=0,r)=\theta_0$. Such a solution in a pipe is described by Burmeister (1993). Here the expressions for local Nusselt number resulted from this solution are presented:

$$Nu_w\left(\tilde{x}\right)=\frac{\sum_{n=1}^{\infty}\exp\left(-4\gamma_n^2\,\tilde{x}\right)}{\sum_{n=1}^{\infty}\dfrac{\exp\left(-4\gamma_n^2\,\tilde{x}\right)}{\gamma_n^2}}, \quad \theta_w=\text{const.}\left(\text{isothermal}\right) \tag{3.19}$$

$$Nu_w\left(\tilde{x}\right)=\frac{8}{1-8\sum_{n=1}^{\infty}\dfrac{\exp\left(-4\gamma_n^2\,\tilde{x}\right)}{\gamma_n^2}}, \quad q_w=\text{const.}\left(\text{isoflux}\right) \tag{3.20}$$

In Equations (3.19) and (3.20), the eigenvalues are given by $J_0\left(\gamma_n\right)=0$ and $\tilde{x}=x/\left(D\,Re_D\,Pr\right)$ is the nondimensional distance from the inlet. Importantly, in the above expression for $\tilde{x}$, the Prandtl number is based on the "effective" thermal conductivity and heat capacity of the fluid. As expected, in the limit of $\tilde{x}\to\infty$, the wall Nusselt numbers given by Equations (3.19) and (3.20) tends to the already mentioned fully developed values of 5.78 and 8, respectively. In both isothermal and isoflux cases, Nu_w is virtually equal to its asymptotic value once $\tilde{x}=O\left(10^{-1}\right)$, which means that

$$\frac{X_T}{D\,Re_D\,Pr}\sim 0.1 \tag{3.21}$$

The dimensionless group $X_T/\left(D\,Re_D\,Pr\right)$ is similar to what is commonly used for thermal development in laminar duct flows without porous medium. Notably, the same order of magnitude as given by Equation (3.21) also holds for the thermal development length of a laminar clear-fluid flow in a pipe (Bejan, 2013). However, one should recall that k_{eff} in a porous medium can be much larger than that of a clear fluid when the solid material is highly conductive. A larger k_{eff} leads to a smaller Pr, and thus, a shorter thermal development length.

Thermal development of the Darcy–Forchheimer flow in geometries other than circular pipes can be tackled by the solution of the Graetz problem for slug flows in those geometries. A full description of such solutions is given by Muzychka et al. (2010). The Graetz methodology can similarly be used for thermal development of Darcy–Brinkman flows as long as the slug velocity profile is replaced by those obtained from the Darcy–Brinkman equation, for example, those in Figure 3.4. Nield et al. (2003a, 2004) reported such solutions in pipes and parallel plate channels for both isothermal and isoflux boundary conditions. The effect of axial conduction is studied by Hooman et al. (2003) in pipes and Nield et al. (2003b) in parallel plate channels. Moreover, the effect of thermal dissipation is investigated by Nield et al. (2003b) and Hooman and Gurgenci (2007). Hooman and Haji-Sheikh (2007) and Hooman et al. (2007) studied the problem comprehensively in a rectangular duct. Thermal development in ducts is also solved for non-Newtonian fluids (Nield and Kuznetsov, 2005) and for rarefied gases (Kuznetsov and Nield, 2009).

3.4.3 Ducts Partially Filled with Porous Media

Fully filled porous ducts, while offering considerable heat transfer enhancement, significantly increase the pressure drop. This has motivated studies on partially filled ducts. In this section, we focus on a specific class of geometries, that is, pipes or parallel plate channels with porous layers lining the wall(s). Such a geometry can be described by a single geometric parameter e, defined as the porous layer

Forced Convection in Porous Media

thickness divided by the pipe radius or channel half height (thus $e = 0$ and 1 correspond to a non-porous and a fully porous duct, respectively). The problem has been investigated analytically by Poulikakos and Kazmierczak (1987) and Chikh et al. (1995) and numerically by Alkam et al. (2002) and Yang et al. (2012) among others. Overall, these studies show that use of a porous layer can lead to an enhancement in heat transfer, but this enhancement is merely a result of an increased k_{eff}. Indeed, for the case of $k_{eff} = k_f$ the porous layer tends to weaken the wall heat transfer coefficient since it suppresses the flow next to the wall, acting similar to an insulation layer. In this case, the wall Nusselt number decreases with an increase in e until it reaches a minimum value, beyond which it grows abruptly to reach the value for the fully porous channel (e.g., the ones in Table 3.1). For $Da \sim 10^{-4}$, the minimum occurs approximately at $e = 0.85$ and increases further with a decrease in Da. It should be noted that Da has a more direct impact on wall heat transfer for partially filled ducts compared to fully filled ducts. It stems from the fact that Da controls the flow portion entering the porous region; a more permeable porous layer (large Da) causes a higher Nusselt number. As mentioned before, large values of effective thermal conductivity compensate the negative effect of flow suppression on heat transfer. Roughly speaking, as long as $k_{eff} / k_f \geq O(10)$, the thermal conductivity effect is dominant, and the heat transfer coefficient increases monotonically with e.

3.5 Local Thermal Non-Equilibrium

There are a number of physical situations in which LTE is no more valid. Highly transient problems are prone to such a failure, as the relaxation time for reaching thermal equilibrium can be longer than the time scales imposed at the boundaries. For instance, when the inlet temperature undergoes a rapid change, if the two phases have significantly different heat capacities or internal thermal resistances, the temperature of one phase may lag behind the other.

LTE can be erroneous also in steady problems under certain circumstances. A classic example is the entrance region of a packed bed with a high-speed hot gas (Kuwahara et al., 2001). Amiri and Vafai (1994) extracted an error map for use of LTE in channels filled with packed porous media and showed that the error increases with a decrease in Darcy number and an increase in particle Reynolds number. Using the fluid-to-solid thermal conductivity ratio and a modified Biot number $Bi = \left(h_{sf} \, a_{sf} \, D^2 \right) / \left(8 k_s \right)$, Lee and Vafai (1999) identified three distinct regimes for convection in porous media, in which either of fluid conduction, solid conduction, or interfacial heat transfer are dominant. Based on their analysis, LTE is flawed when both the Biot number and thermal conductivity ratio are small. Other authors (e.g., Kim and Jang, 2002; Khashan et al., 2005) also proposed their criteria for validity of LTE, in all of which h_{sf} appears as a decisive parameter.

Convection heat transfer has been studied in porous ducts for fully developed (e.g., Nield, 1998; Yang et al., 2011) and developing (e.g., Jiang and Ren, 2001; Nield et al., 2002) flows and also in partially porous ducts (e.g., Forooghi et al., 2011; Xu et al., 2011) under LTNE. These studies either use h_{sf} as a parameter—in which case a realistic range of values is desired—or rely on empirical correlations for its calculation. In either case, an accurate knowledge of this quantity is essential. In view of the above, determination of the heat transfer coefficient between the solid structure und the flowing fluid h_{sf} for open-celled foams is discussed in the following.

Data from the literature show that h_{sf} can be reliably determined using transient experimental methods, where the gas stream flowing through the foam is either heated or cooled abruptly. The former can be realized by switching a valve and the latter by switching off a hot flame (e.g., combustion of hydrogen). Care must be taken to ensure that these processes do not cause any fluctuation in the volume flow. Based on the heating or cooling characteristics, to be determined by temperature sensors in the solid and fluid, h_{sf} can be determined by adapting the solution of Equations (3.7) and (3.8) to the measured temperature field. This is possible because h_{sf} is the only unknown quantity (see Dietrich, 2013 for details). Experimental data on the heat transfer coefficient for ceramic as well as metal open-celled foams have been published by various authors and shown to follow the trend of a power function with an increase in the flow velocity. Consequently, the data are usually correlated using a Nusselt–Reynolds approach according to Equation (3.22), partly without ($n = 0$, e.g., see Younis and Viskanta, 1993), but mostly with consideration of the Prandtl number ($n = 1/3$, e.g., see Schlegel et al., 1993; Calmidi and Mahajan, 2000; Giani et al., 2005; Mancin et al., 2010, 2011; or Dietrich, 2013).

$$Nu_{sf} = C\,Re^m\,Pr^n \tag{3.22}$$

The constants C and m are determined using geometrical considerations or adapted to the experimental data. The characteristic length for the Reynolds number is defined individually by each author. Dietrich (2013) used the hydraulic diameter for example. A significant influence by neither the porosity nor the cell density on h_{sf} is revealed. The experimental values reported by Dietrich (2013) are in the range of $40-50\,\mathrm{W/(m^2 K)}$ for velocities in the range of $0.5-5$ m/s.

Furthermore, Dietrich (2010, 2013) reported that for open-celled foams there is an analogy between momentum and heat transport similar to the Generalized Lévêque Equation (GLE) described in the literature, for instance, for packed beds of spherical particles. This makes it possible to infer h_{sf} from the pressure drop data, which is highly beneficial since an experimental determination of the former requires complex setups while the latter can be measured comparatively easily. The analogy is expressed in the form of a Nusselt–Hagen approach according to Equation (3.23), whereby, on the basis of theoretical considerations, the Nusselt number must depend on the third root of the Hagen number (dimensionless pressure drop). Using Equation (3.24), the fitting constants C_{Re} and C_{geo} can accommodate the effect of geometrical differences among different open-celled foam types. Employing these constants, the correlation in Equation (3.23) reproduces the trends of the experimental data very well. A root mean square deviation (RMSD) of 22% can be achieved, which can be considered a high-quality correlation in this context.

$$Nu_{sf} = C\,Hg^{1/3}\,Pr^{1/3} = 0.45 C_{Re}\,C_{geo}\,Hg^{1/3}\,Pr^{1/3} \tag{3.23}$$

$$C_{Re} = \left(\frac{Re_{d_h}+1}{Re_{d_h}+1000}\right)^{0.25} \quad \text{and} \quad C_{geo} = \left(\frac{d_h/l}{(d_h/l)_{mean}}\right)^{1.5} \tag{3.24}$$

As described in Dietrich (2013), l is a microscopic characteristic length of the open-celled foam structure, which can be achieved using light microscopy. The denominator in the expression for C_{geo} is the arithmetic mean considering all investigated open-celled foam samples used in the experiments and is determined to be $(d_h/l)_{mean} = 1.67$.

3.6 Conclusion

In this chapter, we gave an overview of the problem of forced convection in porous media focusing on heat transfer in ducts fully or partially filled with porous media. The importance of accurate determination of effective thermal conductivity and solid-to-fluid heat transfer coefficient, in the LTE and LTNE frameworks respectively, were highlighted, and recent experimental results on their calculation in ceramic open-celled foams were presented.

NOMENCLATURE

a surface area per unit total volume
Bi modified Biot number; $(h_{sf}\,a_{sf}\,D^2)/(8 k_s)$
c, c_p specific heat (for solid), specific heat at constant pressure (for fluid)
c_F form-drag constant (Forchheimer coefficient)
d_h pore hydraulic diameter; $4\psi/a_{sf}$
d_{mix} mixing length as the characteristic length for Péclet approach
D diameter of pipe
Da Darcy number; $2K/D$ (for pipe)

Forced Convection in Porous Media

e	dimensionless porous layer thickness
E_R	Rosseland extinction coefficient
h	convection heat transfer coefficient
Hg	Hagen number; $(|\Delta p| / \Delta L) \cdot d_h^3 / (\rho v^2)_f$
J_0	Bessel function of the first kind and zero order
k	thermal conductivity
K	permeability
l	characteristic length of the microscopic foam structure
L	axial distance along duct, channel, or open-celled foam sample
Nu_w	wall Nusselt number; $(h_w D) / k_f$
Nu_{sf}	interfacial Nusselt number; $(h_{sf} d_h) / k_f$
p	pore pressure
Pe	Péclet number defined by Equation (3.17)
Pr	Prandtl number; $(c_p \mu) / \lambda$
q	heat flux per unit area
r	radial coordinate (perpendicular to the flow direction)
Re_{d_h}	pore hydraulic Reynolds number; $(u \cdot d_h) / (\psi \cdot v_f)$
Re_D	Reynolds number based on macroscopic scale D; $(u_b \cdot D) / v_f$
t	time
T	absolute temperature
$\boldsymbol{u}$	superficial (Darcy) velocity vector
u	superficial (Darcy) velocity in axial direction
x	axial coordinate (flow direction)
$\tilde{x}$	dimensionless axial coordinate; $x / (D\, Re_D\, Pr)$
X_T	thermal development distance
γ_n	eigenvalue used in Equations (3.19) and (3.20)
θ	temperature
μ	dynamic viscosity
v	kinematic viscosity
ξ	friction factor
ρ	density
σ	Stefan–Boltzmann constant
ψ	Porosity; pore (void) volume per unit total volume

Subscripts

b	bulk value
eff	effective value
f	fluid
s	solid
sf	related to solid–fluid interface
w	value at the wall

REFERENCES

Abu-Hijleh, B. 1997. Convection heat transfer from a laminar flow over a 2D backward facing step with asymmetric and orthotropic porous floor segments. *Numer. Heat Transf. Part A* 35: 325–335.

Alkam, M. K., Al-Nimr, M. A. and Hamdan, M. O. 2002. On forced convection in channels partially filled with porous substrates. *Heat Mass Transfer* 38: 337–342.

Amiri, A. and Vafai, K. 1994. Analysis of dispersion effects and nonthermal equilibrium, non-Darcian, variable porosity incompressible flow through porous media. *Int. J. Heat Mass Transf.* 30: 939–954.

Ávila-Marín, A. L. 2011. Volumetric receivers in solar thermal power plants with central receiver system technology: A review. *Sol. Energy* 5: 891–910.

Baillis, D., Raynaud, M. and Sacadura, J. F. 1999. Spectral radiative properties of open-cell foam insulation. *J. Thermophys. Heat Transf.* 13: 292–298.

Bejan, A. 2013. *Convection Heat Transfer.* Hoboken, NJ: John Wiley & Sons.

Bianchi, E., Heidig, T., Visconti, C. G., Groppi, G. and Freund, H. J. 2012. An appraisal of the heat transfer properties of metallic open-cell foams for strongly exo-/endo-thermic catalytic processes in tubular reactors. *Chem. Eng. J.* 198–199: 512–528.

Boomsma, K. and Poulikakos, D. 2001. On the effective thermal conductivity of a three-dimensionally structured fluid-saturated metal-foam. *Int. J. Heat Mass Transf.* 44: 827–836.

Buonomo, B., Lauriat, G., Manca, O. and Nardini, S. 2016. Numerical investigation on laminar slot-jet impinging in a confined porous medium in local thermal non-equilibrium. *Int. J. Heat Mass Transf.* 98: 484–492.

Burmeister, L. C. 1993. *Convective Heat Transfer.* New York: John Wiley & Sons.

Calmidi, V. V. and Mahajan, R. L. 2000. Forced convection in high porosity metal foams. *ASME J. Heat Transf.* 122: 557–565.

Chikh, S., Boumedien, S., Bouhadef, S. and Lauriat, G. 1995. Analytical solution of non-Darcian forced convection in an annular duct partially filled with a porous medium. *Int. J. Heat Mass Transf.* 38: 1543–1551.

Chumpia, A. and Hooman, K. 2015. Performance evaluation of tubular aluminum foam heat exchangers in single row arrays. *Appl. Therm. Eng.* 83: 121–130.

Della Torre, A., Montenegro, G., Tabor, G. R. and Wears, M. L. 2014. CFD characterization of flow regimes inside open cell foam substrates. *Int. J. Heat Fluid Flow* 50: 72–82.

Dietrich, B. 2010. *Thermische Charakterisierung von keramischen Schwammstrukturen für verfahrenstechnische Apparate.* PhD Thesis. Karlsruhe, Germany: KIT Scientific Publishing.

Dietrich, B. 2012. Pressure drop correlation for ceramic and metal open-celled foams. *Chem. Eng. Sci.* 74(1): 192–199.

Dietrich, B. 2013. Heat transfer coefficients for solid sponges—Experimental results and correlation. *Int. J. Heat Mass Transf.* 61: 627–637.

Dietrich, B., Fischedick, T., Heissler, S., Weidler, P. G., Wöll, C. and Kind, M. 2014. Optical parameters for characterization of thermal radiation in ceramic open-celled foams—Experimental results and correlation. *Int. J. Heat Mass Transf.* 79: 655–665.

Dietrich, B., Schabel, W., Kind, M. and Martin, H. 2009. Pressure drop measurements of ceramic open-celled foams—Determining the hydraulic diameter. *Chem. Eng. Sci.* 64(16): 3633–3640.

Dietrich, B., Schell, G., Bucharsky, E. C. et al. 2010. Determination of the thermal properties of ceramic open-celled foams. *Int. J. Heat Mass Transf.* 53(1): 198–205.

Dybbs, A. and Edwards, R.V. 1984. A new look at porous media fluid mechanics—Darcy to turbulent. In *Fundamentals of Transport Phenomena in Porous Media*, eds. Bear, J. and Corapcioglu, M. Y. NATO ASI Series, Series E, 82: pp. 199–256. Dordrecht, Netherlands: Springer.

Ejlali, A., Ejlali, A., Hooman, K. and Gurgenci, H. 2009. Application of high porosity metal foams as air-cooled heat exchangers to high heat load removal systems. *Int. Commun. Heat Mass Transf.* 36: 674–679.

Feng, S. S., Kuang, J. J., Wen, T., Lu, T. J. and Ichimiya, K. 2014. An experimental and numerical study of finned metal foam heat sinks under impinging air jet cooling. *Int. J. Heat Mass Transf.* 77: 1063–1074.

Fischedick, T., Kind, M. and Dietrich, B. 2015. High temperature effective thermal conductivity of ceramic open-celled foams with stagnant fluid—Experimental results and correlation including thermal radiation. *Int. J. Therm. Sci.* 96: 1–11.

Fischedick, T., Kind, M. and Dietrich, B. 2017. Radial effective thermal conductivity of ceramic open-celled foams up to high temperatures—experimental results and correlation. *Int. J. Therm. Sci.* 114: 98–113.

Forooghi, P., Abkar, M. and Saffar-Avval, M. 2011. Steady and unsteady heat transfer in a channel partially filled with porous media under thermal non-equilibrium condition. *Transp. Porous Med.* 86: 177–198.

Fu, W. S. and Huang, H. C. 1996. Thermal performances of different shape porous blocks under an impinging jet. *Int. J. Heat Mass Transf.* 40: 2261–2272.

Gauthier, S., Nicolle, A. and Baillis, D. 2008. Investigation of the flame structure and nitrogen oxides formation in lean porous premixed combustion of natural gas/hydrogen blends. *Int. J. Hydrogen Energy* 33: 4893–4905.

Giani, L., Groppi, G. and Tronconi, E. 2005. Heat transfer characterization of metallic foams. *Ind. Eng. Chem. Res.* 44: 9078–9085.

Gnielinski, V. 1976. New equations for heat and mass transfer in turbulent pipe and channel flow. *Int. Chem. Eng.* 16: 359–368.

Hendricks, T. J. and Howell, J. R. 1996. Absorption/scattering coefficients and scattering phase functions in reticulated porous ceramics. *ASME J. Heat Transf.* 118: 79–87.

Hooman, K. and Gurgenci, H. 2007. Effects of viscous dissipation and boundary conditions on forced convection in a channel occupied by a saturated porous medium. *Transp. Porous Med.* 68: 301–319.

Hooman, K. and Haji-Sheikh, A. 2007. Analysis of heat transfer and entropy generation for thermally developing Brinkman-Brinkman forced convection problem in a rectangular duct with isoflux walls. *Int. J. Heat Mass Transf.* 50: 4180–4194.

Hooman, K., Haji-Sheikh, A. and Nield, D. A. 2007. Thermally developing Brinkman-Brinkman forced convection in rectangular ducts with isothermal walls. *Int. J. Heat Mass Transf.* 50: 3521–2533.

Hooman, K., Ranjbar-Kani, A. A. and Ejlali, A. 2003. Axial conduction effects on thermally developing forced convection in a porous medium: Circular tube with uniform wall temperature. *Heat Transf. Res.* 34: 34–40.

Hsu, P. and Howell, J. R. 1992. Measurements of thermal conductivity and optical properties of porous partially stabilized zirconia. *Exp. Heat Transf.* 5: 293–313.

Huang, P. C. and Vafai, K. 1994. Analysis of forced convection enhancement in a channel using porous blocks. *J. Thermophys. Heat Transf.* 8: 563–573.

Jang, J. Y. and Chen, J. L. 1992. Forced convection in a parallel plate channel partially filled with a high porosity medium. *Int. Commun. Heat Mass Transf.* 19: 263–273.

Jiang, P. X. and Ren, Z. P. 2001. Numerical investigation of forced convection heat transfer in porous media using a thermal non-equilibrium model. *Int. J. Heat Fluid Flow* 22: 102–110.

Joseph, D. D., Nield, D. A. and Papanicolaou, G. 1982. Nonlinear equation governing flow in a saturated porous medium. *Water Resources Res.* 18: 1049–1052.

Khashan, S. A., Al-Amiri, A. M. and Al-Nimr, M. A. 2005. Assessment of the local thermal non-equilibrium condition in developing forced convection flows through fluid-saturated porous tubes. *Appl. Therm. Eng.* 25: 1429–1445.

Kim, S. J. and Jang, S. P. 2002. Effects of the Darcy number, the Prandtl number, and the Reynolds number on local thermal non-equilibrium. *Int. J. Heat Mass Transf.* 45: 3885–3896.

Kuznetsov, A. V. and Nield, D. A. 2009. Thermally developing forced convection in a porous medium occupied by a rarefied gas: Parallel plate channel or circular tube with walls at constant heat flux. *Transp. Porous Med.* 76: 345–362.

Kuznetsov, A. V., Xiong, M. and Nield, D. A. 2003. Thermally developing forced convection in a porous medium: Circular duct with walls at constant temperature, with longitudinal conduction and viscous dissipation effects. *Transp. Porous Med.* 53: 331–345.

Kuwahara, F., Shirota, M. and Nakayama, A. 2001. A numerical study of interfacial convective heat transfer coefficient in two-energy equation model for convection in porous media. *Int. J. Heat Mass Transf.* 44: 1153–1159.

Lee, D. Y. and Vafai, K. 1999. Analytical characterization and conceptual assessment of solid and fluid temperature differentials in porous media. *Int. J. Heat Mass Transf.* 31: 423–435.

Liu, S. and Masliyah, J. H. 2005. Dispersion in porous media. In *Handbook of Porous Media*, ed. K. Vafai, pp. 81–140. Boca Raton, FL: Taylor & Francis Group.

Mahjoob, S. and Vafai, K. 2008. A synthesis of fluid and thermal transport models for metal foam heat exchangers. *Int. J. Heat Mass Transf.* 51: 3701–3711.

Mancin, S,. Zilio, C., Cavallini, A. and Rossetto, L. 2010. Heat transfer during air flow in aluminum foams. *Int. J. Heat Mass Transf.* 53: 4976–4984.

Mancin, S., Zilio, C., Rossetto, L. and Cavallini, A. 2011. Heat transfer performance of aluminum foams. *ASME J. Heat Transf.* 133: 060904.

Meinicke, S., Möller, C.-O., Dietrich, B., Schlüter, M. and Wetzel, T. 2017. Experimental and numerical investigation of single-phase hydrodynamics in glass sponges by means of combined μPIV measurements and CFD simulation. *Chem. Eng. Sci.* 160: 131–143.

Muzychka, Y. S. Walsh, E. and Walsh, P. 2010. Simple models for laminar thermally developing slug flow in noncircular ducts and channels. *ASME J. Heat Transf.* 132: 111702.

Nield, D. A. 1998. Effects of local thermal nonequilibrium in steady convective processes in a saturated porous medium: Forced convection in a channel. *J. Porous Media* 1: 181–186.

Nield, D. A. and Bejan, A. 2017. *Convection in Porous Media*. Cham, Switzerland: Springer International Publishing AG.

Nield, D. A. and Kuznetsov, A. V. 2005. Thermally developing forced convection in a channel occupied by a porous medium saturated by a non-newtonian fluid. *Int. J. Heat Mass Transf.* 48: 1214–1218.

Nield, D. A., Kuznetsov, A. V. and Xiong, M. 2002. Effect of local thermal non-equilibrium on thermally developing forced convection in a porous medium. *Int. J. Heat Mass Transf.* 45: 4949–4955.

Nield, D. A., Kuznetsov, A. V. and Xiong, M. 2003a. Thermally developing forced convection in a porous medium: Parallel-plate channel or circular tube with walls at constant heat flux. *J. Porous Media* 6: 203–212.

Nield, D. A., Kuznetsov, A. V. and Xiong, M. 2003b. Thermally developing forced convection in a porous medium: Parallel plate channel with walls at uniform temperature, with axial conduction and viscous dissipation effects. *Int. J. Heat Mass Transf.* 46: 643–651.

Nield, D. A., Kuznetsov, A. V. and Xiong, M. 2004. Thermally developing forced convection in a porous medium: Parallel-plate channel or circular tube with isothermal walls. *J. Porous Med.* 7: 19–27.

Odabaee, M., Hooman, K. and Gurgenci, H. 2011. Metal foam heat exchangers for heat transfer augmentation from a cylinder in cross-flow. *Transp. Porous Med.* 86: 911–923.

Orihuela, M. P., Shikh Anuar, F., Ashtiani Abdi, I., Odabaee, M. and Hooman, K. 2018. Thermohydraulics of a metal foam-filled annulus. *Int. J. Heat Mass Transf.* 117: 95–106.

Poulikakos, D. and Kazmierczak, M. 1987. Forced convection in a duct partially filled with a porous material. *ASME J. Heat Transf.* 109: 653–662.

Rashidi, S., Tamayol, A., Valipour, M. S. and Shokri, N. 2013. Fluid flow and forced convection heat transfer around a solid cylinder wrapped with a porous ring. *Int. J. Heat Mass Transf.* 63: 91–100.

Rosseland, S. 1931. Astrophysik, auf atomtheoretischer Grundlage. In *Schriftenreihe Struktur der Materie in Einzeldarstellungen*, pp. 11, Berlin, Germany: Springer-Verlag.

Scheidegger, A. E. 1958. The physics of flow through porous media. *Soil Sci.* 6: 355.

Schlegel, A., Benz, P. and Buser, S. 1993. Wärmeübertragung und Druckabfall in keramischen Schaumstrukturen bei erzwungener Strömung. *Wärme- und Stoffübertragung* 28: 259–266.

Twigg, M. V. and Richardson, J. T. 2002. Theory and applications of ceramic foam catalysts. *Chem Eng. Res. Des.* 80: 183–189.

Wang, T., Luan, W., Wang, W. and Tu, S. T. 2014. Waste heat recovery through plate heat exchanger based thermoelectric generator system. *Appl. Energy* 136: 860–865.

Xu, H. J., Qu, Z. G. and Tao, W. Q. 2011. Analytical solution of forced convective heat transfer in tubes partially filled with metallic foam using the two-equation model. *Int. J. Heat Mass Transf.* 54: 3846–3855.

Yang, C., Ando, K. and Nakayama, A. 2011. A local thermal non-equilibrium analysis of fully developed forced convective flow in a tube filled with a porous medium. *Transp. Porous Med.* 89: 237–249.

Yang, C., Nakayama, A. and Liu, W. 2012. Heat transfer performance assessment for forced convection in a tube partially filled with a porous medium. *Transp. Porous Med.* 89: 237–249.

Younis, L. B. and Viskanta, R. 1993. Experimental determination of the volumetric heat transfer coefficient between stream of air and ceramic foam. *Int. J. Heat Mass Transfer* 36: 1425–1434.

Zeghondy, B., Iacona, E. and Taine, J. 2006. Experimental and RDFI calculated radiative properties of a mullite foam. *Int. J. Heat Mass Transf.* 49: 3702–3707.

Zhao, C. Y., Lu, T. J., Hudson, H. P. and Jackson, J. D. 2004. The temperature dependence of effective thermal conductivity of open-celled steel alloy foams. *Mat. Sci. Eng. A* 367: 123–131.

Section II

Advanced Topics of Convection in Porous Media

4

Convective Heat Transfer of Nanofluids in Porous Media

Bernardo Buonomo, Davide Ercole, Yasser Mahmoudi, Oronzio Manca, and Sergio Nardini

CONTENTS

4.1 Introduction 55
4.2 Natural Convection of Nanofluids in Porous Media 55
4.3 Forced Convection of Nanofluids in Porous Media 61
4.4 Mixed Convection of Nanofluids in Porous Media 69
4.5 Governing Equations 73
    4.5.1 Single Phase Model 73
    4.5.2 Mixture Model 74
4.6 Comparison among Different Configurations 77
4.7 Conclusions 81
Nomenclature 82
References 83

4.1 Introduction

Convective heat transfer enhancement is always a current topic in thermal engineering applications related to the improvement of energy system performance, electronic cooling, nuclear reactors, solar energy systems, aerospace automotive, process industry, and so on. One of the techniques to realize the convective heat transfer augmentation is by means of solid materials, such as nanoparticles or porous media with higher thermal conductivity with respect to the traditional base fluids. The employment of such materials also allows a significant increase of heat transfer contact area inside the fluids, and the heat transfer passes from a surface exchange to a bulk or mass exchange. The coupling of nanoparticles and porous media with high thermal conductivity is an interesting and promising solution. Many research activities have been developed in the convective heat transfer of nanofluids in porous media as reviewed by Nield and Kuznetsov (2014), Kasaeian et al. (2017), Xu et al. (2019), and Khanafer and Vafai (2019). This chapter provides a review of the recent researches on natural, forced, and mixed convection as well as the governing equations related to the convective heat transfer. Some results in terms of comparison are provided in the section Comparisons among Different Configurations, and some remarks are pointed out in the Conclusions.

4.2 Natural Convection of Nanofluids in Porous Media

Recent numerical and experimental studies on natural convection of nanofluids in porous media are given in Table 4.1. Chamkha and Ismael (2014) have studied a vertical porous cavity partially filled with $Cu-H_2O$ nanofluid. The wall close to the porous media is set to a high temperature while the opposite

TABLE 4.1

Natural Convection Studies

Article	Num./Exp.	Model	Regime	Configuration System	Particle's Nature	Base Fluid	dp [nm]	%V	Ra	Da	Porous Medium	e	Output
Chamkha and Ismael (2014)	Num.	Two-phase	Laminar		Cu	H_2O	–	0–1	10^3–10^6	10^{-7}–1	Glass beads	0.398	Nu, the stream function, vorticity, and dimensionless temperature
Ali Agha et al. (2014)	Num.	–	–		Cu Ag	–	–	Variable	–	–	Homogeneous	0.105–0.154	Nu, effect of magnetic field on the velocity profile, temperature profile and concentration
Sheremet et al. (2015a)	Num.	Two-phase mixture	–		–	H_2O	1–100	Variable	30–500	10^{-4}–10^{-1}	–	0.8	Nu, Three-dimensional velocity, temperature, and nanoparticle volume fraction fields
Zargartalebi et al. (2015)	Num.	Two-phase	Laminar		Al_2O_3	H_2O	100	–	–	–	Metal foam	–	Nu, volume fraction of nanoparticles, thermal conductivity, dynamic viscosity of nanofluid

(Continued)

TABLE 4.1 (*Continued*)

Natural Convection Studies

Article	Num./Exp.	Model	Regime	Configuration System	Particle's Nature	Base Fluid	dp [nm]	%V	Ra	Da	Porous Medium	ε	Output
Ismael and Chamkha (2015)	Num.	Single-phase	Laminar		Cu	H_2O	–	0–5	10^3–10^6	10^{-7}–10^{-1}	Glass beads	0.398	Nu, temperature and velocity, stream function
Dastmalchi et al. (2015)	Exp.	Two-component Non-homogeneous	–		Al_2O_3	H_2O	33	0–4	10–10^4	–	Aluminum foam	0.1–0.5	Nu, effect of temperature difference, effects of porosity, streamlines and isotherms
Sheremet et al. (2015b)	Num.	Two-phase	Laminar		Cu	H_2O	–	0–5	10–1000	–	Aluminum foam	0.1–0.9	Variation of Nusselt number, streamlines, isotherms for the nanofluid
Sheremet and Pop (2015)	Num.	Single-phase	Laminar		Cu Ag	H_2O	–	–	50–500	–	Metal foam	–	Nu, distribution of streamlines, isotherms and isoconcentrations

(Continued)

TABLE 4.1 (*Continued*)

Natural Convection Studies

Article	Num./ Exp.	Model	Regime	Configuration System	Particle's Nature	Base Fluid	dp [nm]	%V	Ra	Da	Porous Medium	ε	Output
Ghalambaz et al. (2016)	Num.	Non-homogeneous	–		Al$_2$O$_3$ Cu	H$_2$O	–	–	10–1000	–	Metallic foam	–	Nu, viscous dissipation, radiation effect
Ghasemi and Siavashi (2017)	Num.	Single-phase	Laminar		Cu	H$_2$O	100	0–12	10^3–10^6	0.1	–	0.9	Nuavg, streamlines, isotherms, V, T v vert, Nul, Nuh, Sfrictional, Sthermal, Smagn, Stotal
Toosi and Siavashi (2017)	Num.	Two-phase mixture	Laminar		Cu	H$_2$O	25	0–4	10^3–10^6	0.0001–0.1	–	0.4–0.9	Nuavg, streamlines, isotherms, Nulocal, horiz. and vert. velocity component in the middle of enclosure
Umavathi et al. (2017)	Num.	Single-phase	Laminar		Cu, diamond, TiO$_2$, Ag, SiO$_2$	H$_2$O	–	0–2	–	10^{-4}–10^2	–	–	V and T profile, heat transfer, rate of heat transfer

(*Continued*)

TABLE 4.1 (Continued)

Natural Convection Studies

Article	Num./Exp.	Model	Regime	Configuration System	Particle's Nature	Base Fluid	dp [nm]	%V	Ra	Da	Porous Medium	ε	Output
Xu and Xing (2017)	Num.	Single-phase	Laminar		Al_2O_3	H_2O	–	1–5	10^2–10^7	10^{-7}–10^{-3}	Copper alloy	0.4–0.9	Nuavg, T and v profile and distrib, streamlines, isotherms, average velocity
Sheikholeslami et al. (2017)	Num.	Single-phase	–		Fe_3O_4	H_2O	47	0–4	$10^3,10^4,10^5$	0.01–100	–	–	Nuavg, Nu, streamlines, isotherms, distribution and gradient of temperature
Mehryan et al. (2017)	Exp.	Single-phase	Laminar		Al_2O_3, Al_2O_3+Cu	H_2O	–	0–2	1–10^3	–	Glass ball, Aluminum metal foam	0.3–0.9	V, Nuavg, Nulocal, isotherms, streamlines
Izadi et al. (2018)	Num.	Single-phase	–		Copper, diamond, silicon-dioxide	H_2O	1–100	0–1	10^3–10^6	10^{-6}–10^{-3}	Homogeneous, isotropic	0.1–0.9	Nuavg, streamlines, temp distribution, concentration

Num. = numerical; Exp. = experimental.

wall is set to a lower temperature. The results have shown that the addition of the nanofluid increases the heat transfer even with a porous media at lower permeability, in particular for a Rayleigh number lower than 10^5 there is a critical thickness X_p of the porous media at which the Nusselt number is highest, while for a Rayleigh number higher than 10^5 the Nusselt number decreases rapidly.

Agha et al. (2014) have accomplished an analysis of the boundary layer for a vertical plane plate inside a porous media filled with a nanofluid (Cu and Ag-H_2O) in addition to a uniform magnetic field and a radiant heat flow. The results have shown that the magnetic parameters reduced the velocity profile values while there was an increment of temperature. Moreover, they affirmed that the magnetic parameters and the radiant heat flow had a valuable effect on the Nusselt number. Sheremet et al. (2015a) have studied a constant heat transfer through a three-dimensional porous tank with nanofluid for different heated walls. By the results, they showed that the heat transfer flow has a vortex feature. Moreover, for higher Rayleigh numbers, there is an increment of the convective fluid flow, and therefore the nanoparticle distribution is more uniform inside the cavity. The model is not valid anymore for a lower Rayleigh number. Furthermore, there is a proportion between the Rayleigh number and the Nusselt Number. Zargartalebi et al. (2015) have focused on the heat flux in the boundary layer on a straight plate inside a porous media. The plate is heated, and the Al_2O_3 nanoparticles move along the plate. The results have shown that it is difficult to assess the augment of the heat transfer due to the nanoparticles for reduced Nusselt number. Moreover, the Lewis number does not have a great impact on the velocity and temperature profiles. Ismael and Chamkha (2015) have studied a square cavity with different vertical layers, formed by porous media or just nanofluid. The layer with the porous media is filled with the same nanofluid (Cu-H_2O). By the results, the natural convection is improved when the permeability of the porous media is very low or when the thickness layer is major than 0.5 and the Rayleigh number is minor of 10^4. Therefore, for lower a Rayleigh number, the permeability is not effective; moreover, the temperature gradient increases with the Darcy number. Dastmalchi et al. (2015) have studied the flow field and the heat transfer in a square cavity filled with the aluminum metal foam where the Al_2O_3-H_2O nanofluid flows through. By the results, the volume concentration of nanoparticles is not uniform, and it depends by the temperature field, and the Nusselt number diminishes more than 80% with higher porosity. Therefore, the natural convection is dominant when the porosity is lower, while for high porosity the conduction regime is dominant. Sheremet et al. (2015b) have studied a square cavity filled with Cu-H_2O nanofluid in porous media. They found that heat transfer is higher for a higher Rayleigh number and lower for a decreasing nanoparticles volumetric fraction value. Based on the results, it can be stated that by adding solid nanoparticles in the base fluid, the convective regime is suppressed. Moreover, by increasing the concentration of the nanoparticles there is a decrease of the Nusselt number for both solid and liquid phase. Sheremet and Pop (2015) analyzed the fluid flow and heat transfer of a nanofluid in a porous medium through two concentric, horizontal cylinders. The results have shown that the distribution of the nanoparticles is not uniform for low values of Rayleigh number while for the high ones the nanoparticle distribution becomes homogeneous. At the maximum value of the Rayleigh number in the considered range (50–500), two vortices are formed in each half of the cylinder, one counterclockwise and one clockwise. Moreover, as the Rayleigh number increases, there is also an increase of Nusselt number. Ghalambaz et al. (2016) have studied the influence of viscous dissipation and the radiative effects on the heat transfer of a nanofluid in a square cavity filled with metal foam. The vertical walls of the cavity have different temperature values while the upper and lower walls are adiabatic. The results show that the Nusselt number changes according to the hot or cold wall due to the presence of viscous dissipation effects; in particular, it decreases at the hot wall and increases at the cold one. Because of the same effects, the distribution of the nanoparticles concentration is also significant, which results to be low close to the hot wall and high close to the cold wall. Instead, the radiative effects reduce the temperature gradient but increase the heat diffusion in the porous medium, and therefore they improve the heat transfer. Ghasemi and Siavashi (2017) have studied the natural convection in the magnetohydro-dynamics (MHD) magnetofluid-dynamics field of a Cu-H_2O nanofluid in a porous square cavity. They found that the magnetic field always has a worsened impact on the Nusselt number, but this impact can still be controlled through the nanofluid concentration and the porous medium thermal conductivity. For $Ra = 10^4$, a magnetic instability and sinusoidal variation of the Nusselt number with the Hartmann number was observed. Toosi and Siavashi (2017) carried out a numerical study on the natural convection of copper–water nanofluids in a square cavity partially filled with a porous medium. The horizontal walls are adiabatic while the vertical ones are subjected to constant hot and cold temperature, respectively. It has been shown that for $Ra = 10^3$, as the volume fraction of the particles increases, the number of Nusselt number increases. For different Rayleigh numbers, there is an optimal concentration that maximizes the Nusselt number. For low Rayleigh numbers, the combination of nanofluid and porous medium improves heat transfer performance. The employment of the nanofluid can have positive or negative effects based on the number of Darcy. The Nusselt number for Rayleigh greater than 10^4 is reduced with the introduction of nanoparticles. Umavathi et al. (2017) have studied the natural convection and the heat transfer of nanofluids in a rectangular vertical tube filled with a porous media. The temperature is instead represented with smooth curves covering the entire cavity that are symmetrical with respect to the horizontal line. Five types of nanofluids were employed: Cu, diamond, TiO_2, Ag, and SiO_2. It has been found that heat transfer increases thanks to the nanoparticles. The optimal speed was obtained with the silver particles, while the optimal temperature was obtained with the SiO_2 particles. About the copper, TiO_2 and diamond particles, the maximum values of speed and temperature were obtained with the first ones. Xu and Xing (2017) have employed the Lattice Boltzmann model to study the natural convection of nanofluids in a porous medium. By the results, the Nusselt number increases for higher Darcy number, Rayleigh number, porosity, and thermal conductivity. When the Darcy number is less than 10^{-5}, the effect of convection can be ignored and only conduction exists. Increasing the porosity and thermal conductivity of the porous medium can significantly improve heat transfer by natural convection, but the effect of porosity on heat transfer is less incident with respect to the Darcy number. Sheikholeslami et al. (2017) have studied the influence of a non-uniform magnetic field on heat transfer, through the impact of the Lorentz force on the flow of the Fe_3O_4-H_2O nanofluid in a porous cavity. The results show that the temperature gradient decreases with the increase of the Hartmann number, while it increases with the Darcy and Rayleigh numbers. Furthermore, by selecting the shape of the nanoparticles, the maximum number of Nusselt is obtained. Mehryan et al. (2017) have studied the natural convection of hybrid nanofluids, new types of working fluids designed with advanced thermophysical properties. They have investigated a $Al_2O_3 + Cu$-H_2O hybrid nanofluid in a cavity filled with a porous medium made by glass balls or aluminum metallic foam. By the results, the average number of Nusselt is a decreasing function of the volumetric fraction of the nanoparticles reducing the heat transfer phenomenon. This reduction is greater for hybrid nanofluids than single ones. Moreover, when the solid matrix is aluminum foam, the Nusselt number does not depend by the porosity, whereas when the solid matrix is made of glass balls, the variation of the Nusselt number with the porosity is significant. Izadi et al. (2018) have carried out a numerical study for different nanofluids in a natural convection regime. The nanofluids under investigation are water–copper, water–diamond, and water–silicon–carbon dioxide in a porous medium between two horizontal eccentric cylinders. The Buongiorno model was assumed to take into account the particle concentration. The results show that the probability of collision of the particles decreases as the number of Lewis increases, while the number of Nusselt decreases. The maximum heat transfer flow was detected in water–diamond nanofluid while the minimum heat transfer flow was detected in water–diamond nanofluid and water–silicon–carbon dioxide.

4.3 Forced Convection of Nanofluids in Porous Media

This section provides a detailed literature review on the recent experimental and numerical studies on forced convection of nanofluids in porous media. A summary of the literature review is given in Table 4.2. Matin and Pop (2013) have studied heat transfer forced convection in a porous horizontal channel in which the Cu-H_2O nanofluid flows. The upper wall of the channel is adiabatic while the lower wall is subject to a constant heat flow. Furthermore, it is assumed that the first-order catalytic reaction takes place on the walls. The results show that a high value of the Darcy number leads to high values of permeability of the porous medium, so the nanofluid can flow more easily through the channel. However, it is important to note that by inserting the nanoparticles in the base fluid, the density increases, and therefore it is more difficult to flow through in the porous medium. Instead, the temperature profiles show that the dimensionless temperature decreases as the Darcy number increases but increases with the increase of the particle volumetric fraction. Also important is the distribution of the

TABLE 4.2

Forced Convection Studies

Article	Num./ Exp.	Model	Regime	Configuration System	Particle's Nature	Base Fluid	dp [nm]	%V	Re	Da	Porous Medium	ε	Output
Matin and Pop (2013)	Exp.	Two-phase	Laminar		Cu	H_2O	–	0–1	–	0.05–1	–	–	Nusselt number, velocity, temperature and concentration distributions.
Servati et al. (2014)	Num.	Two-phase	–		Al_2O_3	H_2O	–	0–7	18.5	–	Aluminum foam	0.74	Nu, magnetic field effect, average temperature and velocity
Xu et al. (2015)	Num.	Single-phase	Laminar		Al_2O_3 TiO_3	H_2O	–	0–2	1000	–	Copper foam	0.9–0.98	Nu, velocity and temperature field
Zhang et al. (2015)	Num.	Two-component mixture	Laminar Turbulent		Al_2O_3	H_2O	1–100	0–6	–	0.05 0.1 0.15 0.2	Metal foam	0.9	Nu, velocity, temperature and nanoparticle distributions and heat transfer characteristics

(Continued)

TABLE 4.2 (*Continued*)

Forced Convection Studies

Article	Num./ Exp.	Model	Regime	Configuration System	Particle's Nature	Base Fluid	dp [nm]	%V	Re	Da	Porous Medium	ε	Output
Ting et al. (2018)	Num.	Two-phase	–		Al_2O_3	H_2O	5	2	150	0.1	Aluminum foam	0.95	Nu, entropy generation, temperature distribution, effect of nanoparticles suspensions
Ting et al. (2015)	Num.	Two-phase mixture	Laminar		Al_2O_3	H_2O	5 100	1–4	120	0.1	Metal foam homogeneous, isotropic	0.95	Nu, temperature distributions, effect of nanoparticle suspension
Siavashi et al. (2018b)	Num.	Two-phase mixture	Laminar		Al_2O_3	H_2O	38	0–5	100–2000	0.1 0.01 0.001 0.0001	Metallic foam	–	Nu, temperature field, PN, head loss and entropy generations
Baqaie Saryazdi et al. (2016)	Num.	Two-component mixture	Laminar		Al_2O_3	H_2O	40	5	300–1200	–	Aluminum foam	0.85–0.95	Nu, temperature and velocity profile, wall shear stress

(*Continued*)

TABLE 4.2 (Continued)

Forced Convection Studies

Article	Num./Exp.	Model	Regime	Configuration System	Particle's Nature	Base Fluid	dp [nm]	%V	Re	Da	Porous Medium	ε	Output
Mashaei et al. (2016)	Num.	Single-phase	Laminar		Al_2O_3	H_2O	–	0–6	–	–	–	0.3–0.9	Pressure gradient, thermal conductivity, viscosity, wall temperature distribution, axial and radial v, η, hr, Ur
Amani et al. (2017)	Exp.	Single-phase	Laminar		Fe_3O_4	Di-H_2O	30	1–2	200–1000	–	Metal foam	0.8	Nu, f, Δp
Bayomy and Saghir (2017)	Exp.	Single-phase	Laminar		γ-Al_2O_3	H_2O	50	0.1–0.6	210–631	–	ERG aluminum foam	–	Nuavg, Nulocal, Dimensionless T surface, T distribution, v e T profile
Tayebi et al. (2018)	Num.-Exp.	Mixture	Laminar		Al_2O_3,TiO_2	H_2O	10–20, 0–15	0,0.1, 0.2, 0.3	180–1800	–	Copper-filled porous media	0.93	Rise in fluid temperature, collector efficiency

(*Continued*)

TABLE 4.2 (Continued)

Forced Convection Studies

Article	Num./Exp.	Model	Regime	Configuration System	Particle's Nature	Base Fluid	dp [nm]	%V	Re	Da	Porous Medium	ε	Output
Welsford et al. (2018)	Num.	Single-phase	Laminar		Al_2O_3+MEPCM	H_2O	–	3, 3 + 4, 3 + 20	197, 1801	2.924×10^{-5}, 0.8259×10^{-3}, 2.065×10^{-4}	ERG aluminum foam	0.9	Cp, h, T variation, hm/hf
Siavashi et al. (2018a)	Num.	Single-phase	Laminar		–	Air	–	0,6,12	1000–10,000	0.2×10^{-3} -2×10^{-4}	Metal foam	0.9118	Nuavg, T and v profile, stream function, shear stress
Ting et al. (2015)	Num.	Single-phase	Laminar		Al_2O_3	H_2O	1	0–4	Variable	10^{-4}–10^0	Silicon porous material	0.9	β', Nu/Pe
Siavashi et al. (2015)	Num.	Two-phase mixture	Laminar		Al_2O_3	H_2O	38	0–1	500–2000	–	Nickel foam	0.93–0.98	Nu, h, temperature and velocity profile
Zargartalebi and Azaiez (2018)	Num.	Fluid phase	Variable		Uniformly	–	–	Cost	0–120	–	–	0.8	Nu, Nuavg, T distribution, Nu ratio, gradient temp, HTIR
Ameri and Syed Eshaghi (2018)	Exp.	Single-phase	Laminar		Fe_3O_4	H_2O	30	1–2	250–750	–	Copper foam	0.8	Nu, h, Exergy efficiency, ΔP, PEC, FIP
Siavashi and Joibary (2018)	Num.	Two-phase mixture	Laminar		Al_2O_3	H_2O	38	3	500–2000	10^{-4} 10^{-2} 10^{-1}	Aluminum foam	0.8	Nu, PEC, pressure drop, effect of different Darcy numbers

Num. = numerical; Exp. = experimental.

concentration that increases with the increase in the number of Darcy. Servati et al. (2014) have analyzed the forced convection in a partially filled channel with an aluminum metallic foam, and an applied uniform vertical magnetic field was present. The effects of the volumetric fraction variation of the Al_2O_3 nanoparticles on the heat transfer were investigated. The nanofluid is forced to flow into the channel while the upper and lower walls are maintained at a constant temperature. The Reynolds number is kept constant at 18.5. The results show that as the nanoparticles volumetric fraction increases, there is an increase of the average temperature and the average velocity at the outlet of the channel. It is also notable that the maximum velocity occurs at the center of the channel and that the high thermal conductivity of the porous medium leads to a more effective temperature distribution. A higher velocity gradient near the solid walls is evident and leads to a major temperature gradient. Since the magnetic field does not alter the velocity inside the porous medium, its effects on heat transfer are negligible in this central area where the metal foam is present. Xu et al. (2015) have investigated a channel filled with a homogeneous and isotropic copper metal foam in which a nanofluid flows, whose nanoparticles are made by Al_2O_3 or TiO_3. The results show that the velocity profile of the nanoparticles in the porous medium is very uniform and that the developed analytical solution complies with the numerical solution presented by the Brinkman model. Furthermore, the temperature distribution of the fluid and solid phases is shown for a porosity equal to 0.98. By the results, the solid-phase temperature is slightly higher than the fluid-phase temperature. It is also shown that as the Reynolds number increases, the Nusselt number gradually increases due to the local convective heat transfer between fluid and solid. It is also important to note that as the volumetric fraction of the nanoparticles increases, the Nusselt number gradually increases, and since the thermal conductivity of Al_2O_3 is higher than TiO_3, the Nusselt number is also greater. Zhang et al. (2015) have investigated a numerical study to examine the velocity, temperature, particle distribution, and heat transfer characteristics associated with the thermal and mechanical dispersion of the nanoparticles, considering a homogeneous metallic foam saturated with the Al_2O_3-H_2O nanofluid. The results show a high value of the heat transfer rate 80 times higher than the case of the base fluid without porous foam. Ting et al. (2018b) have investigated the effects of the heat generation on the entropy generation on Al_2O_3-H_2O nanofluid that flows through an asymmetrically heated microchannel filled with an aluminum metal foam. On the upper and lower walls of the microchannel there are two distinct constant heat flows. The results show that the thermal asymmetries significantly influence the temperature distributions and therefore the irreversibility of heat transfer in the system. In the case of heat generation without the metal foam, the generation of entropy can be reduced by 10% by imposing the symmetrical heating of the walls. Furthermore, there is an optimal Reynolds number ($Re = 22$) in which a further reduction of entropy generation occurs. In the case of heat generation with metal foam, the generation of entropy is minimized when the ratio of the wall heat flow is 3/4, but the optimal Reynolds number vanishes. The results also show that the suspension of the nanofluid is favorable when the Reynolds number and the diameter of the nanoparticles are lower, causing a decrease in the entropy generation of 42%. Ting et al. (2015) have carried out the same analyses as in the previous study but were concerned with the effects of viscous dissipation and nanoparticle dimensions on the thermal characteristics of the nanofluid. From the analysis, it was found that in the case of heat generation in the solid phase, there is a bifurcation of the heat flow on the walls of the microchannel due to the inversion of the temperature gradient compared to the case without a solid phase, in which the fluid temperature is lower. Furthermore, the heat transfer of the nanofluid is greater in the case of symmetrical channel heating, with an increase in the average heat transfer coefficient of 47% compared to the non-symmetrical heating case. Siavashi et al. (2018b) have studied the simultaneous application of the porous medium and the addition of Al_2O_3 nanoparticles in order to improve the heat transfer in a tube partially or completely filled with aluminum metal foam in the presence of a constant heat flow. Various configurations were examined based on the position of the porous layer (internal or external to the tube) and the permeability. The results show that for both configurations, the number of Nusselt increases with the increase of the Reynolds number and for the volumetric fraction of the particles. Moreover, for the first case, a larger Nusselt number is obtained with respect to the second case. Furthermore, the ratio between the increase of heat transfer and pressure loss is higher for higher permeability value ($Da = 0.1$ and 0.01). The ratio reaches a maximum value when $Da = 0.001$, whereas when $Da = 0.0001$, the ratio starts to decrease. Baqaie Saryazdi et al. (2016) have analyzed the flow and heat transfer of the Al_2O_3-H_2O nanofluid flowing through a circular tube filled with a porous medium. The distribution of the nanoparticles is assumed to be non-homogeneous within the tube. The results show that the Nusselt number and the shear stress increase as the volume fraction of the particles increases, and the heat transfer coefficient is higher for an increment of the nanoparticles concentration or with high values of the Reynolds number. The velocity profile, on the other hand, decreases with the increase of the solid phase and consequently also of the temperature phase. Moreover, as the porosity increases, permeability increases, causing viscous and inertial effects to decrease while the number of Nusselt decreases. Mashaei et al. (2016) have performed a numerical analysis on the forced convection of the Al_2O_3-H_2O nanofluid in an annular section filled with a porous medium. The results show that the addition of the particles leads to a reduction in terms of velocity. Although the use of the nanofluid instead of the base fluid increases the pressure drop in the structure, the magnitude of this increase is not quantifiable. Furthermore, the temperature of the pipe wall decreases with the introduction of the nanofluid, even if a greater uniformity of the wall temperature can be obtained with the increase in the concentration of the particles. The effectiveness of the use of nanofluid in the porous medium on the thermal performance of the heat pipe becomes more significant with the increase in the heat load, which has different effects on the pressure drop, depending on the concentration of the particles. In conclusion, the best thermohydraulic performances of the use of nanofluid in thermal ducts are detected at the average level of particle concentration (i.e., 4%) and the highest thermal load. Amani et al. (2017) have experimentally studied the characteristics of convective heat transfer and pressure drop of the Fe_3O_4-H_2O nanofluid in a tube filled with a metallic porous foam. Experimental observations reveal that the increase of the volumetric fraction and the Reynolds number improves the Nusselt number. Furthermore, the improvement is more pronounced for high Reynolds numbers due to the addition of the particles. A slight increase in pressure drop was observed using the nanofluid compared to pure water, due to the increase in viscosity of the fluid for the dispersion of the particles in the water. The performance index aims to consider the effects of raising the Nusselt number and the pressure drop simultaneously. In this study, it was observed that the improvement of heat transfer dominates the pressure losses in term of performance index. Bayomy and Saghir (2017) have carried out an experimental survey on the flow of an Al_2O_3-H_2O nanofluid in a porous sink and a comparison with the numerical results of an analogous numerical model. The aim of the study was to investigate the characteristics related to heat transfer and thermal performance of an aluminum foam heat sink for the Intel Core i7 processor. The results show that low volumetric percentages of the nanoparticles lead to an improvement in the local and global Nusselt number. The maximum improvement of the heat transfer rate was reached at 0.2% while at 0.3% a sudden drop of the positive effects compared to pure water was obtained. The presence of the channel filled with aluminum metal foam leads to an improvement of 20% compared to an empty channel. The numerical results relating to the surface temperature and the Nusselt number are in good agreement with the experimental results with a maximum relative error of 2% and 3%, respectively. The empirical correlations of the mean number of Nusselt have been developed on the basis of experimental data. Tayebi et al. (2018) have carried out a numerical survey on the improvement of efficiency in a direct absorption parabolic collector occupied by a porous medium and filled with a nanofluid. The purpose of the study was to investigate the effect of metal foams and nanofluids on the thermal performance of the collector. The nanofluids used were Al_2O_3 and TiO_2/H_2O while the porous medium was copper. The results show that the maximum efficiency of 34.51% is achieved with the TiO_2-H_2O nanofluid due to the high extinction coefficient of titanium dioxide nanoparticles. Furthermore, using a porous medium with a high absorption coefficient and dispersion coefficient, which could absorb more incoming radiation and transfer heat to the fluid, may lead to an improvement of the efficiency of the collector. The efficiency can increase further by increasing the volume flow rate and the volume fraction of the nanoparticles. Welsford et al. (2018) studied the role of a metal foam in the heat storage in the presence of a ternary fluid consisting of a nanofluid and a micro-encapsulated phase change material (MEPCM). Different values of heat flux have been applied to the outer wall of a porous tube with the fluid entering at different flow rates. The results reveal that by adding a metal foam there is an improvement of the heat transfer between the tube wall and the fluid. Heat transfer improves further by adding nanoparticles to the fluid. The heat storage capacity of the proposed fluid also increases with the addition of particles in the MEPCM material. The results revealed that 20% of MEPCM and

3% of nanoparticles in water represent an ideal balance for the conservation and improvement of heat, with an improvement of 6.7% in the case of water flowing in a porous tube. Siavashi et al. (2018a) have performed a numerical study on the flow and heat transfer of an impinging jet consisting of air and nanofluid through a porous and cylindrical heat sink. The cooling fluid flows smoothly through the porous foam on the hot disk. The results indicate that the addition of nanoparticles, the increase in thermal conductivity, and the reduction of the aspect ratio increase the thermal performance of the heat sink. Moreover, as the thermal conductivity increases, the average Nusselt number increases. This increase is also achieved by adding metal particles to the base fluid and increasing their volume fraction. Moreover, as the number of Darcy increases, the heat transfer can also increase moderately. Ting et al. (2015) analyzed the flow of nanofluids in microporous channels using the principle of field synergy. They studied the effects of the presence of the porous medium on the field synergy of the Al_2O_3-H_2O nanofluid. The results show that the introduction of the porous medium into the microchannel reduces the synergy angle and increases the number of synergies in the field, which in turn increases the convection performance of the system. Moreover, as the number of Darcy decreases and the number of Biot increases, it can improve the coordination between the heat field and the flow field. The synergy field can be further intensified by suspending the nanoparticles in the fluid. Siavashi et al. (2015) have performed a numerical study using porous media with a variable porosity [gradient porous media (GPM)] and multi-layer porous media (MLPM) with optimized properties to maximize the heat transfer and minimize the pressure drop. The fluids flow in a tube filled with GPM or MLPM, and they are simulated with Ansys-Fluent. The goal is to maximize the energy performance ratio (EPR). Two cases are examined: gradual increase or reduction of the dimensions of the nanoparticles (Case 1) and reduction of the porosity (Case 2) of the layers with respect to the radius. The results show that the optimal arrangement for Case 1 and Case 2 gives the EPR equal to 0.845 and 0.789, respectively. Furthermore, simultaneous optimization of the two cases provides a higher EPR. Zargartalebi and Azaiez (2018) have performed a numerical study to generate heat in a miniature electronic device by adopting nanofluid-based microchannel heat sinks (MCHSs) incorporated with fins. The effect of nanoparticles on the MCHS was studied. The results show that the influence of nanoparticles on heat transfer depends on both the size of the fins and the flow regime. For low Reynolds numbers, as the fins increase, the effect of nanoparticles is very evident. While at high Reynolds numbers, due to the reduced size of the fins, an increase in size produces a greater effect of the nanoparticles. This happens up to a specific fin dimension, above which the effect is reversed. The performance of the MCHS improves as the size of the fins decreases and/or the number of Reynolds increases. Ameri and Eshaghi (2018) have accomplished an experimental study introducing a new system that uses the Fe_3O_4-H_2O nanofluid in a porous medium in the presence of a constant magnetic field in the absorber tubes of a flat plate collector (FPC). The nanofluid is used in two concentrations, and it flows in a porous medium with a porosity of 0.8 in a constant magnetic field. A comparison is made between two systems, the new one and a regular FPC, which uses water as working fluid and a porous medium in the pipes. The results reveal that the Nusselt number of the new system is 1.36 times that of the latter system. Furthermore, the overall heat-loss coefficient is reduced in the new system, while the thermal efficiency, as well as the pumping power and the heat transfer coefficient, is affected by an increment. However, the latter system has greater exergetic efficiency. In conclusion, the new system can be used wherever space availability and greater thermal efficiency are crucial. Siavashi and Joibary (2018) have performed a numerical study on the performance analysis of a shell-and-tube heat exchanger with both walls filled with a porous medium in which the Al_2O_3-H_2O nanofluid flows. Nanofluid and porous media are simultaneously applied to improve the heat transfer of the exchanger with a minimum increase in power. Results are shown in terms of EPR, heat transfer rate, and pressure drop. First, it is assumed that the inner and outer tubes were filled with porous media having the same Darcy number. For a low value of Darcy number, there are two optimal situations to maximize the EPR: the first with the filling of 60% of the inner tube and the second with the total filling of both tubes. As the Reynolds number increases, the first optimal situation disappears and the second prevails. The same results are verified for higher values of Darcy number. Second, different combinations of the Darcy number for internal and external tubes were analyzed for various thickness combinations of the porous layers. It has been found that as Darcy and Reynolds vary, there are three optimal situations to maximize the EPR: the first with partial filling of the inner tube, the second with the total filling of both tubes, and the third with the partial

4.4 Mixed Convection of Nanofluids in Porous Media

Recent studied on mixed convection of nanofluids in porous media are conducted using numerical approaches as summarized in Table 4.3. Hajipour and Dehkordi (2012) have analyzed the heat transfer by mixed convection of a nanofluid in a vertical channel partially filled with a porous medium. The walls of the channel are kept at a constant temperature with the right wall at a hot temperature and the left wall at a cold temperature. The results show that there is an increment in the temperature and the velocity of the nanofluid with the increase of the viscous dissipation and the concentration of the nanoparticles. Moreover, as the permeability of the porous medium decreases (Darcy number), the velocity of the fluid is reduced. About the Nusselt number, as the viscous dissipation increases, the heat transfer rate from the nanofluid to the cold wall ($Nu_1 > 0$) increases while from the hot wall to the nanofluid ($Nu_2 > 0$) decreases until the direction of heat transfer changes and becomes from the nanofluid to the hot wall. Hajipour et al. (2014) have performed a numerical analysis on the mixed convection of the Al_2O_3-H_2O nanofluid at the inlet of a vertical channel partially filled with a porous medium whose walls are heated differently. The behavior of flow and thermal fields were analyzed under the influence of viscous and inertial dissipative effects. The results indicate that the presence of the nanoparticles in the fluid base increases the temperature along the channel. Furthermore, the temperature of the nanofluid is increased when the thickness of the thermal boundary layer is reduced. Finally, they have also obtained that the Nusselt number of the left (cold) wall increases while that of the right (warm) wall decreases so the variation of the Nusselt number of the right wall is reversed. Matin and Ghanbari (2014) have considered the flow and heat transfer by mixed convection of a nanofluid flowing in a vertical channel filled with a homogeneous and isotropic porous medium whose walls are heated differently. From the analysis, they obtained relationships according to the ratio between the Grashof number and the Reynolds number, and in particular in the case of Gr/Re < 1, the heat transfer takes place by forced convection. As Gr/Re increases, the effects of buoyancy forces arise, and the heat transfer regime becomes of mixed convection, in which there is a circulatory flow along the channel so that the intensity of the inverted flow increases. Furthermore, the volumetric fraction is maximum near the cold wall and along the channel decreases until it reaches a minimum value near the hot wall. Finally, as the number of Brinkmann increases, the number of Nusselt on the cold wall increases while the number on the hot wall decreases. Furthermore, as the number of Brinkmann increases, the heat transfer along the channel is improved. Zahmatkesh and Naghedifar (2017) have numerically studied the mixed oscillatory convection in the cooling, with an impinging jet, of a horizontal surface immersed in a porous medium saturated by a nanofluid. The purpose of this study is to observe how the governing parameters can alter the oscillations and the subsequent heat exchange. It has been observed that the final constant or oscillatory response of the flow depends on the Reynolds number values, the Grashof number, and the Darcy number while it is not influenced by the porosity of the medium and the volumetric fraction of the nanoparticles. Hussain et al. (2017) have performed a numerical analysis on the entropy generation of a mixed convective flow in a sloped channel with a cavity filled by a porous medium in which the Al_2O_3-H_2O nanofluid flows. The temperatures on the right wall and at the entrance of the cavity are fixed while the remaining walls are thermally insulated. It has been found that with an increase in the channel inclination angle up to 135° the maximum temperature gradient occurs, and an improvement in the temperature distribution in the channel leads to an increase in heat transfer. The opposite effect occurs for an angle greater than or less than 135°. Moreover, as the angle of inclination increases, there is also an improvement in the average Nusselt number and an increase in entropy. These increments are proportional with the porosity. Biswas et al. (2018) have performed an analysis on heat transfer and pumping power in a porous heated cavity saturated with a Cu-H_2O nanofluid. The nanofluid is injected into the cavity from the middle up to the adiabatic upper wall, and it is discharged through the cold side walls. The results reveal that heat transport is heavily influenced by parameters such as Richardson number, porosity, Darcy number, Reynolds number, volumetric fraction of the nanoparticles, and aspect ratio. By the results, there is an

TABLE 4.3

Mixed Convection Studies

Article	Num./Exp.	Model	Regime	Configuration System	Particle's Nature	Base Fluid	dp [nm]	%V	Ra	Re	Da	Porous Medium	ε	Output
Hajipour and Molaei Dehkordi (2012)	Num.	Two-component mixture	Laminar		–	–	–	Variable	–	–	0.25	Isotropic and homogeneous	0.9	Velocity and temperature profiles and expressions for the Nusselt number values
Hajipour et al. (2014)	Num.	Two-component mixture	Laminar		Al_2O_3	H_2O	–	0–4	–	–	0.25	–	0.8	Nu, velocity profile, temperature profile
Matin and Ghanbari (2014)	Num.	Two-component mixture	Laminar		–	–	–	Const	–	–	0.1	Homogeneous and isotropic	–	Nu, velocity and temperature distributions, pressure drop

(Continued)

TABLE 4.3 (*Continued*)

Mixed Convection Studies

Article	Num./Exp.	Model	Regime	Configuration System	Particle's Nature	Base Fluid	dp [nm]	%V	Ra	Re	Da	Porous Medium	ε	Output
Zahmatkesh and Ali Naghedifar (2017)	Num.	Single-phase	Laminar		Al_2O_3	H_2O	–	0–8	–	1–1000	10^{-6}–10^{-2}	–	0.5–0.9	Nuavg, Nulocal, T distribution, streamlines
Hussain et al. (2017)	Num.	Single-phase	Laminar		Al_2O_3	H_2O	–	0–4	–	10–200	10^{-6}–10^{-3}	–	0.2–0.8	Isotherms, streamlines, Nuavg, St, Sht, Sff, Stavg
Biswas et al. (2018)	Num.	Single-phase	Laminar		Cu	H_2O	1	1–5	10–10^{3}	10–300	10^{-7}–10^{-4}	–	0.1–1	Streamlines, isotherms, Nu, Pd, T distribution
Asiaei et al. (2018)	Num.	Two-phase	Laminar		Cu	H_2O	25	0–4	–	–	10^{-4}–10^{-2}	Porous foam	0.8	Nu, streamlines, isothermal lines, entropy

Num. = numerical; Exp. = experimental.

increment of the pumping power for Reynolds number above 100 and Darcy number below the value of 10^{-5}. The increase in heat transfer using Cu-H_2O nanofluid is at most 15% when the flow domain is filled with porous medium. The results show the usual tendency for greater heat transfer with greater particle concentration. Asiaei et al. (2018) have numerically studied the mixed convection of Cu-H_2O nanofluid in a two-sided casing with a cover filled with multilayered porous foam. The results indicate that the use of multilayer porous material can limit flow vortexes closing to the coating walls and could increase heat flow up to 17% compared to the case where a homogeneous porous material is used with the highest permeability. Moreover, an increase in the volumetric fraction of the nanoparticles decreases the generation of entropy for low numbers of Richardson, while producing the opposite effect for high numbers of Richardson. Kuznetsov and Bubnovich (2012) have developed a theory on the biothermal convection of a nanofluid in a porous layer in which two different types of mobile microorganisms are suspended. Their goal is to investigate the combined effect of these two species of microorganisms, nanoparticles, and the variation of vertical temperature on the hydrodynamic stability of the porous layer. It is obtained that, since the microorganisms are heavier than the fluid base (H_2O), they act as a destabilizing factor. Moreover, if a species of microorganisms is present, the system is less sensitive to the concentration of the second species. Furthermore, the effect of nanoparticles depends on their distribution, if heavier on the bottom or on the top, while the effects of temperature distribution are destabilizing if heated from the bottom and stabilizing if it cools from the bottom. Kuznetsov (2012a) carried out a study on the biothermal convection of a nanofluid in a horizontal porous medium heated by the bottom. The main reason for bioconvection is to improve mass transfer in microvolumes, but before being implemented in microdevices it must be well analyzed. The results show that the effects of microorganisms on the stability of suspension could depend on the number of Peclet. Ghaziani and Hassanipour (2016) have accomplished an experimental analysis of the heat transfer of a nanofluid (Al_2O_3-H_2O) flowing in a rectangular channel filled with metallic foam. The high thermal conductivity and the mixing of metal foams (porous media) lead to an increase in the heat transfer rate, which increases further thanks to the use of nanoparticles suspended in the fluid. Kuznetsov (2012b) carried out a study on a new type of nanofluid that contains, in addition to the particles, oxidized mobile microorganisms. It has shown that the stability of these nanofluids is controlled by three agents: the distribution of nanoparticles, the stratification of density induced by the vertical temperature gradient, and the upward movement of microorganisms. Mahdi et al. (2015) presented a review of the expected effects for different convective and heat transfer schemes, using an aluminum metal foam in which a nanofluid flows. The nanoparticles used are Al_2O_3, TiO_3, Cu, and SiO_2 while the base fluid is H_2O. Previous studies have shown that convection heat transfer increases in the presence of the porous medium due to its thermal conductivity leading to a significant increase of the heat transfer coefficient. Furthermore, the heat transfer improves better if in the porous medium there is a nanofluid, formed by base fluid and nanoparticles, which has a high thermal conductivity, which also depends on the type of nanofluid used. Miguel (2015) has experimentally analyzed the flow of the Al_2O_3-H_2O nanofluid in a porous cylinder formed by many capillary tubes to obtain the exact viscosity, permeability, and inertial parameters of the porous cylinder. The experiment was completed in two phases. In the first phase, the physical properties of the nanofluid, which consisted of deionized water and different volumetric fraction of the nanoparticles, were measured. In the second phase, the nanofluid was injected into the cylinder, whose porosity was 0.249, to evaluate permeability and inertial parameters. Jouybari et al. (2017) carried out an experimental study on the effect of a porous medium and nanofluid on the thermal performance of a porous flat solar collector. The nanofluid used was SiO_2-H_2O. With the use of nanofluids, the thermal efficiency improves up to 8.1%, but the combination of nanofluid and porous media causes an unwanted increase in pressure drop. In order to take into account both the improvement of heat transfer and the pressure drop, the EPR is used separately for the nanofluid and the porous medium. It has been observed that with increasing of the volumetric fraction of the nanoparticles, the EPR relative to the nanofluids can increase up to 1.34 while as the concentration of the nanofluid increases and the flow rate decreases, the EPR relative to the porous medium increases to the value of 0.92. Moreover, the effect of the nanoparticle dimensions was studied, and the results show that the slope of the efficiency curve decreases as the size decreases. Ghaziani and Hassanipour (2012) experimentally analyzed the performance of a heat sink with a porous medium in which coolant nanofluids are present.

Convective Heat Transfer of Nanofluids in Porous Media 73

The nanofluid used was a mixture of deionized water and alumina nanoparticles at three different concentrations. The experimental section is a rectangular mini-channel filled with a metal foam electrically heated in order to provide a constant heat flow. The effects of the volumetric fraction of the particles on the heat transfer coefficient were studied. The results show that the particles combined with the porous foam have a significant effect on the heat transfer rate, which is maximal at an optimum value of the foam permeability. Moreover, due to the fixed concentration of the nanoparticles, the smaller ones are more effective about improving the heat transfer. Xu et al. (2019) conducted a review on conduction, convection, thermal radiation, and heat transfer due to the phase change of nanofluids in porous media. The motivation of this review paper was to stimulate researchers to pay attention to heat transfer enhancement techniques through the use of nanofluids in porous media. These show different characteristics for engineering applications, including the promotion of heat conduction, reduction of drag friction, transfer of boiling heat, freezing, and the increase of the heat transfer rate in the thermal storage process. Future work can be carried out on turbulence, on the interphase force model, on the non-Newtonian effect, and on the sliding phenomenon for nanofluids in porous metals.

4.5 Governing Equations

In the porous media, the governing equations in the macroscopic scale for the momentum and energy of nanofluids are determined by applying the volume averaging process on a representative elementary volume (REV) at the microscopic equations, as illustrated by Amiri and Vafai (1994), Whitaker (1999) and Nield and Bejan (2006).

4.5.1 Single Phase Model

The following hypotheses are assumed:

- Homogeneous porous media with uniform properties
- For drag model the Darcy–Forchheimer assumption is applied
- For heat transfer between the two phases, local thermal non-equilibrium (LTNE) assumption is applied
- Laminar, transient, and incompressible flow
- The effect of buoyancy is considered
- Effects of pressure work and viscous dissipation are neglected
- The LTNE model for two phases
- The viscous dissipation and the work done by the pressure forces are neglected
- The thermal dispersion and thermal tortuosity are neglected under these hypotheses, and the volume-averaged governing equations of continuity, momentum, and energy for fluid and solid phases are:
  - Continuity equation

$$\nabla \cdot \langle \mathbf{u} \rangle = 0 \tag{4.1}$$

  - Momentum equation

$$\rho_{nf} \left(\frac{1}{\varepsilon} \frac{\partial \langle \mathbf{u} \rangle}{\partial t} + \frac{1}{\varepsilon^2} \langle \mathbf{u} \rangle \cdot \nabla \langle \mathbf{u} \rangle \right) = -\nabla \langle p \rangle^f + \rho_{nf} \mathbf{g} + \frac{\mu_{nf}}{\varepsilon} \nabla^2 \langle \mathbf{u} \rangle$$
$$- \frac{\mu_{nf}}{K} \langle \mathbf{u} \rangle - \frac{C_F}{K^{1/2}} \rho_{nf} |\langle \mathbf{u} \rangle| \langle \mathbf{u} \rangle \tag{4.2}$$

74 Convective Heat Transfer in Porous Media

- Energy equation for the liquid phase:

$$\varepsilon\left(\rho c_p\right)_{nf}\frac{\partial\langle T_f\rangle^f}{\partial t}+\left(\rho c_p\right)_{nf}\langle\mathbf{u}\rangle\cdot\nabla\langle T_f\rangle^f=\varepsilon k_{nf}\nabla^2\langle T_f\rangle^f+h_{sf}a_{sf}\left(\langle T_s\rangle^s-\langle T_f\rangle^f\right) \tag{4.3}$$

- Energy equation for the solid phase:

$$\left(1-\varepsilon\right)\rho_s c\frac{\partial\langle T_s\rangle^s}{\partial t}=\left(1-\varepsilon\right)k_s\nabla^2\langle T_s\rangle^s-h_{sf}a_{sf}\left(\langle T_s\rangle^s-\langle T_f\rangle^f\right) \tag{4.4}$$

where ε, K, and C_F are the porosity, permeability, and inertia coefficients, respectively. The presence of the last term in both energy equations is responsible of the heat exchange between the two phases thanks to the LTNE assumption.

If the local thermal equilibrium (LTE) assumption is considered, it is necessary that a single energy equation can be derived applying the equivalence of the temperature functions:

$$\langle T_f\rangle^f=\langle T_s\rangle^s=\langle T\rangle.$$

Therefore, by adding Equations (4.3) and (4.4). It is obtained:

$$\left(\rho c\right)_e\left(\frac{\partial\langle T\rangle}{\partial t}+\langle\mathbf{u}\rangle\cdot\nabla\langle T\rangle\right)=k_e\nabla^2\langle T\rangle \tag{4.5}$$

where

$$\left(\rho c\right)_e=\left(1-\varepsilon\right)\rho_s c_s+\varepsilon\left(\rho c_p\right)_{nf} \tag{4.6a}$$

$$k_e=\left(1-\varepsilon\right)k_s+\varepsilon k_{nf} \tag{4.6b}$$

are the heat capacity per unit volume and effective thermal conductivity of the porous medium, respectively. In the following, the volume-average symbol < > is dropped for convenience.

4.5.2 Mixture Model

In the two-phase mixture model, the presence of particles is modeled by adding a new term in the momentum equation of the fluid phase and solving a new equation about the concentration particles.

Applying the hypotheses of the single-phase model, the macroscopic governing equations are:

- Continuity equation

$$\nabla\cdot\left(\rho_m\langle\mathbf{u}_m\rangle\right)=0 \tag{4.7}$$

- Momentum equation

$$\frac{1}{\varepsilon}\frac{\partial\left(\rho_m\langle\mathbf{u}_m\rangle\right)}{\partial t}+\frac{1}{\varepsilon^2}\nabla\cdot\left(\rho_m\langle\mathbf{u}_m\rangle\langle\mathbf{u}_m\rangle\right)=-\nabla\langle p\rangle^f+\rho_m\,\mathbf{g}+\frac{1}{\varepsilon}\nabla^2\left(\mu_m\langle\mathbf{u}_m\rangle\right)$$

$$+\nabla\cdot\left(\sum_{k=1}^n\varphi_k\rho_k\langle\mathbf{u}_{dr,k}\rangle\langle\mathbf{u}_{dr,k}\rangle\right)-\frac{\mu_m}{K}\langle\mathbf{u}_m\rangle-\frac{C_F}{K^{1/2}}\rho_m\left|\langle\mathbf{u}_m\rangle\right|\langle\mathbf{u}_m\rangle \tag{4.8}$$

Convective Heat Transfer of Nanofluids in Porous Media

Volume fraction equation for nanoparticles:

$$\nabla \cdot \left(\varphi_p \rho_p \langle \mathbf{u}_m \rangle \right) = -\nabla \cdot \left(\varphi_p \rho_p \langle \mathbf{u}_{dr,p} \rangle \right) \tag{4.9}$$

Energy equation for mixture phase (fluid and nanoparticle):

$$\varepsilon \frac{\partial}{\partial t} \left[\sum_{k=1}^{n} \varphi_k \rho_k c_{pk} \langle T_f \rangle^f \right] + \nabla \cdot \left(\sum_{k=1}^{n} \varphi_k \rho_k c_{pk} \langle \mathbf{u}_k \rangle \langle T_f \rangle^f \right) = \varepsilon \nabla \cdot \left[k_m \nabla \langle T_f \rangle^f \right]$$
$$+ h_{sf} a_{sf} \left(\langle T_s \rangle^s - \langle T_f \rangle^f \right) \tag{4.10}$$

Energy equation for the solid phase (porous):

$$(1-\varepsilon)\rho_s c \frac{\partial \langle T_s \rangle^s}{\partial t} = (1-\varepsilon) k_s \nabla^2 \langle T_s \rangle^s - h_{sf} a_{sf} \left(\langle T_s \rangle^s - \langle T_f \rangle^f \right) \tag{4.11}$$

The mass-average mixture properties are:

$$\langle \mathbf{u}_m \rangle = \frac{\sum_{k=1}^{n} \varphi_k \rho_k \langle \mathbf{u}_k \rangle}{\rho_m}; \qquad \rho_m = \sum_{k=1}^{n} \varphi_k \rho_k; \qquad k_m = \sum_{k=1}^{n} \varphi_k k_k; \qquad \mu_m = \sum_{k=1}^{n} \varphi_k \mu_k \tag{4.12}$$

where n is the number phase, and φ_k is the volume fraction of the k-phase.

The drift velocity $\langle \mathbf{u}_{dr,k} \rangle$ for the phase k in the momentum equation (4.8) is given by:

$$\langle \mathbf{u}_{dr,k} \rangle = \langle \mathbf{u}_k \rangle - \langle \mathbf{u}_m \rangle \tag{4.13}$$

The relative velocity (or slip velocity) $\langle \mathbf{u}_{pf} \rangle$, defined as velocity of the nanoparticles "p" relative to the velocity of the fluid "f," is given by:

$$\langle \mathbf{u}_{pf} \rangle = \langle \mathbf{u}_p \rangle - \langle \mathbf{u}_f \rangle \tag{4.14}$$

The relation between drift velocity and slip velocity is:

$$\langle \mathbf{u}_{dr,p} \rangle = \langle \mathbf{u}_{pf} \rangle - \langle \mathbf{u}_m \rangle \tag{4.15}$$

The slip velocity is taken by (Manninen et al. 1996):

$$\langle \mathbf{u}_{pf} \rangle = \frac{\rho_p d_p^2}{18 \mu_f f_{\text{drag}}} \left(\frac{\rho_p - \rho_m}{\rho_p} \right) \left(\mathbf{g} - \langle \mathbf{u}_m \rangle \cdot \nabla \langle \mathbf{u}_m \rangle \right) \tag{4.16}$$

The drag force coefficient f_{drag} is calculated by Schiller and Naumann relations (Schiller and Naumann 1935):

$$f_{\text{drag}} = \begin{cases} 1 + 0.15 Re_p^{0.678} & Re_p \leq 1000 \\ 0.0183 Re_p & Re_p > 1000 \end{cases} \tag{4.17}$$

A weighted averaging process with the volume concentration between the nanoparticles and the base fluid is accomplished to find the thermal properties of the nanofluid. For the density, a simple mass balance is applied:

$$\rho_{nf} = \rho_{bf}\left(1-\phi\right) + \rho_p\phi \tag{4.18}$$

For the specific heat, the simple weighted average with the concentration is not corrected:

$$c_{p,nf} = c_{p,bf}\left(1-\phi\right) + c_{p,p}\phi \tag{4.19}$$

It is better to use an analytical model by assuming thermal equilibrium between nanoparticles and the base fluid (Khanafer et al. 2015):

$$c_{p,nf} = \frac{\left(1-\phi\right)\rho_{bf}c_{p,bf} + \rho_n c_{p,n}\phi}{\rho_{bf}\left(1-\phi\right) + \rho_n\phi} \tag{4.20}$$

For the thermal conductivity and the dynamic viscosity, in literature there are many models to simulate the interaction between the nanoparticles and the base fluid. Some of them are shown in Table 4.4 for the dynamic viscosity and Table 4.5 for thermal conductivity.

TABLE 4.4

Some Models in Literature to Simulate the Dynamic Viscosity of Nanofluid

References	Dynamic Viscosity [Pa·s]
Einstein (1906)	$\mu_{nf} = \mu_{bf}\left(1+2.5\phi\right)$
Brinkman (1952)	$\mu_{nf} = \dfrac{\mu_{bf}}{(1-\phi)^{2.5}}$
Lundgren (1972)	$\mu_{nf} = \mu_{bf}\left(1+2.5\phi + \dfrac{25}{4}\varphi^2\right)$
Corcione (2011)	$\mu_{nf} = \mu_{bf}\dfrac{1}{1-34.87\,(d_p/d_{bf})^{-0.3}\phi^{1.03}}\quad d_{bf} = 0.1\left(\dfrac{6M}{N\pi\rho_{bf}}\right)^{1/3}$

TABLE 4.5

Some Models in Literature to Simulate the Thermal Conductivity of Nanofluid

References	Thermal Conductivity [W m^{-1} k^{-1}]
Maxwell (1954)	$k_{nf} = k_{bf}\left[\dfrac{k_p + 2k_{bf} + 2(k_p - k_{bf})\phi}{k_p + 2k_{bf} - 2(k_p - k_{bf})\phi}\right]$
Hamilton-Crosser model (1962)	$k_{nf} = k_{bf}\left[\dfrac{k_p + 2k_{bf} + 2(k_{bf} - k_p)\phi}{k_p + 2k_{bf} - (k_{bf} - k_p)\phi}\right]$
Maiga et al. (2004)	$k_{nf} = k_{bf}\left(4.97\varphi^2 + 2.72\phi + 1\right)$
Corcione (2011)	$k_{nf} = k_{bf}\left(1 + 4.4\,\mathrm{Re}_p^{0.4}\,\mathrm{Pr}_{bf}^{0.66}\left(\dfrac{T}{T_{fr}}\right)^{10}\left(\dfrac{k_p}{k_{bf}}\right)^{0.03}\varphi^{0.66}\right); \mathrm{Re} = \dfrac{2\rho_{bf}k_{bf}T}{n\mu_{bf}^2 d_p}$

Convective Heat Transfer of Nanofluids in Porous Media 77

The parameters of the porous media, permeability K, and drag coefficient C_F, can be assessed using the Calmidi and Mahajan relations (Calmidi 1998):

$$K = 0.00073(1-\varepsilon)^{-0.224}\left(\frac{d_f}{d_p}\right)^{-1.11} d_p^2 \tag{4.21}$$

$$C_F = 0.00212(1-\varepsilon)^{-0.132}\left(\frac{d_f}{d_p}\right)^{-1.63} \tag{4.22}$$

These relations are valid for metal foam as a porous media, where d_f and d_p are the ligament diameter and pore diameter, respectively. These diameters are in relation with the parameters of the porous media by means of the following relations (Calmidi 1998):

$$\frac{d_f}{d_p} = 1.18\sqrt{\frac{1-\varepsilon}{3\pi}}\left(\frac{1}{1-e^{-(1-\varepsilon)/0.04}}\right) \tag{4.23}$$

$$d_p = \frac{0.0224}{\omega} \tag{4.24}$$

where ω is the pore density of the metal foam, that is, the number of pores across a linear inch. To estimate, α_{sf} and h_{sf} are adopted by the following correlations (Bhattacharya et al. 2002):

$$\alpha_{sf} = \frac{3\pi d_f}{\left(0.59 d_p\right)^2}(1-e^{-(1-\varepsilon)/0.04}) \tag{4.25}$$

$$h_{sf} = \begin{cases} \left(0.75 Re_{d_f}^{0.4} Pr_f^{0.37}\right)\left(\dfrac{k_f}{d_f}\right), & 1 \le Re_{df} \le 40 \\[2ex] \left(0.51 Re_{d_f}^{0.5} Pr_f^{0.37}\right)\left(\dfrac{k_f}{d_f}\right), & 40 \le Re_{df} \le 1000 \\[2ex] \left(0.26 Re_{d_f}^{0.6} Pr_f^{0.37}\right)\left(\dfrac{k_f}{d_f}\right), & 1000 \le Re_{df} \le 2\times10^5 \end{cases} \tag{4.26}$$

where Pr_f is the Prandtl number referred to the fluid phase, and Re_{df} is the local Reynolds number referred to the ligament diameter:

$$Re_{d_f} = \frac{\rho_f u_0 d_f}{\mu_f} \tag{4.27}$$

4.6 Comparison among Different Configurations

The following several results available from literature are reported in terms of average Nusselt number in order to compare different configurations to underline the general findings of heat transfer with nanofluids in porous media.

The average Nusselt number as a function of the Rayleigh number is reported in Figure 4.1 for natural convection both in a porous enclosure filled with a nanofluid (Sheremet et al. 2015a) and in porous medium through two concentric, horizontal cylinders (Sheremet and Pop 2015) with a nanofluid using Buongiorno's mathematical model.

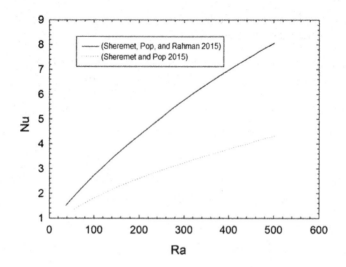

FIGURE 4.1 Natural convection in a porous enclosure.

For the two investigated configurations and same nanofluid (Ag-H$_2$O, Cu-H$_2$O), Nu-Ra profiles are very similar, but at the same value of the Rayleigh number, the average Nusselt in the cubic enclosure is rather higher than in the horizontal annulus filled with porous medium.

Figure 4.2 shows the average Nusselt number values as functions of volume particle concentration φ in Cu-H$_2$O nanofluid for the configurations reported in Chamkha and Ismael (2014), Ismael and Chamkha (2015), and Izadi et al. (2018), at $Ra = 10^5$ and $Ra = 10^6$.

It is worth noting that, in the case of natural convection in a cavity, the Nusselt number can decrease with the volumetric concentration for both values of the reported Rayleigh number as reported in (Ismael and Chamkha 2015).

A comparison among average Nusselt numbers as a function of volumetric concentration for $Da = 10^{-1}$ is reported in Chamkha and Ismael (2014), Ismael and Chamkha (2015), and Toosi and Siavashi (2017) and presented in Figure 4.3. The values of Nu are nearly independent with φ for the configuration investigated in Chamkha and Ismael (2014) and is about 8.0. For the configuration of article Toosi and Siavashi (2017), the average Nusselt number increases slightly with φ and assumes values in the range of 2.2–2.4. Greater values are reached in Ismael and Chamkha (2015) where average the Nusselt number values are in the range of 10.0–12.0 but decrease with the volumetric concentration.

As far as the forced convection is concerned, a comparison among average Nusselt number values as functions of Reynolds number for $\varphi = 0.01$ and 0.02 and for a porous medium radius equal to 0.05 m is shown in Figure 4.4. The presented results are extracted from Zargartalebi and Azaiez (2018), where the nanofluid is Al$_2$O$_3$-water and Amani et al. (2017), where the nanofluid is Fe$_3$O$_4$-water. The average Nusselt number increases as the Reynolds number increases for both nanofluids, but the greater values are for Al$_2$O$_3$-water nanofluid.

Convective Heat Transfer of Nanofluids in Porous Media

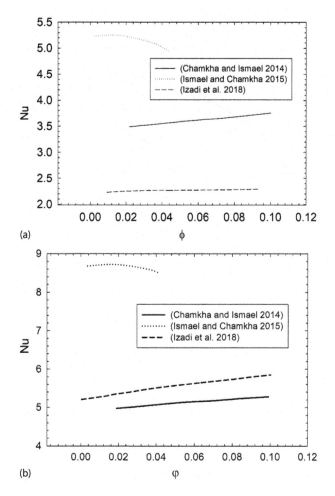

FIGURE 4.2 Profile Nu-c for: (a) $Ra = 10^5$; and (b) $Ra = 10^6$.

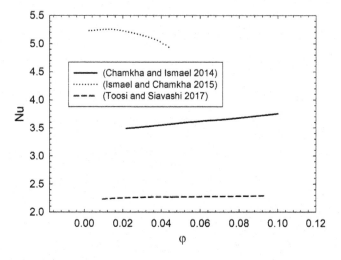

FIGURE 4.3 Profile Nu-φ for $Da = 10^{-1}$.

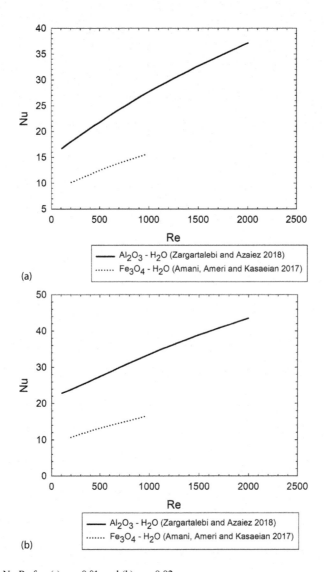

FIGURE 4.4 Profile Nu-Re for: (a) $\varphi = 0.01$; and (b) $\varphi = 0.02$.

The average Nusselt number as a function of volumetric particle concentration, which are presented in the papers Ghaziani and Hassanipour (2012) and Kuznetsov (2012a) for mixed convection, is reported in Figure 4.5. The average Nusselt numbers, which are obtained for mixed convention in a square geometry with bottom heating and filled porous medium and Cu-water, investigated in Asiaei et al. (2018) in Figure 4.5b, are greater than the ones for mixed convention of Cu-water nanofluid inside a two-sided lid-driven enclosure with an internal heater, filled with multi-layered porous foams, investigated in Kuznetsov and Bubnovich (2012) Figure 4.5a.

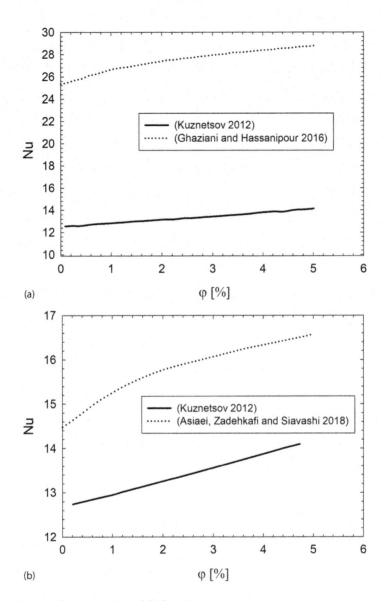

FIGURE 4.5 Profile Nu-φ for: (a) $Ri = 0.1$; and (b) $Ri = 1$.

4.7 Conclusions

The application of two passive techniques to enhance the heat transfer, such as nanofluids combined with porous media, has been significantly employed in convection as reviewed recently (Nield and Kuznetsov 2014; Kasaeian et al. 2017; Xu et al. 2019; Khanafer and Vafai 2019) and highlighted in the present chapter. Basic configurations have been mainly investigated by means of numerical models, but more experimental activities should be also carried out to validate and confirm the theoretical models and results.

As recently underlined by Khanafer and Vafai (2019), some more research activities should be oriented toward turbulent and local non-equilibrium convective heat transfer of nanofluids in porous media. Furthermore, the effect of magnetic fields can be an additive active technique to improve the convective heat transfer in several engineering applications.

NOMENCLATURE

C_F	inertial drag factor (–)
c_p	specific heat (Jkg^{-1}K^{-1})
d_p	pore diameter (m)
d_f	ligament diameter (m)
f_{drag}	drag force coefficient (–)
g	gravity acceleration vector (ms^{-2})
h_{sf}	local heat transfer coefficient (Wm^{-2}K^{-1})
k	thermal conductivity (Wm^{-1}K^{-1})
K	permeability (m^2)
n	number of phases (–)
p	relative pressure (Pa)
Pr_f	Prandtl number [$(\mu c_p)/k$]
Re_{df}	ligament reynolds number [$(\rho V d_f)/\mu$]
t	time (s)
T	temperature (K)
u	velocity vector (ms^{-1})

Greek Letters

α_{sf}	surface area density (m^{-1})
ε	porosity (–)
φ	volume fraction (–)
μ	dynamic viscosity (kgm^{-1}s^{-1})
ρ	density (kgm^{-3})
ω	pore density

Subscripts

dr	drift
e	effective
f	fluid phase
k	k-phase
p	nanoparticles
pf	nanoparticles-fluid
m	mass-average mixture
nf	nanofluid
s	solid phase

Superscripts

f	fluid phase
s	solid phase

REFERENCES

Ali Agha, Hamza, Mohamed Najib Bouaziz, and Salah Hanini. 2014. "Free Convection Boundary Layer Flow from a Vertical Flat Plate Embedded in a Darcy Porous Medium Filled with a Nanofluid: Effects of Magnetic Field and Thermal Radiation." *Arabian Journal for Science and Engineering* 39 (11): 8331–8340. doi:10.1007/s13369-014-1405-z.

Amani, Mohammad, Mohammad Ameri, and Alibakhsh Kasaeian. 2017. "The Experimental Study of Convection Heat Transfer Characteristics and Pressure Drop of Magnetite Nanofluid in a Porous Metal Foam Tube." *Transport in Porous Media* 116 (2): 959–974. doi:10.1007/s11242-016-0808-6.

Ameri, Mohammad, and Mahyar Syed Eshaghi. 2018. "Exergy and Thermal Assessment of a Novel System Utilizing Flat Plate Collector with the Application of Nanofluid in Porous Media at a Constant Magnetic Field." *Thermal Science and Engineering Progress* 8: 223–235. doi:10.1016/j.tsep.2018.08.004.

Amiri, Abdalla M., and Kambiz Vafai. 1994. "Analysis of Dispersion Effects and Non-Thermal Equilibrium, Non-Darcian, Variable Porosity Incompressible Flow through Porous Media." *International Journal of Heat and Mass Transfer.* doi:10.1016/0017-9310(94)90219-4.

Asiaei, Sasan, Ali Zadehkafi, and Majid Siavashi. 2018. "Multi-Layered Porous Foam Effects on Heat Transfer and Entropy Generation of Nanofluid Mixed Convection Inside a Two-Sided Lid-Driven Enclosure with Internal Heating." *Transport in Porous Media.* doi:10.1007/s11242-018-1166-3.

Baqaie Saryazdi, A., Farhad Talebi, Taher Armaghani, and Ioan Pop. 2016. "Numerical Study of Forced Convection Flow and Heat Transfer of a Nanofluid Flowing inside a Straight Circular Pipe Filled with a Saturated Porous Medium." *European Physical Journal Plus* 131 (4). doi:10.1140/epjp/i2016-16078-6.

Bayomy, Ayman M., and Ziad Saghir. 2017. "Experimental Study of Using γ-Al$_2$O$_3$–water Nanofluid Flow through Aluminum Foam Heat Sink: Comparison with Numerical Approach." *International Journal of Heat and Mass Transfer* 107: 181–203. doi:10.1016/j.ijheatmasstransfer.2016.11.037.

Bhattacharya, Anandaroop., Varaprasad V. Calmidi, and Roop L. Mahajan. 2002. "Thermophysical Properties of High Porosity Metal Foams." *International Journal of Heat and Mass Transfer* 45 (5): 1017–1031. doi:10.1016/S0017-9310(01)00220-4.

Biswas, Nirmalendu, Nirmal K. Manna, Priyankan Datta, and Pallab Sinha Mahapatra. 2018. "Analysis of Heat Transfer and Pumping Power for Bottom-Heated Porous Cavity Saturated with Cu-Water Nanofluid." *Powder Technology* 326: 356–369. doi:10.1016/j.powtec.2017.12.030.

Brinkman, H. C. 1952. "The viscosity of concentrated suspensions and solutions." *The Journal of Chemical Physics* 20 (4): p. 571. doi: 10.1063/1.1700493.

Calmidi, Varaprasad V. 1998. *Transport Phenomena in High Porosity Fibrous Metal Foams.* University of Colorado, Boulder, CO.

Chamkha, Ali J., and Muneer A. Ismael. 2014. "Natural Convection in Differentially Heated Partially Porous Layered Cavities Filled with a Nanofluid." *Numerical Heat Transfer; Part A: Applications* 65 (11): 1089–1113. doi:10.1080/10407782.2013.851560.

Corcione, M. 2011. "Empirical correlating equations for predicting the effective thermal conductivity and dynamic viscosity of nanofluids." *Energy Conversion and Management* 52 (1): 789–793. doi: 10.1016/j.enconman.2010.06.072.

Dastmalchi, Majid, Ghanbar A. Sheikhzadeh, and Ali A. Abbasian Arani. 2015. "Double-Diffusive Natural Convective in a Porous Square Enclosure Filled with Nanofluid." *International Journal of Thermal Sciences* 95: 88–98. doi:10.1016/j.ijthermalsci.2015.04.002.

Einstein, A. 1906. *Annalen der Physik* 324 (2): 289–306. doi: 10.1002/andp.19063240204.

Ghalambaz, Mohammad, Mahmoud H. Sabour, and Ioan Pop. 2016. "Free Convection in a Square Cavity Filled by a Porous Medium Saturated by a Nanofluid: Viscous Dissipation and Radiation Effects." *Engineering Science and Technology, an International Journal* 19 (3): 1244–1253. doi:10.1016/j.jestch.2016.02.006.

Ghasemi, Kasra, and Majid Siavashi. 2017. "MHD Nanofluid Free Convection and Entropy Generation in Porous Enclosures with Different Conductivity Ratios." *Journal of Magnetism and Magnetic Materials* 442: 474–490. doi:10.1016/j.jmmm.2017.07.028.

Ghaziani, Navid O., and Fatemeh Hassanipour. 2012. "Experimental Analysis of Nanofluid Slurry Through Rectangular Porous Channel." Volume 7: *Fluids and Heat Transfer, Parts A, B, C, and D* 7: 713. doi:10.1115/IMECE2012-89758.

Ghaziani, Navid O., and Fatemeh Hassanipour. 2016. "Convective Heat Transfer of Al_2O_3 Nanofluids in Porous Media." *Journal of Heat Transfer* 139 (3): 032601. doi:10.1115/1.4034936.

Hajipour, Mastaneh, and Asghar Molaei Dehkordi. 2012. "Analysis of Nanofluid Heat Transfer in Parallel-Plate Vertical Channels Partially Filled with Porous Medium." *International Journal of Thermal Sciences* 55: 103–113. doi:10.1016/j.ijthermalsci.2011.12.018.

Hamilton, R. L., and O.K. Crosser. 1962. "Thermal conductivity of heterogeneous two-component systems." *Industrial and Engineering Chemistry Fundamentals* 1 (3): 187–191. doi: 10.1021/i160003a005.

Hussain, Shafqat T., Khalid Mehmood, Muhammad Sagheer, and Arfa A. Farooq. 2017. "Entropy Generation Analysis of Mixed Convective Flow in an Inclined Channel with Cavity with Al_2O_3-Water Nanofluid in Porous Medium." *International Communications in Heat and Mass Transfer* 89: 198–210. doi:10.1016/j.icheatmasstransfer.2017.10.009.

Hussain, S., K. Mehmood, M. Sagheer, and A. Farooq. 2017. "Entropy Generation Analysis of Mixed Convective Flow in an Inclined Channel with Cavity with Al_2O_3-Water Nanofluid in Porous Medium." *International Communications in Heat and Mass Transfer* 89: 198–210. doi:10.1016/j.icheatmasstransfer.2017.10.009.

Ismael, Muneer A., and Ali J. Chamkha. 2015. "Conjugate Natural Convection in a Differentially Heated Composite Enclosure Filled with a Nanofluid." *Journal of Porous Media* 18 (7): 699–716. doi:10.1615/JPorMedia.v18.i7.50.

Izadi, Mohsen, Sara Sinaei, S. A.M. Mehryan, Hakan F. Oztop, and Nidal Abu-Hamdeh. 2018. "Natural Convection of a Nanofluid between Two Eccentric Cylinders Saturated by Porous Material: Buongiorno's Two Phase Model." *International Journal of Heat and Mass Transfer* 127: 67–75. doi:10.1016/j.ijheatmasstransfer.2018.07.066.

Jouybari, H. Javaniyan, S. Saedodin, A. Zamzamian, M. Eshagh Nimvari, and Somchai Wongwises. 2017. "Effects of Porous Material and Nanoparticles on the Thermal Performance of a Flat Plate Solar Collector: An Experimental Study." *Renewable Energy* 114: 1407–1418. doi:10.1016/j.renene.2017.07.008.

Kasaeian, Alibakhsh, Reza Daneshazarian Azarian, Omid Mahian, Lioua Kolsi, Ali J. Chamkha, Somchai Wongwises, and Ioan Pop. 2017. "Nanofluid flow and heat transfer in porous media: A review of the latest developments." *International Journal of Heat and Mass Transfer* 107: 778–791. doi:10.1016/j.ijheatmasstransfer.2016.11.074.

Khanafer, Khalil, and Kambiz Vafai. 2019. "Applications of nanofluids in porous medium: A critical review." *Journal of Thermal Analysis and Calorimetry* 135 (2): 1479–1492. doi: 10.1007/s10973-018-7565-4.

Khanafer, Khalil, Fatemeh Tavakkoli, Kambiz Vafai, and Abdalla AlAmiri. 2015. "A Critical Investigation of the Anomalous Behavior of Molten Salt-Based Nanofluids." *International Communications in Heat and Mass Transfer* 69: 51–58. doi:10.1016/j.icheatmasstransfer.2015.10.002.

Kuznetsov, Andrey V. 2012a. "Nanofluid Bioconvection in a Horizontal Fluid-Saturated Porous Layer." *Journal of Porous Media* 15 (1): 11–27. doi:10.1615/JPorMedia.v15.i1.20.

Kuznetsov, Andrey V. 2012b. "Nanofluid Bioconvection in Porous Media: Oxytactic Microorganisms." *Journal of Porous Media* 15 (3): 233–248. doi:10.1615/JPorMedia.v15.i3.30.

Kuznetsov, Andrey V., and Valeri Bubnovich. 2012. "Investigation of Simultaneous Effects of Gyrotactic and Oxytactic Microorganisms on Nanofluid Bio-Thermal Convection in Porous Media." *Journal of Porous Media* 15 (7): 617–631. doi:10.1615/JPorMedia.v15.i7.20.

Lundgren, T. S. 1972. "Slow flow through stationary random beds and suspensions of spheres." *Journal of Fluid Mechanics* 51 (2): 273–299. doi: 10.1017/S002211207200120X.

Mahdi, Raed Abed, H. A. Mohammed, K. M. Munisamy, and N. H. Saeid. 2015. "Review of Convection Heat Transfer and Fluid Flow in Porous Media with Nanofluid." *Renewable and Sustainable Energy Reviews* 41: 715–734. doi:10.1016/j.rser.2014.08.040.

Maiga, Sidi El Bécaye, Cong Tam Nguyen, Nicolas Galanis, Gilles Roy. 2004. "Heat transfer behaviours of nanofluids in a uniformly heated tube." *Superlattices Microstructures* 35: 543–557. doi: 10.1016/j.spmi.2003.09.012.

Manninen, M, Taivassalo, V, Kallio, S. 1996. "On the Mixture Model for Multiphase Flow." *Technical Research Centre of Finland: VTT Publications* 288. doi:10.17660/ActaHortic.2017.1164.18.

Mashaei, P. R., M. Shahryari, and S. Madani. 2016. "Numerical Hydrothermal Analysis of Water-Al2O3nanofluid Forced Convection in a Narrow Annulus Filled by Porous Medium Considering Variable Properties: Application to Cylindrical Heat Pipes." *Journal of Thermal Analysis and Calorimetry* 126 (2): 891–904. doi:10.1007/s10973-016-5550-3.

Matin, Meisam Habibi, and Behzad Ghanbari. 2014. "Effects of Brownian Motion and Thermophoresis on the Mixed Convection of Nanofluid in a Porous Channel Including Flow Reversal." *Transport in Porous Media* 101 (1): 115–136. doi:10.1007/s11242-013-0235-x.

Matin, Meisam Habibi, and Ioan Pop. 2013. "Forced Convection Heat and Mass Transfer Flow of a Nanofluid through a Porous Channel with a First Order Chemical Reaction on the Wall." *International Communications in Heat and Mass Transfer* 46: 134–141. doi:10.1016/j.icheatmasstransfer.2013.05.001.

Maxwell, J. C. 1954. *A Treatise on Electricity and Magnetism*. 3rd ed. Dover, New York.

Mehryan, Seyed A. M., Farshad M. Kashkooli, Mohammad Ghalambaz, and Ali J. Chamkha. 2017. "Free Convection of Hybrid Al_2O_3-Cu Water Nanofluid in a Differentially Heated Porous Cavity." *Advanced Powder Technology* 28 (9): 2295–2305. doi:10.1016/j.apt.2017.06.011.

Miguel, Antonio F. 2015. "Experimental Study on Nanofluid Flow in a Porous Cylinder: Viscosity, Permeability and Inertial Factor." *Trans Tech Publications* 362: 47–57. doi:10.4028/www.scientific.net/DDF.362.47.

Nield, Donald, and Adrian Bejan. 2006. *Convection in Porous Media*. 3rd ed. Springer, New York.

Nield, D.A., and A. V. Kuznetsov. 2014. "Forced convection in a parallel-plate channel occupied by a nanofluid or a porous medium saturated by a nanofluid", *International Journal of Heat and Mass Transfer* 70: 430–433. doi: 10.1016/j.ijheatmasstransfer.2013.11.016.

Schiller, Links, and Z. Naumann. 1935. *A Drag Coefficient Correlation*. *Zeitschrift des Vereins Deutscher Ingenieure* 77: 318–320. doi:10.1016/j.ijheatmasstransfer.2009.02.006.

Servati V., Ata A., Koroush Javaherdeh, and Hamid Reza Ashorynejad. 2014. "Magnetic Field Effects on Force Convection Flow of a Nanofluid in a Channel Partially Filled with Porous Media Using Lattice Boltzmann Method." *Advanced Powder Technology* 25 (2): 666–675. doi:10.1016/j.apt.2013.10.012.

Sheikholeslami, Mohsen, Davood D. Ganji, and Rasoul Moradi. 2017. "Heat Transfer of Fe_3O_4–water Nanofluid in a Permeable Medium with Thermal Radiation in Existence of Constant Heat Flux." *Chemical Engineering Science* 174: 326–336. doi:10.1016/j.ces.2017.09.026.

Sheremet, Mikhail A., and Ioan Pop. 2015. "Free Convection in a Porous Horizontal Cylindrical Annulus with a Nanofluid Using Buongiorno's Model." *Computers and Fluids* 118: 182–190. doi:10.1016/j.compfluid.2015.06.022.

Sheremet, Mikhail A., Ioan Pop, and Mohammad M. Rahman. 2015a. "Three-Dimensional Natural Convection in a Porous Enclosure Filled with a Nanofluid Using Buongiorno's Mathematical Model." *International Journal of Heat and Mass Transfer* 82: 396–405. doi:10.1016/j.ijheatmasstransfer.2014.11.066.

Sheremet, Mikhail A., Ioan Pop, and Roslinda M. Nazar. 2015b. "Natural Convection in a Square Cavity Filled with a Porous Medium Saturated with a Nanofluid Using the Thermal Nonequilibrium Model with a Tiwari and Das Nanofluid Model." *International Journal of Mechanical Sciences* 100: 312–321. doi:10.1016/j.ijmecsci.2015.07.007.

Siavashi, Majid, and Seyed Mohammad Miri Joibary. 2018. "Numerical Performance Analysis of a Counter-Flow Double-Pipe Heat Exchanger with Using Nanofluid and Both Sides Partly Filled with Porous Media." *Journal of Thermal Analysis and Calorimetry*. doi:10.1007/s10973-018-7829-z.

Siavashi, Majid, Hamed Rasam, and Aliakbar Izadi. 2018a. "Similarity Solution of Air and Nanofluid Impingement Cooling of a Cylindrical Porous Heat Sink." *Journal of Thermal Analysis and Calorimetry*. doi:10.1007/s10973-018-7540-0.

Siavashi, Majid, Hamid Reza Talesh Bahrami, and Ehsan Aminian. 2018b. "Optimization of Heat Transfer Enhancement and Pumping Power of a Heat Exchanger Tube Using Nanofluid with Gradient and Multi-Layered Porous Foams." *Applied Thermal Engineering* 138: 465–474. doi:10.1016/j.applthermaleng.2018.04.066.

Siavashi, Majid, Hamid Reza Talesh Bahrami, and Hamid Saffari. 2015. "Numerical Investigation of Flow Characteristics, Heat Transfer and Entropy Generation of Nanofluid Flow Inside an Annular Pipe Partially or Completely Filled with Porous Media Using Two-Phase Mixture Model." *Energy* 93: 2451–2466. doi:10.1016/j.energy.2015.10.100.

Tayebi, Rashid, Sanaz Akbarzadeh, and Mohammad Sadegh Valipour. 2018. "Numerical Investigation of Efficiency Enhancement in a Direct Absorption Parabolic Trough Collector Occupied by a Porous Medium and Saturated by a Nanofluid." *Environmental Progress & Sustainable Energy* 1–14. doi:10.1002/ep.13010.

Ting, Tiew Wei, Yew Mun Hung, and Ningqun Guo. 2015. "Viscous Dissipative Nanofluid Convection in Asymmetrically Heated Porous Microchannels with Solid-Phase Heat Generation." *International Communications in Heat and Mass Transfer* 68: 236–247. doi:10.1016/j.icheatmasstransfer.2015.09.003.

Ting, Tiew Wei., Yew Mun Hung, Mohd S. B. Osman, and Peter N. Y. Yek. 2018. "Heat and Flow Characteristics of Nanofluid Flow in Porous Microchannels." *International Journal of Automotive and Mechanical Engineering* 15 (2): 5238–5250. doi:10.15282/ijame.15.2.2018.7.0404.

Toosi, Mohammad Hesam, and Majid Siavashi. 2017. "Two-Phase Mixture Numerical Simulation of Natural Convection of Nanofluid Flow in a Cavity Partially Filled with Porous Media to Enhance Heat Transfer." *Journal of Molecular Liquids* 238: 553–569. doi:10.1016/j.molliq.2017.05.015.

Umavathi, Jawali C., Odelu Ojjela, and Kuppalappalle Vajravelu. 2017. "Numerical Analysis of Natural Convective Flow and Heat Transfer of Nanofluids in a Vertical Rectangular Duct Using Darcy-Forchheimer-Brinkman Model." *International Journal of Thermal Sciences* 111: 511–524. doi:10.1016/j.ijthermalsci.2016.10.002.

Welsford, Christofer, Ayman M. Bayomy, and Ziad Saghir. 2018. "Role of Metallic Foam in Heat Storage in the Presence of Nanofluid and Microencapsulated Phase Change Material." *Thermal Science and Engineering Progress* 7: 61–69. doi:10.1016/j.tsep.2018.05.003.

Whitaker, Stephen. 1999. *The Method of Volume Averaging. Theory and Applications of Transport in Porous Media.* Kluwer Academic Publishers, Dordrecht, the Netherlands. doi:10.1016/j.cbpc.2006.06.008.

Xu, Huijin J., and Zhanbin B. Xing. 2017. "The Lattice Boltzmann Modeling on the Nanofluid Natural Convective Transport in a Cavity Filled with a Porous Foam." *International Communications in Heat and Mass Transfer* 89: 73–82. doi:10.1016/j.icheatmasstransfer.2017.09.013.

Xu, Huijin, Liang Gong, Shanbo Huang, and Minghai Xu. 2015. "Flow and Heat Transfer Characteristics of Nanofluid Flowing through Metal Foams." *International Journal of Heat and Mass Transfer* 83: 399–407. doi:10.1016/j.ijheatmasstransfer.2014.12.024.

Xu, Huijin J., Zhanbin B. Xing, Fuqiang Wang, and Ziming Cheng. 2019. "Review on Heat Conduction, Heat Convection, Thermal Radiation and Phase Change Heat Transfer of Nanofluids in Porous Media: Fundamentals and Applications." *Chemical Engineering Science* 195: 462–483. doi:10.1016/j.ces.2018.09.045.

Zahmatkesh, Iman, and Seyyed Ali Naghedifar. 2017. "Oscillatory Mixed Convection in the Jet Impingement Cooling of a Horizontal Surface Immersed in a Nanofluid-Saturated Porous Medium." *Numerical Heat Transfer; Part A: Applications* 72 (5): 401–416. doi:10.1080/10407782.2017.1376961.

Zargartalebi, Hossein, Aminreza Noghrehabadi, Mohammad Ghalambaz, and Ioan Pop. 2015. "Natural Convection Boundary Layer Flow over a Horizontal Plate Embedded in a Porous Medium Saturated with a Nanofluid: Case of Variable Thermophysical Properties." *Transport in Porous Media* 107 (1): 153–170. doi:10.1007/s11242-014-0430-4.

Zargartalebi, Mohammad, and Jalel Azaiez. 2018. "Heat Transfer Analysis of Nanofluid Based Microchannel Heat Sink." *International Journal of Heat and Mass Transfer* 127: 1233–1242. doi:10.1016/j.ijheatmasstransfer.2018.07.152.

Zhang, Wenhao, Wenhao Li, and Akira Nakayama. 2015. "An Analytical Consideration of Steady-State Forced Convection within a Nanofluid-Saturated Metal Foam." *Journal of Fluid Mechanics* 769: 590–620. doi:10.1017/jfm.2015.131.

5

Pore-Network Simulation of Drying of Heterogeneous and Stratified Porous Media

Hassan Dashtian, Nima Shokri, and Muhammad Sahimi

CONTENTS

5.1 Introduction..87
5.2 The Correlated and Anisotropic Pore Network Model 90
5.3 Pore-Network Simulation 91
5.4 Results and Discussion 92
    5.4.1 Isotropic Porous Media..93
    5.4.2 Stratified Porous Media 95
    5.4.3 The Saturation Distribution 96
    5.4.4 The Drying Rates..97
    5.4.5 Effect of the Type of the Correlations 98
5.5 Summary...99
Acknowledgments...99
References..99

HIGHLIGHTS

- Development of three-dimensional pore-network model of drying of heterogeneous and stratified porous media
- Demonstrating the effect of long-range correlations and stratification, hitherto unexplored
- Describing the efficiency of drying in porous media in terms of the structure of vapor-filled cluster, hitherto not studied

---

5.1 Introduction

Drying of porous materials is a phenomenon that is important to numerous problems in science and technology (Or et al. 2013), ranging from agriculture and soil physics (Aydin et al. 2008; Ben-Noah and Friedman 2018), to remediation and recovery of soil contaminated with hydrocarbons (Nadim et al. 2000; Shokri et al. 2010; Jambhekar et al. 2015; Hosseini and Alfi 2016; Soltanian et al. 2016; Lu et al. 2017), to recovery of volatile hydrocarbons from oil reservoirs, cosmetics, building restoration, and such material processing as the production of food, wood, paper, textiles, pharmaceuticals, and washing powders. In addition, water evaporation disturbs the availability, transport, and partitioning of nutrients, as it controls the transfer of heat, air, and humidity between the atmosphere and subsurface. Thus, understanding evaporation and drying of porous media and factors that affect them has been a long-standing problem and studied for decades.

87

Drying of porous materials belong to the much broader class of fluid flow and transport processes in heterogeneous porous media, which are controlled to a large degree by the pore-size distribution (PSD), pore connectivity, wettability, and several other factors. Patterns of flow paths in disordered porous media are usually highly complex, with the complexity being mostly due to the spatial heterogeneity of the pore space (Helmig and Schulz 1997; Sahimi 2011; Blunt 2017). In the case of drying, the evaporation flux depends not only on the morphology of the pore space but also on the nature of the evaporating fluid (Shokri-Kuehni et al. 2017). Although evaporation is a very complex phenomenon, it generally consists of two stages with a transition between the two. Stage-1 evaporation is controlled mostly by the external conditions and occurs over a short time. It represents the period when the saturated zone at the bottom of the porous medium is connected to the surface via capillary-induced liquid flow and is limited by the evaporative demand—the external condition. The duration of the first stage is influenced by the PSD, the correlations between the pore sizes, the pore connectivity, and transport properties of porous media. When connection of the liquid with the surface begins to break down, it marks the end of Stage-1 evaporation and the onset of the transition period. The transition period, which lasts much longer than Stage-1 evaporation (Shokri et al. 2008a, 2008b; Lehmann and Or 2009; Chauvet et al. 2009), ends when all the connections of the liquid are disrupted from the surface, which marks the onset of Stage-2 evaporation, and is limited by vapor diffusion through porous media.

In addition to the PSD, pore connectivity, and the correlations between the sizes of the pores, stratification is a key feature of sedimentary rock that occurs on the scale of micrometer up to hundreds of meters. Stratification is caused by the variety of processes, such as weathering, deposition, cementation, and compaction that together form sedimentary rock. Small-scale stratification is due to alternately operating depositional processes. Through experimental and numerical studies, Zhang et al. (2013) showed that such small-scale stratifications have a dominant effect on the capillary pressure field, CO_2/brine saturation patterns, the flow regime, and the apparent relative permeability of sandstone. Shorki et al. (2010) developed a model to describe the drying of stratified porous media and verified the model with experimental data. They showed that the presence of interfaces in stratified porous media affects the fluid-phase distribution, the drying rates, and other aspects of evaporation in porous formations. Vorhauer et al. (2017) showed that the impact of nonuniform distribution of evaporation rate becomes increasingly less important with an increasing number of porous layers. The transition from thin porous media, in which the drying process is highly dependent on the local structure of the external mass transfer, to thick porous media in which the drying process is much less sensitive to the details of the external mass transfer at the surface, is progressive under isothermal conditions.

Evaporation, similar to all multiphase flow phenomena in porous media, is also influenced by the interplay between capillarity and viscous and gravitational forces (Or 2008; Yiotis et al. 2015; Pegler et al. 2016) in which wettability also plays a prominent role (Shokri et al. 2009). The combined effect of the main factors gives rise to ramified preferential flow paths and the spatial distribution of the fluids. Whether heterogeneity contributes positively or negatively to a given phenomenon in porous media depends on the nature of the process considered. For example, whereas in order to have efficient separation of fluid mixtures by membranes and filters, relative uniformity in the size of the pores is desired, high heterogeneity contributes greatly (Sahimi and Goddard 1986; Sahimi and Arbabi 1992) to controlling propagation of fractures in rock. In any case, the importance of the PSD of the pore space of both natural and synthetic porous media to the phenomena that occur there has been recognized and studied; see for example, Muljadi et al. (2016), Alim et al. (2017), Fantinel et al. (2017), and Liu et al. (2012, 2017) for the most recent studies of the effect of the PSD on fluid flow.

Laboratory-scale porous media usually contain extended correlations between the pore sizes. Large-scale porous media, on the other hand, contain extended correlations between the permeability and porosity of their various zones, first demonstrated by Hewett (1986) for the porosity distribution, and by Neuman (1994) for the permeabilities. Since then, numerous studies have demonstrated (for reviews see Molz et al. 1997; Sahimi 2011) that large-scale porous formations exhibit long-range correlations, with the extent of the correlations being on the order of the length scale over which they are studied.

The existence of such correlations motivated several studies of their effect on fluid flow and transport in porous media. Aside from early work of Hewett and co-workers (Hewett 1986; Hewett and Behrens 1990) who included long-range correlations in the spatial distribution of the porosity in order to simulate fluid flow and transport in oil reservoirs, Sahimi (1994, 1995) appears to be the first who studied the effect of long-range correlations in the spatial distribution of the permeability on fluid flow and dispersion in porous media. Knackstedt et al. (1998) demonstrated excellent agreement between their capillary pressure data and a model of the same phenomenon in a porous medium with extended correlations, while Knackstedt et al. (2001a, 2001b) studied the effect of the same type of correlations on two-phase flow and residual saturations in rock, and Dashtian et al. (2015) investigated the same effect on the resistivity logs of oil reservoirs. Weitz and co-workers (Datta and Weitz 2013; Alim et al. 2017) used confocal microscopy to study the spatial fluctuations in fluid flow through a 3D porous medium and demonstrated that pore-scale correlations in the flow are determined by the geometry of the medium.

Evaporation and drying in porous materials are also strongly affected by the morphology of a pore space. The spatiotemporal dynamics of a fluid, liquid, or vapor distribution and the drying front are controlled by two sets of factors. One set includes the morphological properties of the porous medium and, if they are present, the fractures (Prat 2002; Fantinel et al. 2017; Borgman et al. 2017). The second set includes dynamic factors, such as capillarity-driven flow (Scherer 1990; Shokri et al. 2008b; Le et al. 2009), pore-scale evaporation, and vapor diffusion (Miri et al. 2015; Lehmann and Or 2009; Vorhauer et al. 2015; Hajirezaie et al. 2017), and formation of liquid films on the pores' surfaces (Yiotis et al. 2001; Chauvet et al. 2009; Lehmann and Or 2009).

Evaporation from a pore space is a displacement process in which the liquid phase in the pore space recedes as a result of vaporization into a vapor phase. Thus, drying may be viewed as a drainage process in which the nonwetting vapor displaces the evaporating wetting liquid. Therefore, in the absence of viscous effects, that is, when the interface between the liquid and vapor moves slowly, the receding interface is described by invasion percolation (IP) (Shaw 1987). Taking advantage of the link between drying and the IP, pore-network (PN) models, a powerful tool of studying the link between pore-scale phenomena and the macroscopic properties of porous media, have been developed for studying evaporation and drying in porous media (Prat 1995, 2002, 2011; Laurindo and Prat 1998; Bray and Prat 1999; Yiotis et al. 2001; Surasani et al. 2008). The models range from simple to relatively advanced PNs, although most of them have been developed for 2D porous media (Yiotis et al. 2001; Rahimi et al. 2016; Le et al. 2017). Three-dimensional models have also been developed more recently (Le et al. 2009; Kharaghani et al. 2012; Yiotis et al. 2015; Attarimoghadam et al. 2017), including a recent model (Vorhauer et al. 2017) for thin porous media.

The goal of this paper is to study the effect of heterogeneity of porous media on evaporation and drying in porous media. The particular aspects of the heterogeneity that we study include the effect of the correlations between the sizes of the pores, as well as anisotropy induced by stratification that is prevalent in large-scale porous formations. Despite its obvious significance, the effect on evaporation and drying of the spatial correlations between the sizes of the pores has not been studied before. We report the results of a study in a 3D model of porous media that includes the effects of the correlations, the PSD, and anisotropy caused by stratification. In a recent study (Dashtian et al. 2018), we demonstrated that evaporation of brine and the resulting salt precipitation in porous media are strongly influenced by the correlations between the sizes of the pores. Borgman et al. (2017) also included such effects in their experimental and modeling studies of 2D porous media, showing that increasing the correlation length in porous media prolongs the liquid connectivity to the surface and, hence, increases the drying rates, although the type of correlations that they simulated is different from what we consider in our work.

The rest of this paper is as follows. The correlated PN model is described in the next section. In Section 5.3, the PN simulation of evaporation and drying is described. The results are presented and discussed in Section 5.4, while the paper is summarized in the last section.

5.2 The Correlated and Anisotropic Pore Network Model

As pointed out earlier, analyzing a vast amount of data, Neuman (1994) showed that the permeability distribution of large-scale porous formations follows the statistics of the fractional Browning motion (FBM), a nonstationary stochastic process that is characterized by its power spectrum $S(\omega)$ that for a d-dimensional medium is given by (Mandelbrot and van Ness 1968),

$$S(\omega) = \frac{a(d)}{\left(\omega_x^2 + \omega_y^2 + \omega_z^2\right)^{H+d/2}}, \tag{5.1}$$

where $a(d)$ is a d-dependent constant, and H is the Hurst exponent. Following Sahimi and Mukhopadhyay (1996) and Ansari-Rad et al. (2012), we generalize the power spectrum by rewriting it as

$$S(\omega) = \frac{b(d)}{\left(\beta_x \omega_x^2 + \beta_y \omega_y^2 + \beta_z \omega_z^2\right)^{H+d/2}}, \tag{5.2}$$

where $b(d)$ is another d-dependent constant, and the βs are the anisotropy parameters. Equation (5.2) reduces to the standard power spectrum for an isotropic FBM, Equation (5.1), in the limit $\beta_x = \beta_y = \beta_z == 1$. The Hurst exponent H characterizes the nature of the correlations. For $H > 0.5$ ($H < 0.5$) one has positive (negative) large-scale correlations, while $H = 0.5$ represents a random walk process with no correlation between the successive increments of the FBM. To generate anisotropy induced by stratification, we set $\beta_x/\beta_z \neq 1$. Figure 5.1 presents examples of the 3D heterogeneous PNs generated by the model.

Knackstedt et al. (1998) demonstrated that the extended correlations at core scale are also well approximated by a FBM. Thus, in this study we utilize the FBM to generate correlated PSDs. We ignore the effect of liquid films in the throats and pore bodies, so that the former may be assumed to be cylindrical. The PN that we utilize is, therefore, a cubic network in which each bond represents a pore throat with an effective radius r, selected from the FBM, while assuming that the throats' length is constant, although it poses no particular difficulty to select the throats' length from a statistical distribution.

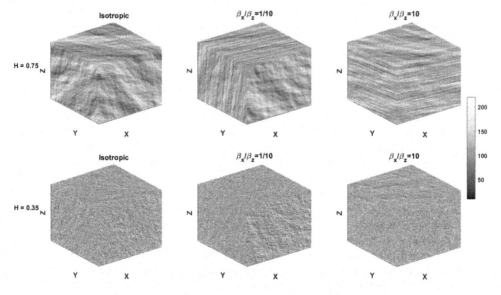

FIGURE 5.1 Examples of pore networks with correlated heterogeneity and stratification.

5.3 Pore-Network Simulation

Yiotis et al. (2001) showed that both advection and capillarity affect the drying patterns and rates. Thus, we assume slow (laminar) flow of the liquid in the PN. Mass transfer in the vapor phase is by diffusion, for which the driving force is the concentration gradient alongside the drying front. The PN's top surface is assumed to be a layer whose throats are initially empty and through which air (or another gas) flows into the pore space, while the other sides are sealed. The liquid flows in the pore space, while the vapor diffuses to the boundary layer and induces further evaporation.

Advancement of the drying front in the pore space is simulated by the IP processes in which a gas (the nonwetting phase as the agent for drying) invades the liquid-saturated porous medium throat-by-throat according to the lowest capillary pressure needed for entering a throat at the interface between the vapor and liquid. The capillary pressure for entering a throat of radius r_{ij} between pore bodies i and j at the liquid–vapor interface is given by

$$P_c = \frac{2\sigma}{r_{ij}}, \tag{5.3}$$

where σ is the surface tension. At every step of the simulation, we identify the state of the pores, which is determined by the filling state of the pore throats connected to them: if at least one neighboring throat does not contain the liquid, it is a vapor-filled pore; otherwise, it is considered a liquid-filled pore. A throat can be filled with the liquid, fully or partially, or contain its vapor that is leaving it through a path of pores and throats that are also filled by the vapor. A vapor-filled pore can be either at equilibrium or at an unknown vapor pressure, to be determined. In reality, equilibrium vapor pressure prevails only at the menisci between liquid and the vapor, but we assume that a pore is at equilibrium vapor pressure if at least one of its neighboring throats still contains the liquid, even if it is only par-tially so. Therefore, diffusion of the vapor between menisci of partially filled throats and the adjoin-ing vapor-filled pore happens without any resistance. Our preliminary numerical simulation indicated that the assumption leads to a slight—about 5%—overestimate of the drying rate but simplifies the computations greatly.

Once the states of pores and throats are identified, the second stage of the simulation, namely, model-ing of flow of the liquid throughout the PN, begins. After one or a few steps of the IP that advance the drying front, we solve the governing flow equations. The liquid flow rate $Q_{ij}^{(l)}$ in a throat ij that connects pores i and j is given by the Hagen–Poiseuille equation

$$Q_{ij}^{(l)} = \frac{\pi r_{ij}^4}{8\nu^{(l)} l_{ij}} \left[P_i^{(l)} - P_j^{(l)} \right], \tag{5.4}$$

where the liquid viscosity $\nu^{(l)}$ is assumed to be constant, as we simulate an isothermal process. Here, $P_i^{(l)}$ is the liquid pressure in pore i, and l_{ij} is the length of the throat ij. Since there is no liquid accumulation in the pores, we must have

$$\sum_{\{ij\}} \rho^{(l)} Q_{ij}^{(l)} = 0, \tag{5.5}$$

with $\rho^{(l)}$ being the liquid's mass density. The sum is over the set of the throats $\{ij\}$ that are connected to pore i. Substituting for $Q_{ij}^{(l)}$ and writing the mass balance for every interior node of the PN in the liquid phase results in a set of linear equations, $\mathbf{GP} = \mathbf{b}$, where G is the conductance matrix that depends only on the morphology of the network, $\mathbf{P}$ is the nodal pressure vector to be calculated, and $\mathbf{b}$ is a vector related to the external boundary conditions. We solve the set of the equations by the conjugate-gradient method.

For the vapor, we use the equation for evaporation in the so-called Stefan tube (Bird et al. 2007) in order to compute its mass flow rate in the throats. Ignoring the pressure in the vapor-filled throats, vapor mass conservation at the pores must hold, implying that,

$$\sum_{\{ij\}} \rho^{(v)} Q_{ij}^{(v)} = \sum_{\{ij\}} S_{ij} \frac{D^{(v)}}{l_{ij}} \frac{PM^{(v)}}{RT} \ln\left[\frac{P - P_i^{(v)}}{P - P_j^{(v)}}\right] = 0, \qquad (5.6)$$

where $D^{(v)}$, $M^{(v)}$, and $\rho^{(v)}$ are, respectively, the diffusion coefficient, molecular weight, and mass density of the vapor, P is the total pressure in the vapor phase, and S_{ij} is the cross-sectional area of throat ij. We used $\rho^{(v)} \approx 8.12 \times 10^{-4}$ gr/cm³ and $D^{(v)} \approx 2.6 \times 10^{-5}$ m²s⁻¹, both corresponding to 298 K. The vapor concentration at the vapor–liquid interface is equal to the equilibrium concentration, which is zero outside the boundary layer of the PN. Writing down Equation (5.6) for all the vapor-filled pores and throats results in another set of equations that govern the pressures in such pores, which is again solved by the conjugate-gradient method. The pressure at the liquid–vapor interface is assumed to be equal to the equilibrium vapor pressure that we took it to be, $P^* = 2340$ Pa. At a meniscus between the liquid and the vapor, the Kelvin equation describes the dependence of the parameters on the vapor pressure:

$$\ln\left[\frac{P^{(v)}}{P^*}\right] = -\frac{2M^{(v)}\sigma}{RT r_{ij} \rho^{(l)}} = -\frac{P_c M^{(v)}}{RT \rho^{(l)}}, \qquad (5.7)$$

where R is the gas constant. Once the set of the equations is solved, the vapor flux at any particular stage of the drying is computed. We used $100 \times 100 \times 100$ PNs and averaged the results over 10 independent realizations.

5.4 Results and Discussion

Figure 5.1 presents examples of the 3D models generated by the FBM, indicating clearly that the correlations and stratification are relevant. The degree of stratification and the resulting anisotropy both depend on the Hurst exponent H and the anisotropy parameters β_i. The corresponding PSDs are shown in Figure 5.2. In Table 5.1, we show the properties of the PNs that we used to simulate drying.

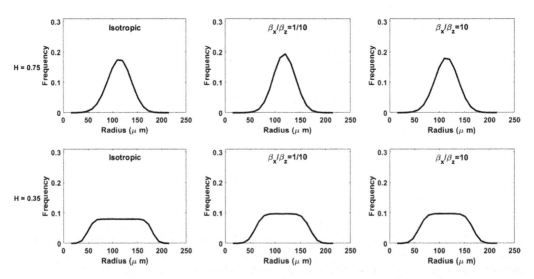

FIGURE 5.2 Size distributions of the pore throats corresponding to Figure 5.1.

TABLE 5.1
Pore Size Distribution Properties of Different Cases

H	β_x/β_z	β_y/β_z	Mean Throat Radius (µm)
0.35	1	1	120.0
0.35	1/10	1/10	119.3
0.35	10	10	116.8
0.75	1	1	112.2
0.75	1/10	1/10	123.4
0.75	10	10	117.8

5.4.1 Isotropic Porous Media

We first studied drying in isotropic porous media and varied the Hurst exponent H. It should be pointed out that the shapes and statistics of the clusters of vapor-filled pores in isotropic PNs are strongly dependent on the Hurst exponent H. To show this clearly, we present examples of the clusters of vapor-filled pores in 2D PNs (although our main simulations were carried out with 3D PNs) for $H = 0.2, 0.5$, and 0.9, and compare them with the case in which there are correlations between the sizes of the pores. As Figure 5.3 indicates, increasing the Hurst exponent increases the compactness of the vapor-filled clusters, implying that the liquid evaporates in most of the throats and pores. Thus, drying in such pore space is efficient in the sense that not much liquid is left behind. Indeed, it has been shown (Knackstedt et al. 2000) that, in both 2D and 3D, such IP clusters are compact for $0.5 < H \leq 1$, but they have a fractal structure for $0 < H < 0.5$ with a fractal dimension that depends continuously on H. A fractal structure implies that drying is not done efficiently, as it leaves many liquid-filled pores behind that are distributed in clusters of various sizes.

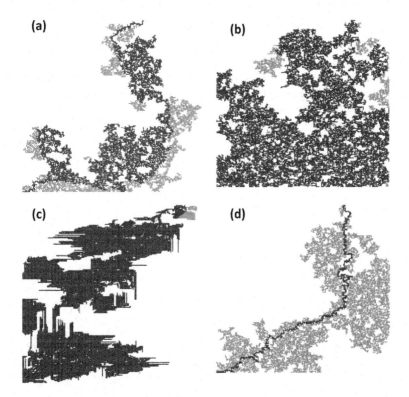

FIGURE 5.3 Structure of the vapor-filled clusters in 2D pore networks for Hurst exponent H of (a) 0.2, (b) 0.5, and (c) 0.9. The structure in (d) is random without any correlations.

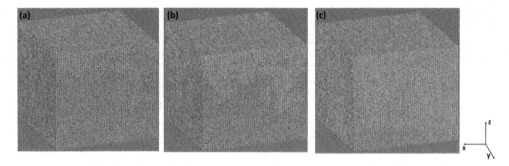

FIGURE 5.4 The phase distributions in an isotropic pore network with the Hurst exponent $H = 0.75$, when vapor saturation is (a) 0.2, (b) 0.4, and (c) 0.9. The top surface of the network is open.

The qualitative insight provided by Figure 5.3 is corroborated by the simulation in 3D PNs. Figure 5.4 shows the evolution of the spatial distributions of the liquid and vapor phases with the $H = 0.75$, implying long-range positive correlations in the throat-size distribution. The drying front invades from the open top surface toward the bottom. Figure 5.4a represents the early stage of drying in which its rate is essentially constant. Close to the open surface at the top, the throats empty (dry out) at a faster rate than those deeper in the pore space. Invasion of the PN by the drying gas eventually disconnects the liquid phase. As drying proceeds, the evaporated liquid is transferred to the vapor phase, building up its concentration. This also affects the rate at which clusters of the liquid pores and throats dry out. The clusters that are formed near the open boundary of the PN are emptied faster because they are subjected to faster evaporation. Figure 5.4b and c present the phase distributions at the later stages of drying. As the figures indicate, the sizes of the liquid clusters closer to the open boundary are smaller than those that are formed at the bottom of the PN.

Figure 5.5 shows the corresponding evolution in the PN for which $H = 0.35$, manifesting strong differences with those shown in Figure 5.4. In this case, the sizes of the liquid clusters are smaller than those for the higher H. Positive correlations ($H > 0.5$) that generate compact clusters (see the discussion of Figure 5.3) augment the probability of connectivity of the throats with similar sizes and, hence, the invading gas phase has better accessibility to the throats with minimum capillary pressure—the largest throats—through the PN. As described by Borgman et al. (2017) and Dashtian et al. (2018), the larger throats will dry out first, whereas the smaller pores remain saturated with the liquid. This further indicates the role of the underlying morphology of a porous medium on evaporation of the liquid in its pore space.

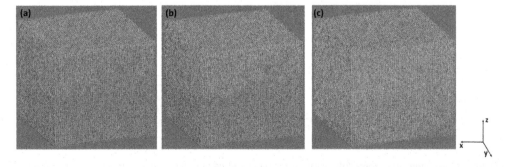

FIGURE 5.5 (a–c) Same as in Figure 5.4, but with the Hurst exponent $H = 0.35$.

5.4.2 Stratified Porous Media

We varied β_x/β_z in Equation (5.2) (see Figure 5.1) in order to generate stratification and, hence, anisotropy in the PNs by setting $\beta_x/\beta_y = 1$. With $\beta_x/\beta_z > 1$ the strata are parallel to the xy planes, whereas $\beta_x/\beta_z < 1$ generates layers that are more or less parallel to the xz planes; see Figure 5.1. Once again, the effect of the correlations on the structure of the vapor-filled pores is quite strong. Figure 5.6 presents two examples of such clusters in 2D. In this case, the strata are more or less parallel to the direction of the movement of the interface. Once again, highly connected and compact clusters of vapor-filled pores are formed that leave behind a significant amount of liquid.

The structure of 2D clusters of vapor-filled pores shown in Figure 5.6 is reproduced by our simulations in 3D PNs. In Figures 5.7 and 5.8 we show the resulting fluid phase distributions in 3D with $\beta_x/\beta_z = 1/10$ for, respectively, $H = 0.75$ and $H = 0.35$. There are evident differences with the isotropic PNs (Figures 5.3 through 5.5). In this case, the direction of the moving interface is more or less parallel to the strata. As a result, the interface moves downward in a sweep-like manner, the clear 2D examples of which are shown in Figure 5.6.

Figures 5.9 and 5.10 show the evolution of the fluid phase distribution for $\beta_x/\beta_z = 10$ for, respectively, $H = 0.75$ and 0.35. The macroscopic direction of the interface motion is perpendicular to the strata. Thus, as the two figures indicate, the drying front prefers to first invade the throats that are parallel to the open surface in the planes of the strata, and then expand downward toward the bottom of the porous medium. In addition, Figures 5.7 through 5.10 indicate that the presence of the strata modifies the liquid phase distribution through preferential drying of the throats in a certain direction.

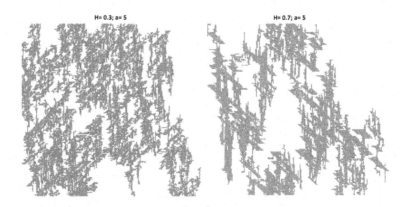

FIGURE 5.6 The structure of the vapor-filled clusters in 2D stratified pore networks in which the strata are more or less parallel to the macroscopic direction of motion of vapor-liquid interface, with $a = \beta_z/\beta_x$.

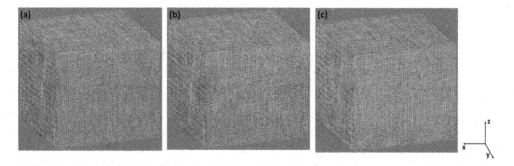

FIGURE 5.7 The phase distributions in an anisotropic pore network with the Hurst exponent $H = 0.75$ and $\beta_x/\beta_z = 1/10$, when vapor saturation is (a) 0.2; (b) 0.4, and (c) 0.9.

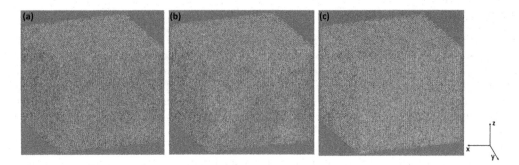

FIGURE 5.8 (a–c) Same as in Figure 5.7, but with the Hurst exponent $H = 0.35$.

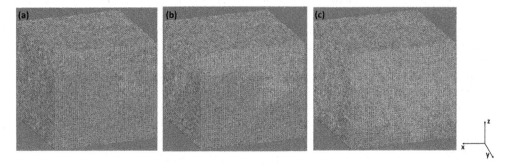

FIGURE 5.9 The phase distributions in an anisotropic pore network with the Hurst exponent $H = 0.75$ and $\beta_x/\beta_z = 10$, when vapor saturation is (a) 0.2; (b) 0.4, and (c) 0.9.

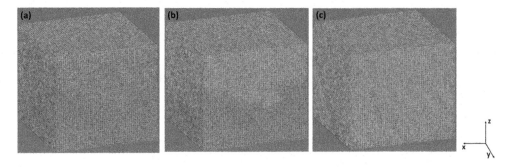

FIGURE 5.10 (a–c) Same as in Figure 5.9, but with the Hurst exponent $H = 0.35$.

5.4.3 The Saturation Distribution

We further quantified the effect of the correlations and stratifications by computing the saturation S of the liquid phase at various depths throughout the PN. Four distinct times were selected for plotting the saturations, averaged over the xy planes. Figure 5.11 presents the transversely averaged saturations in the same PNs as before. The blue areas indicate the range of variations over multiple realizations. A comparison between Figure 5.11a–c and 5.11d–f indicates the stronger effect of positive correlations ($H = 0.75$) on the saturation distributions. On the other hand, PNs with negative long-range correlations ($H = 0.35$) exhibit a narrower range of variations in the saturations over multiple realizations. This is surprising since negative correlations lead to more heterogeneous PNs, implying that the variations of the saturation in such networks should be larger than those in PNs with positive correlations. Note also that when the strata are perpendicular to the open end surface of the porous medium (i.e., when $\beta_x/\beta_z = 1/10$), there is

Pore-Network Simulation of Drying of Heterogeneous and Stratified Porous Media

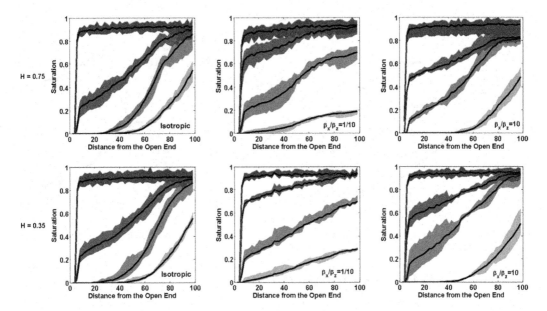

FIGURE 5.11 Evolution of the transversely averaged liquid saturations in the six pore networks of Figure 5.1. Black curves show the average saturations, while the shaded areas indicate the variations over multiple realizations.

no region in which the saturation is zero, as the liquid phase stays connected to the surface. The reason is that if the interface enters a stratum with relatively large pore throats, it quickly advances through them, hence leaving behind a significant portion of the pore space with liquid-filled throats that span the PN from top to the bottom. Such a phenomenon has been observed in the experiments of Shokri et al. (2010) and Shokri-Kuehni et al. (2018).

We also note that there exist three distinct regions in the saturation distributions. The first region represents the dried portion of the PN with zero saturation. The second region is associated with a transition zone in which the saturation is $0 < S < 1$, while the third region presents the PN regions with $S \approx 1$, that is, those in which the drying front has not arrived yet. Generally speaking, at each time step, the slope of the saturation curve, $\partial S/\partial z$ in the pore space with stratification, $\beta_x/\beta_z \neq 1$, is smaller than those in isotropic porous media, $\beta_x/\beta_z = 1$.

5.4.4 The Drying Rates

The invasion of the PN by the drying front increases the resistance imposed by the emptied larger throats in the dried regions. The resistance slows down the evaporation and, therefore, vapor diffusion dominates the drying. As discussed in the Introduction, the evaporation process is typically divided into two stages, with a transition region in between. The two stages, together with the transition zone, are all manifested by Figure 5.12, where we present the dependence of the drying rate on the saturation. Note that the transition zone with positive correlations ($H = 0.75$) and the strata perpendicular to the open end surface ($\beta_x/\beta_z = 1/10$) has the highest evaporation rate.

To better understand the results, consider flow of the liquid through the PN between two points separated by several throats. If the throats' sizes between the two points are positively correlated, there will be smaller resistant against the flow. In the PN in which the strata are parallel to the xy direction—Figure 5.11c and f—the falling rate is slower and, thus, this stage lasts until almost the end of the process.

The trends of the evaporation rate and saturation profiles produced by the PN simulation are similar to those reported by experimental studies (Shokri et al. 2008a, 2008b, 2010). Possible deviation from experimental data are presumably due to not including such effects as condensation, adsorption (Bakhshian and Sahimi, 2017; Bakhshian et al. 2018), thin liquid films, or nonisothermal conditions. These issues will be taken up in a future paper.

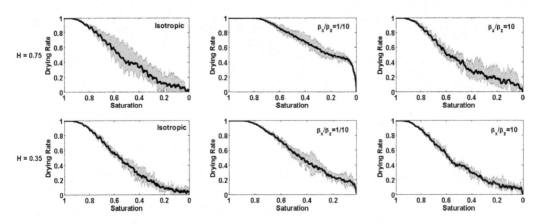

FIGURE 5.12 Evolution of the average drying rates (black curves), averaged over 10 realizations. The gray areas indicate variations over multiple realizations.

5.4.5 Effect of the Type of the Correlations

As described earlier, the threshold capillary pressure for invading a pore throat is a key factor in the shape of the drying patterns. Since the threshold capillary pressure is smaller for larger pores, the invading gas tends to enter larger pores first. On the other hand, positive correlations ($H > 0.5$) implies that large pores are neighbors to similarly large pores, whereas negative correlations ($H < 0.5$) generates PNs in which large pores are next to small ones. Thus, we expect the type of correlations to strongly influence the drying patterns.

Therefore, in addition to the results described and discussed so far, we define a parameter χ_L that represents the fraction of the larger-than-average invaded pore throats, and computed it at three water saturations (or time steps) for a range of the Hurst exponent H. Figure 5.13 depicts the results. At low saturations, χ_L remains essentially constant for all the H values with $\chi_L \approx 0.58$. This is because at such low saturation the liquid phase is disconnected and, therefore, long-range correlations between the pore sizes are not relevant, or represent at most a secondary effect. The same is not, however, true at higher saturations for which χ_L also seems to be more or less constant for $H < 0.5$, that is, the regime in which the throats' sizes are negatively correlated. But, χ_L seemingly undergoes a transition for $H > 0.5$ when the pores are positively correlated and, therefore, larger pores with effective radius larger than the average pore size are clustered together.

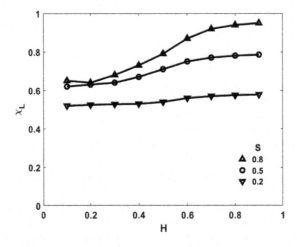

FIGURE 5.13 The fraction χ_L of the pore throats with a size larger than the average size of the invaded pores for three saturations S and a range of the Hurst exponent H.

5.5 Summary

This paper reported on pore-network simulation of evaporation and drying of a porous medium, initially saturated by a liquid. Using large 3D pore networks, we reported on the result of a study of the effect of extended correlations between the sizes of the pore throats, as well as the stratification of the pore space, on evaporation and drying, demonstrating their strong influence on the drying pattern, the liquid saturation profile, and the drying rate, effects that to our knowledge had not been studied before. In particular, evaporation is maximum if the strata are more or less parallel to the direction of macroscopic motion of the interface between the vapor and liquid phases, as demonstrated convincingly by Figure 5.6. This is because the liquid remains connected to the external surface of the pore space and, therefore, evaporation happens in the throats near the surface, hence leading to higher drying rates and longer periods of constant drying rate.

ACKNOWLEDGMENTS

This work was supported in part by the Petroleum Research Fund, administered by the American Chemical Society.

REFERENCES

Alim, K., Parsa, S., Weitz, D.A., Brenner, M.P., 2017. Local pore size correlations determine flow distributions in porous media. *Phys. Rev. Lett.* 119, 144501.

Ansari-Rad, M., Vaez Allaei, S.M., Sahimi, M., 2012. Nonuniversality of roughness exponent of quasi-static fracture surfaces. *Phys. Rev.* E 85, 021121.

Attarimoghadam, A., Kharaghani, A., Tsotsas, E., Prat, M., 2017. Kinematics in a slowly drying porous medium: Reconciliation of pore network simulations and continuum modeling. *Phys. Fluids* 29, 022102.

Aydin, M., Yano, T., Evrendilek, F., Uygur, V., 2008. Implications of climate change for evaporation from bare soils in a Mediterranean environment. *Envir. Monitor. Assess.* 140, 123.

Bakhshian, S., Sahimi, M., 2017. Adsorption-induced swelling of porous media. *Int. J. Greenh. Gas Con.* 57, 1.

Bakhshian, S., Shi, Z., Sahimi, M., Tsotsis, T.T., Jessen, K., 2018. Image-based modeling of gas adsorption and deformation in porous media. *Sci. Rep.* 8, 8249.

Ben-Noah, I., Friedman, S.P., 2018. Review and evaluation of root respiration and of natural and agricultural processes of soil aeration. *Vadose Zone J.* 17, 170119.

Bird, R.B., Stewart, W.E., Lightfoot, E.N., 2007. *Transport Phenomena*, 2nd ed. Wiley, New York.

Blunt, M.J., 2017. *Multiphase Flow in Permeable Media: A Pore-Scale Perspective.* Cambridge University Press, Cambridge, MA.

Borgman, O., Fantinel, P., Lühder, W., Goehring, L., Holtzman, R., 2017. Impact of spatially correlated pore-scale heterogeneity on drying porous media. *Water Resour. Res.* 53, 5645.

Bray, Y. L., Prat, M., 1999. Three-dimensional pore network simulation of drying in capillary porous media. *Int. J. Heat Mass Transfer* 42, 4207.

Chauvet, F., Duru, P., Geoffroy, S., Prat, M., 2009. Three periods of drying of a single square capillary tube. *Phys. Rev. Lett.* 103, 124502.

Dashtian, H., Shokri, N., Sahimi, M., 2018. Pore-network model of evaporation-induced salt precipitation in porous media: The effect of correlations and heterogeneity. *Adv. Water Resour.* 112, 59.

Dashtian, H., Yang, Y., Sahimi, M., 2015. Nonuniversality of the Archie exponent due to multifractality of resistivity well logs. *Geophys. Res. Lett.* 42, 655.

Datta, S.S., Weitz, D.A., 2013. Drainage in a model stratified porous medium. *Europhys. Lett.* 101, 14002.

Fantinel, P., Borgman, O., Holtzman, R., Goehring, L., 2017. Drying in a microfluidic chip: Experiments and simulations. *Sci. Rep.* 7, 15572.

Hajirezaie, S., Wu, X.; Peters, C.A., 2017. Scale formation in porous media and its impact on reservoir performance during water flooding. *J. Natural Gas Sci. Eng.* 39, 188.

Helmig, R., Schulz, P., 1997. *Multiphase Flow and Transport Processes in the Subsurface.* Springer, Berlin, Germany.

Hewett, T.A., 1986. Fractal distributions of reservoir heterogeneity and their influence on fluid transport. SPE Annual Technical Conference and Exhibition 15386, New Orleans, LA.

Hewett, T.A., Behrens, R.A., 1990. Conditions affecting the scaling of displacements in heterogeneous permeability distribution. *SPE Form. Eval.* 5, 217.

Hosseini, S.A., Alfi, M., 2016. Time-lapse application of pressure transient analysis for monitoring compressible fluid leakage. *Greenhouse Gases: Sci. Technol.* 6, 352.

Jambhekar, V. A., Helmig, R., Schröder, N., Shokri, N., 2015. Free-flow-porous-media coupling for evaporation-driven transport and precipitation of salt in soil. *Transp. Porous Media* 110, 251.

Kharaghani, A., Metzger, T., Tsotsas, E., 2012. An irregular pore network model for convective drying and resulting damage of particle aggregates. *Chem. Eng. Sci.* 75, 267.

Knackstedt, M.A., Marrink, S.J., Sheppard, A.P., Pinczewski, W.V., Sahimi, M., 2001b. Invasion percolation on correlated and elongated lattices: Implications for the interpretation of residual saturations in rock cores. *Transp. Porous Media* 44, 465.

Knackstedt, M.A., Sahimi, M., Sheppard, A.P., 2000. Invasion percolation with long-range correlation: First-order phase transition and nonuniversal scaling properties. *Phys. Rev. E* 61, 4920.

Knackstedt, M.A., Sheppard, A. P., Sahimi, M., 2001a. Pore network modelling of two-phase flow in porous rock: The effect of correlated heterogeneity. *Adv. Water Resour.* 24, 257.

Knackstedt, M.A., Sheppard, A.P., Pinczewski, W.V., 1998. Simulation of mercury porosimetry on correlated grids: Evidence for extended correlated heterogeneity at the pore scale in rocks. *Phys. Rev. E* 58, R6923.

Laurindo, J.B., Prat, M., 1998. Numerical and experimental network study of evaporation in capillary porous media. Drying rates. *Chem. Eng. Sci.*, 53, 2257.

Le, D., Hoang, H., Mahadevan, J., 2009. Impact of capillary-driven liquid films on salt crystallization. *Transp. Porous Media* 80, 229.

Le, K.H., Kharaghani, A., Kirsch, C., Tsotsas, E., 2017. Discrete pore network modeling of superheated steam drying. *Drying Technol.* 35, 1584.

Lehmann, P., Or, D., 2009. Evaporation and capillary coupling across vertical textural contrasts in porous media. *Phys. Rev. E* 80, 046318.

Liu, M., Zhang, S., Mou, J., 2012. Effect of normally distributed porosities on dissolution pattern in carbonate acidizing. *J. Pet. Sci. Eng.* 94–95, 28.

Liu, P., Yao, J., Couples, G.D., Ma, J., Iliev, O., 2017. 3-D modelling and experimental comparison of reactive flow in carbonates under radial flow conditions. *Sci. Rep.* 7, 17711.

Lu, J., Darvari, R., Nicot, J.P., Mickler, P., Hosseini, S.A., 2017. Geochemical impact of injection of Eagle Ford brine on Hosston sandstone formation: Observations of autoclave water rock interaction experiments. *Appl. Geochem.* 84, 26.

Mandelbrot, B.B., van Ness, J.W., 1968. Fractional Brownian motions, fractional noises and applications. *SIAM Rev.* 10, 422.

Miri, R., van Noort, R., Aagaard, P., Hellevang, H., 2015. New insights on the physics of salt precipitation during injection of CO_2 into saline aquifers. *Int. J. Greenhouse Gas Control.* 43, 10.

Molz, F.J., Liu, H.H., Szulga, J., 1997. Fractional Brownian motion and fractional Gaussian noise in subsurface hydrology: A review, presentation of fundamental properties, and extensions. *Water Resour. Res.* 33, 2273.

Muljadi, B.P., Blunt, M.J., Raeini, A.Q., Bijeljic, B., 2016. The impact of porous media heterogeneity on non-Darcy flow behaviour from pore-scale simulation. *Adv. Water Resour.* 95, 329.

Nadim, F., Hoag, G.E., Liu, S., Carley, R.J, Zack, P., 2000. Detection and remediation of soil and aquifer systems contaminated with petroleum products: An overview. *J. Pet. Sci. Eng.* 26, 169.

Neuman, S.P., 1994. Generalized scaling of permeabilities: Validation and effect of support scale. *Geophys. Res. Lett.* 21, 1944.

Or, D., 2008. Scaling of capillary, gravity and viscous forces affecting flow morphology in unsaturated porous media. *Adv. Water Resour.* 31, 1129.

Or, D., Lehmann, P., Shahraeeni, E., Shokri, N., 2013. Advances in soil evaporation physics A review. *Vadose Zone J.* 12, 4.

Pegler, S.S., Huppert, H.E., Neufeld, J.A., 2016. Stratified gravity currents in porous media. *J. Fluid Mech.* 791, 329.

Prat, M., 1995. Isothermal drying on non-hygroscopic capillary-porous materials as an invasion percolation process. *Int. J. Multiphase Flow* 21, 875.

Prat, M., 2002. Recent advances in pore-scale models for drying of porous media. *Chem. Eng. J.* 86, 153.

Prat, M., 2011. Pore network models of drying, contact angle, and film flows. *Chem. Eng. Technol.* 34, 1029.

Rahimi, A., Metzger, T., Kharaghani, A., Tsotsas, E., 2016. Interaction of droplets with porous structures: Pore network simulation of wetting and drying. *Drying Technol.* 34, 1129.

Sahimi, M., 1994. Long-range correlated percolation and flow and transport in heterogeneous porous media. *J. de Physique I 4*, 1263.

Sahimi, M., 1995. Effect of long-range correlations on transport phenomena in disordered media. *AICHE J.* 41, 229.

Sahimi, M., 2011. *Flow and Transport in Porous Media and Fractured Rock*, 2nd ed. Wiley-VCH, Weinheim.

Sahimi, M., Arbabi, S., 1992. Percolation and fracture in disordered solids and granular media: Approach to a fixed point. *Phys. Rev. Lett.* 68, 608.

Sahimi, M., Goddard, J.D., 1986. Elastic percolation models for cohesive mechanical failure in heterogeneous systems. *Phys. Rev. B* 33, 7848.

Sahimi, M., Mukhopadhyay, S., 1996. Scaling properties of a percolation model with long- range correlations. *Phys. Rev. E* 54, 3870.

Scherer, G.W., 1990. Theory of drying. *J. Am. Ceramic Soc.* 73, 3.

Shaw, T.M., 1987. Drying as an immiscible displacement process with fluid counter flow. *Phys. Rev. Lett.* 59, 1671.

Shokri, N., Lehmann, P., Or, D., 2008a. Effects of hydrophobic layers on evaporation from porous media. *Geophys. Res. Lett.* 35, L19407.

Shokri, N., Lehmann, P., Or, D., 2009. Characteristics of evaporation from partially-wettable porous media. *Water Resour. Res.*, 45, W02415.

Shokri, N., Lehmann, P., Or, D., 2010. Evaporation from layered porous media. *J. Geophys. Res: Solid Earth.* 115, B06204.

Shokri, N., Lehmann, P., Vontobel, P., Or, D., 2008b. Drying front and water content dynamics during evaporation from sand delineated by neutron radiography. *Water Resour. Res.* 44, 1944.

Shokri-Kuehni, S.M.S., Norouzirad, M., Webb, C., Shokri N., 2017. Impact of type of salt and ambient conditions on saline water evaporation from porous media. *Adv. Water Resour.*, 105, 154–161.

Shokri-Kuehni, S.M.S., Bergstad, M., Sahimi, M., Webb, C., Shokri N., 2018. Iodine k-edge dual energy imaging reveals the influence of particle size distribution on solute transport in drying porous media. *Sci. Rep.*, 10, 10731, London: Nature Publishing Group..

Soltanian, M.R., Amooie, M.A, Cole, D.R., Graham, D.E., Hosseini, S.A., Hovorka, S., Pfiffner, S.M., Phelps, T.J., Moortgat, J., 2016. Simulating the Cranfield geological carbon sequestration project with high-resolution static models and an accurate equation of state. *Int. J. Greenhouse Gas Control.* 54, 282.

Surasani, V.K., Metzger, T., Tsotsas, E., 2008. Consideration of heat transfer in pore network modelling of convective drying. *Int. J. Heat Mass Transfer* 51, 2506.

Vorhauer, N., Tsotsas, E., Prat, M., 2017. Drying of thin porous disks from pore network simulations. *Drying Technol.* 36, 651.

Vorhauer, N., Wang, Y.J., Kharaghani, A., Tsotsas, E., Prat, M., 2015. Drying with formation of capillary rings in a model porous medium. *Transp. Porous Media* 110, 197.

Yiotis, A. G., Salin, D., Yortsos, Y.C., 2015. Pore network modeling of drying processes in macroporous materials: Effects of gravity, mass boundary layer and pore microstructure. *Transp. Porous Media* 110, 175.

Yiotis, A.G., Stubos, A.K., Boudouvis, A.G., Yortsos, Y.C., 2001. A 2-D pore-network model of the drying of single-component liquids in porous media. *Adv. Water Resour.* 24, 439.

Zhang, Y., Kogure, T., Chiyonobu, S., Lei, X., Xue, Z., 2013. Influence of heterogeneity on relative permeability for CO_2/brine: CT observations and numerical modeling. *Energy Procedia* 37, 4647.

6

Wicking of Liquids under Non-Isothermal and Reactive Conditions: Some Industrial Applications

Mohammad Amin Faghihi Zarandi, and Krishna M. Pillai

CONTENTS

6.1 Introduction.. 103
6.2 History of Heat Transfer in Wicking... 104
    6.2.1 Propellant Management Devices (PMD).. 104
    6.2.2 Sheet Drying 105
    6.2.3 Porous Burners... 106
    6.2.4 Heat Pipes 108
6.3 Effect of Heat Transfer on the Wicking-Process Governing Equations................................. 109
    6.3.1 Governing Equations in Propellant Management Devices 109
    6.3.2 Governing Equations in Sheet Drying 111
    6.3.3 Governing Equations in Porous Burners 113
        6.3.3.1 Models Independent of Porous-Medium Processes.................................... 113
        6.3.3.2 Models Including Porous-Medium Processes 115
        6.3.3.3 Consequences of Including Porous-Medium Processes 116
    6.3.4 Governing Equations in Heat Pipes 116
6.4 Conclusions.. 119
Nomenclature... 119
References... 121

6.1 Introduction

The industrial applications of porous media involving heat transfer have become very widely spread. There are numerous applications that deal with heat transfer studies in porous media including heat pipes (Faghri 1995; Milani Shirvan et al. 2017; Yeo and No 2018), porous media burners (Mujeebu et al. 2009a; Ellzey and Goel 1995; Olayiwola 2015; Zhang et al. 2017), and porous evaporators (Canbazoglu et al. 2016; Donnarumma et al. 2015; Mosthaf et al. 2014). These are going to be reviewed in this chapter.

Wicking is defined as the process of imbibition of a wetting liquid into a dry porous medium where the liquid displaces air in the pores under the action of capillary suction pressure as the driving force under isothermal conditions (Masoodi and Pillai 2012a; Zarandi et al. 2018). The topic becomes interesting and challenging as the heat transfer conjoins with the wicking process in porous media. There have been different modeling methods for the two processes, but their combination raises the difficulty due to the complexity of modeling both processes simultaneously after considering the mutual coupling effects. This complexity is caused by (a) intricacy of the pore structure and presence of two or more flowing phases necessitate upscaling for modeling processes at the lab scale, (b) occurrence of multiple phase changes, (c) presence of different types of heat transfer phenomena, and (d) accompaniment of the mass transfer aspects (e.g., due to the use of liquid mixtures), which leads to difficulties regarding investigating this phenomenon theoretically, numerically, and experimentally. Hence, for different scenarios, by simplifying and confining the problem aspects, researchers have conducted many comprehensive and reliable studies. In this chapter, we are first going to have a thorough review of the main industrial applications involving the combined effects of heat transfer and wicking. Then, in Section 6.4, we will describe the equations employed to model these coupled processes in the four considered applications.

6.2 History of Heat Transfer in Wicking

Investigating the coupling of heat transfer and wicking has had a long history in literature. In this section, we present previous instances of some the most-applied applications of the wicking process conjugated with heat transfer. In all the considered applications, heat transfer plays an important role in wicking compared to the case of the "pure" wicking process under isothermal conditions. In all these applications, the fluid mechanics is not the only science governing the phenomena—addition of heat transfer accompanied by phase change and reactive mass-transfer phenomena renders them complicated and of "multiphysics" in nature.

6.2.1 Propellant Management Devices (PMD)

One of the most complicated and advanced set of studies has been done in propellant management devices (PMDs) or liquid acquisition devices (LADs) (Tam and Jaekle 2016; Tam et al. 2002, 2016). PMDs are static metallic structures inside spacecraft propellant tanks that use only the surface tension to ensure gas-free liquid delivery to the tank outlet (Hartwig 2016). (Figure 6.1 shows the reason for using a PMD.) With no moving parts, they do not break down and hence are inherently reliable. PMDs are used in satellites, solar system probes, rockets, and the Space Shuttle. They are used mostly for propellants but can also be used in water-supply systems, thermal-management systems, cryogenic systems and, in fact, in any space-bound liquid system.

As shown in Figure 6.2, PMDs are designed to ensure a constant and consistent propellant flow from the main fuel tank to a specific location (Fries et al. 2007, 2008; PMD Technology Capabilities). PMD contains galleries with porous screen windows, which are made of a metal weave. The essential application of weaves is to prevent the gas-phase from entering the liquid phase when the pressure falls below the critical bubble-point pressure (Fries et al. 2008). Hence, the weave must be completely saturated with propellant as the evaporation process occurs. On the other hand, in microgravity, the surface tension force becomes significant and dominant, which determines the location and orientation of fuel as it imbibes into the wicks.

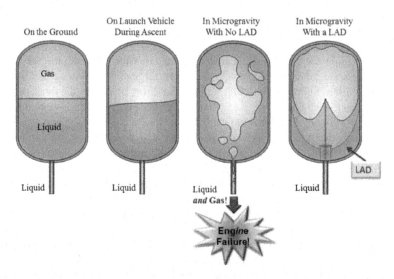

FIGURE 6.1 The schematic showing the necessity of PMD (or LAD) usage. (From Tam, W.H. et al., *AIAA Paper*, 4137, 2002, 2002.)

Wicking of Liquids under Non-Isothermal and Reactive Conditions: Some Industrial Applications 105

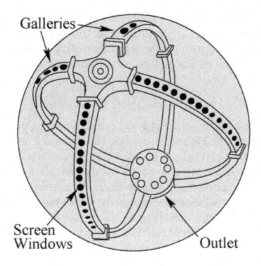

FIGURE 6.2 An example of a propellant management device (PMD). (From Fries, N. et al., Wicking of perfectly wetting liquids into a metallic mesh, in *Proceedings of the 2nd International Conference on Porous Media and its Applications in Science and Engineering*, 2007; Dodge, F.T., *The New Dynamic Behavior of Liquids in Moving Containers*, Southwest Research Inst, San Antonio, TX, 2000.)

This topic as a conjugate heat-transfer and wicking problem has been investigated under different subjects. During the wicking and evaporation process, due to the surface tension, formation of interfaces between gas and liquid face occurs; the pressure difference across such interfaces can be estimated by the Laplace equation, which is a function of the surface tension as well as the local radii of curvature.

There have been some studies on the wicking process observed in weaves on using the conventional (wicking) governing equations and conducting experiments to evaluate the direct effect of evaporation (Fries et al. 2008). Some other studies developed specific analytical models for the wicking process affected by evaporation, which were validated through experiments (Jaekle 1970, 1995; Jaekle and Jaekle 1997; Symons 1974). The developed models can be used in different applications of wicking rate prediction, PMD design, etc.

6.2.2 Sheet Drying

Another example of the coupled wicking and heat-transfer processes is in paper and sheet-drying industries. This application is not confined to paper and sheets and sometimes includes water and ink/dye transport in textiles, body fluids absorption by medical clothing and sanitary sheets, and moisture migration and drying process in porous building surfaces like tiles and claddings. In these applications, the wicking process and the subsequent evaporation process are intensely important in terms of determining the rate of evaporation, evaporation uniformity, etc. In this regard, there have been some investigations that just focused on the effect of the porous structure on both the wicking and evaporation processes (Camplisson et al. 2015; Hashemi and Douglas 2001; Hashemi and Murray Douglas 2003; Hashemi et al. 2003; McDonald 2006; Saricam and Kalaoğlu 2014; Yanılmaz and Kalaoğlu 2012). In these types of studies, the researchers have been trying to assess the effect of the medium (in terms of the parameters such as porosity, fabric or paper thickness, tightness, and pore size) and the wicking-liquid properties (such as viscosity and surface tension.). In addition, some researchers have developed the transport-process equations for drying (evaporation) in fabrics and paper materials (Bandyopadhyay et al. 2002; Chatterjee et al. 1997; Horas and Toso 1992; Zhou et al. 2008). This specific type of evaporation deals with unsteady diffusion, which makes it more complicated.

Generally, there are two major approaches in studying this process: (1) as the diffusion-driven drying (Peysson et al. 2011; Widjaja and Harris 2008; Zou and Kim 2014), and (2) as the flow-through drying (Allerton et al. 1949; Huostila and Haapsaari 1980; Mahadevan et al. 2007). In diffusion-driven drying, it is assumed that a gas phase flows over the external surface of a porous medium wetted by a liquid, and this liquid gets evaporated into this gas. In the second approach, the wetting liquid evaporates due to the flow of a gas forced through the porous medium. One of the major differences between these two methods is that in diffusion-driven drying the evaporation of the wetting liquid is into an unsaturated gas, but in the flow-through drying the injected gas is fully saturated (Mahadevan et al. 2007).

6.2.3 Porous Burners

One of the most complex heat-transfer processes associated with wicking process is in the case of porous burners or other similar applications. Porous burners have attracted much attention due to their many benefit. Of the many advantages, the higher burning rates and low emission of pollutants are the most significant ones. In this type of application, the imbibed liquid gets burnt at the end of a porous medium after wicking through it. Different sets of complex processes happen during this combination of combustion, evaporation, and imbibition phenomena. Figure 6.3 shows a sample of the porous-burner flame composition and regions.

The whole process starts with the imbibition of liquid fuel as it wicks into a porous medium. This "absorption" occurs under the effect of capillary, gravity, viscous, and inertia forces. Masoodi and Pillai (2012b) have done a comprehensive study on different cases in which one or more of these forces are negligible. As the wicking process commences, different levels of complexity can come to the picture as a result of studying the interface between different phases. Multiphase flow occurs due to distinct causes, which have been studied thoroughly (Brennen 2005; Cleary et al. 2017; Gibou et al. 2018; Wang et al. 2016). Two major reasons behind phase change are the heat transfer and the immobile or trapped phase of a fluid, which could be another type of fluid (other than the wicking liquid) in a porous medium. Whatever the causes that lead to the multiphase flow, scientists have been investigating the porous burners through two different approaches. The **first** is introducing an interphase between the two-phase region (where both liquid and vapor exist in the pores) and the single-phase region (where only the liquid phase exists in the pores) and studying this phenomenon by tracking this interface and employing different physics in the two regions. The **second** approach involves considering three different regions with only-liquid phase, both liquid–vapor phase, and only-vapor phase present in the pores without considering and tracking the interface (Benard et al. 2005; Kaviany and Tao 1988; Raju and James 2007, 2008). Figure 6.4 shows a schematic on these two methods of multiphase flow modeling: in Figure 6.4a there is no distinguishable boundaries between the phases, which is in contrast with Figure 6.4b where three distinct regions with clearly distinguishable boundaries are shown.

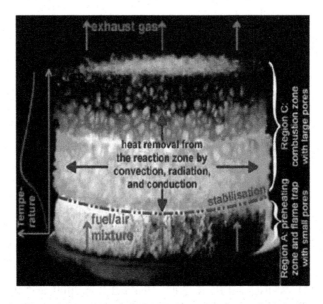

FIGURE 6.3 Schematic of a two-layer premixed combustor. (From Mossbauer, S. et al., *Clean Air*, 3, 185–198, 2002; Mujeebu, M.A. et al., *J. Environ. Manag.*, 90, 2287–2312, 2009b.)

Wicking of Liquids under Non-Isothermal and Reactive Conditions: Some Industrial Applications

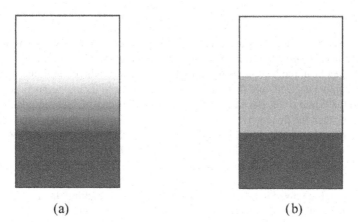

FIGURE 6.4 Schematic of the two ways to investigate multiphase flow in porous burners: (a) interface approach and (b) three-region approach.

The overall mechanism of porous burners is that the fuel gets imbibed toward the top of the porous medium from its base, and the fuel is evaporated in the process, thus providing vaporized fuel for combustion and flame generation. The heat produced through such a combustion causes drying of the porous medium, thus leading to a capillary-pressure-based suction force for consistent imbibition of fuel and its evaporation. This cycle continues till either the whole of the fuel gets consumed by the combustion, or till the point where the heat generated is not adequate to evaporate the fuel and to create strong-enough capillary pressure to suck the fuel into the porous medium against opposing forces such as gravity.

Another example of combustion in porous media involves the torches used for lighting or for insect-repellent purposes. In every torch, there is a porous wick, often made of glass fibers, which is used for sucking fuel (oil) toward the combustion zone at the wick top. Figure 6.5 show an example of such a torch and its parts.

As it was explained formerly, the combustion of evaporated fuel at the top of the wick provides the driving force for imbibition of fuel into the rest of the wick, which is inside the canister. Though the model combining the wicking and combustion processes are very complicated and have not been attempted effectively, efforts to model pure wicking are quite advanced (Zarandi and Pillai 2018; Zarandi et al. 2018).

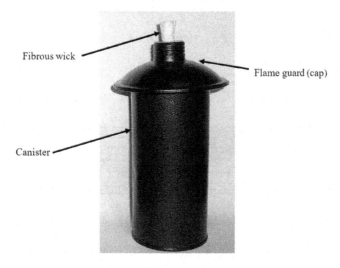

FIGURE 6.5 The main parts of a tiki torch that is used atop a bamboo stick and is used for providing lighting during camping and backyard parties.

6.2.4 Heat Pipes

Since the 1960s, heat pipes have been used in a wide range of applications where heat needs to be carried away quickly from heat-generation sites. The idea is quite simple—unlike metal fins, which use solid metal to conduct heat, the heat pipes use liquids to carry heat where the liquid is evaporated at the heat pipe's high-temperature end, and then the vapors move over to its low-temperature end where they condense back to the liquid state. Because of the high-heat capacity of such liquids, the heat pipes are much more effective than the fins in carrying heat. Some very high-tech applications involve using these devices (Alkam and Al-Nimr 1998; Faghri 1995; Imke 2004; Nazari and Toghraie 2017). For example, the flat plate heat pipes (FPHPs) (Annamalai and Dhanabal 2010) and mini heat pipes (MHP) (Florez et al. 2011) are the most utilized heat pipes in aerospace and satellite industries for cooling electronic equipment.

Thin porous materials are used in heat pipes to facilitate through wicking the movement of liquid from the cold end to the hot end. Adding porous materials causes impressive improvements in the performance of heat pipes, such as increasing the heat transfer capacity of the device and smaller mass-flow rates to transport equivalent heat flux. Like the other above-mentioned devices, the liquid flowing through the porous structure is engaged in wicking as well as phase change/heat transfer processes. These coupled processes in porous media has led to the emergence of the new type of heat pipes, those which deal with the two-phase systems. Such heat pipes are more beneficial since they are smaller and lighter and, more importantly, they work with lower temperature gradients or lower-mass flow rates to provide the same amount of heat transfer compared to the single-phase systems (Nemec 2017).

As shown in Figure 6.6, a two-phase system includes a container with porous structure (wick) on the inner surface. The working liquid gets evaporated by absorbing heat from the evaporator section, and then the evaporated vapor flows through the (open) central core of the system. After flowing through this adiabatic section, the evaporated working fluid flows into the condenser, and then by releasing heat it changes back to the liquid state. Here the use of wicking material in heat pipes provides two benefits. First, the heat released due to vapor condensation gets dissipated through the porous part. Second, the condensed liquid starts moving toward the hot end under the influence of capillary pressure, and the liquid is "pumped" back into the evaporator.

An estimate of the capillary pressure can be obtained through the formula

$$P_{cmax} = \frac{2\gamma}{r_c}$$

To perform correctly, the essential condition for every heat pipe must be

$$P_{cmax} \geq \Delta P_v + \Delta P_l + \Delta P_g$$

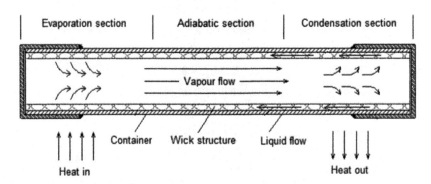

FIGURE 6.6 Schematic of a two-phase heat pipe system. (From Nemec, P., Porous Structures in Heat Pipes, in *Porosity-Process, Technologies and Applications*, IntechOpen, 2017.)

As the heat flux is one of the most important quantities in evaluating the performance of a heat pipe, it is of interest to investigate the effect of porous-media structure on this quantity. The total heat flux of heat pipe is given by

$$Q = \dot{m}_{max}.L$$

where the mass flow rate is obtained using Darcy's law as

$$\dot{m}_{max} = \frac{\rho_l \gamma}{\mu_l} \frac{KA}{l} \left(\frac{2}{r_c} - \frac{\rho_l g l}{\gamma} \sin\theta \right)$$

This equation shows the effect of porous-medium properties, such as permeability, K, contact angel, θ, cross-sectional area of porous wick, A, and capillary radius, r_c, on the mass flow rate of the liquid and subsequently the total heat flux in the heat pipe.

Investigations on the effect of porous-wick parameters in heat pipes has shown that as the porosity of the wick in a heat pipe increases, the heat pipe performance gets enhanced. In addition, the higher permeability of the wick causes an increase in the heat flux. But increasing permeability has an optimum point beyond which the increment in permeability may lead to dry out of the heat-pipe evaporation section and thus decrease the total heat-pipe performance (Nemec 2017). Also, as the thickness of the porous medium increases, the heat-pipe performance increases. Like permeability, increasing the wick thickness works in an optimum range of working temperature for the heat pipe. For example, in the temperature region of 30°C–60°C, this increment causes better performance, but it decreases the heat-pipe performance in the operating temperature region of 80°C–130°C (Nemec 2017).

6.3 Effect of Heat Transfer on the Wicking-Process Governing Equations

We described in the earlier sections all the examples and related details of the wicking process coupled with heat transfer. In this section, we give details of the most cited and applicable mathematical models for analyzing the described phenomena. In every mathematical model, there are some assumptions to make in order to make it less complex and more practical without compromising accuracy. Hence, the following to-be-discussed methods are categorized based on different conditions and assumptions.

6.3.1 Governing Equations in Propellant Management Devices

PMDs can be found in numerous sizes and configurations. In the space applications, the PMDs can be classified as being used either as a communication-device type application or as a control-device type (see www.pmdtechnology.com) (PMD Technology Capabilities). The most important part of a PMD in the former is that it provides gas-free propellant delivery, and the galleries play this important role of creating internal flow paths (Tam et al. 2008). In a gallery, metallic porous elements are the main parts that prevent gas penetration, as the propellant flows through the gallery. Bubble point is the pressure or head at which the wetted porous element will prevent gas penetration into the combustion chamber, which depends upon the size of the pores in the porous element and the fluid properties. In order to prevent gas from entering entrances, the bubble point has to be greater than the sum of the dynamic loads, the viscous losses, and the hydrostatics. Hence, it is important to obtain this pressure. Figure 6.7 elaborates the physics of the galleries.

One can estimate the pressure head of all the loads (Jaekle and Jaekle 1997) as

$$\Delta H_{\text{Hydrostatics}} = a \Delta Z$$

$$\Delta H_{\text{porous elements flow losses}} = f\left(u_{pe}, v, \text{Porous element}\right)$$

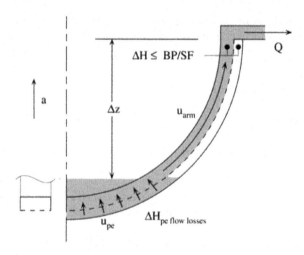

FIGURE 6.7 Schematic of a gallery in PMD. (From Jaekle, Jr D. and Jaekle, Jr D., Propellant management device conceptual design and analysis-Galleries, in *33rd Joint Propulsion Conference and Exhibit*, 2811, 1997.)

$$\Delta H_{\text{arm flow losses}} = f(Re) \frac{L}{D_{eq}} \frac{u_{arm}^2}{2}$$

$$\Delta H_{\text{dynamic head}} = \frac{u_{arm}^2}{2}$$

As mentioned earlier, the bubble point has to be greater than all the loads. Hence, the following equation should be solved to obtain a safety factor in order to ensure the prevention of gas flow.

$$\frac{\text{Bubble Point}}{\text{Safety Factor}} \geq a\Delta Z + \frac{u_{arm}^2}{2} + f(u_{pe}, v, pe) + f(Re) \frac{L}{D_{eq}} \frac{u_{arm}^2}{2}$$

Here, a is the acceleration, u_{arm} is the velocity in the gallery arm, u_{pe} is the velocity in porous elements, and D_{eq} is the equivalent diameter.

Another important parameter is the *wicking velocity under zero gravity conditions* (Symons 1974). During the wicking process, the capillary pressure as the pressure difference across the liquid–vapor interface as a function of the surface tension and the curvatures of the surface can be defined using the Young and Laplace equation:

$$\Delta P_c = \sigma \left(\frac{1}{R_1} + \frac{1}{R_2} \right)$$

Here, σ is the surface tension while R_1 and R_2 are principal radii of curvature of the liquid–vapor interface. If the pores are assumed to be capillary tubes, this expression for pressure reduces to

$$\Delta P_c = \frac{2\sigma}{R}$$

The radius of curvature in a tube of circular cross section can be defined as

$$R = \frac{D}{2\cos\theta}$$

with D being the capillary-tube diameter and θ being the contact angle. In reality, the pores in the screen through which the liquid rises are not circular in cross section. Hence, in order to consider the effect of pore shape, a coefficient, ξ, is employed and the use of which results in a new equation for capillary pressure:

$$\Delta P_c = \frac{\xi \sigma \cos \theta}{D_s}$$

Here, D_s is the pore size diameter. As an example, $\xi = 4$ for circular pores.

In order to include the effect of gravity, the corresponding pressure differential to be considered is

$$\Delta P_g = \rho g h \sin \omega$$

where h is the height of liquid in capillary tube, and ω indicates the capillary tube orientation.

The other important factor is the pressure drop associated with viscous losses over the height of liquid in the porous element, which can be found through Darcy's law as

$$\Delta P_f = \frac{V_w A_c \, \mu h}{K A_{c,pe}}$$

where V_w is the wicking velocity (Darcy's velocity), K is the permeability, A_c is the cross-sectional area for wicking flow, and $A_{c,pe}$ is the total cross-sectional area of the porous element.

Force balance indicates that

$$\Delta P_c = \Delta P_f + \Delta P_g$$

Hence

$$\frac{\xi \sigma \cos \theta}{D_s} = \frac{V_w A_c \, \mu h}{K A_{c,pe}} + \rho g h \sin \omega$$

By assuming the contact angel to be zero and neglecting the gravity effect, the wicking velocity can be expressed as

$$V_w = \frac{\xi K D_s}{D_s} \frac{A_{c,pe}}{A_c} \frac{\sigma}{\mu L}$$

Symons (1974) developed this above-described model to obtain the rate of wicking in screens containing porous structures, and he validated his model through experiments in zero and non-zero gravity conditions. He found that the developed theoretical model accurately predicts on the wicking process the effect of important parameters, such as distance from the liquid source, liquid properties, and gravity.

6.3.2 Governing Equations in Sheet Drying

In this area, there have been a number of theoretical models based on a wide range of assumptions, such as uniform temperature, uniform moisture content, or constant thermal conductivity (Harrmann and Schulz 1990; Lee and Hinds 1981). All these assumptions were made in order to make the complex heat and mass transfer phenomenon simpler. One of the main deficiencies of these models is ignoring the internal transport mechanisms, which affects the drying process (Lu and Shen 2007). However, some researchers have considered structural changes within the porous media during drying in their models. Considering the porous structure change is very important as it affects moisture migration and the drying process (Rogers and Kaviany 1992; Turner and Ilic 1990). Generally, most of the proposed drying models have been based on the volume-averaging approach developed by Whitaker (1977).

The following model is composed of the following conservation equations based on Whitaker's volume-averaging approach, which upscales the heat and mass transfer processes in the paper sheet during drying. Before presenting the governing equations, let us state some necessary assumption as follows:

1. Negligible shrinkage and deformation of paper sheet during the drying process
2. Thermodynamic equilibrium in all phases
3. One-dimensional drying along the paper thickness
4. Ideal gases behavior for air and vapor within the pores
5. Negligible inertia and viscous dissipation

The governing equations for drying in a sheet can be stated as follows.

Moisture mass conservation

$$\frac{\partial\left(\varepsilon_l\rho_l + \varepsilon_v\rho_v\right)}{\partial t} + \frac{\partial\left(\varepsilon_l\rho_l u_l + \varepsilon_v\rho_v u_v\right)}{\partial x} = 0$$

Fluid mixture mass conservation

$$\frac{\partial\left(\varepsilon_l\rho_l + \varepsilon_g\rho_g\right)}{\partial t} + \frac{\partial\left(\varepsilon_l\rho_l u_l + \varepsilon_g\rho_g u_g\right)}{\partial x} = 0$$

where the velocities u_l and u_g can be found through Darcy's law as

$$u_l = -\frac{K_{rl}K}{\mu_l\varepsilon_l}\left(\frac{\partial P_l}{\partial x} - \rho_l g\right)$$

$$u_g = -\frac{K_{rg}K}{\mu_g\varepsilon_g}\left(\frac{\partial P_g}{\partial x} - \rho_g g\right)$$

While the air and vapor velocities are given by

$$u_a = u_g - \frac{\rho_g D M_{m\cdot a}M_{m\cdot v}}{\rho_a M_{m\cdot g}{}^2}\frac{\partial}{\partial x}\left(\frac{P_a}{P_g}\right)$$

$$u_v = u_g - \frac{\rho_g D M_{m\cdot a}M_{m\cdot v}}{\rho_v M_{m\cdot g}{}^2}\frac{\partial}{\partial x}\left(\frac{P_v}{P_g}\right)$$

As the gaseous phase pressure given by Dalton's law,

$$P_g = P_a + P_v$$

where the air pressure follows the ideal gas law:

$$P_a = \frac{\rho_a RT}{M_a}$$

Here, M_m is the molar mass, D is the diffusion coefficient, and subscripts a, v, g represent air, vapor, and gas, respectively.

The porosity of each phase can be found from the saturation as follows.

$$\varepsilon_l = \varepsilon S$$

$$\varepsilon_a = \varepsilon_v = \varepsilon_g = \varepsilon(1-S)$$

$$\varepsilon_s = 1 - \varepsilon$$

Energy conservation

$$\frac{\partial}{\partial t}\left(\varepsilon_l \rho_l h_l + \varepsilon_v \rho_v h_v + \varepsilon_g \rho_g h_g + \varepsilon_s \rho_s h_s\right) + \frac{\partial}{\partial x}\left(\varepsilon_l \rho_l h_l u_l + \varepsilon_v \rho_v u_v h_v + \varepsilon_a \rho_a h_a u_a\right)$$

$$= \frac{\partial}{\partial x}\left(K_{\mathit{eff}} \frac{\partial T}{\partial t}\right)$$

Lu and Shen (2007) used this model for predicting paper drying in a running machine, and they acquired an extremely satisfying (less than 4% error) match between the numerical and experimental results. As one of their interesting results, they found that temperature and gas pressure are not uniformly distributed across the paper thickness, but are higher in the middle than on the two boundaries; however, the moisture distribution is uniform (Lu and Shen 2007). In addition, they concluded that as long as the paper is in touch with rolling parts of the drying machine (cylinders), the temperature and gas pressure increase, and the average drying rate is low. However, once the paper strip loses this touch and undergoes free movement, the temperature and gas pressure become low but cause the average drying rate to be high.

6.3.3 Governing Equations in Porous Burners

We categorized these methods based on whether they include the porous-media effects directly in the modeling process or not. In all the following models, we have some common assumptions:

- The medium is a cylindrical wick as shown in Figure 6.8.
- The wicking process occurs along both the axial, z, and the radial, r, directions.
- Radiative heat transfer is not negligible.

6.3.3.1 Models Independent of Porous-Medium Processes

In this kind of modeling method, the governing equations for flame on top and around porous media (such as continuity, momentum, energy, and species equations) get solved without considering the effect of porous medium on fluid velocity and the other velocity-related properties. Hence, instead of having a wick with porous structure and the associated flow-and-transport complexities, it is assumed to be solid.

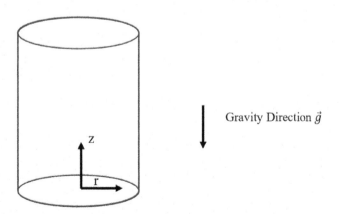

FIGURE 6.8 The schematic of a cylindrical wick and the flow directions.

In such problems, the wick geometry is assumed to be radially symmetric and thus amenable to cylindrical coordinates. Also, it is assumed that there is just one phase of fluid (fuel) issuing from the boundaries of the wick geometry, and which acts as a boundary condition for solving the flame equations.

We will present the governing equations for flames for one typical case. A comprehensive investigation has been done by Alsairafi et al. on diffusive flame around a cylindrical wick (Alsairafi et al. 2004), the governing equations for which are as follows.

Continuity equation:

$$\frac{\partial}{\partial z}(\rho u_z) + \frac{1}{r}\frac{\partial}{\partial r}(r\rho u_r) = 0$$

Radial momentum equation:

$$\frac{\partial}{\partial z}(\rho u_z u_r) + \frac{1}{r}\frac{\partial}{\partial r}(r\rho u_r u_r)$$

$$= -\frac{\partial p}{\partial r} + \frac{\partial}{\partial z}\left(\mu\frac{\partial u_r}{\partial z}\right) + \frac{1}{r}\frac{\partial}{\partial r}\left(2r\mu\frac{\partial u_r}{\partial r}\right) + \frac{\partial}{\partial z}\left(\mu\frac{\partial u_z}{\partial r}\right) - \frac{2\mu u_r}{r^2} - \frac{\partial}{\partial r}\left(\frac{2}{3}\mu\left(\frac{\partial u_r}{\partial r} + \frac{u_r}{r} + \frac{\partial u_z}{\partial z}\right)\right)$$

Axial momentum equation:

$$\frac{\partial}{\partial z}(\rho u_z u_z) + \frac{1}{r}\frac{\partial}{\partial r}(r\rho u_r u_z)$$

$$= -\frac{\partial p}{\partial z} + \frac{\partial}{\partial z}\left(\frac{4}{3}\mu\frac{\partial u_z}{\partial z}\right) + \frac{1}{r}\frac{\partial}{\partial r}\left(r\mu\frac{\partial u_z}{\partial r}\right) + \frac{1}{r}\frac{\partial}{\partial r}\left(r\mu\frac{\partial u_r}{\partial z}\right) - \frac{1}{r}\frac{\partial}{\partial z}\left(\frac{2}{3}\mu\left(\frac{\partial(ru_r)}{\partial r}\right)\right) - \rho g$$

Species equation:

$$\frac{\partial}{\partial z}(\rho u_z Y_i) + \frac{1}{r}\frac{\partial}{\partial r}(r\rho u_r Y_i) = \frac{\partial}{\partial z}\left(\rho D_i\frac{\partial Y_i}{\partial x}\right) + \frac{1}{r}\frac{\partial}{\partial r}\left(r\rho D_i\frac{\partial Y_i}{\partial r}\right) + \dot{R}_i$$

Energy equation:

$$\frac{\partial}{\partial z}(\rho u_z T) + \frac{1}{r}\frac{\partial}{\partial r}(r\rho u_r T)$$

$$= \frac{\partial}{\partial z}\left(\frac{\lambda}{C_p}\frac{\partial T}{\partial z}\right) + \frac{1}{r}\frac{\partial}{\partial r}\left(r\frac{\lambda}{C_p}\frac{\partial T}{\partial r}\right) + \frac{\lambda}{C_p^2}\left(\frac{\partial C_p}{\partial z}\frac{\partial T}{\partial z} + \frac{\partial C_p}{\partial r}\frac{\partial T}{\partial r}\right)$$

$$- \frac{1}{C_p}\left(\sum_i \int_{T_0}^{T}\left(C_{p,i}dT\dot{R}_i + Q\dot{R}_F\right)\right) + \sum_i\left(\frac{\rho C_{p,i}D_i}{C_p}\left(\left(\frac{\partial Y_i}{\partial z}\frac{\partial T}{\partial z} + \frac{\partial Y_i}{\partial r}\frac{\partial T}{\partial r}\right)\right)\right)$$

$$- \frac{1}{C_p}\nabla\cdot\vec{q_r}$$

Here, Q is the heat released per unit mass of burned fuel, $\dot{R}_F$ is the reaction rate defined from Arrhenius equation (Alsairafi et al. 2004), and q_r is the radiative heat-flux vector defined as

$$\nabla\cdot\vec{q_r} = k\left[4\sigma T^4 - G\right]$$

where G is the incident radiation.

Wicking of Liquids under Non-Isothermal and Reactive Conditions: Some Industrial Applications 115

As it was mentioned earlier, radiation is important. Hence, the radiative heat from the flame can be predicted through

$$\left(\vec{\Omega}.\nabla\right)I\left(\vec{r},\vec{\Omega}\right)+\kappa\left(\vec{r}\right)I\left(\vec{r},\vec{\Omega}\right)=\kappa\left(\vec{r}\right)I_b\left(\vec{r},\vec{\Omega}\right)$$

where $\vec{\Omega}$ is the solid angle in terms of orientation, I is the radiation intensity, and κ is the absorption coefficient. By assuming a complete combustion, κ can be estimated from the types of combustion products in terms of their material, phase, etc. (e.g. H_2O (l) or H_2O (g), etc.) (Abu-Romia and Tien 1967; Alsairafi et al. 2004; Mahadevan et al. 2006).

6.3.3.2 Models Including Porous-Medium Processes

The second type of modeling methods includes the effect of the flow and transport processes occurring inside the (wick) porous medium in the set of governing equations. Hence, although the number of to-be-solved equations is more than that in the previous method, this method seems more reliable and accurate. In addition, we account for the two phases (liquid and vapor) of the fuel. (That is, the model accounts for the phase-change phenomena encountered by the fuel inside the wick where the liquid fuel is evaporated to vapor form under the influence of heat.) There have been various studies using this method to investigate the wicking process coupled with heat transfer (Mahadevan et al. 2006; Raju and T'ien 2008).

The governing equations for flame remain unchanged, and hence are the same as listed in the last section. However, we now list the additional governing equations used to model processes inside the porous wick (or the "burner"). The continuity equation will be considered for both the phases in the axial and radial directions while the momentum equation used is Darcy's Law. Hence, the governing equations in the porous domain are as follows.

Continuity equation:

$$\frac{\partial}{\partial z}\left(\rho_l u_{z,l}\right)+\frac{\partial}{\partial z}\left(\rho_g u_{z,g}\right)+\frac{1}{r}\frac{\partial}{\partial r}\left(r\rho_l u_{r,l}\right)+\frac{1}{r}\frac{\partial}{\partial r}\left(r\rho_g u_{r,g}\right)=0$$

Momentum equations:

$$u_{z,l}=-\frac{K_{rl}K}{\mu_l}\left(\frac{\partial P_l}{\partial z}-\rho_l g\right)$$

$$u_{z,g}=-\frac{K_{rg}K}{\mu_g}\left(\frac{\partial P_g}{\partial z}-\rho_g g\right)$$

$$u_{r,l}=-\frac{K_{rl}K}{\mu_l}\left(\frac{\partial P_l}{\partial r}\right)$$

$$u_{r,g}=-\frac{K_{rg}K}{\mu_g}\left(\frac{\partial P_g}{\partial r}\right)$$

Here, K_r is the relative permeability, and it can be found by the Brooks and Corey (1966) equation as a function of saturation of wicking fluid:

$$K_r=S^n$$

In the same way, the capillary pressure can be estimated through different models for finding pressure as a function of saturation (Landeryou et al. 2005; Leverett 1941; Van Genuchten 1980):

$$P_c(S) = P_g - P_l$$

Energy equation:

$$\frac{\partial}{\partial z}(\rho_l C_{p,l} T u_{z,l}) + \frac{\partial}{\partial z}((C_{p,l} T - h_{fg,l})\rho_g u_{z,g}) + \frac{1}{r}\frac{\partial}{\partial r}(r\rho_l C_{p,l} T u_{r,l})$$
$$+ \frac{1}{r}\frac{\partial}{\partial r}(r\rho_g (C_{p,l} T - h_{fg,l}) u_{r,g}) = \frac{\partial}{\partial z}\left(k_{eff}\frac{\partial T}{\partial z}\right) + \frac{1}{r}\frac{\partial}{\partial r}\left(rk_{eff}\frac{\partial T}{\partial r}\right)$$

6.3.3.3 Consequences of Including Porous-Medium Processes

Inclusion of porous-media physics resulted in enhancing the results obtained after assuming the wick (burner) to be solid. For example, Raju and T'ien (2008) investigated the effect of some of the porous-media properties, and they demonstrated that the permeability affects the flame structure and the burning rate. As the permeability decreases, the resistance increases, and the height to which the fuel rises in the wick decreases. In addition, it was observed that by increasing the diameter of the "burning" porous medium, the flame diameter, flame height, and the total burning rate increase.

In another investigation, M. P. Raju and James (2007) demonstrated that more liquid fuel is evaporated than the vaporized fuel entering into the flame region. The reason behind this discrepancy is that a small portion of the evaporated fuel condenses inside the wick.

Hence, one can surmise that the inclusion of porous-media physics enhances the accuracy and completeness of the combustion simulation.

6.3.4 Governing Equations in Heat Pipes

In most applications of porous media in heat pipes, the working fluids are subjected to phase-change processes, such as condensation and evaporation, and which undergo flow and transport phenomena due to temperature and pressure gradients (Udell 1985). Therefore, we consider a general case involving evaporation and condensation in an inclined plane as shown in Figure 6.9.

While developing this particular model, the following assumptions are made:

- Heat and mass transfer processes are one dimensional and steady state.
- Flow of both liquid and vapor phases is governed by viscosity, gravity, and capillary forces.

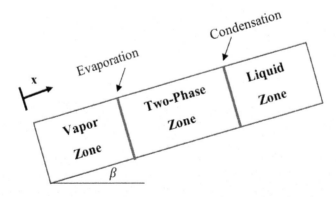

FIGURE 6.9 A representative porous-medium system in a heat pipe subjected to evaporation and condensation.

Wicking of Liquids under Non-Isothermal and Reactive Conditions: Some Industrial Applications 117

- Darcy's law is the governing equation for flow of liquid and vapor phases.
- Negligible temperature gradient in the two-phase zone.
- Condensation and evaporation happen at the boundaries of the two-phase zone.

Continuity equation:

$$\dot{m}_l + \dot{m}_v = 0$$

Energy equation:

If we assume that the temperature gradient in the two-phase zone is negligible, the energy equation reduces to

$$Q = \dot{m}_v h_{fg}$$

Momentum equations:

On using Darcy's law, the mass flow rate for the liquid and vapor phases can be obtained as

$$\dot{m}_l = -\frac{\rho_l K K_{rl}}{v_l} \frac{d}{dx}\left(\frac{P_l}{\rho_l} + gx\sin\beta\right)$$

$$\dot{m}_v = -\frac{\rho_v K K_{rv}}{v_v} \frac{d}{dx}\left(\frac{P_v}{\rho_v} + gx\sin\beta\right)$$

where β is defined as the orientation of the heat pipe, and K_r is the relative permeability.

Capillary pressure is defined as the difference between the vapor and liquid pressures as

$$P_c = P_v - P_l$$

On substituting the energy equation in the momentum equations, we obtain

$$\frac{dP_l}{dx} = +\frac{\mu_l Q}{h_{fg}\rho_l K K_{rl}} - \rho_l\, g\sin\beta$$

$$\frac{dP_v}{dx} = -\frac{\mu_v Q}{h_{fg}\rho_v K K_{rv}} - \rho_v\, g\sin\beta$$

Now on using these two equations in the capillary-pressure equation, we get

$$\frac{dP_c}{dx} = -\frac{Q}{h_{fg}K}\left(\frac{\mu_v}{\rho_v K_{rv}} + \frac{\mu_l}{\rho_l K_{rl}}\right) + \left(\rho_l - \rho_v\right)g\sin\beta$$

In order to include the effect of saturation, Leverett (1941) proposed a model for capillary pressure as a function of saturation:

$$f(s) = \frac{P_c}{\gamma}\left(\frac{K}{\varepsilon}\right)^{\frac{1}{2}}$$

A typical expression for the function f is

$$f = 1.417(1-s) - 2.120(1-s)^2 + 1.263(1-s)^3$$

Now, one can define the following dimensionless parameters (Udell 1985):

$$\Psi = \frac{Q\mu_v}{\rho_v\, K\, h_{fg}\left(\rho_l - \rho_v\right)g}$$

$$\Lambda = \frac{x\left(\rho_l - \rho_v\right)g}{\gamma}\left(\frac{K}{\varepsilon}\right)^{\frac{1}{2}}$$

$$\Omega = \frac{\rho_v\mu_l}{\mu_v\rho_l}$$

Here Ψ is taken as the ratio of vapor pressure gradient to the hydrostatic pressure gradient, while Λ is the scaling of x by the capillary-pressure height. Using these variables, the gradient of capillary pressure can be expressed as

$$\frac{df}{d\Lambda} = \sin\beta - \Psi\left(\frac{1}{K_{rv}} - \frac{\Omega}{K_{rl}}\right)$$

Similarly, the gradient of saturation with respect to Λ can be formulated as

$$\frac{ds}{d\Lambda} = \frac{\sin\beta - \Psi\left(\dfrac{1}{K_{rv}} - \dfrac{\Omega}{K_{rl}}\right)}{\acute{f}}$$

Since the length of the two-phase zone is Λ_1, the boundary conditions are defined as

$$s \;=\; 0 \text{ at } \Lambda = 0$$

$$s \;=\; 1 \text{ at } \Lambda = \Lambda_1$$

The thermal characteristics can be obtained through the vapor and liquid pressures. Therefore, on referencing the pressures with respect to the saturation pressure P_o and then scaling it with a characteristic capillary pressure, the vapor and liquid pressures are defined in non-dimensional forms as

$$\overline{P}_v = \frac{\left(P_v - P_o\right)}{\gamma}\left(\frac{K}{\varepsilon}\right)^{\frac{1}{2}}$$

$$\overline{P}_l = \frac{\left(P_l - P_o\right)}{\gamma}\left(\frac{K}{\varepsilon}\right)^{\frac{1}{2}}$$

By using the saturation-gradient equation with respect to Λ described earlier, the derivatives with respect to saturation of the vapor and liquid pressures can be rendered as

$$\frac{d\overline{P}_v}{ds} = \frac{-\Psi\,\acute{f}}{K_{rv}\left\{\sin\beta - \Psi\left(\dfrac{1}{K_{rv}} + \dfrac{\Omega}{K_{rl}}\right)\right\}}$$

$$\frac{d\overline{P}_l}{ds} = \frac{\Psi\dfrac{\acute{f}}{K_{rl}} - \acute{f}\sin\beta}{\left\{\sin\beta - \Psi\left(\dfrac{1}{K_{rv}} + \dfrac{\Omega}{K_{rl}}\right)\right\}}$$

After effecting all these manipulations, the temperature within the two-phase zone as a function of vapor and liquid pressures is given by

$$T = \frac{T_0\left(1 + \dfrac{P_c}{h_f \rho_l}\right)}{1 - T_0 R / h_{fg} Ln\left(\dfrac{P_v}{P_o}\right)}$$

Udell applied this method and investigated the effect of gravity, capillary pressure, viscosity, and phase changes on heat and mass transfer at various orientations of a heat-pile-like system (1985). He evaluated his theoretical result by conducting sets of experiments. Udell found that in media with high values of permeability, the thermal conductivity of the two-phase zone is several orders of magnitude greater than that of the single phase one, and because of which the heat transfer rate is enhanced significantly.

Further work has been done to model heat pipes under different conditions. In one such case of using heat pipes inside the drills used for machining metals, it was observed that not much is lost by ignoring the details of wicking of cooling liquids inside the heat pipes (Jen et al. 2002).

6.4 Conclusions

In the current chapter, we discussed four industrial examples where heat transfer is coupled with the wicking-type two-phase flows in porous media. In three out of the four cases considered, numerical and theoretical models based on detailed physics of wicking/two-phase flow were found to exist where heat transfer had to be incorporated. In the sole case of porous burners and candle wicks, the heat transfer had to incorporate both the convective and radiative effects in the presence of chemical reactions during the combustion of fuel.

In all the considered cases, Darcy's law was found to be applicable where the permeability of the considered porous medium plays an important role. However, in the three cases of sheet drying, porous burners, and heat pipes, simultaneous presence of two phases (liquid and vapor/gas) led to the use of the generalized form of Darcy's law where additional properties including the relative permeability and capillary pressure as a function of local saturation became also important. The phase change of the liquids implied that enthalpy of change is used to compute the mass undergoing such a change. In some applications like porous burners, the investigations become more complicated due to high temperatures and complex fluid mechanics, and hence diverse aspects had to be considered due to the coupled nature of various phenomena.

In the final analysis, when heat transfer (with or without chemical reaction) equations are added to the basket of relations employed to model the wicking process involving two-phase flows in porous media, a higher accuracy in the prediction of results is achieved under non-isothermal conditions, even though it makes the model more complex and difficult to solve. Such theoretical and simulation models help engineers in predicting the behavior of systems involving non-isothermal wicking, and thus paves a way toward their subsequent optimization and improvement.

NOMENCLATURE

a acceleration
A cross-sectional area of porous wick
C specific heat
C_p specific heat of the mixture
$C_{p,i}$ specific heat of species i
C_I inertia coefficient
D_i diffusion coefficient of species i

D_s	pore size diameter
D_{eq}	equivalent diameter
G	incident radiation
g	gravity
h_{fg}	latent heat
I	radiation intensity
K	permeability
K_r	relative permeability
k	thermal conductivity
k_{eff}	effective thermal conductivity
L	characteristic length
l	length of the pipe
M	molecular weight
M_m	molar mass
$\dot{m}_{max}$	maximum mass flow
n	power of saturation
P	pressure
$\overline{P}$	non-dimensional pressure
P_c	capillary pressure
P_o	saturation pressure
ΔP_l	pressure drop in the wick structure
ΔP_v	pressure drop in the vapor
ΔP_g	pressure drop due gravity
Q	heat flux
q_r	radiative heat flux
r_c	capillary radius
$\dot{R}_F$	reaction rate
$\dot{R}_i$	sink or source term of species i
Re	Reynold's number
$\vec{r}$	position vector
S	saturation
s	effective saturation
T	temperature
t	time
y	mole fraction
Y_i	mass fraction of species i
u	velocity

Subscripts

a	air phase
arm	gallery arm
g	gas phase
l	liquid phase
V	vapor phase
pe	porous element
r	radial direction
x	axial direction
z	vertical direction

Wicking of Liquids under Non-Isothermal and Reactive Conditions: Some Industrial Applications 121

Greek Letters

ε	Porosity
σ	Stefan–Boltzmann constant
$\sigma_\kappa, \sigma_\epsilon, \sigma_T$	turbulence model constant for k, ϵ, and T
β	heat pipe orientation
κ	absorption coefficient
λ	gas thermal conductivity
ξ	coefficient for pore-shape type
ρ	density
γ	surface tension
θ	contact angel
μ	viscosity
ν	kinematic viscosity
$\vec{\Omega}$	orientation in terms of solid angle

REFERENCES

Abu-Romia M., and C. L. Tien, "Appropriate mean absorption coefficients for infrared radiation of gases," *Journal of Heat Transfer*, vol. 89, no. 4, pp. 321–327, 1967.

Alkam M. K., and M. A. Al-Nimr, "Transient non-Darcian forced convection flow in a pipe partially filled with a porous material," *International Journal of Heat and Mass Transfer*, vol. 41, no. 2, pp. 347–356, 1998.

Allerton J., L. Brownell, and D. Katz, "Through-drying of porous media," *Chemical Engineering Progress*, vol. 45, no. 10, pp. 619–635, 1949.

Alsairafi A., S.-T. Lee, and J. S. T'ien, "Modeling gravity effect on diffusion flames stabilized around a cylindrical wick saturated with liquid fuel," *Combustion Science and Technology*, vol. 176, no. 12, pp. 2165–2191, 2004.

Annamalai M., and S. Dhanabal, "Experimental studies on porous wick flat plate heat pipe," *International Refrigeration and Air Conditioning Conference at Purdue*, P. 1045, 2010.

Bandyopadhyay A., B. Ramarao, and S. Ramaswamy, "Transient moisture diffusion through paperboard materials," *Colloids and Surfaces A: Physicochemical and Engineering Aspects*, vol. 206, no. 1–3, pp. 455–467, 2002.

Benard J., R. Eymard, X. Nicolas, and C. Chavant, "Boiling in porous media: Model and simulations," *Transport in Porous Media*, vol. 60, no. 1, pp. 1–31, 2005.

Brennen C. E., *Fundamentals of Multiphase Flow*. Cambridge University Press, Cambridge, England, 2005.

Brooks R. H., and A. T. Corey, "Properties of porous media affecting fluid flow," *Journal of the Irrigation and Drainage Division*, vol. 92, no. 2, pp. 61–90, 1966.

Camplisson C. K., K. M. Schilling, W. L. Pedrotti, H. A. Stone, and A. W. Martinez, "Two-ply channels for faster wicking in paper-based microfluidic devices," *Lab on a Chip*, vol. 15, no. 23, pp. 4461–4466, 2015.

Canbazoglu F., B. Fan, A. Kargar, K. Vemuri, and P. Bandaru, "Enhanced solar evaporation of water from porous media, through capillary mediated forces and surface treatment," *AIP Advances*, vol. 6, no. 8, p. 085218, 2016.

Chatterjee S., B. Ramarao, and C. Tien, "Water-vapour sorption equilibria of a bleached-kraft paperboard- a study of the hysteresis region," *Journal of Pulp and Paper Science*, vol. 23, pp. J366–373, 1997.

Cleary P. W., J. E. Hilton, and M. D. Sinnott, "Modelling of industrial particle and multiphase flows," *Powder Technology*, vol. 314, pp. 232–252, 2017.

Dodge F. T., *The New Dynamic Behavior of Liquids in Moving Containers*. Southwest Research Inst, San Antonio, TX, 2000.

Donnarumma D., G. Tomaiuolo, S. Caserta, Y. Gizaw, and S. Guido, "Water evaporation from porous media by dynamic vapor sorption," *Colloids and Surfaces A: Physicochemical and Engineering Aspects*, vol. 480, pp. 159–164, 2015.

Ellzey J. L., and R. Goel, "Emissions of co and no from a two stage porous media burner," *Combustion Science and Technology*, vol. 107, no. 1–3, pp. 81–91, 1995.

Faghri A., *Heat Pipe Science and Technology*. CRC Press, Boca Raton, FL, 1995.

Florez J., G. Nuernberg, M. Mantelli, R. Almeida, and A. Klein, "Effective thermal conductivity of layered porous media," in *Proceedings of 10th International Heat Pipe Symposium*, Taipei, Taiwan, 2011.

Fries N., K. Odic, and M. Dreyer, "Wicking of perfectly wetting liquids into a metallic mesh," in *Proceedings of the 2nd International Conference on Porous Media and its Applications in Science and Engineering*, 2007.

Fries N., K. Odic, M. Conrath, and M. Dreyer, "The effect of evaporation on the wicking of liquids into a metallic weave," *Journal of Colloid and Interface Science*, vol. 321, no. 1, pp. 118–129, 2008.

Gibou F., D. Hyde, and R. Fedkiw, "Sharp interface approaches and deep learning techniques for multiphase flows," *Journal of Computational Physics*, vol. 380, pp. 442–463, 2019.

Harrmann M., and S. Schulz, "Convective drying of paper calculated with a new model of the paper structure," *Drying Technology*, vol. 8, no. 4, pp. 667–703, 1990.

Hartwig J. W., "A detailed historical review of propellant management devices for low gravity propellant acquisition," in *52nd AIAA/SAE/ASEE Joint Propulsion Conference*, p. 4772, 2016.

Hashemi S., and W. M. Douglas, "Paper drying: A strategy for higher machine speed. I. Through air drying for hybrid dryer sections," *Drying Technology*, vol. 19, no. 10, pp. 2487–2507, 2001.

Hashemi S., and W. Murray Douglas, "Moisture nonuniformity in drying paper: Measurement and relation to process parameters," *Drying Technology*, vol. 21, no. 2, pp. 329–347, 2003.

Hashemi S., M. Roald, and W. Murray Douglas, "Mechanism of through air drying of paper: Application in hybrid drying," *Drying Technology*, vol. 21, no. 2, pp. 349–368, 2003.

Horas J. A., and J. P. Toso, "Diffusion in glassy polymers: A model using a homogenization method and the effective medium theory," *Journal of Polymer Science Part B: Polymer Physics*, vol. 30, no. 2, pp. 127–131, 1992.

Huostila M., and T. Haapsaari, *Transferring a Web from a Pick-Up Fabric to a Flow-Through Drying Wire*. U.S. Patent 4,194,947, issued March 25, 1980.

Imke U., "Porous media simplified simulation of single- and two-phase flow heat transfer in microchannel heat exchangers," *Chemical Engineering Journal*, vol. 101, no. 1, pp. 295–302, 2004.

Jaekle D. E., "Propellant management device conceptual design and analysis-Sponges," in *29th Joint Propulsion Conference and Exhibit*, p. 1970, 1993.

Jaekle Jr D., "Propellant management device conceptual design and analysis-Traps and Troughs," in *31st Joint Propulsion Conference and Exhibit*, p. 2531, 1995.

Jaekle Jr D., and Jr D. Jaekle, "Propellant management device conceptual design and analysis-Galleries," in *33rd Joint Propulsion Conference and Exhibit*, p. 2811, 1997.

Jen et al. T. C., "Investigation of heat pipe cooling in drilling applications. Part I: Preliminary numerical analysis and verification," *International Journal of Machine Tools and Manufacture*, vol. 42, pp. 643–652, 2002.

Kaviany M., and Y. Tao, "A diffusion flame adjacent to a partially saturated porous slab: Funicular state," *Journal of Heat Transfer*, vol. 110, no. 2, pp. 431–436, 1988.

Landeryou M., I. Eames, and A. Cottenden, "Infiltration into inclined fibrous sheets," *Journal of Fluid Mechanics*, vol. 529, pp. 173–193, 2005.

Lee P., and J. Hinds, "Optimizing Dryer Performance: Modeling Heat and Mass Transfer Within a Moist Sheet of Paper or Board," in *IN 1981 ENG. CONF*, 1981, pp. 117–124.

Leverett M., "Capillary behavior in porous solids," *Transactions of the AIME*, vol. 142, no. 01, pp. 152–169, 1941.

Lu T., and S. Q. Shen, "Numerical and experimental investigation of paper drying: Heat and mass transfer with phase change in porous media," *Applied Thermal Engineering*, vol. 27, no. 8, pp. 1248–1258, 2007.

Mahadevan J., M. M. Sharma, and Y. C. Yortsos, "Flow-through drying of porous media," *AIChE Journal*, vol. 52, no. 7, pp. 2367–2380, 2006.

Masoodi R., and K. M. Pillai, *Wicking in Porous Materials: Traditional and Modern Modeling Approaches*. CRC Press, Boca Raton, FL, 2012a.

Masoodi R., and K. M. Pillai, "Traditional theories of wicking: Capillary models," *Wicking in Porous Materials: Traditional and Modern Modeling Approaches*, pp. 31–53, 2012b.

McDonald P. E., *Wicking In Multi-Ply Paper Structures With Dissimilar Plies*. PhD thesis, Georgia Institute of Technology, 2006.

Milani Shirvan K., S. Mirzakhanlari, S. A. Kalogirou, H. F. Oztop, and M. Mamourian, "Heat transfer and sensitivity analysis in a double pipe heat exchanger filled with porous medium," *International Journal of Thermal Sciences*, vol. 121, pp. 124–137, 2017.

Mossbauer S., O. Pickenacker, K. Pickenacker, and D. Trimis, "Application of the porous burner technology in energy- and heat-engineering," *Clean Air*, vol. 3, no. 2, pp. 185–198, 2002.

Mosthaf K., R. Helmig, and D. Or, "Modeling and analysis of evaporation processes from porous media on the REV scale," *Water Resources Research*, vol. 50, no. 2, pp. 1059–1079, 2014.

Mujeebu M. A., M. Z. Abdullah, M. Z. A. Bakar, A. A. Mohamad, and M. K. Abdullah, "Applications of porous media combustion technology—A review," *Applied Energy*, vol. 86, no. 9, pp. 1365–1375, 2009a.

Mujeebu M. A., M. Z. Abdullah, M. A. Bakar, A. Mohamad, R. Muhad, and M. Abdullah, "Combustion in porous media and its applications–a comprehensive survey," *Journal of Environmental Management*, vol. 90, no. 8, pp. 2287–2312, 2009b.

Nazari S., and D. Toghraie, "Numerical simulation of heat transfer and fluid flow of Water-CuO Nanofluid in a sinusoidal channel with a porous medium," *Physica E: Low-Dimensional Systems and Nanostructures*, vol. 87, pp. 134–140, 2017.

Nemec P., "Porous Structures in Heat Pipes," in *Porosity-Process, Technologies and Applications*, IntechOpen, 2017.

Olayiwola R. O., "Modeling and simulation of combustion fronts in porous media," *Journal of the Nigerian Mathematical Society*, vol. 34, no. 1, pp. 1–10, 2015.

Peysson Y., M. Fleury, and V. Blazquez-Pascual, "Drying rate measurements in convection-and diffusion-driven conditions on a shaly sandstone using nuclear magnetic resonance," *Transport in Porous Media*, vol. 90, no. 3, pp. 1001–1016, 2011.

PMD Technology Capabilities, http://www.pmdtechnology.com/

Raju M. P., and S. James, "Heat and mass transports in a one-dimensional porous wick driven by a gasphase diffusion flame," *Journal of Porous Media*, vol. 10, no. 4, pp. 327–342, 2007.

Raju M. P. , and S. James, "Two-phase flow inside an externally heated axisymmetric porous wick," *Journal of Porous Media*, vol. 11, no. 8, pp. 701–718, 2008.

Raju M., and J. T'ien, "Modelling of candle burning with a self-trimmed wick," *Combustion Theory and Modelling*, vol. 12, no. 2, pp. 367–388, 2008.

Rogers J., and M. Kaviany, "Funicular and evaporative-front regimes in convective drying of granular beds," *International Journal of Heat and Mass Transfer*, vol. 35, no. 2, pp. 469–480, 1992.

Saricam C., and F. Kalaoğlu, "Investigation of the wicking and drying behaviour of polyester woven fabrics," *Fibres & Textiles in Eastern Europe*, vol. 105, no. 3, pp. 73–78,2014.

Symons E. P., "Wicking of liquids in screens," NASA Technical Report, NASA-TN-D-7657, E-7781, 1974.

Tam W., I. Ballinger, and D. E. Jaekle, "Review and history of ATK space systems commerce-the past 15 years," Internal Report. *ATK Space Systems Commerce Division*, California, 2008.

Tam W., P. Behruzi, D. Jaekle, and G. Netter, "The evolutionary forces and the design and development of propellant management devices for space flight in Europe and the United States," in *Space Propulsion 2016 Conference*, Rome, Italy, 2016.

Tam W., and D. Jaekle Jr, "The evolution of a family of propellant tanks containing propellant management devices," 2016: SP2016 3124654. *Paper Presented at the Space Propulsion 2016 Conference*, Rome, Italy, 2016.

Tam W. H., J. Kuo, and D. E. Jaekle Jr, "Design and manufacture of an ultra-lightweight propellant management device," *AIAA Paper*, vol. 4137, p. 2002, 2002.

Turner I., and M. Ilic, "Convective drying of a consolidated slab of wet porous material including the sorption region," *International Communications in Heat and Mass Transfer*, vol. 17, no. 1, pp. 39–48, 1990.

Udell K. S., "Heat transfer in porous media considering phase change and capillarity—The heat pipe effect," *International Journal of Heat and Mass Transfer*, vol. 28, no. 2, pp. 485–495, 1985.

Van Genuchten M. T., "A closed-form equation for predicting the hydraulic conductivity of unsaturated soils 1," *Soil Science Society of America Journal*, vol. 44, no. 5, pp. 892–898, 1980.

Wang Z.-B., R. Chen, H. Wang, Q. Liao, X. Zhu, and S.-Z. Li, "An overview of smoothed particle hydrodynamics for simulating multiphase flow," *Applied Mathematical Modelling*, vol. 40, no. 23, pp. 9625–9655, 2016.

Whitaker S., "Simultaneous heat, mass, and momentum transfer in porous media: A theory of drying," *Advances in Heat Transfer*, vol. 13, pp. 119–203, 1977, Elsevier.

Widjaja E., and M. T. Harris, "Numerical study of vapor phase-diffusion driven sessile drop evaporation," *Computers & Chemical Engineering*, vol. 32, no. 10, pp. 2169–2178, 2008.

Yanılmaz M., and F. Kalaoğlu, "Investigation of wicking, wetting and drying properties of acrylic knitted fabrics," *Textile Research Journal*, vol. 82, no. 8, pp. 820–831, 2012.

Yeo D. Y., and H. C. No, "Modeling film boiling within chimney-structured porous media and heat pipes," *International Journal of Heat and Mass Transfer*, vol. 124, pp. 576–585, 2018.

Zarandi M. A. F., and K. M. Pillai, "Spontaneous imbibition of liquid in glass fiber wicks, Part II: Validation of a diffuse-front model," *AIChE Journal*, vol. 64, no. 1, pp. 306–315, 2018.

Zarandi M. A. F., K. M. Pillai, and A. S. Kimmel, "Spontaneous imbibition of liquids in glass-fiber wicks. Part I: Usefulness of a sharp-front approach," *AIChE Journal*, vol. 64, no. 1, pp. 294–305, 2018.

Zhang X., H. Li, L. Zheng, Z. Chen, and C. Qin, "Combustion characteristics of porous media burners under various back pressures: An experimental study," *Natural Gas Industry B*, vol. 4, no. 4, pp. 264–269, 2017.

Zhou L., G. Wu, and J. Liu, "Modeling of transient moisture equilibrium in oil-paper insulation," *IEEE Transactions on Dielectrics and Electrical Insulation*, vol. 15, no. 3, pp. 872–878, 2008.

Zou J., and F. Kim, "Diffusion driven layer-by-layer assembly of graphene oxide nanosheets into porous three-dimensional macrostructures," *Nature Communications*, vol. 5, p. 5254, 2014.

7

Thermal Effect on Capillary Imbibition in Porous Media

Jianchao Cai, Wei Wei, and Yasser Mahmoudi

CONTENTS

7.1 Introduction 125
7.2 Basic Imbibition Model 126
7.3 Effect of Temperature on Liquid Viscosity 128
7.4 Effect of Temperature on the Contact Angle 130
7.5 Imbibition in Porous Media with Thermal Effect 132
7.6 Conclusions 134
Nomenclature 134
Acknowledgments 135
References 135

7.1 Introduction

The static and dynamic problems connected with spontaneous capillary imbibition and wicking, wetting, and spreading of wetting liquid have been the subject of intense analytical and experimental studies for more than a century due to their applications in multiple disciplines, including engineering, environmental and applied sciences. Spontaneous imbibition or wicking of wetting liquid is derived by capillary pressure to displace non-wetting liquid saturated in a capillary tube and porous media (Alava et al. 2004; Masoodi and Pillai 2012). This phenomenon is ubiquitous in many applications, for example, water suction into porous matrix from fractures in wetting reservoirs. Measuring and characterizing the infiltration rate of wetting liquid into a capillary tube or porous medium is the general method to understand the capillary imbibition mechanism.

The imbibition process is controlled and influenced by the characterization of porous media, liquid properties, and the interaction between liquid and porous media (Li and Horne 2001; Cai et al. 2010b). These factors include permeability, pore structure and connectivity, matrix size and shape, liquid viscosity and initial saturation, wettability, interface tension, etc. Generally, the capillary imbibition properties are usually studied at room temperature condition. However, some cases are non-isothermal capillary-driven flow, such as the enhancing heavy oil recovery by introducing heat into reservoirs. If the wicking process in porous media is under the non-isothermal condition, liquid viscosity, wettability, and interface tension will be influenced by temperature and will not be constant in general studies. Thus, the available traditional theory on imbibition should be retested. This chapter focuses on the thermal effect on capillary imbibition in porous media based on experimental and theoretical analysis.

7.2 Basic Imbibition Model

The earliest theoretical model for capillary imbibition was proposed by Lucas (1918) and Washburn (1921) based on the Hagen–Poiseuille equation and the assumption of the capillary bundle model, named as the Lucas–Washburn (LW) equation. From the LW equation, the spontaneous imbibition of a wetting liquid in a capillary is expressed as

$$l_s = \sqrt{\frac{r\sigma\cos\theta}{2\mu}t}, \tag{7.1}$$

where l_s is the penetration height, σ is the surface tension, r is the capillary radius, μ is the fluid viscosity, θ is the contact angle, and t is the imbibition time. From the LW equation [Equation (7.1)], the rise height of the wetting liquid $l_s(t)$ versus time t shows a power law relation with exponent 0.5, which is defined as the imbibition time exponent k. When deriving Equation (7.1), the contact angle between the wetting liquid and solid is assumed to be constant, and the gravity factor is neglected during the capillary imbibition process.

If the gravity factor is included, the velocity of liquid capillary rise in a capillary tube is give as (Cai et al. 2010a)

$$\frac{dl_s}{dt} = \frac{r^2}{8\mu l_s}\left(\frac{2\sigma\cos\theta}{r} - \rho g l_s\right), \tag{7.2}$$

where ρ is the liquid density, and g is the gravitational acceleration. Due to the hydrostatic pressure, the capillary-driven process initially rises rapidly and then slowly approaches to the equilibrium height l_e, which is calculated by (Cai et al. 2010a):

$$l_e = \frac{2\sigma\cos\theta}{r\rho g}. \tag{7.3}$$

Integrating Equation (7.2) with the initial condition $l(0) = 0$ yields

$$t = -\frac{16\mu\sigma\cos\theta}{r^3\rho^2 g^3}\left[\ln\left(1 - \frac{l_s}{l_e}\right) + \frac{l_s}{l_e}\right]. \tag{7.4}$$

If the hydrostatic pressure is negligible compared to the capillary pressure, the second term on the right-hand side of Equation (7.2) can be neglected, and Equation (7.1) is obtained. Equations (7.1), (7.2), and (7.4) are validated to straight channel, while if the capillary flow channel is tortuous, these equations are modified as follows (Cai et al. 2010a):

$$l_s = \sqrt{\frac{r\sigma\cos\theta}{2\mu}}\frac{\sqrt{t}}{\tau}, \tag{7.5}$$

$$\frac{dl}{dt} = \frac{\lambda^2}{32\mu l}\left(\frac{4\sigma\cos\theta}{\lambda} - \rho g l_s\right), \tag{7.6}$$

$$t = -\frac{A_h}{B_h^2}\ln\left(1 - \frac{B_h}{A_h}l_s\right) - \frac{l_s}{B_h}, \tag{7.7}$$

where $l_s \equiv l_s(t)$ is the height/elevation of fluid at time t, l is the tortuous length, $\tau = l/l_s$ is the tortuosity, $A_h = \lambda\sigma\cos\theta/(8\mu\tau^2)$, and $B_h = \rho g\lambda^2/(32\mu\tau^2)$.

Thermal Effect on Capillary Imbibition in Porous Media 127

Besides the parameter tortuosity, fractal dimension may be a fundamental and the best parameter to describe the tortuousness of tortuous streamtubes in porous media (Majumdar 1992). If using the scaling correction $l = (2r)^{1-D_T} l_s^{D_T}$ (it indicates that a liquid traveling in a fractal streamtube travels farther than a straight-line distance) (Yu and Cheng 2002), the capillary rise velocity is calculated by

$$\frac{dl_s}{dt} = \frac{(2r)^{2D_T-1}\sigma\cos\theta}{8\mu D_T l_s^{2D_T-1}} - \frac{(2r)^{2D_T}\rho g}{32\mu D_T l_s^{2D_T-2}}, \tag{7.8}$$

where D_T is the tortuosity fractal dimension. D_T lies between 1 and 3 in three-dimensional spaces, representing the extent of convolutedness of streamtubes. For a straight capillary path, $D_T = 1$, a higher value of D_T corresponds to a highly tortuous capillary. If $D_T = 3$, it means a highly tortuous line is so irregular that it fills a three-dimensional space (Wheatcraft and Tyler 1988; Yu et al. 2009).

For short capillary rise time, the second term of the right-hand side of Equation (7.8) is neglected. Integrating this equation under the initial condition $l_s(0) = 0$ yields

$$l_s = \left(\frac{\sigma\cos\theta}{4\mu(2r)^{1-2D_T}}\right)^{\frac{1}{2D_T}} t^{\frac{1}{2D_T}}. \tag{7.9}$$

From Equation (7.9), for the tortuous capillary described by fractal geometry, the height of capillary rise against time follows $l_s \sim t^{1/2D_T}$, only for straight capillary $(D_T = 1)$, we have $l_s \sim t^{1/2}$. Based on the tortuous capillary model and fractal geometry, Cai and Yu (2011) further discussed the effect of tortuosity on the capillary imbibition in porous media. For D_T in the range of 1 to 3, this, $1/6 < k < 1/2$, which indicates the classical LW, is the special case of fractal imbibition time exponent proposed by Cai and Yu (2011).

For imbibition in porous media, Handy (1960) proposed an equation to characterize the process of spontaneous water imbibition into gas-saturated rocks by neglecting gravity and assuming piston-like displacement of the imbibition process and infinite mobility of the gas phase, that is,

$$V = A\sqrt{\frac{P_c K\phi S_{wf}}{\mu}}\sqrt{t}, \tag{7.10}$$

where V is the volume of wetting fluid imbibed, A is the cross-section area of the core, ϕ is the porosity, μ is the viscosity, and t is the time; K, P_c, S_{wf} are the effective permeability, the capillary pressure, and the wetting phase saturation, respectively. In Equation (7.10), permeability and capillary pressure cannot be determined separately from a spontaneous water imbibition test (Li and Horne 2001). In addition, how the influence of pore structure and the liquid–solid interaction are not explicitly described in Equation (7.10).

Based on the fractal characters of porous media, Cai et al. (2010b) proposed an analytical expression for characterizing the spontaneous co-current imbibition process of the wetting fluid into gas-saturated porous media. In their model, the imbibition liquid mass is expressed as a function of the fractal dimensions for pores and for tortuous capillaries (D_f, D_T), the minimum and maximum hydraulic diameter of pores $(\lambda_{min}, \lambda_{max})$, porosity (ϕ), liquid viscosity and density, (ρ, μ), interface tension (σ), and contact angle (θ), that is,

$$W = A\phi\rho\left[\frac{D_f(2-D_f)}{(2+D_T-D_f)(D_T+D_f-1)}\frac{\sigma\cos\theta}{4\mu}\frac{\lambda_{max}^{D_T}}{\lambda_{min}^{D_T-1}}\frac{1-\beta^{2+D_T-D_f}}{1-\phi}\right]^{1/2} t^{1/2}, \tag{7.11}$$

where parameter $\beta = \lambda_{min}/\lambda_{max}$. Although many factors influencing imbibition in porous media are well included in Equation (7.11), the temperature factor is not taken into account, which has an important effect on imbibition properties, and also specifically includes liquid viscosity and contact angle.

7.3 Effect of Temperature on Liquid Viscosity

In the oil recovery of naturally fractured reservoirs, the spontaneous imbibition of water from fracture to porous matrix is an important recovery mechanism. The spontaneous imbibition of water is strongly influenced by the viscosity of oil in the pores (Babadagli 1996). The viscosities of liquids at reservoir temperature are necessary to consider when evaluation of fluids flow in an oil reservoir. Generally, the spontaneous imbibition rate decreases with the increase of the oil viscosity. Specially for heavy oil contained in a porous matrix, the imbibition efficiency will be improved if the oil viscosity is reduced. Introducing heat into the reservoir could reduce oil viscosity. Babadagli (1996) studied the temperature effects on the spontaneous imbibition efficiency of porous matrix that contained heavy oil. From Figure 7.1, the oil viscosity reduces sharply with the increase in temperature.

If the oil viscosity at a desired temperature is predicted by an empirical correlation or an analytical model, it will be very useful in the thermal recovery process of heavy oil reservoirs. Considering a linear relationship between $\log_{10} \mu_{OD}$ and $\log_{10} \mu$ at a dissolved gas value, Beggs and Robinson (1975) developed an empirical viscosity correlation for dead and gas-free crude oil as a function of gravity and temperature as:

$$\mu_{OD} = 10^X - 1, \tag{7.12}$$

$$X = 10^{3.0324 - 0.02023\gamma_o} T^{-1.163}, \tag{7.13}$$

where μ_{OD} is the viscosity of gas-free oil at temperature T, and γ_o is the oil gravity (° American Petroleum Institute). The live oil viscosity is calculated by (Beggs and Robinson 1975):

$$\mu = E\mu_{OD}^F, \tag{7.14}$$

$$E = 10.715(R_s + 100)^{-0.515}, \tag{7.15}$$

$$F = 5.44(R_s + 150)^{-0.338}, \tag{7.16}$$

where R_s is the dissolved gas–oil ratio.

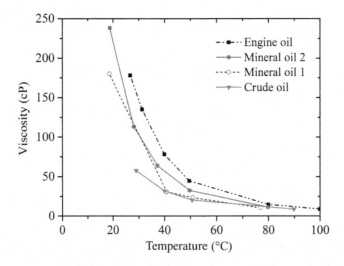

FIGURE 7.1 The change of oil viscosity versus temperature. The viscosity of this oil samples is in the range of 2.19 and 238 cP at room temperature. (From Babadagli, T., *J. Pet. Sci. Eng.*, 14, 197–208, 1996.)

The above empirical expressions are easy to use and give fair precision over a wide range of temperature, oil gravity, and dissolved gas. However, Egbogah and Ng (1990) further examined these correlations and pointed out a significant deviation of viscosity between the prediction of the Beggs and Robinson correlation and the measurement, and modified the Beggs and Robinson correlation by introducing pour-point characterization of oil and multiple-regression analysis as well as new viscosity data. Egbogah and Ng (1990) rewrote Equation (7.12) as:

$$\log_{10}\log_{10}(\mu_{OD}+1) = \chi_1 + \chi_2 T_p + \chi_3 \gamma + (\chi_4 + \chi_5 T_p + \chi_6 \gamma)\log_{10}(T-T_p), \qquad (7.17)$$

where χ_1 to χ_6 are constants, γ is the specific gravity, and T_p is the pour-point temperature that is the lowest temperature at which the oil could flow. A plot of $\log_{10}\log_{10}(\mu_{OD}+1)$ versus $\log_{10}(T-T_p)$ from Equation (7.17) shows a series of straight lines with these slopes against pour-point temperatures. The modified correction (Equation 7.17) also is verified by measurement of about 400 oil systems and shown a substantial improvement over the correction by Beggs and Robinson (1975).

Seeton (2006) reviewed the historical development of viscosity–temperature relations in years 1886–1974 and developed a new viscosity–temperature equation and corresponding chart to extend the range of the current viscosity–temperature charts, which extends the temperature and viscosity range for hydrocarbons and has the ability to extend to a low-viscosity regime of halocarbons and low-temperature fluids. The proposed viscosity–temperature relation is expressed in a generalized form as:

$$\ln(\ln(\mu + 0.7 + e^{-\mu} K_0(\mu + 1.244067))) = H - G \ln T, \qquad (7.18)$$

where H and G are constants, and the zero-order-modified Bessel function of the second kind K_0 is calculated by

$$K_0(x) = \int_0^\infty \frac{\cos(xt)}{\sqrt{t^2+1}}\,dt, \qquad (7.19)$$

Compared to other available correlations, Equation (7.18) can fit the measured viscosity data (Vogel et al. 1998) well over the entire temperature range (from triple point to critical point) as shown in Figure 7.2 (reproduced from Seeton (2006). With a modification to temperature scaling, Equation (7.18) could fit liquid metal viscosity data too. However, it cannot accurately linearize the viscosity with respect to the temperature of liquids with strong molecular bonding, liquids with molecular structure consisting of long coils, and liquid mixtures in which one liquid precipitates out of solution (Seeton 2006).

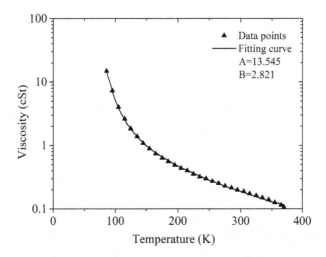

FIGURE 7.2 The fitting of propane viscosity data by Equation (7.18). (From Seeton, C.J., *Tribol. Lett.*, 22, 67–78, 2006.)

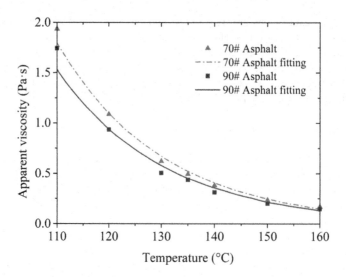

FIGURE 7.3 Change of apparent viscosity of warm-mix asphalt versus temperature.

Zhang et al. (2015) measured and analyzed the dynamic viscosities and the apparent viscosities of two matrix asphalt binders at 60°C and above 110°C, respectively. They developed the apparent viscosity in terms of temperature and additive dosage as:

$$\mu_A = \beta_1(1+\beta_2 C)e^{-\beta_3(1+\beta_4 C)T}, \tag{7.20}$$

where μ_A is the apparent viscosity above 110°C, C is the dosage of Sasobit®, and β_1 to β_4 are the regression parameters. From Figure 7.3 (replotted with data from Zhang et al. (2015)), the apparent viscosity exponentially decreases with the increase of test temperature. Cleary, Equation (7.20) can fit the viscosity data well, and the regression formulas for asphalt binders 70# and 90#, respectively, are expressed as

$$\mu_A = 1070(1+0.0356C)e^{-0.056(1+0.0381C)T}, \tag{7.21}$$

$$\mu_A = 812.1(1+0.0356C)e^{-0.056(1+0.0148C)T}. \tag{7.22}$$

If a fitting constant $m = \beta_3(1+\beta_4 C)$ is defined, Equation (7.20) can be used to evaluate the construction temperature field. The higher m value means the more susceptibility of temperature.

7.4 Effect of Temperature on the Contact Angle

The contact angle is also one complex factor influencing liquid wetting dynamic behavior under different temperatures. Experimental and theories for the liquid wetting process under the effect of temperature have been performed by several researchers. Zhang et al. (1995) measured contact angle on water-swollen cross-linked poly gel and showed unique behavior with the change of temperature. The advancing contact angle of water versus temperature is plotted in Figure 7.4. At the beginning, the contact angle is 42° when the poly gel contains about 90% water at 25°C. With the increasing the temperature, the contact angle first decreases and then increases sharply and reaches to 90°. Rechecking Figure 7.4 in detail, the arrows indicate the temperature change direction.

The temperature dependence of the contact angle is usually studied through experiments. Adamson (1973) presented a more fundamental model to predict contact angle trends with respect to temperature, which is related the molecular surface adsorption to liquid–solid contact angle, that is,

$$\cos\theta = 1+D(T_\infty - T)^{a/(b-a)}, \tag{7.23}$$

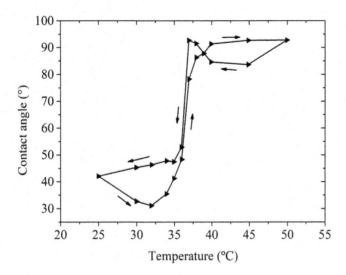

FIGURE 7.4 The change on advancing contact angle of water on cross-link poly gel swollen with water versus temperature. (From Zhang, J. et al., *Langmuir*, 11, 2301–2302, 1995.)

where T_∞ indicates a pseudo-critial temperature, or the temperature at which the contact angle reaches to zero, D is an intergration constant, and a and b are temperature-independent constants from a balance of intermolecular forces.

Bernardin et al. (1997) presented an experimental investigation of the temperature dependence of advancing contact angle of water on an aluminum surface, and contact angles were measured with the sessile drop technique for surface temperatures ranging from 25°C to 170°C and pressures from 101.3kPa to 827.4 kPa. Two distinct temperature-dependent regimes were observed. In the lower temperature regime (<120°C), a relatively constant contact angle of 90° was observed, and the contact angle decreased in a fairly linear manner in the high-temperature regime (>120°C) as shown in Figure 7.5. They also developed an empirical correlation to describe the temperature dependence behavior of advancing contact angle, that is,

$$\begin{aligned} \theta &= 90° & T \leq 120°C \\ \theta &= 157.4 - 0.55T & T > 120°C \end{aligned} \quad (7.24)$$

It is also suggested that the temperature dependance of the contact angle presented by Equation (7.24) is valid for other surface conditions.

de Ruijter et al. (1998) studied experimentally the effect of temperature on the change of contact angle of partially wetting drops for squalene on poly. They further analyzed the problem using a molecular kinetic model, a hydrodynamic model, and a combined model by molecular kinetic model and hydrodynamic model. The equilibrium contact angle resulted by the above models are shown in Figure 7.6 (data are from de Ruijter et al. (1998)). Similar to the phenomenon seen in liquid–solid system, the equilibrium contact angle linearly decreases with the increase of temperature.

For the development of a high-temperature contact angle measurement in recent 20 years, Eustathopoulos et al. (2005) summarized and reviewed the present state of knowledge and future perspectives in which experimental results are obtained by different versions of the sessile drop method and by various procedures. In the studies of wicking in porous media and spreading on a surface at micro- and nanoscale, high-resolution techniques for characterizing the topological features of surfaces are helpful. In addition, an automatic system for data acquisition and image analysis for simultaneously measuring contact angle and surface tension is also desirable.

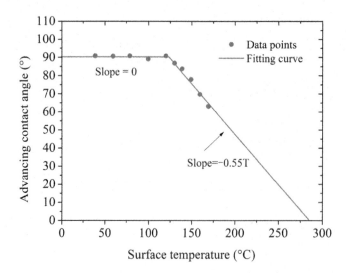

FIGURE 7.5 The linear fitting of advancing contact angle versus temperature on a polished aluminum surface. (From Bernardin, J.D. et al., *Int. J. Heat Mass Transfer*, 40, 1017–1033, 1997.)

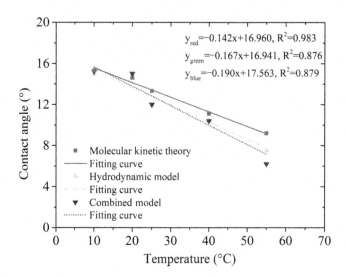

FIGURE 7.6 Variation of the equilibrium contact angle versus temperature.

7.5 Imbibition in Porous Media with Thermal Effect

For oil recovery from fractured carbonate reservoirs, reservoir temperature plays an important role. Phase behavior of the oil–water mixture, wettability, interface tension, oil viscosity, and imbibition rate are influenced by reservoir temperature. Capillary imbibition in fractured oil reservoirs with temperature effect has been reported in literature for water-wet (Babadagli 1996), oil-wet (Gupta and Mohanty 2010), and mixed-wet (Strand et al. 2008) reservoirs, which are briefly analyzed as follows:

1. Imbibition in water-wet reservoirs: Babadagli (1996) investigated temperature effects on capillary imbibition mechanism in fractured water-wet oil reservoirs. Three-dimension spontaneous imbibition experiments under static conditions and at different temperatures (between 20°C and 90°C) are conducted using Berea sandstone samples (cylindrical plugs of 2.5 cm in

diameter and 7.5 cm in length) taken from the same block. Permeability of sample to brine is determined to be 400 mD and porosity ranges from 19% to 21%. Rock samples are saturated with oil without water prewetting. In each set of tests, different types of liquid pairs representing a wide range of oil/water viscosity ratios and interface tensions were used. A temperature increase resulted in a decrease in oil viscosity (see Figure 7.1) and interface tension between oil and brine. The reduction in oil viscosity significantly increased the capillary imbibition rate. The validity of the scaling law is also studied with the temperature effect to be involved.

2. Imbibition in oil-wet reservoirs: Surfactant treatments are usually used to enhance oil recovery from oil-wet carbonate reservoirs by water flooding. Gupta and Mohanty (2010) investigated the effect of temperature in the range of (25°C–90°C) on surfactant imbibition in oil-wet fractured carbonates and on surfactant stability as well as phase behavior of the surfactant–oil–water mixture. The final contact angle decreases with an increase of temperature for all the surfactants. An increase of temperature leads to reduction of viscosity and contact angle, which increases oil relative permeability and accordingly enhances the oil-recovery rate. The interface tension and wettability are modeled as functions of local surfactant concentration based on their tested data. It is found that the oil recovery rate increases with the increase of temperature. For the recoverable oil, it is found that it almost has 80% recovery at 90°C in 50 days compared to 40% recovery at 50°C. The primary reason of this improvement is the reductions of oil viscosity and contact angle.

3. Imbibition in mixed-wet reservoirs: The high temperature also plays an important role for chemical reactions taking place at the chalk surface and for active ion diffusion in mixed-wet chalk matrix (Strand et al. 2008). The rock–liquid interactions and symbiotic interactions (wettability modification) in seawater can improve the water wetness of chalk at high temperature. Hence, the oil recovery is increased by the capillary imbibition of seawater. Strand et al. (2008) studied enhanced oil recovery using seawater both spontaneous imbibition and viscous flooding of formation water and seawater at different temperatures in the range of 90°C–120°C. In their study for 110°C and 120°C, an increased oil recovery by spontaneous imbibition of seawater was observed with the increase of temperature. The wettability alteration by seawater was found to be slow at low temperature (<100°C), and enhanced oil recovery from a naturally fractured oil-wet chalk reservoir is difficult.

4. Imbibition in cellulosic materials: Spontaneous imbibition in cellulosic materials (such as paper) is also a very active research field. However, the thermal field during an isothermal imbibition process is always ignored. Terzis et al. (2017) focused on the heat release property for the wetting front of capillary rising in cellulosic materials, and experimentally found that temperature at the wetting front is temporarily increased. Imbibition experimenters with several liquids and two types of filter paper demonstrate a significant temperature rise at the wetting front. The temperature rise is related to the energetics of imbibition compounds. The imbibition penetration height with time follows Washburn-like behavior, where the imbibition time exponent is not 0.5 (Terzis et al. 2017).

Songok et al. (2014) investigated the temperature effect on water imbibition/absorption into paper. They found that the increase in water imbibition rate with temperature is controlled by the molecular processes occurring in front of the advancing liquid front. A non-linear relationship between the imbibition length and $t^{1/2}$ is found, and the classical LW equation is inadequate to characterize water imbibition into paper both at short times and as a function of temperature. The LW equation underpredicts the liquid imbibition rate at short times. To account for this inconsistency and temperature dependence on imbibition, Songok et al. (2014) considered dynamic and temperature-dependent factors affecting the advancing liquid front at pore level and used a linearized molecular equation to fit the experiment data of water imbibition into groundwood containing base paper (Popescu et al. 2008; Martic et al. 2002):

$$\frac{dl_s}{dt} = \frac{\sigma K^0 \lambda^3}{k_B T} (\cos\theta - \cos\theta_d), \tag{7.25}$$

where k_B is the Boltzmann constant, K^0 is the molecular displacement frequency, λ is the average molecular displacement length, θ is the static contact angle, and θ_d is the dynamic contact angle. Compared to the LW equation, a clear improvement of fitting [Equation (7.25)] to experimental data shows that retardation of water imbibition as short times is due to the dynamic contact angle effect.

7.6 Conclusions

In this chapter, the thermal effect on spontaneous imbibition of a wetting liquid in porous media is analyzed by theoretical, numerical, and experimental methods. First, for imbibition with no thermal effect, the basic equation for capillary rise in single capillary and spontaneous imbibition in porous media is introduced, and the influence factors on imbibition are given from the fractal theory. Second, the temperature effect on liquid viscosity and contact angle is discussed from experimental result and empirical corrections. At last, the imbibition properties in water-wet, oil-wet, and mixed-wet reservoirs and in cellulosic materials are presented. This chapter helps to understand the complex imbibition mechanism in porous media with the thermal effect.

NOMENCLATURE

C	dosage of Sasobit
D_f	pore fractal dimension
D_T	tortuosity fractal dimension
g	gravitational acceleration
k	imbibition time exponent
K	effective permeability
K_0	the zero-order-modified Bessel function
K^0	molecular displacement frequency,
k_B	is the Boltzmann constant
l	tortuous length
l_e	equilibrium height
l_s	penetration straight height
M	imbibition liquid mass
P_c	capillary pressure
r	capillary radius
R_s	dissolved gas–oil ratio
S_{wf}	wetting-phase saturation
t	imbibition time
T	temperature
T_p	pour-point temperature
T_∞	pseudo-critical temperature
V	volume of wetting fluid imbibed

Greek Letters

β	parameter, $\left(\beta = \lambda_{min} / \lambda_{max} \right)$
β_1 to β_4	regression parameter used in Equation (7.20)
χ_1 to χ_6	constant used in Equation (7.17)
ϕ	porosity
γ	specific gravity
γ_o	oil gravity (°API)
λ	average molecular displacement length

λ_{max}	maximum hydraulic diameter
λ_{min}	minimum hydraulic diameter
μ	fluid viscosity
μ_A	apparent viscosity
μ_{OD}	viscosity of gas-free oil at temperature T
θ	contact angle
θ_d	dynamic contact angle
ρ	liquid density
σ	surface tension
τ	tortuosity

ACKNOWLEDGMENTS

This project was supported by the National Natural Science Foundation of China (Nos. 41722403, 41572116) and the Hubei Provincial Natural Science Foundation of China (No. 2018CFA051).

REFERENCES

Adamson A. W., Potential distortion model for contact angle and spreading. Ii. Temperature dependent effects, *J. Colloid Interface Sci.* 44 (1973) 273–281.

Alava M., M. Dubé, M. Rost, Imbibition in disordered media, *Adv. Phys.* 53 (2004) 83–175.

Babadagli T., Temperature effect on heavy-oil recovery by imbibition in fractured reservoirs, *J. Pet. Sci. Eng.* 14 (1996) 197–208.

Beggs H. D., J. R. Robinson, Estimating the viscosity of crude oil systems, *J. Pet. Technol.* 27 (1975) 1140–1141.

Bernardin J. D., I. Mudawar, C. B. Walsh, et al., Contact angle temperature dependence for water droplets on practical aluminum surfaces, *Int. J. Heat Mass Transfer* 40 (1997) 1017–1033.

Cai J. C., B. M. Yu, A discussion of the effect of tortuosity on the capillary imbibition in porous media, *Transp. Porous Media* 89 (2011) 251–263.

Cai J. C., B. M. Yu, M. F. Mei, et al., Capillary rise in a single tortuous capillary, *Chin. Phys. Lett.* 27 (2010a) 054701.

Cai J. C., B. M. Yu, M. Q. Zou, et al., Fractal characterization of spontaneous co-current imbibition in porous media, *Energy Fuel.* 24 (2010b) 1860–1867.

de Ruijter M., P. Kölsch, M. Voué, et al., Effect of temperature on the dynamic contact angle, *Colloids Surf. A* 144 (1998) 235–243.

Egbogah E. O., J. T. Ng, An improved temperature-viscosity correlation for crude oil systems, *J. Pet. Sci. Eng.* 4 (1990) 197–200.

Eustathopoulos N., N. Sobczak, A. Passerone, et al., Measurement of contact angle and work of adhesion at high temperature, *J. Mater. Sci.* 40 (2005) 2271–2280.

Gupta R., K. Mohanty, Temperature effects on surfactant-aided imbibition into fractured carbonates, *SPE J.* 15 (2010) 588–597.

Handy L. L., Determination of effective capillary pressures for porous media from imbibition data, *Pet. Trans. AIME* 219 (1960) 75–80.

Li K. W., R. N. Horne, Characterization of spontaneous water imbibition into gas-saturated rocks, *SPE J.* 6 (2001) 375–384.

Lucas R., Rate of capillary ascension of liquids, *Kolloid-Zeitschrift* 23 (1918) 15–22.

Majumdar A., Role of fractal geometry in the study of thermal phenomena, *Annu. Rev. Heat Transfer* 4 (1992) 51–110.

Martic G., F. Gentner, D. Seveno, et al., A molecular dynamics simulation of capillary imbibition, *Langmuir* 18 (2002) 7971–7976.

Masoodi R., K. M. Pillai, *Wicking in Porous Materials: Traditional and Modern Modeling Approaches.* 1st ed. 2012: CRC Press, Boca Raton, FL.

Popescu M. N., J. Ralston, R. Sedev, Capillary rise with velocity-dependent dynamic contact angle, *Langmuir* 24 (2008) 12710–12716.

Seeton C. J., Viscosity–temperature correlation for liquids, *Tribol. Lett.* 22 (2006) 67–78.

Songok J., P. Salminen, M. Toivakka, Temperature effects on dynamic water absorption into paper, *J. Colloid Interface Sci.* 418 (2014) 373–377.

Strand S., T. Puntervold, T. Austad, Effect of temperature on enhanced oil recovery from mixed-wet chalk cores by spontaneous imbibition and forced displacement using seawater, *Energy Fuel.* 22 (2008) 3222–3225.

Terzis A., E. Roumeli, K. Weishaupt, et al., Heat release at the wetting front during capillary filling of cellulosic micro-substrates, *J. Colloid Interface Sci.* 504 (2017) 751–757.

Vogel E., C. Kuechenmeister, E. Bich, et al., Reference correlation of the viscosity of propane, *J. Phys. Chem. Ref. Data* 27 (1998) 947–970.

Washburn E. W., Dynamics of capillary flow, *Phys. Rev.* 17 (1921) 273–283.

Wheatcraft S. W., S. W. Tyler, An explanation of scale-dependent dispersivity in heterogeneous aquifers using concepts of fractal geometry, *Water Resour. Res.* 24 (1988) 566–578.

Yu B. M., P. Cheng, A fractal permeability model for bi-dispersed porous media, *Int. J. Heat Mass Transfer* 45 (2002) 2983–2993.

Yu B. M., J. C. Cai, M. Q. Zou, On the physical properties of apparent two-phase fractal porous media, *Vadose Zone J.* 8 (2009) 177–186.

Zhang J., R. Pelton, Y. Deng, Temperature-dependent contact angles of water on poly(n-isopropylacrylamide) gels, *Langmuir* 11 (1995) 2301–2302.

Zhang J., F. Yang, J. Pei, et al., Viscosity-temperature characteristics of warm mix asphalt binder with Sasobit®, *Constr. Build. Mater.* 78 (2015) 34–39.

8

Convection in Bi-Disperse Porous Media

Arunn Narasimhan

CONTENTS

8.1 Introduction	137
8.2 Bi-Disperse Porous Medium	137
8.3 Forced Convection in BDPM	138
8.3.1 Laminar Flow through a BDPM Channel	139
8.3.2 Conjugate and Mixed-Forced Convection in BDPM	142
8.3.3 Studies Related to Interfaces in BDPM	143
8.4 Natural Convection in BDPM	144
8.4.1 Natural Convection in BDPM Bounded by One or Two Walls	144
8.4.2 Natural Convection inside a BDPM Enclosure	145
8.4.3 Further Studies on Natural Convection inside BDPM Enclosures	146
8.5 Convection with Change of Phase and Mass Transfer in BDPM	147
8.6 Forced Convection in Tri-Disperse Porous Media	148
8.7 Natural Convection in Tri-Disperse Porous Media	148
8.8 Conclusion	149
References	149

8.1 Introduction

We begin with a definition of the bi-disperse porous medium (BDPM), which is also discussed in a subsection in (Narasimhan 2012). Excellent introduction and thorough documentation of research literature related to convection in BDPM are available in (Nield and Bejan 2013). To avoid repetition, this chapter collects and discusses research literature that was published mostly after 2010. The few exceptions cited from previous years are only to provide the expository context. However, we include research summaries on convection in tri-disperse porous media (TDPM) not documented elsewhere, yet.

8.2 Bi-Disperse Porous Medium

A BDPM is defined by Chen et al. (2000a) as clusters of large particles that are themselves agglomerates of smaller particles. Figure 8.1 shows a schematic of channel flow through such a BDPM. This was claimed as a form of porous medium by (Nield and Bejan 2013) in which the solid constituent of the parent porous medium is porous in itself, with interconnected pores. The pores of the two constituent porous media of a BDPM differ in length scale by at least an order of magnitude.

It is customary to denote the macropores as the f-phase (fissure) and the rest of the structure as the p-phase (pore) in a BDPM. The fluid is common to both pores and is present in the entire f-phase and some of the p-phase. A different approach to BDPM is that it is a porous medium with fractures, resulting in the f-phase being the "fracture phase" and the p-phase the "porous phase."

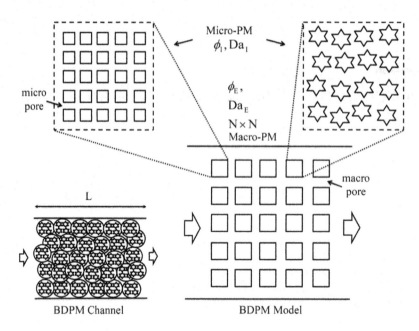

FIGURE 8.1 Schematic of a bi-disperse porous medium channel flow.

The resolution levels of porosity can be extended to tri- and higher orders of dispersion, with progressively finer porous structures, eventually allowing visualization of porous media geometry as fractals (Yu and Li 2001).

Some natural examples of BDPM are cooked rice and cauliflower. In the first case, water soaks into the raw rice when being boiled, making it a porous medium with pore scales much smaller than the pores that are formed between the grains of rice due to the passage of water vapor through the bulk. The porous flowers of the cauliflower are in turn made of finer florets. The adsorbents and capillary wicks of heat pipes were treated by (Kuznetsov and Nield 2013) as BDPM. A coal pile is a BDPM that is formed by piling small coal particles that are themselves porous. The row of plants in a field or garden and an array of servers in data centers could be analyzed as BPDM to understand dispersal of seeds in the former and for designing thermal management solutions for the latter. Aquifers, absorbent pellets, catalysts, proton exchange membrane (PEM) fuel cells, and mircoreactors can all be modeled as BDPM (Narasimhan and Reddy 2011b).

The higher area-to-volume ratio in bi- and higher-disperse porous media results in interesting mass flow and thermal properties. Perhaps the earliest reported study to examine mass flow through BDPM was by Heeter and Liapis (1997), even before the formulation of the term BDPM. They developed a quantitative method to estimate the pore diameter that would allow intraparticulate fluid flow in the microporous region of bi-disperse porous structure of particles packed in a column for chromatographic applications. The value of pore diameter required to obtain specified values of the velocity of supercritical fluid flowing through the column and specified values of the intraparticle Peclet number were obtained by this method.

8.3 Forced Convection in BDPM

Pioneering work by Nield and Kuznetsov (2004, 2005) modeled the heat transfer and fluid flow in BDPM. Assuming local thermal equilibrium (LTE) within the microporous media, a volume-averaged two-velocity two-temperature model was proposed to simulate steady-state fluid flow and heat transfer in BDPM. They extended the Brinkman model for a monodisperse porous medium by coupling a pair of equations for the dimensionless variables v_f and v_p.

Convection in Bi-Disperse Porous Media

$$G = \left(\frac{\mu}{K_f}\right)v_f^* + \varsigma\left(v_f^* - v_p^*\right) - \tilde{\mu}_f \nabla^2 v_f^* \tag{8.1}$$

$$G = \left(\frac{\mu}{K_p}\right)v_p^* + \varsigma\left(v_p^* - v_f^*\right) - \tilde{\mu}_p \nabla^2 v_p^* \tag{8.2}$$

In the above equations, tt is the negative of the applied pressure gradient, μ is the fluid viscosity, K_f and K_p are the permeabilities of the two phases, and ζ is the coefficient for momentum transfer between the two phases. The quantities $\tilde{\mu}_f$ and $\tilde{\mu}_p$ are the respective effective viscosities. The effect of form (quadratic, Forchheimer) drag was neglected, and the hydrodynamic interaction between the two phases was modeled by a simple expression. This model has since been used in many papers on forced, natural, and mixed convection. A detailed discussion on the works of Nield and Kuznetsov (2004, 2005) in the area of BDPM has been presented by Straughan (2009).

8.3.1 Laminar Flow through a BDPM Channel

Porous medium channel flows have found several engineering applications. Extending this configuration to a BDPM case, Narasimhan and Reddy (2011a) studied forced convection cooling of heat-generating porous blocks arranged inside a channel separated mutually by fissures to form the f-phase of a BDPM as described earlier and depicted in Figure 8.1. The influence of the thermophysical and hydraulic properties of the heat-generating porous blocks forming the p-phase of the BDPM on the overall forced convection effect is investigated using numerical simulations.

In a subsequent work, Narasimhan et al. (2012) modeled heat-generating electronics as BDPM and studied forced convection cooling as a function of micro and macroscale porosities and permeabilities of the medium. The BDPM channel was considered as a two-dimensional porous block array, and in turn made of arrays of regularly arranged square electronic components distributed uniformly. Fluid flow through the micro and macropores of the BDPM channel was visualized as removing the heat generated by the electronics components.

The convecting fluid and the porous matrix were assumed to be in LTE, and 2D versions of the mass, momentum, and energy conservation equations were used to model fluid flow and heat transfer. The governing partial differential equations were represented in non-dimensional form as follows.

$$x^* = \frac{x}{H}, \, y^* = \frac{y}{H}, \, u^* = \frac{u}{u_{in}}, \, v^* = \frac{v}{u_{in}}, \, \theta = \frac{T - T_{in}}{Q_s''' H^2 / k_f},$$

$$p^* = \frac{\left(p - p_{ref}\right)}{\rho u_{in}^2}, \, Pr = \frac{\nu}{\alpha}, \, Da_I = \frac{K_I}{H^2}, \, Re_H = \frac{\rho u_{in} H}{\mu}, \, \gamma = \frac{k_{eff}}{k_f} \tag{8.3}$$

in which, p_{ref} and T_{in} were the reference pressure and temperature, respectively. Here they were taken as the atmosphere pressure and inlet temperature, respectively.

The non-dimensional governing equations for mass, momentum in the x^*, and y^* directions and heat transport in porous media were derived using the above terms as:

$$\frac{\partial u^*}{\partial x^*} + \frac{\partial v^*}{\partial y^*} = 0 \tag{8.4}$$

$$u^*\frac{\partial u^*}{\partial x^*} + v^*\frac{\partial u^*}{\partial y^*} = -\phi_I^2 \frac{\partial p^*}{\partial x^*} + \frac{\phi_I}{Re_H}\left(\frac{\partial^2 u^*}{\partial x^{*2}} + \frac{\partial^2 u^*}{\partial y^{*2}}\right)$$

$$- \left[\frac{\phi_I^2}{Re_H Da_I} + \frac{c_F \phi_I^{1/2}}{\sqrt{Da_I}}\sqrt{v^{*2} + u^{*2}}\right]u^*, \tag{8.5}$$

$$u^* \frac{\partial v^*}{\partial x^*} + v^* \frac{\partial v^*}{\partial y^*} = -\phi_I^2 \frac{\partial p^*}{\partial y^*} + \frac{\phi I}{Re_H}\left(\frac{\partial^2 v^*}{\partial x^{*2}} + \frac{\partial^2 v^*}{\partial y^{*2}}\right) \quad (8.6)$$

$$-\left[\frac{\phi_I^2}{Re_H Da_I} + \frac{c_F \phi_I^{1/2}}{\sqrt{Da_I}}\sqrt{v^{*2}+u^{*2}}\right]v^*,$$

$$u^* \frac{\partial \theta}{\partial x^*} + v^* \frac{\partial \theta}{\partial y^*} = \frac{\gamma}{Re_H Pr}\left(\frac{\partial^2 \theta}{\partial x^{*2}} + \frac{\partial^2 \theta}{\partial y^{*2}}\right) + \frac{(1-\phi_I)}{Re_H Pr}, \quad (8.7)$$

where γ (k_{eff}/k_f) is the ratio of effective thermal conductivity of the porous medium to the thermal conductivity of a flowing fluid.

Figure 8.2 shows sample steady-state results from the solution of the above governing equations with associated boundary conditions in a BDPM channel as shown in Figure 8.1, with arrays of heat-generating porous blocks cooled by a convecting fluid.

Comparing the top and bottom streamlines in Figure 8.2 one can see the flow through the porous blocks of the bottom and relative lack of it at the top, owing to the internal permeability Da_I variation. This effect is the pertinent bi-dispersion feature that causes the BDPM to behave very different or similar to that of a monodisperse porous medium flow, provided that even at the limit of $Da_I \to 0$, the f-phase can be treated as a porous medium.

The width and length ($W^* = L^*$) for the BDPM channel are shown in Figure 8.2. The individual block size D^* and macropore width δ could be determined once the macropore volume fraction and the number of porous blocks were known. D^* and δ were determined using the following equations:

$$\phi_E = 1 - N^2 D^{*2}, \quad (8.8)$$

and

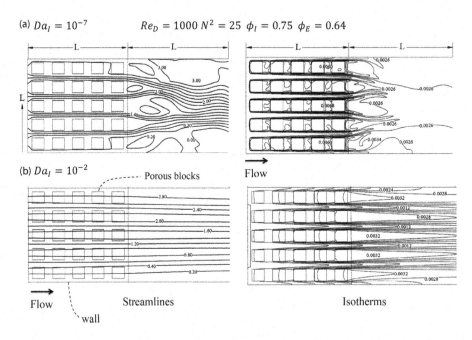

FIGURE 8.2 Streamlines and isotherms for forced convection inside a BDPM channel configuration with 5 × 5 porous block array, (a) $Da_I = 10^{-7}$ and (b) $Da_I = 10^{-2}$.

Convection in Bi-Disperse Porous Media 141

$$\delta = \frac{1 - D^* N}{N},$$

(8.9)

The results from the solution of the conservation equations were analyzed using the non-dimensional variables, pressure drop Δp^*, and Nusselt number Nu, defined as:

$$\Delta p^* = \frac{1}{W^*} \left[\int_0^{W^*} p^* dy^* \Big|_{x^* = L^*} - \int_0^{W^*} p^* dy^* \Big|_{x^* = 2L^*} \right],$$

(8.10)

and

$$Nu = \frac{\sum_{i=1}^{N^2} \int_0^{D^*} \int_0^{D^*} Nu_{x^* y^*} dx^* dy^*}{\left(N^2 D^{*2} \right)},$$

(8.11)

where,

$$Nu_{x^* y^*} = \frac{Q_s''' H^2}{\left(T_{x,y} - T_{in} \right) k_f} = \frac{1}{\theta_{x^* y^*}}.$$

(8.12)

$$\chi = \frac{Nu}{\Delta p^* Re^3},$$

(8.13)

The Reynolds number Re was based on the hydraulic diameter of the channel ($Re = Re_H \times (D_h/H)$). For given values of fluid ($Pr = 0.7$), BDPM channel volume ($W^* \times L^* \times 1$) and temperature difference, the dissipated heat was proportional to Nu, which in turn was dependent on the area ($W^* \times L^*$) of the BDPM channel. The pumping power was found to be proportional to $\Delta p^* Re^3$.

Increasing internal permeability in a BDPM was found to cause a reduction in the pressure drop across the BDPM channel and an increase in forced convection. However, the effects of internal permeability on in Nu and Δp^* were not linear. When the internal permeability was in the order of 10^{-7}, convection was restricted mostly to the macropores around the microporous blocks, and the porous medium was mono-disperse. On increasing permeability, bi-dispersion effects appeared. Permeation of fluids into the micro-pores due to low viscous drag resulted in uniform distribution of streamlines as convection increased, resulting in a fall in the local maximum temperature. The introduction of bi-dispersion into a porous medium resulted in nearly fourfold enhancement of heat transfer and an eightfold decrease in pressure drop. In terms of the application envisaged, viz., electronics cooling, judicious spacial distribution of heat-generating electronics that form the microporous blocks by choosing appropriate permeability can result in good thermohydraulic designs.

Figure 8.3 shows the results of modeling-forced convection inside a BDPM channel for the thermal management of electronics like, say, arrays of computers in a typical data center. The individual blocks in the arrays can be viewed as local computer clusters with fixed heat generation, distributed either as arrays of small local clusters or as a single bulk cluster at the center of the cooling configuration. The forced convection results are, as expected, heavily influenced by the local distribution characterized by the internal permeability, Da_I, variation. The external permeability, Da_E, and porosity, Φ_E, control the influence of different clustering arrangements as highlighted by the flows and temperature distribution in Figure 8.3.

Bi-dispersion effects may also be induced by varying the macropore volume fraction and number of blocks in the BDPM channel. For a fixed internal permeability, $Da_I = 10^{-3}$, as macropore volume fraction increases, there is a corresponding increase in the effective permeability of the BDPM channel. This reduces the macropore-related viscous drag, and therefore, for a fixed $Re = 1000$, there is a reduction in Δp^* with increase in volume fraction. When the volume fraction falls below 0.2, a limit is reached

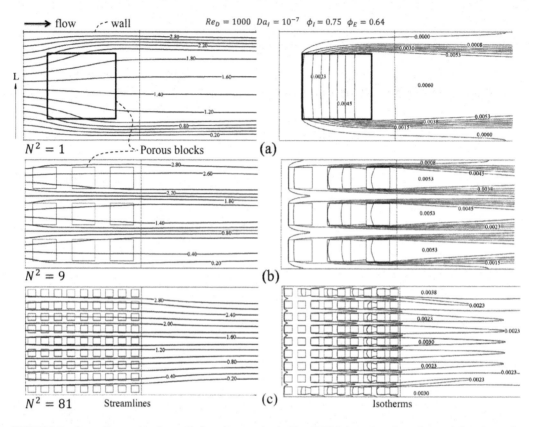

FIGURE 8.3 Streamlines and isotherms for forced convection inside a BDPM channel configuration modeling electronics blocking, (a) $N^2 = 1$, (b) $N^2 = 9$ and (c) $N^2 = 81$.

when the macropore channel size approaches the micropore size. When macropore and micropore become equal in length, the BDPM becomes monodisperse. It is shown by Narasimhan et al. (2012) that introduction of bi-dispersion by increasing the volume fraction provides larger heat-transfer enhancement (two times) and large reduction (eight times) in pressure drop compared to a monodisperse porous medium channel.

Translating the above research to practical applications, modeling heat-generating electronics such as data centers as BDPM can help in arriving at important design decisions for heat management without tedious multi-scale computational fluid dynamics (CFD) computations. For a fixed number of heat-generating electronics components (heat generation fixed), clustering in bulk or in progressively smaller local clusters of large arrays (depicted by the increase in N^2) provides a method for optimization using the First Law efficiency of the BDPM configuration. There is an operative window in terms of N^2 as shown in Narasimhan et al. (2012) for deriving maximum First-Law efficiency (heat transfer versus corresponding pump power) for such BDPM configurations. The geometry and heat distribution of such a bi-disperse system should be suitably tuned by changing the bi-dispersion parameters, such as the microporous medium φ_I and Da_I or the macroporous medium Da_E by varying φ_E and N^2.

8.3.2 Conjugate and Mixed-Forced Convection in BDPM

The conjugate problem of forced convection in a plane channel filled with saturated BDPM, coupled with conduction in plane slabs bounding the channel was analyzed by Nield and Kuznetsov (2004). The local Nusselt number was found to depend strongly on the BDPM thermal conductivity ratio but only moderately dependent on the velocity ratio in the two bi-disperse pore scales. The critical Rayleigh number was found to be a function of the BDPM interface momentum transfer, permeability ratio, thermal

Convection in Bi-Disperse Porous Media

conductivity and capacity ratios, volume fractions, and interface heat transfer coefficients. An increase in value of the Peclet number was found to reduce the rate of exponential decay in the downstream direction.

Lattice Boltzmann simulations were performed by Hoef et al. (2005) for low Reynolds number flow past mono and BDPM to measure the permeability and drag force. New drag force relations were developed using 58 parameter sets for both monodisperse and polydisperse systems, based on the Carman–Kozeny equations. For binary systems with large diameter ratios (1:4), the individual drag force on a particle was found to differ by up to a factor of 5 compared to other prediction models.

The above two-temperature and two-velocity model of BDPM was used by Kumari and Pop (2009) to study the mixed convection boundary layer flow past a horizontal circular cylinder embedded in a BDPM. Expressions for flow and heat transfer characteristics were derived in terms of an interphase momentum parameter, thermal conductivity ratio, thermal diffusivity ratio, permeability ratio, modified thermal capacity ratio, and buoyancy parameter. The partial differential equations governing the flow and heat transfer in the f-phase (the macropores) and the p-phase (the remainder of the structure) were solved using the Keller-box method.

The two-velocity two-temperature formulation and Darcy's law was used by Kuznetsov and Nield (2010) to analyze forced convection in a parallel-plate channel partly occupied by a BDPM and partly by a fluid in an asymmetrical distribution when the walls are subject to an uniform heat flux. The relationships between the Nusselt number on conductivity ratio, velocity ratio, volume fraction, internal heat exchange parameter, and the position of the porous-fluid interface were investigated. In asymmetric heating, the Nusselt number was found to become singular at a certain value of asymmetric heating parameter that depends on the dimensionless interfacial position coordinate. This does not point to infinite heat transfer but is a result of the characteristic temperature difference in the equation defining the Nusselt number taking a zero value. Redefining the Nusselt number by changing the temperature difference can eliminate this singularity.

In Ajay Kumar et al. (2013), forced convection cooling in electronic chip was analyzed with and without porous media. The conventional heat sink model was compared with BDPM heat sinks and forced convection without heat sink. The cooling of BDPM heat sink was found to be 67% better than the other two cases. It was shown that the use of BDPM heat sinks not only improves thermal management in electronic devices, but also results in weight savings because BDPM has inherently more voids than solid heat sinks.

More recently, forced convective heat transfer of a circular pipe filled with a BDPM was studied by Wang et al. (2017). As expected, the temperature distribution changes significantly with bi-dispersion intensity. There is a critical value of the effective thermal conductivity ratio below which the temperature in the p-phase is higher than that in the f-phase; beyond this critical value, bi-dispersivity results in the reversal of this trend. The authors have also discussed the effects of thermophysical parameters on the Nusselt number. The BDPM Nusselt number was found to fall to zero beyond the critical conductivity ratio. This is in contrast to the behavior of monodisperse porous media.

In another recent study, Wang and Li (2018) analyzed forced convective heat transfer in a parallel-plate channel filled with BDPM after incorporating viscous dissipation, assuming a constant wall heat flux condition. The flow was assumed to be hydrodynamically and thermally fully developed and is described by a two-velocity one-temperature formulation. The analysis shows that bi-dispersivity plays a significant role in balancing the heat flux at the wall and the heat generated by viscous dissipation. For both cases with and without viscous dissipation, the Nusselt numbers were found to become independent of bi-dispersivity with an increase in the Darcy number ratio.

8.3.3 Studies Related to Interfaces in BDPM

In an interesting work, Hooman et al. (2015) obtained the momentum transfer coefficient between the f and p phases of a BDPM and mapped the experimental data to the theoretical model proposed by Nield and Kuznetsov (2005). The BDPM was of plate-fin heat exchangers. The configuration is similar to that experimentally investigated in Narasimhan et al. (2014); the solid fins were replaced by a simple porous medium. The pressure drops that were experimentally measured for such heat exchangers were used to obtain the overall permeability that was linked to porosity and permeability of each phase and the interfacial momentum transfer between the two phases. Numerical values were obtained for the momentum transfer coefficient for three different fin-spacing values considered in the heat exchanger experiments. Both form drag and the momentum exchange coefficient were found to be homogeneous, constant, and velocity independent over the range of parameters considered.

A new model was proposed in Nield (2016) for flow in BDPM, involving unidirectional flow in a stack of channels with alternating fluid and porous phases, with the Beavers–Joseph boundary condition imposed at the interphase boundaries. A simple analytical expression was derived for the increase of effective permeability caused by the replacement of a monodisperse medium by a bi-disperse one.

The macroscopic properties of two-dimensional random periodic packs of polydisperse cylinders were numerically studied by Matsumura and Jackson (2014). Unsteady, two-dimensional Navier–Stokes equations were solved on a staggered Cartesian grid to study the effects of porosity, polydispersivity, and Reynolds numbers on the macroscopic permeability. For small Reynolds numbers, permeability could be correlated to the underlying microstructure using the mean shortest Delaunay edge. Appropriate scaling resulted in the collapse of the polydisperse cylinders to monodisperse values. A modified Forchheimer equation could characterize flow for large Reynolds numbers.

8.4 Natural Convection in BDPM

Natural convection heat transfer in porous media has been extensively studied over the years for a number of engineering applications, such as in electronics cooling, heat exchangers, and various thermal systems. Its extension to the bi-disperse case is natural.

8.4.1 Natural Convection in BDPM Bounded by One or Two Walls

The first report on natural convection in BDPM was by Rees et al. (2008), who extended the Cheng–Minkowycz study on vertical natural convection in porous medium to the BDPM case. The boundary-layer equations were expressed in terms of modified volume fraction, modified thermal conductivity ratio, and a third parameter incorporating both thermal and BDPM properties. BDPM was found to differ from the classical local thermal non-equilibrium (LTNE) model for a regular porous medium in that LTNE is non-generic that requires an asymptotic analysis that matches inner and outer solutions.

The classical Rayleigh–Benard theory of porous medium was extended by Nield and Kuznetsov (2006) and was used to study the onset of convection to BDPM. Four new dimensionless parameters were introduced—volume fraction, permeability ratio, modified thermal capacity ratio, and interphase momentum transfer parameter—and their effects were analyzed. Linear stability analysis resulted in an expression for the critical Rayleigh number as a function of the Darcy number and the four additional parameters. The classical Cheng–Minkowycs study of convection past a vertical plate embedded in a porous medium was extended by Nield and Kuznetsov (2008) to the BDPM case and obtained expressions for velocity and temperature in terms of a geometrical parameter, an interphase momentum transfer parameter, thermal diffusivity ratio, permeability ratio, thermal conductivity ratio, and interphase heat transfer.

Natural convection from an inclined wavy plate in a BDPM with uniform wall temperature was studied by Cheng (2013b). The governing equations were derived using a two-velocity two-temperature formulation, and the Prandtl coordinate transformation was used to transform the wavy surface into a regular plane. Order-of-magnitude analysis was used to simplify the equations, and the cubic spline collocation method was used to solve the boundary layer governed equations. They found that an increase in the modified thermal conductivity ratio and permeability ratio enhances the natural convection heat transfer of the inclined plate in BDPM. The thermal non-equilibrium effects were found to be significant for low values of the interphase heat transfer parameter. With an increase in the dimensionless amplitude, there was enhancement in fluctuations of the local Nusselt number for the f-phase and the p-phase with the streamwise coordinate. Similarly, an increase in the modified thermal conductivity ratio or the permeability ratio enhanced the free convection of the vertical truncated cone in BDPM (Cheng 2013a).

Mixed convective heat transfer in the vertical plate BDPM case with constant wall temperature was reported to be similar to natural convection (Cheng 2015a). As before, an increase in the mixed convection parameter, the modified thermal conductivity ratio, or the permeability ratio could enhance the mixed convection heat transfer. As in the natural convection study by the same researcher, the thermal non-equilibrium effects between the *f*-phase and the *p*-phase assumed significance when the interphase heat transfer parameter or the mixed convection parameter was small.

8.4.2 Natural Convection inside a BDPM Enclosure

The bi-dispersion effects on the steady-state natural convection inside a BDPM enclosure were investigated by Narasimhan and Reddy (2009) using distributed arrays of porous blocks. The configuration was similar to that described in the case with forced convection (Narasimhan and Reddy 2011a), with all sides bounded by adiabatic walls. Figure 8.4 shows the sample results obtained by solving appropriate governing equations using numerical methods.

Natural convection was induced with a temperature gradient across the sidewalls. The finite volume method was used to obtain the fluid flow and heat transfer in the macropore scale, but in the microscale, the volume-averaged formulation was used considering the LTE. That is, the BDPM was assumed to be detached from the enclosure walls. In this work, bi-dispersion effects in natural convection were studied by varying the internal Darcy number, Da_I, of the microporous blocks and the external Darcy number, Da_E, of the macropores. At the monodisperse limit, at which the control volume forming the microphase porous blocks are solid ($Da_I \to 0$), the convection flow was found to be restricted to the macropores. As the thermal conductivity ratio between the solid and fluid constituent was set as $\gamma = 1$, the wall-to-wall convection heat transfer across the enclosure was strongly influenced by the nature of the fluid flow—whether happening only in the fissure phase or also through the porous phase. The wall-to-wall heat transfer was found to be dominated by the combined convection flow for a BDPM, as seen in the middle pair of streamlines and isotherms in Figure 8.4. There was little convection heat transfer because of the severely restricted flow at $Da_I \to 0$ Increase in Da_I resulted in increased bi-dispersivity that allowed flow through the microporous blocks.

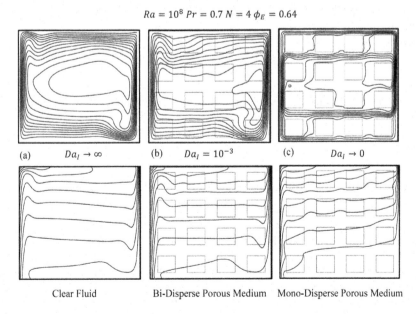

FIGURE 8.4 Streamlines and isotherms for natural convection inside an enclosure with: (a) clear fluid; (b) bi-disperse porous medium (BDPM); and (c) monodisperse porous medium (MDPM). (Adapted from Fig. 2 of Narasimhan, A., and B.V.K. Reddy, *ASME J. Heat Transf.*, 132, 012502, 2009.)

146   *Convective Heat Transfer in Porous Media*

Bi-dispersion effects were seen between the limits of $Da_I \to 0$, which denotes a reduction of the enclosure to a monodisperse porous medium, and $Da_I \to \infty$, where the enclosure is solid. With increase in Da_I, there was more flow through the micropores, resulting in increase of Nu between $Da_I \to 0$ and $Da_I \to \infty$ for all Ra. An increase in Ra reduced the bi-dispersivity effects on Nu.

A correlation was developed in Narasimhan and Reddy (2009) to predict the average Nusselt number, Nu_h, as a function of Rayleigh number, Ra_φ, and external Darcy number, Da_E, defined as

$$Ra_\phi = 0.144 Ra^{1.208} \quad \text{and} \quad Da_E = 1.005 Da_{eq}^{1.208}, \tag{8.14}$$

where

$$Ra = \frac{g\beta(T_h - T_{ref})L^3}{v\alpha}, \tag{8.15}$$

and Da_{eq} is calculated by using

$$Da_E = \frac{KE}{L^2} = \frac{1}{180}\frac{\phi_E^3 D^{*2}}{(1-\phi_E)^2}, \tag{8.16}$$

where Da_E is the non-dimensional representation of the permeability of the solid-block-filled enclosure when treated as a porous medium.

The correlation was written as

$$Nu = 0.577\left(Ra_\varphi Da_E\right)^{0.5} \quad 10 \le Ra_\varphi Da_E \le 200,$$
$$Nu = 0.37 Ra_\phi^{0.25} Da_E^{0.07} \quad 200 < Ra_\varphi Da_E \le 10^5, \tag{8.17}$$

where the Da_E was separately modeled as a function of a number of blocks, N, and the internal Darcy number, Da_I, and was written as follows:

$$Da_E = \frac{4.1 \times 10^{-3}}{N^2} + 0.007 Da_I^{0.32}. \tag{8.18}$$

The above correlation was found to be valid for fixed macropore volume fraction ($\varphi_E = 0.64$) and micropore volume fraction ($\varphi_I = 0.5$), within the parameter range: $3 \le N \le 8$ and $0 \le Da_I \le 10^{-7}$. It predicted the Nu_h data set is generated by numerical simulations with a correlation coefficient of 0.98, within $\pm 15\%$ and $\pm 9\%$ for $10 \le Ra_\varphi Da_E \le 200$ and $200 < Ra_\varphi Da_E \le 10^5$, respectively.

8.4.3 Further Studies on Natural Convection inside BDPM Enclosures

Resonance of natural convection inside such a BDPM enclosure was investigated by Narasimhan and Reddy (2011b), as an extension of a similar problem investigated by Lage and Bejan (1993) for clear fluid and porous-medium enclosures. The BDPM enclosure was made from uniformly spaced, disconnected, square, porous blocks that formed the microporous medium saturated with a fluid of $Pr = 0.7$ and was subjected to time periodic heat flux at a side wall, with the opposite wall kept isothermal while the top and bottom walls were adiabatic.

Natural convection resonance was observed in the BDPM enclosure for $Ra = 10^8$ and 10^7, and the resonance heating frequency f_r was observed to increase with Ra, Da_I, and Da_E. However, f_r of BDPM enclosure was found to be always less than that of the corresponding clear fluid enclosure limit (at $Da_I \to \infty$)—compare the left-most and middle cases of Figure 8.4.

For a fixed N^2 and φ_p, the effect of bi-dispersion, an internal permeability, Da_p, increase, raised the resonance frequency, f_r, from the monodisperse porous medium limit, $Da_l \to 0$. The increase in f_r is due to the additional flow through the porous blocks of the BDPM, enhancing the convection. At the clear fluid limit, $Da_l \to \infty$, the resonance frequency was found to reach a maximum. The effect of bi-dispersion is understood to vary the resonance frequency, f_r, between the monodisperse porous medium and clear fluid enclosure limiting values. Predictions of f_r using a modified scale analysis incorporating BDPM effects were also provided by Narasimhan and Reddy (2011b).

A differentially heated enclosure filled with BDPM was simulated by Imani and Hooman (2017) at a pore scale encompassing both micro and microporous structures. In addition, the BDPM was modeled as being attached to the enclosure walls in addition to detachment. The effects of geometrical and thermophysical parameters on streamlines and isotherms as well as the hot wall average Nusselt number in both attached and detached BDPM were studied. The pore-scale results were then examined using the thermal lattice Boltzmann method to establish the validity of assumption LTE within the microporous media of a BDPM invoked by the two-velocity two-temperature model. Their main conclusions were as follows:

- When the BDPM is detached from the enclosure walls, the local thermal equilibrium condition within the microporous media is justified for all ranges of all parameters.
- The effects of solid–fluid thermal conductivity ratio on the hot wall average Nusselt number assume importance at lower Raleigh numbers.
- The microporous parameters on the hot wall average Nusselt number have significant effects only when the permeability of the microporous medium is comparable to that of the macropores.
- When the BDPM is attached to the walls, there is departure from the LTE condition within the microporous media for higher values of the Rayleigh number, microporous porosity and solid–fluid thermal conductivity ratio, and lower values of the macropores volume fraction.
- An increase in the solid–fluid thermal conductivity ratio decreases the hot wall average Nusselt number for the macropores and increases that of the micropores. The heat transfer characteristics are influenced more in the attached case than the detached, by the microporous geometrical and thermophysical parameters.
- An increase in microporosity increases the hot wall average Nusselt numbers for both micropores and macropores; however, an increase in macropore volume fraction increases the hot wall average Nusselt number of the macropores and decreases that of the micropores in the BDPM.

Steady Darcy free convection in a square cavity filled with a porous medium has been extended to BDPM. Modified partial differential equations that include dimensionless stream function and temperature were solved numerically using a finite-difference method for a few values of the governing parameters for Rayleigh numbers of 10^2 and 10^3. In Revnic et al. (2009), the interphase heat transfer parameter and the modified conductivity ratio parameter were found to influence fluid flow and heat transfer characteristics for free convection. For small values of γ, the heat transfer results are similar to that of heat conduction, while convection appears for $\gamma \geq 1$ as shown in Figure 8.4.

8.5 Convection with Change of Phase and Mass Transfer in BDPM

Few works have appeared so far, investigating convection inside a BDPM with associated change of phase effects. Lin and coworkers developed a mathematical model for evaporative heat transfer in a loop heat pipe containing different kinds of sintered nickel wicks with monoporous and bi-disperse structures and ammonia as the working fluid (Lin et al. 2011). Distinct heat transfer characteristics were observed, and the bi-disperse wick was seen to decrease the thickness of the vapor blanket region that induces thermal resistance and lower heat transfer capacity of the evaporator.

The heat and mass transfer associated with phase change has been modeled using a mixed pore network model. The mass and heat transfer in two kinds of wicks, characterized by biomodal pore size distribution, one, a monoporous structure and the other, a bi-dispersed capillary structure, were studied by

Mottet and Prat (2016) using simulations and experiments. The conductance and overheating of the casting were studied, and it was found that the bi-dispersed wick had higher thermal performance than the monoporous wick. The liquid–vapor phase distribution, vapor saturation, and the vapor mass flow rate were used to explain the superiority of bi-disperse wick over monodisperse. From the results, one can observe that the temperature difference to sustain identical heat flux is about 50% lower for bi-disperse wicks compared to monodisperse wicks.

Experiments were conducted by Chen et al. (2000b) on boiling in a BDPM channel and compared the results with a monodisperse porous medium. The BDPM have a better boiling heat transfer co-efficient for flow boiling at high heat flux with low-flow resistance compared with the monodispersed porous material when the pore diameter is the same as micropore diameter of BDPM. The pressure drop along the bi-dispersed porous channel is lower by nearly 10% than in the monodispersed porous channel when the micropore diameter (80 μm) of the bi-dispersed porous sample is the same as the pore diameter of monodispersed porous sample. However, no significant change is seen in wall temperature gradient due to bi-dispersivity.

A framework based on pore network modeling was presented in Sadeghi et al. (2017) for simulation of reactive transport in a porous catalyst with a hierarchy of porosity. Steady-state reactive transport was studied in a nanoporous catalyst particle interlaced with macropores that result from the use of pore-formers. The influences of structural features such as macroporosity, pore size ratio, and the particle size, as well as transport properties on the net reaction rate inside the particle, were analyzed. They showed that depending on the transport properties, increasing the macroporosity does not necessarily improve the catalytic activity of the particle. Particles with lower pore-size ratios were also found to be more kinetically active. The study showed the influence of microstructure on reactivity of hierarchical porous catalyst particles.

8.6 Forced Convection in Tri-Disperse Porous Media

A TDPM contains three levels of porosity, viz., macroporosity, mesoporosity, and microporosity. A three-velocity three-temperature model for forced convection in TDPM was formulated by Nield and Kuznetsov (2011b). The Nusselt number values were given as functions of conductivity ratios, velocity ratios, volume fractions, and internal heat exchange parameters. It was shown that for the case of an LTE, the Nusselt number tends to its corresponding value in a monodisperse porous medium. For the constant wall heat flux situation, the Nusselt number behavior could become singular, which is explained by the fact that the wall temperature becomes equal to the bulk mean temperature of the fluid. In a related study, the same authors have described a theory of mass, momentum, and heat transfer in a TDPM (Kuznetsov and Nield 2011). They accounted for the coupling among the three scales of porosity by introducing momentum and interphase heat transfer coupling coefficients. The classical Rayleigh–Benard problem on the onset of convection in a horizontal layer uniformly heated from below, for TDPM was solved using the Darcy law. Linear stability analysis resulted in an expression for the critical Rayleigh number as a function of three volume fractions, two permeability ratios, two thermal capacity ratios, two thermal conductivity ratios, two interphase heat transfer parameters, and two interphase momentum transfer parameters.

As with the BDPM case, Cheng (2015b) studied the natural convection about a vertical cone embedded in a TDPM, and reported that increasing the three modified thermal conductivity ratios or the two permeability ratios tends to increase the natural convection heat transfer rate. They also reported that the thermal non-equilibrium phenomena among the three phases are influenced by the streamwise coordinate and the two interphase heat transfer parameters.

8.7 Natural Convection in Tri-Disperse Porous Media

The first study on natural convection in TDPM was reported in Nield and Kuznetsov (2011a). The three phases were called the fluid phase, monodisperse porous phase, and bi-disperse porous phase. In their model, the three phases were considered as forming a hierarchy, in which, there is coupling between the

Convection in Bi-Disperse Porous Media 149

first and second phases, and between the second and third phases, but not between the first and third phases. The Darcy model was used for simplicity. The leading edge alone was considered because this sets the scene for downstream events. The analysis involved eight BDPM parameters comprising two geometrical parameters, two interphase momentum transfer parameters, two porosity-modified thermal diffusivity ratios, and two permeability ratios. An interesting finding of this study was that there was no dramatic difference between the case of a TDPM and that of a BDPM, unlike for the forced convection in a channel as reported in Kuznetsov and Nield (2011), in which, there could be qualitatively different behavior according to the conditions employed.

Finite-element methods were used by Ghalambaz et al. (2017) to study natural convection in a square enclosure filled with TDPM. Isotherms, streamlines, and Nusselt number values were presented for three levels of porosity. The influence of various governing parameters on convection heat transfer was reported.

8.8 Conclusion

A discussion on the research literature on convection in BDPM published roughly after 2010 has been provided. Convection research in TDPM has also been briefly discussed.

Bi-dispersivity by definition involves pores of two length scales and associated porous medium properties. The distinction of these two pore scales when rendered very strong, that is, the smaller of the scales differing by more than three orders of magnitude compared to the larger length scale, results in convection patterns and heat transfer results not very different from that of convection in monodisperse porous media. Usually such porous media, although BDPM by definition, are primarily governed by the larger pore scale and associated porous medium properties. To achieve a combined influence of the two length scales of a BDPM on convection, it is essential to keep the two length scales different within the range 10^{-1} and 10^{-3}. It appears from the research literature published so far that the influence of the smaller length scale porous medium is maximum and apparent when the two pore scale lengths of the parent BDPM differ only by an order.

LTNE modeling of convection in bi- and tri-disperse media and interface phenomena across multiple pore scales requires research attention. While applications of the BDPM approach to thermal management of electronics and wicks of heat pipes have expanded the convection possibilities to thermal control during heat generation, a unified theory of convection inside simple BDPM configurations as an extension of the well-established monodisperse porous medium theory is found required. Such a theory should provide a connection for the available results with the established homogeneous theory and even a generalization for convection in multi-scale and heterogeneous porous media. Change of phase heat transfer inside bi- and tri-disperse porous media calls for attention due to the interesting possibilities that arise due to the differing length scales of the porous phases.

Unless corroborative experiments are feasible and conducted, much of the convection theory for bi- and tri-disperse porous media shall remain impractical for shaping engineering applications.

REFERENCES

Ajay Kumar, P.V., A.C.G. Chandran, and P.M. Kamath. 2013. "Heat transfer enhancement of electronic chip cooling using porous medium." In *Proceedings of the 22th National and 11th International ISHMT-ASME Heat and Mass Transfer Conference*, IIT Kharagpur, India.

Chen, Z.Q., P. Cheng, and C.T. Hsu. 2000a. "A theoretical and experimental study on stagnant thermal conductivity of bi-dispersed porous media." *International Communications in Heat and Mass Transfer* 27 (5): 601–610. doi:10.1016/S0735-1933(00)00142-1.

Chen, Z.Q., P. Cheng, and T.S. Zhao. 2000b. "An experimental study of two phase flow and boiling heat transfer in bi-dispersed porous channels." *International Communications in Heat and Mass Transfer* 27 (3): 293–302. doi:10.1016/S0735-1933(00)00110-X.

Cheng, C.-Y. 2013a. "Free convective boundary-layer flow over a vertical truncated cone in a bidisperse porous medium." In *Proceedings of the World Congress on Engineering*, London, UK.

Cheng, C.-Y. 2013b. "Natural convection heat transfer from an inclined wavy plate in a bidisperse porous medium." *International Communications in Heat and Mass Transfer* 43: 69–74. doi:10.1016/j.icheatmasstransfer.2013.01.001.

Cheng, C.-Y. 2015a. "Mixed convection heat transfer from a vertical plate embedded in a bidisperse porous medium." In *Proceedings of the World Congress on Engineering*, London, UK.

Cheng, C.-Y. 2015b. "Natural convection heat transfer about a vertical cone embedded in a tridisperse porous medium." *Transport in Porous Media* 107 (3): 765–779. doi:10.1007/s11242-015-0466-0.

Ghalambaz, M., H. Hendizadeh, H. Zargartalebi, and I. Pop. 2017. "Free convection in a square cavity filled with a tridisperse porous medium." *Transport in Porous Media* 116 (1): 379–392. doi:10.1007/s11242-016-0779-7.

Heeter, G.A., and A.I. Liapis. 1997. "Estimation of pore diameter for intraparticle fluid flow in bidisperse porous chromatographic particles." *Journal of Chromatography A* 761 (1): 35–40. doi:10.1016/S0021-9673(96)00791-1.

Hoef, M. A., Van Der., R. Beetstra, and J.A.M. Kuipers. 2005. "Lattice-Boltzmann simulations of low-Reynolds-number flow past mono—and bidisperse arrays of spheres: Results for the permeability and drag force." *Journal of Fluid Mechanics* 528: 233–254. doi:10.1017/S0022112004003295.

Hooman, K., E. Sauret, and M. Dahari. 2015. "Theoretical modelling of momentum transfer function of bi-disperse porous media." *Applied Thermal Engineering* 75: 867–870. doi:10.1016/j.applthermaleng.2014.10.067.

Imani, G., and K. Hooman. 2017. "Lattice Boltzmann pore scale simulation of natural convection in a differentially heated enclosure filled with a detached or attached bidisperse porous medium." *Transport in Porous Media* 116 (1): 91–113. doi:10.1007/s11242-016-0766-z.

Kumari, M., and I. Pop. 2009. "Mixed convection boundary layer flow past a horizontal circular cylinder embedded in a bidisperse porous medium." *Transport in Porous Media* 77 (2): 287–303. doi:10.1007/s11242-008-9293-x.

Kuznetsov, A.V., and D.A. Nield. 2010. "Forced convection in a channel partly occupied by a bidisperse porous medium: Asymmetric case." *International Journal of Heat and Mass Transfer* 53 (23): 5167–5175. doi:10.1016/j.ijheatmasstransfer.2010.07.046.

Kuznetsov, A.V., and D.A. Nield. 2011. "The onset of convection in a tridisperse porous medium." *International Journal of Heat and Mass Transfer* 54 (15): 3120–3127. doi:10.1016/j.ijheatmasstransfer.2011.04.021.

Kuznetsov, A.V., and D.A. Nield. 2013. "An historical and topical note on convection in porous media." *ASME Journal of Heat Transfer* 135 (6): 061201 (1–10). doi:10.1115/1.4023567.

Lage, J.L., and A. Bejan. 1993. "The resonance of natural convection in an enclosure heated periodically from the side." *International Journal of Heat and Mass Transfer* 36 (8): 2027–2038. http://www.sciencedirect.com/science/article/pii/S0017931005801346.

Lin, F.-C., B.-H. Liu, C.-T. Huang, and Y.-M. Chen. 2011. "Evaporative heat transfer model of a loop heat pipe with bidisperse wick structure." *International Journal of Heat and Mass Transfer* 54 (21): 4621–4629. doi:10.1016/j.ijheatmasstransfer.2011.06.015.

Matsumura, Y., and T.L. Jackson. 2014. "Numerical simulation of fluid flow through random packs of polydisperse cylinders." *Physics of Fluids* 26 (12): 123302. doi:10.1063/1.4903954.

Mottet, L., and M. Prat. 2016. "Numerical simulation of heat and mass transfer in bidispersed capillary structures: Application to the evaporator of a loop heat pipe." *Applied Thermal Engineering* 102: 770–784. doi;10.1016/j.applthermaleng.2016.03.143.

Narasimhan, A. 2012. *Essentials of Heat and Fluid Flow in Porous Media*. 2nd ed. CRC Press, New York.

Narasimhan, A., K. S. Raju, and S. R. Chakravarthy. 2014. "Experimental and numerical determination of interface slip coefficient of fluid stream exiting a partially filled porous medium channel." *Journal of Fluids Engineering* 136 (4): 041201. doi:10.1115/1.4026194.

Narasimhan, A., and B.V.K. Reddy. 2009. "Natural convection inside a bidisperse porous medium enclosure." *ASME Journal of Heat Transfer* 132 (1): 012502. doi:10.1115/1.3192134.

Narasimhan, A., and B.V.K. Reddy. 2011a. "Laminar forced convection in a heat generating bi-disperse porous medium channel." *International Journal of Heat and Mass Transfer* 54 (1): 636–644. doi:10.1016/j.ijheatmasstransfer.2010.08.022.

Narasimhan, A., and B.V.K. Reddy. 2011b. "Resonance of natural convection inside a bidisperse porous medium enclosure." *ASME Journal of Heat Transfer* 133 (4): 042601. doi:10.1115/1.4001316.

Narasimhan, A., B.V.K. Reddy, and P. Dutta. 2012. "Thermal management using the bi-disperse porous medium approach." *International Journal of Heat and Mass Transfer* 55 (4): 538–546. doi:10.1016/j.ijheatmasstransfer.2011.11.006.

Nield, D.A. 2016. "A note on the modelling of bidisperse porous media." *Transport in Porous Media* 111 (2): 517–520. doi:10.1007/s11242-015-0607-5.

Nield, D.A., and A. Bejan. 2013. *Convection in Porous Media.* 4th ed. Springer-Verlag, New York.

Nield, D.A., and A.V. Kuznetsov. 2004. "Forced convection in a bi-disperse porous medium channel: A conjugate problem." *International Journal of Heat and Mass Transfer* 47 (24): 5375–5380. doi:10.1016/j.ijheatmasstransfer.2004.07.018.

Nield, D.A., and A.V. Kuznetsov. 2005. "A two-velocity two-temperature model for a bi-dispersed porous medium: Forced convection in a channel." *Transport in Porous Media* 59 (3): 325–339. doi:10.1007/s11242-004-1685-y.

Nield, D.A., and A.V. Kuznetsov. 2006. "The onset of convection in a bidisperse porous medium." *International Journal of Heat and Mass Transfer* 49 (17): 3068–3074. doi:10.1016/j.ijheatmasstransfer.2006.02.008.

Nield, D.A., and A.V. Kuznetsov. 2008. "Natural convection about a vertical plate embedded in a bidisperse porous medium." *International Journal of Heat and Mass Transfer* 51 (7): 1658–1664. doi:10.1016/j.ijheatmasstransfer.2007.07.011.

Nield, D.A., and A.V. Kuznetsov. 2011a. "The Cheng Minkowycz problem for natural convection about a vertical plate embedded in a tridisperse porous medium." *International Journal of Heat and Mass Transfer* 54 (15): 3485–3493. doi:10.1016/j.ijheatmasstransfer.2011.03.037.

Nield, D.A., and A.V. Kuznetsov. 2011b. "A three-velocity three-temperature model for a tridisperse porous medium: Forced convection in a channel." *International Journal of Heat and Mass Transfer* 54 (11): 2490–2498. doi:10.1016/j.ijheatmasstransfer.2011.02.013.

Rees, D.A.S., A.V. Kuznetsov, and D.A. Nield. 2008. "Vertical free convective boundary-layer flow in a bidisperse porous medium." *ASME Journal of Heat Transfer* 130 (9): 09601. doi:10.1115/1.2943304.

Revnic, C., T. Grosan, I. Pop, and D.B. Ingham. 2009. "Free convection in a square cavity filled with a bidisperse porous medium." *International Journal of Thermal Sciences* 48 (10): 1876–1883. doi:10.1016/j.ijthermalsci.2009.02.016.

Sadeghi, M.A., M. Aghighi, J. Barralet, and J.T. Gostick. 2017. "Pore network modeling of reaction-diffusion in hierarchical porous particles: The effects of microstructure." *Chemical Engineering Journal* 330: 1002–1011. doi:10.1016/j.cej.2017.07.139.

Straughan, B. 2009. "On the Nield–Kuznetsov theory for convection in bidispersive porous media." *Transport in Porous Media* 77 (2): 159–168. doi:10.1007/s11242-008-9307-8.

Wang, K., and P. Li. 2018. "Forced convection in bidisperse porous media incorporating viscous dissipation." *Applied Thermal Engineering* 140: 86–94. doi:10.1016/j.applthermaleng.2018.05.036.

Wang, K., K. Vafai, P. Li, and H. Cen, 2017. "Forced convection in a bidisperse porous medium embedded in a circular pipe." *ASME Journal of Heat Transfer* 139 (10): 102601. doi:10.1115/1.4036574.

Yu, B., and J. Li. 2001. "Some fractal characters of porous media." *Fractals* 9 (3): 365–372. doi:10.1142/S0218348x01000804.

9

Pore Scale Analysis in Forced Convection Heat Transfer in Porous Media

Hasan Celik, Moghtada Mobedi, and Akira Nakayama

CONTENTS

9.1 Introduction.. 153
9.2 Governing Equations and Boundary Conditions 156
     9.2.1 Pore-Scale Governing Equations 156
     9.2.2 Volume-Averaged Governing Equations.. 156
     9.2.3 Determination of Volume-Averaged Transport Properties............................ 157
9.3 Pore-Scale Analysis of Periodic Structures 159
     9.3.1 Selecting of Representative Control Volume 159
     9.3.2 Creation of Fully Developed Heat and Fluid Flow for a Unit Cell Representative
         Control Volume 161
     9.3.3 Validation Study... 162
9.4 Pore-Scale Study of Stochastic Structure 163
     9.4.1 Selection of Representative Control Volume 163
     9.4.2 Steps for Pore-Scale Analysis with Microcomputed Tomography 163
         9.4.2.1 Microscanning the Structures and Obtaining the Images................... 164
         9.4.2.2 Processing the Images and Obtaining an Appropriate Digital 3D Structure...... 164
         9.4.2.3 Generating Mesh and Solving Pore-Scale Governing Equations..................... 164
         9.4.2.4 Application of Volume-Averaged Technique to Determine Transport
             Properties... 166
         9.4.2.5 Validation of the Obtained Results... 166
9.5 Conclusion 169
References... 170

9.1 Introduction

A porous medium is a composite medium that contains voids that might be filled with a fluid. The applications of heat and fluid flow in porous media are faced in many industrial areas, such as the food and crop drying process, chemical reactors, filters, membranes, gas separator adsorbent beds, nuclear reactors, and heat exchangers, as well as applications in nature, such as soil and rocks (Kaviany 1995; Vafai 2015; Nield and Bejan 2013; Ingham and Pop 1998). The mechanism of heat and fluid flow through a porous medium is complex and three dimensional in many applications. The pore-scale velocity and temperature cannot be determined for the entire domain of a porous medium due to the huge number of pores. Thus, some approaches are required to overcome this difficulty. The volume-averaged method is effective in removing the difficulties of the pore-scale approaches. The discontinuity of phases is removed by using volume averaging of the dependent parameters, and as a result, the motion and energy equations are solved for a continuum domain. The volume-averaged method allows the use of all conventional solution methods, employing for solving single-phase heat and fluid flow governing equations. The main difficulty of the volume-averaged method is the transport properties arising during the integration of equations. Almost all thermophysical parameters take effective form (such as effective thermal capacitance, effective stagnant thermal conductivity), and furthermore, new transport properties appear (such as permeability or thermal dispersion conductivity). Numerous porous structures exist in industry and nature having distinctive effective and transport properties. The determination of these properties, which highly depends on the structure of porous media and heat and flow character, is the main drawback of volume-averaged method.

Recent developments in the field of image processing and X-ray microcomputed tomography facilitate determining the values of volume-averaged transport parameters computationally, which is considerably faster and cheaper than experimental methods. It is challenging to discover the mechanism of heat and fluid flow in pores since a 3D porous medium is a closed box. However, the use of X-ray microtomography enables researchers to observe all details of fluid flow and heat transfer even in the smallest pore of the medium. Furthermore, it enables researchers to determine the effective and volume-averaged transport properties by taking the volume average of dependent variables calculated computationally. All volume-averaged transport properties (such as permeability, thermal dispersion, and interfacial heat transfer coefficient that needed precise experimental methods and devices to be determined) can be calculated accurately by using the results of the pore-scale analysis. As it was mentioned before, one of the main difficulties of the researchers who study on heat and fluid flow in porous media is the large variety of porous structures causing volume-averaged transport properties and have incredibly different values from each other. Fortunately, pore-scale analysis with the support of X-ray microcomputed tomography and advanced simulation technique is a proper remedy for this difficulty, and that is why it has taken the attention of many researchers in recent years.

Two basic classifications of porous media, which are important for the analysis of pore scale heat and fluid flow, are described below.

1. *Open pore and isolated pore*: The connection between the pores is an essential issue for analyzing heat and fluid flow in porous media (Figure 9.1). In a porous medium with open pore, the fluid can flow through the porous media and the volume-averaged continuity and momentum equations should be included into the governing equations, while for a porous medium with isolated pores (closed pores) the continuity and momentum equations vanish since the volume average of the velocity for a closed space is zero.

2. *Periodic and stochastic porous media*: As it can be understood from the name of structures, when a periodicity exists in the structure of a porous medium, it is called periodic porous media. A periodic porous structure might be two or three dimensional. Figure 9.2 shows two- and three-dimensional periodic porous media as well as a stochastic structure porous media. The repetition of the structure in a periodic porous media can be easily observed from Figure 9.2a and b. Pore-scale analysis of a periodic medium is simpler than stochastic one. The main advantage of periodic porous media is the selection of control volume (usually is a

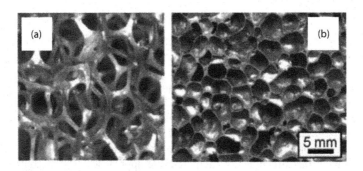

FIGURE 9.1 Two samples of porous media: (a) an open pore porous medium (e.g., open cell metal foam); and (b) an isolated pore porous medium (e.g., closed-cell metal foam). (With kind permission from Taylor & Francis: *Heat Transfer Eng.*, A review of metal foam and metal matrix composites for heat exchangers and heat sink 33, 2012, 991–1009. Han, X., Wang, Q., Park, Y., T'Joen, C., Sommers, A., and Jacobi, A.)

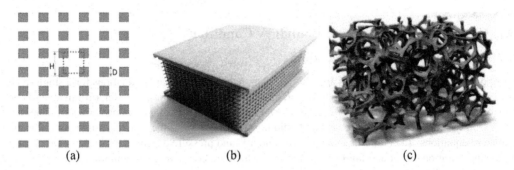

FIGURE 9.2 Samples of periodic and stochastic porous media: (a) 2D periodic; (b) 3D periodic; and (c) stochastic porous media. (With kind permission from Taylor & Francis: *Mech. Adv. Mater. Struc.*, Microstructure-based models for multifunctional material systems 19, 2012, 421–430, Sarzynski, M.D., Schaefer, S., and Ochoa, O.O.)

cell that has small size), which can correctly represent the entire domain. For forced convection fluid flow and heat transfer in a periodic porous media, the concept of the fully developed heat and fluid flow is valid, and therefore the analysis of a representative control volume (cell) is sufficient to discover the mechanism of heat and fluid flow precisely and determine the volume-averaged transport properties.

The velocity and temperature distributions in the pores of a stochastic porous medium can be completely different from each other due to the structure; therefore, a cluster of pores should be selected to represent the entirety of the porous media. The cluster should be sufficiently large to represent the entire domain. The difficulties of pore-scale analysis for a stochastic porous medium can be described as:

- Creation of representative control volume in the computer environment requires special techniques, such as the microcomputed tomography technique. For the generation of representative control volume from the tomography images, additional software and extra efforts are required.
- The representative control volume is the cluster of pores and particles (or ligaments), and this causes the increase of the size of the computational domain and consequently the computational time and cost.
- Precise separation of solids (particles or ligaments) from voids (or fluid) due to discontinuity of images and steep changes of the interface surface is not easy.
- Generation of mesh and matching of the solid and fluid meshes at the interface and requires a special treatment and software.
- Generally, an inlet and an outlet region are added to the representative control volume since the pore scale hydraulically and thermally fully developed concept is not valid, and this makes the computational domain larger.
- Due to the sudden geometrical changes in the pore structure, instabilities in flow may appear during the computational study, causing difficulties on the decision for convergence.

According to the above difficulties, the pore-scale analysis of stochastic porous media requires special attention, treatments, and validations.

In this chapter, first, the pore-scale analysis techniques for periodic structures and considerations for reliable numerical results are described. Then, the pore-scale analysis of the stochastic porous media and methods for checking the reliability of the obtained results are explained in details. The computational methods are risky, and there are many possibilities of occurrence of mistakes in a pore-scale analysis; therefore, authors believe that this chapter of the book helps researchers not only to discover the pore-scale analysis but also to have an idea about the inspection and checking methods for the reliable results.

9.2 Governing Equations and Boundary Conditions

9.2.1 Pore-Scale Governing Equations

The pore-scale governing equations are continuity, momentum, and energy equations for a single-phase heat and fluid flow in a porous media; however, based on the problem, mass transfer, turbulent, and reaction equations as well as radiation effect can also be added into the set of governing equations. The focus of this study is on the single-phase heat and fluid flow; however, the same strategy can be used for other transport equations. To obtain the temperature, velocity, and pressure distributions in the scale of pores, following traditional heat and fluid flow equations, which are continuity, momentum, and energy equations, for the solid and fluid phases should be solved.

$$\vec{\nabla}\cdot\vec{V} = 0$$

$$\vec{V}\left(\vec{\nabla}.\vec{V}\right) = -\frac{1}{\rho}\vec{\nabla}p + v\,\vec{\nabla}^2\vec{V} \tag{9.1}$$

$$(\rho c_p)_f(\vec{V}.\vec{\nabla})T_f = k_f\,\vec{\nabla}^2 T_f$$

$$\vec{\nabla}^2 T_s = 0$$

where $\vec{V}$ is the pore-scale velocity vector, T_s and T_f are the pore scale temperature of the solid and fluid phases. The boundary conditions depend on the structure of the representative control volume, but it is clear that velocity is zero at the interface between solid and fluid, while a continuous heat flux exists at the interface.

9.2.2 Volume-Averaged Governing Equations

The volume average for a representative control volume with volume of V can be defined as,

$$\langle\varphi\rangle = \frac{1}{V}\int_V \varphi\,dV$$

$$\langle\varphi\rangle^x = \frac{1}{V_x}\int_{V_v} \varphi\,dV \tag{9.2}$$

where V is the total volume of representative control volume and V_x is the volume of considered phase in representative elementary volume (REV), such as V_s and V_f (s or f stands for solid or fluid phases, respectively). After applying the above volume-average definitions into the pore-scale continuity, momentum, and energy equations, and doing some mathematical manipulations, the following form of the volume-averaged equations for a steady flow are obtained:

$$\vec{\nabla}\cdot\left\langle\vec{V}\right\rangle = 0$$

$$\frac{1}{\varepsilon^2}\left\langle\vec{u}\right\rangle\cdot\nabla\left\langle\vec{u}\right\rangle = -\frac{1}{\rho_f}\nabla\left\langle p\right\rangle^f + \frac{\mu}{\varepsilon}\nabla^2\left\langle\vec{u}\right\rangle - \frac{\mu}{\rho_f K}\left\langle\vec{u}\right\rangle - \frac{C}{K^{1/2}}\left|\left\langle\vec{u}\right\rangle\right|\left\langle\vec{u}\right\rangle \tag{9.3}$$

$$\varepsilon\rho_f c_{p_f}\left\langle\vec{u}\right\rangle\cdot\nabla\left\langle T\right\rangle^f = \nabla\cdot k_{eff,f}\nabla\left\langle T\right\rangle^f + h_v(\left\langle T\right\rangle^s - \left\langle T\right\rangle^f)$$

$$\nabla\cdot(k_{eff,s})\nabla\left\langle T\right\rangle^f - h_v(\left\langle T\right\rangle^s - \left\langle T\right\rangle^f) = 0$$

where K and C are permeability and inertia coefficients while $k_{eff,f}$, $k_{eff,s}$ and h_v are effective thermal conductivity for solid and fluid and interfacial heat transfer coefficient, respectively. The comparison of these equations with the pore-scale governing equations shows that some parameters in the volume-averaged transport equations take the effective form (such as $k_{eff,f}$ and $k_{eff,s}$), and some new parameters appear, called volume-averaged transport properties (such as h_v, K, C). The effective thermal conductivity for the solid and fluid can be defined as

$$k_{eff,s} = (1-\varepsilon)k_s + k_{tor,s} \tag{9.4}$$

$$k_{eff,f} = \varepsilon k_f + k_{tor,f} + \varepsilon k_{disp} \tag{9.5}$$

As it can be seen, the effective solid thermal conductivity ($k_{eff,s}$) involves the stagnant thermal conductivity and tortuosity effect. For effective fluid thermal conductivity ($k_{eff,f}$), in addition to stagnant and the tortuosity thermal conductivity, the thermal dispersion appears, and consequently fluid effective thermal conductivity ($k_{eff,f}$), is the summation of them ($k_{eff,f}$). Since both the stagnant and tortuosity conductivities do not involve velocity, some researchers prefer to combine the stagnant and tortuosity thermal conductivities. Yang and Nakayama (2010) stated that for the heat and fluid flow in porous media with a high thermal conductivity ratio between solid and fluid phases (such as aluminum–air), the effective thermal conductivity for diffusion term could be defined as (Yang and Nakayama 2010):

$$k_{eff,s} = k_{stag,s} = (1-\varepsilon^*)k_s \tag{9.6}$$

$$k_{eff,f} = k_{stag,f} + k_{disp} = \varepsilon^* k_f + \varepsilon k_{disp} \tag{9.7}$$

where ε^* is the effective porosity, and it can be easily calculated numerically or by using a suggested correlation. The accurate results from the solution of the above volume-averaged equations strongly depend on accurate knowledge of the volume-averaged transport properties. If a local thermal equilibrium exists between the solid and fluid phases, the energy equations can be added to each other, and an equilibrium energy equation can be obtained as

$$\varepsilon \rho_f c_{p_f} \langle \vec{u} \rangle \cdot \nabla \langle T \rangle = \nabla \cdot k_{eff} \nabla \langle T \rangle \tag{9.8}$$

where k_{eff} is the effective thermal conductivity that involves both solid and fluid phases thermal conductivities and defined as

$$k_{eff} = k_{eff,s} + k_{eff,f} \tag{9.9}$$

Equation 9.8 shows that the effective thermal conductivity for the solid and fluid phases are combined, whereas the interfacial heat transfer coefficients are eliminated.

9.2.3 Determination of Volume-Averaged Transport Properties

A brief information on the calculation of effective and volume-averaged transport properties is given below.

- *Stagnant thermal conductivity*: Stagnant thermal conductivity involves both the effect of physical thermal conductivity and tortuosity, and it can be calculated numerically. It should be mentioned that the stagnant thermal conductivity is tensor quantity, but since this work concerns with isotropic porous media, it has a unique value in this study. The effective thermal conductivity in a direction can be numerically calculated by imposing a temperature difference in a direction, and then the effective porosity can be obtained.

$$\varepsilon^* = \frac{k_{eff} - k_f}{k_s - k_f} \tag{9.10}$$

After calculation of effective porosity, the values of $k_{stag,f}$ and $k_{stag,s}$ can be calculated by using Equations (9.6) and (9.7).

- *Permeability and inertia coefficient*: The effect of friction between the surface of the solid and fluid phases, as well as the change of direction of fluid and occurrence of vortices in the pores, are stresses on a fluid, which flows through a porous medium. The last two terms in the momentum equation (Equation 9.3) are the inclusion of the additional effects. Similar to the stagnant thermal conductivity, permeability is a tensor quantity whose value can vary with direction. In this study, the discussion on the permeability is done for an isotropic porous structure; hence, a unique permeability value exists. The permeability and inertia coefficients can be easily calculated by using the pore-scale velocity and pressure drop. It is assumed that the friction effect due to the solid phase in porous media is considerably greater than the clear fluid inertia and clear fluid viscous stresses. Hence, the momentum equation (Equation 9.3) takes the following form (Celik et al. 2019; Mancin et al. 2010; Mancin et al. 2012; Della Torre et al. 2014):

$$\Pi = A + B\,Re \qquad (9.11)$$

where the dimensionless parameters of A, B, and Π are defined as:

$$\Pi = -\frac{d_c^2}{\langle u \rangle \mu_f}\frac{d\langle p \rangle^f}{dx},\ A = \frac{d_c^2}{K};\ B = C\left(\frac{d_c^2}{K}\right)^{1/2},\ Re = \frac{\rho_f \langle u \rangle d_c}{\mu_f} \qquad (9.12)$$

The dimensionless pressure, Darcy velocity, and Re number can be calculated from pore-scale results. The other parameters in the above equations (except K and C) are known, and they can be found from fluid thermophysical properties and the geometry of porous media. The drawing of change of Π respect to Re number provides a diagram shown in Figure 9.3. As can be seen for the small values of Re number, the inertia effect does not play an important role, and $\Pi = A$, while for high values of Re number, the inertia force is significant, and $\Pi = B\,Re$, consequently permeability and inertia coefficients can be found from Figure 9.3. It should be mentioned that the plotting of change of Π respect to Re number also yields a critical Re number indicating the end of the Darcy region.

- *Interfacial heat transfer coefficient*: The interfacial heat transfer coefficient appears in the set of equations when a local thermal non-equilibrium state exits between the solid and fluid phases. This may happen when the temperature at the boundary surfaces of porous media changes

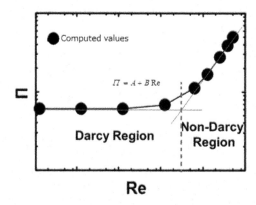

FIGURE 9.3 The change of dimensionless pressure drop with pore-scale Re number. (Thin line shows theoretical expectation and thin dashed line separates the regions.) (Celik et al., 2019, Reprinted from *Journal of Porous Media*, 22(5), H. Celik, M. Mobedi, A. Nakayama, U. Ozkol, A study on numerical determination of permeability and inertia coefficient of aluminum foam using X-Ray micro-tomography technique: Focus on Inspection Methods for Reliability, pp. 511–529, © 2019 with permission from Begell House.)

rapidly or when the solid and fluid phases have significantly different heat capacities and thermal conductivities. In those cases, both the heat transfer equations for fluid and solid phases should be solved to obtain temperature distribution for the entire volume-averaged domain. Based on the pore-scale heat transfer between the solid and fluid phases, the interfacial heat transfer coefficient can be calculated as below (Nakayama et al. 2009):

$$h_{sf} = \frac{\dfrac{1}{V} \displaystyle\int_{A_{sf}} k_f \vec{\nabla} T . \vec{n} \, dA}{\langle T \rangle^s - \langle T \rangle^f} \qquad (9.13)$$

The numerator shows the total volumetric heat transfer from solid to fluid in the representative control volume, while the denominator is the different volume-averaged temperatures between solid and fluid phases. The calculated interfacial heat transfer coefficient from Equation (9.13) is the volumetric heat transfer coefficient. In the most of studies, a uniform temperature is considered for the solid phase, which eliminates the solid heat conduction equation in the governing equations (Ozgumus and Mobedi 2015, 2016; Pedras and Lemos 2008).

- *Thermal dispersion*: There is a local difference between the velocity and temperature of pore scale and the volume-averaged one in a cross section perpendicular to the fluid flow. These local differences in the velocity and temperature causes a difference between the pore scale and volume-averaged heat transport through the cross section. To balance the heat transport between the volume averaged and pore scale, thermal dispersion concept is employed. Thermal dispersion is a tensor quantity, which has non-zero diagonal components for the isotropic porous media. Hence, for an isotropic porous media in which fluid flows in a direction such as *x*, both longitudinal and transverse thermal diffusivity should be calculated. The equation used for calculation of the longitudinal and transverse thermal dispersion for a direction such as *x* can be written as (Nakayama 1995; Nakayama and Kuwahara 1999; Nakayama et al. 2002):

$$k_{dis,XX} = -\frac{1}{\left(\Delta \langle T \rangle_{xx} / L_{ref,X} \right)} \frac{\rho_f C_{pf}}{V} \iint \left(T_f - \langle T \rangle^f \right) \left(u - \langle u \rangle^f \right) dV \qquad (9.14)$$

$$k_{dis,YY} = -\frac{1}{\left(\Delta \langle T \rangle_{yy} / L_{ref,Y} \right)} \frac{\rho_f C_{pf}}{V} \iint \left(T_f - \langle T \rangle^f \right) \left(v - \langle v \rangle^f \right) dV \qquad (9.15)$$

where $\Delta \langle T \rangle$ is the temperature difference in the direction of X or Y of the representative control volume having length of L_{ref}. Hence, the value of longitudinal and transverse thermal dispersion can be found easily if the pore-scale velocity and temperatures for the representative control volume are known.

9.3 Pore-Scale Analysis of Periodic Structures

As it was mentioned before, the pore-scale analysis of the periodic porous media is easier than stochastic since the entire domain can be represented by a representative control volume (or cells). In this section, the necessary information about the representative control volume, boundary conditions, and validation methods for pore-scale analysis of periodic porous media is given.

9.3.1 Selecting of Representative Control Volume

Our literature survey shows that there are two kinds of representative control volumes, which may be called the "unit cell" and "unit layer" selected according to the structure of the porous media and character of heat and fluid flow.

- *Unit cell representative control volume*: In many periodic structures, a unit cell can be selected to represent the entire domain as shown in Figure 9.4 due to the repetition of not only the structure but also heat and fluid flow. The pore-scale governing equations can be solved for the unit cell by considering the concept of hydraulically and thermally fully developed heat and fluid flow. The pore-scale analysis in a unit cell yields all details of the mechanism of heat and fluid flow as well as provides the calculation of all effective and volume-averaged transport properties. An important point that should be considered for selecting a unit cell is the repetition of both structure and flow. In some cases, (particularly three-dimensional structures), the structure may be periodic, but the heat or fluid flow is not repeated in the periodic cells. Hence, the size of the unit cell should be sufficiently increased to satisfy the periodic condition both for structure and flow.
- *Unit layer*: Sometimes the selected unit cell cannot represent heat and fluid flow in the entire domain due to the character of flow. The heat and fluid flow are not repeated in the flow direction (despite periodic structure in flow direction); however, it is periodic in the transverse direction (Figure 9.5). For instance, when a natural or mixed convection heat transfer exists in a periodic structure, the concept of being fully developed cannot be applied, and the unit cell representative control volume cannot be used, although the structure is repeated in the flow direction. For those cases, a layer of the structure can be selected to represent heat and

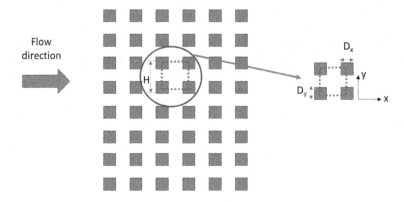

FIGURE 9.4 Schematic view of a periodic porous structure and the selected unit cell.

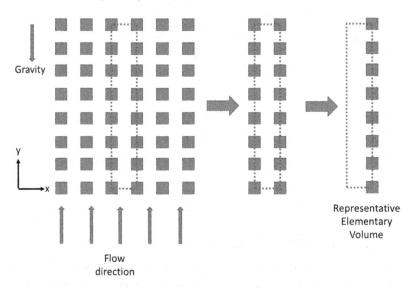

FIGURE 9.5 Schematic view of a periodic porous structure and the selected unit layer.

Pore Scale Analysis in Forced Convection Heat Transfer in Porous Media 161

fluid flow for the entire domain. Although the number of studies used the unit cell model is profoundly greater than the unit layer, there are some examples for the application of the unit layer in the literature. The study by Celik et al. (2017), in which a pore-scale study was done for mixed convection, can be an example for the use of the unit layer. Furthermore, Ucar et al. (2015) performed a study on a random porous media, which was not periodic in the flow direction but in the transverse direction, and they used the unit layer for a pore-scale study of heat and fluid flow. Pathak et al. (2013) used a unit layer for analyzing heat transfer in a periodic porous structure under the unsteady state (oscillatory and pulsated) conditions.

In the unit layer representative control volume approach, the pore-scale equations are solved for the unit layer (which is considerably larger than a unit cell), and run time is considerably longer. Sometimes a dummy region (buffer zone) is added to the inlet and outlet regions to reduce the effect of entrance and outlet boundary conditions on the heat and fluid flow on the flow in the porous region. In the unit layer approach, the pore-scale analysis is done for entire of the layer (including many cells), and the creation of a fully developed heat and fluid flow is not required.

9.3.2 Creation of Fully Developed Heat and Fluid Flow for a Unit Cell Representative Control Volume

The use of a unit cell that can represent heat and fluid flow for the entire domain is possible if a hydraulically and thermally fully developed heat and fluid flow can be created. Some techniques should be used to create a fully developed condition for the unit cell. The literature survey shows that there are two methods that can be used to create hydraulically and thermally fully developed heat and fluid flow, and these methods may be called as the boundary condition iteration and cell location iteration.

- *Boundary condition iteration*: For the creation of hydraulic fully developed fluid flow in a unit cell, a velocity function such as $f(y)$ can be chosen for the inlet boundary, and a zero-diffusion gradient condition can be applied to the outlet of the unit cell. After solving the governing equations for the unit cell under the aforementioned boundary conditions and obtaining the pore-scale velocity and pressure distributions in the REV, the outlet velocity profile is substituted into the inlet boundary. This iterative procedure continues until attaining an identical velocity profile at the inlet and outlet boundaries. A sample of this iteration is shown in Figure 9.6. As can be seen, by increasing the number of iterations, the inlet and outlet velocity profiles approach each other and become almost identical (Ozgumus and Mobedi 2015; Sabet et al. 2018).

  To generate a thermally fully developed boundary condition for a unit cell, an inlet temperature function such as $g(y)$ is used and imposed to the inlet boundary, while the zero-diffusion boundary condition is applied for the outlet. Then the temperature field for the entire domain is obtained by solving the energy equation. The dimensionless temperature profile at the outlet boundary is used to obtain a new temperature profile for the inlet. The iterative process continues until no change in the inlet and outlet dimensionless temperature profile is observed. The dimensionless temperature for the unit cell can be defined as (Ozgumus and Mobedi 2015):

$$\theta(y) = \frac{T(y) - T_s}{T_b - T_s} \tag{9.16}$$

where T_b is the bulk temperature and for an incompressible flow and can be defined as:

$$T_b = \frac{\int_{-(H-D_y)/2}^{(H-D_y)/2} uT \, dy}{\int_{-(H-D_y)/2}^{(H-D_y)/2} u \, dy} \tag{9.17}$$

For definition of the integral limits, Figure 9.4 can be seen.

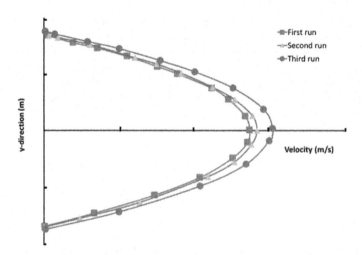

FIGURE 9.6 The change of velocity profiles at the inlet and outlet through the iterative procedure to obtain periodic velocity distribution.

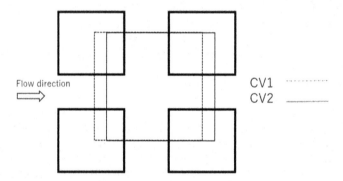

FIGURE 9.7 Cell location iteration for the creation of hydraulically and thermally fully developed condition.

- *Cell location iteration*: Cell location iteration is employed by some researchers such as Nakayama and Kuwahara (1999) and Kuwahara et al. (2001) extensively. In this method, two representative control volumes as CV1 and CV2 are used as shown in Figure 9.7. Solving heat and fluid flow equations for CV1 yields the velocity and temperature profiles for inlet boundary of CV2. Furthermore, solving the same equations under the boundary condition found from CV1 gives the inlet profile of CV2. This procedure can be continued till no change in the velocity profiles can be observed.

9.3.3 Validation Study

The pore-scale analysis for the periodic structure is easier than the stochastic one since the computational domain is smaller and regular in many cases. However, the obtained solution should be validated. The following points should be considered to be ensured of the obtained results:

- Like the other simulation studies, the grid independency should be done.
- There are many 2D and 3D pore-scale studies on the periodic structure porous media in literature, the comparison of the obtained results with some reported studies should be done.
- The velocity and temperature boundary layers at the solid surface (velocity and temperature gradient) should be observed and checked for different Reynolds numbers.

- For the representative control volume in which a unit layer is used, the effect of the outlet dummy region on the results should be inspected. In other words, the length of the outlet dummy region in flow direction should be sufficiently far from the porous region to remove the effect of outlet boundary condition on the results.

9.4 Pore-Scale Study of Stochastic Structure

The stochastic structure is not regular and needs some special treatment. It can be two or three dimensional. However, in literature, it is preferred to call the 2D stochastic structure a random porous structure. There are many studies in literature related to 2D random structure (Wang et al. 2007; Koponen et al. 1998; Xuan et al. 2010). The 2D random porous structure can be generated in a computer environment easily, and the motion and heat flow equations can be solved. There are methods such as Monto Carlo for creating random porous media in the computer environment. Since the structure is random, the studied domain (representative control volume) should be sufficiently large to discover the effect of the randomness on heat and fluid flow. In many pore-scale analyses on the random porous media, concepts such as Voronoi Tessellation or nearest neighbor distance are used to determine the volume-averaged transport parameters (Ucar et al. 2015; Young et al. 2008; Xiao and Yin 2016).

Pore-scale study of 3D stochastic porous media is complicated and needs special techniques. The 3D stochastic structure is a closed box in which the structure and heat and flow patterns cannot be predicted easily. That is why techniques such as the combination of microtomography with the heat and fluid flow simulation were developed to overcome the difficulties. Even though all equations employed to determine volume-averaged transport parameters for the stochastic structure are the same with the periodic structure, the generation of porous media in the computer environment and then the solution of the governing equations for an irregular structure creates trouble (Petrasch et al. 2008; Ranut et al. 2014; Zafari et al. 2015; Diani et al. 2015; Akolkar and Petrasch 2012).

9.4.1 Selection of Representative Control Volume

The important point for the selection of a representative control volume is the size of the volume. As it was mentioned before, the control volume should represent the entire porous media. The result from a small representative volume (with small number of particles or ligaments), or thin volume, may not be reliable since the location of the boundary affects the results considerably. Generally, a cube volume with sufficient side length is suitable as a representative control volume. The representative volume should be selected from a region of porous media and can represent almost the entire domain.

9.4.2 Steps for Pore-Scale Analysis with Microcomputed Tomography

As it was mentioned before, microcomputed tomography is a favorable technology to create a porous structure in a computer environment and then to simulate the heat and fluid flow in the structure. The process of obtaining volume-averaged transport properties by using microcomputed tomography may involve five steps (Celik et al. 2018):

- Microscanning of the structures and obtaining of images,
- Processing of the images and obtaining an appropriate digital 3D structure in the computer environment,
- Generating mesh for the structured domain and solving pore-scale governing equations,
- Determination of macroscopic transport properties by taking the volume average of the obtained velocity and temperature fields, and
- Inspection and validation of results.

The details of these steps are explained in the following subsections.

9.4.2.1 Microscanning the Structures and Obtaining the Images

X-ray-computed microtomography is one of the most widely used non-destructive methods for analyzing the structure of materials. In a microtomography scanner, an X-ray beam cone passes through the sample, and it is collected by a detector. The sample is rotated, providing a series of 2D projection images at different angles. Unfortunately, due to the limitations of computational resources, the precise microtomography images can be obtained only for small size representative elementary volumes (Celik et al. 2019). The size of the representative control volume, which is a cluster of disordered pores and ligaments in a stochastic structure, is significant. It should contain sufficient pores and solid particles (or ligaments) to represent the entire stochastic porous media. Incorrect results are obtained from the small size of representative control volume while a huge one increases computational time, considerably. The distance between the images and the number of pixels is another crucial issue that should be seriously taken into account. An insufficient number of pixel or distance between images causes the creation of a different porous structure from the original one in the computer environment, yielding completely wrong results.

9.4.2.2 Processing the Images and Obtaining an Appropriate Digital 3D Structure

The obtained 2D images are processed by a software (imageJ, MATLAB, etc.), and the images are converted into the black-and-white regions. Processing images in the computer environment is an expensive procedure since the high amount of Random Access Memory (RAM) is required. Figure 9.8 shows the obtained digital structure of the studied 20 PPI metal foam and a close view of cells of the same foam in the study of Celik et al. (2018). The pore-scale analysis with using X-ray tomography has many other advantages, such as observation of cells, pores, and ligaments to understand the structure and measuring their sizes in the computer environment.

9.4.2.3 Generating Mesh and Solving Pore-Scale Governing Equations

For solving the governing equations, the mesh should be created. There are some commercial software or even in-house computer programs for generation of the mesh in the solid and fluid regions. Some software generates rectangular prism voxel while others create a triangular prism voxel. Each of them has advantages and disadvantages with respect to each other. The rectangular prism voxel yields solutions faster while the triangular prism voxel can create the surface of the solid phase smoothly. Figure 9.9 shows a rectangular prism voxel in the fluid region, which is created for aluminum foam with 20 PPM in the study of Celik et al. (2018).

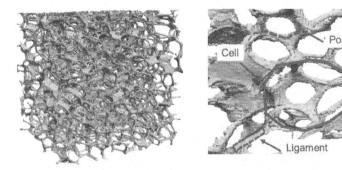

FIGURE 9.8 The digital structure of 20 PPM aluminum foam created in the computer environment and observation of cell, pore, and ligament of the same structure. (With kind permission from Taylor & Francis: *Numerical Heat Transfer, Part A: Applications*, A numerical study on determination of volume averaged thermal transport properties of metal foam structures using X-ray microtomography technique 74, 2018, 1368–1386, Celik, H., Mobedi, M., Nakayama, A., and Ozkol, U.)

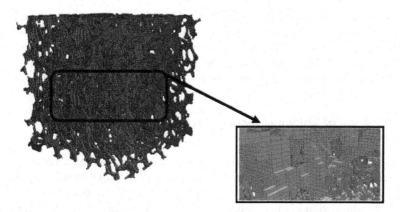

FIGURE 9.9 The rectangular prism voxels for the fluid phase of 20 PPI metal foam (With kind permission from Taylor & Francis: *Numerical Heat Transfer, Part A: Applications*, A numerical study on determination of volume averaged thermal transport properties of metal foam structures using X-ray microtomography technique 74, 2018, 1368–1386, Celik, H., Mobedi, M., Nakayama, A., and Ozkol, U.)

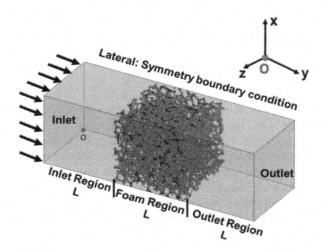

FIGURE 9.10 The digital structure of analyzed porous media with dummy inlet and outlet regions, and boundary conditions (O indicates the origin point). (With kind permission from Taylor & Francis: *Numerical Heat Transfer, Part A: Applications*, A numerical study on determination of volume averaged thermal transport properties of metal foam structures using X-ray microtomography technique 74, 2018, 1368–1386, Celik, H., Mobedi, M., Nakayama, A., and Ozkol, U.)

A commercial software (Fluent, COMSOL, OpenFOAM, etc.) or even in-house computer program can be used to solve the pore-scale equations for the entire representative domain employing the generated mesh. To resemble real flow through the porous media (without forcing or imposing boundary conditions at the inlet and outlet), an entrance and outlet dummy region should be created as shown in Figure 9.10. A uniform fluid flow with constant velocity value and zero-gradient temperature can be applied at the inlet section. The same zero-gradient boundary condition can be used for the velocity and temperature at the outlet, referring to the negligible diffusion transport. After solving the pore-scale equations by using the generated mesh under the aforementioned boundary conditions, the temperature, velocity, and pressure distribution for the entire representative control volume can be obtained. It should be mentioned that for many industrial cases, $k_s \gg k_f$ and ligament or particle size is small. This situation may allow researchers to apply constant temperature assumptions for the surface of the solid phase. Hence, there might be no need to solve the heat conduction equation for the solid phase.

Our numerical experience showed that for low Re number, there was no difficulty in the convergence of the solver due to smooth flow in the pores; however, by increasing the Re number and collision of the fluid particles to the walls of the pores, instabilities may appear. For those cases, the values of some volume-averaged transport parameters can be calculated in different iterations, and by increasing the number of iterations if there are no considerable changes in those volume-averaged values, the solution is converged.

9.4.2.4 Application of Volume-Averaged Technique to Determine Transport Properties

The values of volume-averaged transport properties as permeability, inertia coefficient, interfacial heat transfer coefficient, and thermal dispersion conductivity can be calculated by using equations explained in Section 9.2.2 after obtaining the temperature, pressure, and velocity distributions in the solid and fluid phases. Fortunately, most of the commercial software has options for the numerical integration of a quantity in a domain providing the calculation volume-averaged properties easier. It should be mentioned that the interfacial heat transfer coefficient and the inertia coefficient are scalar quantities whose values are independent of the direction. However, thermal dispersion and permeability are tensor quantities whose values depend on the direction. Hence, if the porous media is anisotropic, the components of tensor should be calculated (Vu et al. 2014; Galindo-Torres et al. 2012).

9.4.2.5 Validation of the Obtained Results

Since the porous media is stochastic, the possibility of error is high, some inspections and checks should be done to be ensured of correctness of the obtained results.

- *Verification for elimination of images*: The high amount of RAM and CPU resources are required to obtain high-quality 3D views. The number of images taken through the domain by the tomography devices affects voxel length (i.e., the distance between two slices). The decrease of distance between the images improves the accuracy and quality; however, it increases a number of images. A high number of images considerably affects the computational resources for processing the images. Our experience showed that some images can be eliminated for the sake of computational resources. But this elimination of images may reduce the accuracy of the obtained numerical results. In order to check the effect of eliminated images, the variation of a volume-averaged transport versus the number of images can be drawn. The value of volume-averaged transport properties does not change considerably after elimination of a reasonable number of images, indicating the existence of a sufficient number of images to obtain accurate results (Ucar et al. 2015; Celik et al. 2019).

- *Verification of grid numbers*: For verification of the grid numbers, the change of volume-averaged transport properties with the number of the voxel can be drawn. The values of volume-averaged transport properties do not change considerably after a voxel number, indicating the sufficient number of the voxels for obtaining accurate results. Furthermore, structure change of a cross section in the domain with number of grids can be observed. The structure of the porous media at that cross section should not change after the sufficient grid number. A sample of grid refinement performed in the study by Celik et al. (2018) is shown in Figure 9.11. The generated structure considerably changes by changing grids from $48 \times 48 \times 48$ to $148 \times 148 \times 148$. As it is seen, both the size and the location of solid phase and voids change. However, the size and location of ligaments by changing grids from $148 \times 148 \times 148$ to $192 \times 192 \times 192$ or $250 \times 250 \times 250$ do not vary, indicating independence of generated structure from the mesh.

- *Check of the flow rate and velocity*: Based on the continuity equation for a steady-state process, the flow rate through a porous media with open pores should remain constant. In order to check the flow rate, the intrinsic velocity and the net flow area of some cross sections perpendicular to the flow can be calculated and plotted in the same diagram as shown in Figure 9.12, which is a sample of the study by Celik et al. (2019). The flow rate should be identical through the flow

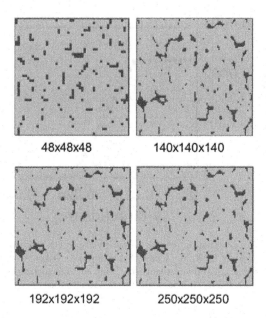

FIGURE 9.11 The change of the generated structure of aluminum foam with PPI = 20 with a number of grids for a cross section (black is the solid region while gray represents a void). (With kind permission from Taylor & Francis: *Numerical Heat Transfer, Part A: Applications*, A numerical study on determination of volume averaged thermal transport properties of metal foam structures using X-ray microtomography technique 74, 2018, 1368–1386, Celik, H., Mobedi, M., Nakayama, A., and Ozkol, U.)

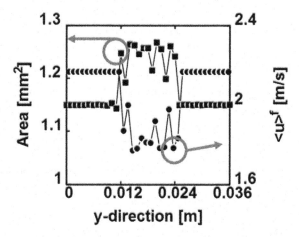

FIGURE 9.12 The change of volume-averaged velocity and cross-sectional area through the domain. (Celik et al., 2019, Reprinted from *Journal of Porous Media*, 22(5), H. Celik, M. Mobedi, A. Nakayama, U. Ozkol, A study on numerical determination of permeability and inertia coefficient of aluminum foam using X-ray micro-tomography technique: Focus on Inspection Methods for Reliability, pp. 511–529, © 2019 with permission from Begell House.)

direction if a unidirectional flow in an isotropic porous media exists. The change of intrinsic velocity with the area of the cross section should be reverse, as seen in Figure 9.12. This reverse change should also be checked to be ensured from the correctness of the solution of conservation of mass and momentum equations.

- *Existence of fully develop region*: A linear pressure gradient exists in a fully developed flow. That is why a linear change of pressure in the flow direction of the volume-averaged porous media domain should be observed, if fully developed flow exists. Figure 9.13 shows a sample

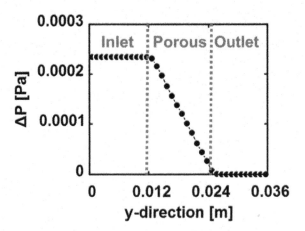

FIGURE 9.13 The change of pressure gradient through a representative control volume. (Celik et al., 2019, Reprinted from *Journal of Porous Media*, 22(5), H. Celik, M. Mobedi, A. Nakayama, U. Ozkol, A study on numerical determination of permeability and inertia coefficient of aluminum foam using X-ray micro-tomography technique: Focus on Inspection Methods for Reliability, pp. 511–529, © 2019 with permission from Begell House.)

of the change of pressure drop through a metal foam with 20 PPI in the study by Celik et al. (2019). A steep linear pressure gradient is observed for the porous region while the change of pressure in the dummy regions is negligible. The plotted diagram proves the existence of a fully developed flow in the porous media region and also provides an accurate value of the pressure gradient through the porous media by using an appropriate curve-fitting method (Celik et al. 2019).

- *Voxels number in narrow throats*: In order to obtain accurate results, sufficient number of voxels in the narrow throat (narrow pores) should exist, and this important point should be checked in order to be sure of the accuracy of the results. The shrinking and expansion of the pores play an essential role in pressure drop and pore-scale velocity profiles. Hence, in order to simulate pore-scale flow accurately, a sufficient number of voxels should exist in the throats. One method for checking the number of voxels in the throat is that the grid distribution in different cross sections can be plotted, and the number of mesh in the pores and throats can be observed in order to be ensured that sufficient grids exist in the throat for obtaining accurate pore-scale results (Celik et al. 2019).

- *Transverse velocity gradient for low and high Reynolds numbers*: There should be a velocity gradient near the solid surface of pores as we know from the flow in channel and pipe, and the velocity gradient near the surface region must become steeper by the increase of Re number. In order to deduce this fact, a sufficient number of voxels should be generated in the pores. Figure 9.14 shows a sample of the velocity profile in the middle line of a cross section of the study by Celik et al. (2019). As can be seen, when the Reynolds number is very small (i.e., $Re = 0.001$), a smooth flow exists in the pore, and a velocity gradient on the region close to the surface can be observed. The gradient near the surface increases by increasing of Re number from 0.001 to 600, and fluctuations due to collide of flow to the anterior ligament appear. The same check should be done for temperature, and the observation of temperature gradient in the region close to the solid surface for the extreme cases should be done (Celik et al. 2018, 2019).

- *Validation with reported studies*: The comparison of the obtained results, such as volume-averaged transport properties, can be done with some experimental or theoretical works in literature, after ensuring all checks explained above. Our experience showed that it is almost impossible to catch the same values of volume-averaged results for two porous media, even with the same porosity and particle (or ligament) size since the topology might be different.

Pore Scale Analysis in Forced Convection Heat Transfer in Porous Media

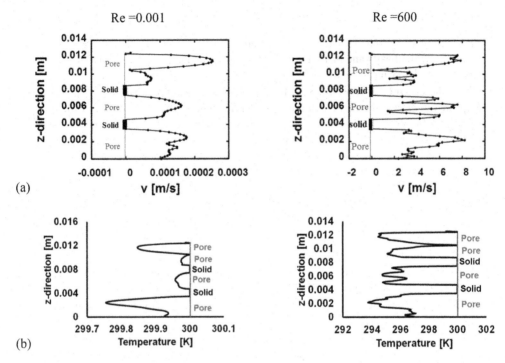

FIGURE 9.14 (a) Velocity and (b) temperature profile in a line of the cross section for two different Re numbers. ((a): Celik et al., 2019, Reprinted from *Journal of Porous Media*, 22(5), H. Celik, M. Mobedi, A. Nakayama, U. Ozkol, A study on numerical determination of permeability and inertia coefficient of aluminum foam using X-ray micro-tomography technique: Focus on Inspection Methods for Reliability, pp. 511–529, © 2019 with permission from Begell House. ((b): With kind permission from Taylor & Francis: *Numer. Heat Transfer, Part A: Appl.*, A numerical study on determination of volume averaged thermal transport properties of metal foam structures using X-ray microtomography technique 74, 2018, 1368–1386, Celik, H., Mobedi, M., Nakayama, A., and Ozkol, U.)

However, the obtained results and reported values must be sufficiently close to each other. For the comparison with the literature, the structural parameters of porous media for which the results are reported should be close to the analyzed porous as much as possible to achieve better validation.

Another important issue about the stochastic porous media is the anisotropic effect. Since the structure is stochastic, the possibility of anisotropic behavior of volume-averaged transport properties may exist. That is why authors recommend the checking of anisotropic effect at least for permeability by determination of permeability for different directions. A small difference between the calculated permeability values in different directions can be accepted, while if the difference is considerable the stochastic structure is anisotropic, and the tensor concept should be included to the study for permeability, stagnant thermal conductivity, and thermal dispersion transport properties.

9.5 Conclusion

Pore-scale analysis both for periodic and stochastic porous media has become popular in recent years. Particularly, the creation of the digital structure of porous media in the computer environment not only yields important information about the structure of the porous media but also enables the application of numerical heat and fluid flow in pore scale. In this study, important information is presented about the pore-scale analysis of both periodic and stochastic porous media. The following remarks can be concluded:

For periodic structure

- The analysis of periodic structure is easier than a stochastic one, and a unit cell or unit layer under the hydraulically and thermally fully developed condition can be selected to observe the heat and fluid flow for the entire domain.
- If both the structure and flow are periodic, a unit cell as a representative control volume can be employed to present the entire domain; however, in some applications, although the structure is periodic but flows (or heat) in the cells are not identical, for those cases a unit layer can be employed to represent the entire domain.
- For the creation of two hydraulically and thermally fully developed methods, the boundary condition iteration and the cell location iteration can be used.

For stochastic structure

- The representative control volume is the cluster of pore and particles (or ligaments, fibers, etc.), and it should be sufficiently large to resemble the entire porous media domain. A small representative domain yields completely incorrect results.
- Microcomputed tomography is a useful technique to generate stochastics porous media in a computer environment, but it requires software and large RAM and computational resources.
- Elimination of an image may be a proper method and can reduce the number of images, but it should be properly checked to prevent the altering of structure.
- During solving governing equations, particularly for high Re number, instabilities may appear, and the program may not be converged. For those cases, the convergence of volume-averaged transport parameters can be used.
- Using dummy inlet and outlet (buffer region) regions plays an important role on the convergence and the accuracy of results. These dummy regions should be absolutely used, and they should be sufficiently long from the digital structure of porous media to remove the effect of inlet and outlet boundary conditions on the results.
- Inspection and verification of the results are important. Inspections and checks on the obtained results should be done. In this study, the suggested inspection, verifications, and validation are as the verification for the elimination of images, verification of grid numbers, check of the flow rate and velocity through porous media, inspection of existence of a fully develop region, check of number of voxels in narrow throats, check of transverse velocity gradient for low and high Reynolds number, and the comparison with reported studies.
- Fortunately, many programs have options for taking volume and surface average. That is why calculation of the volume-averaged transport parameters is not a difficult issue, and accurate results can be obtained if the correct temperature, velocity, and pressure distributions in the porous media can be obtained.

Authors believe that the number of pore-scale studies will increase in the near future since the capacity and ability of software increases, as well as computational resources become cheaper. In the future, the microcomputed tomography technique and simulation will be combined in the same device, which provides all detailed information about porous media as well as yields the volume-averaged transport parameters in a short run time.

REFERENCES

Akolkar, A., and Petrasch, J. 2012. Tomography-Based Characterization and Optimization of Fluid Flow through Porous Media. *Transport in Porous Media* 95(3): 535–550.

Celik, H., M. Mobedi, A. Nakayama, and U. Ozkol. 2018. A Numerical Study on Determination of Volume Averaged Thermal Transport Properties of Metal Foam Structures using X-ray Microtomography Technique. *Numerical Heat Transfer Part A: Applications* 74(7): 1368–1386.

Celik, H., M. Mobedi, A. Nakayama, and U. Ozkol. 2019. A Study on Numerical Determination of Permeability and Inertia Coefficient of Aluminum Foam using X-Ray Micro-Tomography Technique: Focus on Inspection Methods for Reliability. *Journal of Porous Media* 22(5): 511–529.

Celik, H., M. Mobedi, O. Manca, and U. Ozkol. 2017. A Pore Scale Analysis for Determination of Interfacial Convective Heat Transfer Coefficient for Thin Periodic Porous media Undermixed Convection. *International Journal of Numerical Methods for Heat and Fluid Flow* 27(12): 2775–2798.

Della Torre, A., G. Montenegro, G. R. Tabor, and M. L. Wears. 2014. CFD Characterization of Flow Regimes Inside Open Cell Foam Substrates. *International Journal of Heat and Fluid Flow* 50: 72–82.

Diani, A., K. K. Bodla, L. Rossetto, and S. V. Garimella. 2015. Numerical Investigation of Pressure Drop and Heat Transfer through Reconstructed Metal Foams and Comparison against Experiments. *International Journal of Heat and Mass Transfer* 88: 508–515.

Galindo-Torres, S. A., A. Scheuermann, L. Li. 2012. Numerical Study on the Permeability in a Tensorial form for Laminar Flow. *Physical Review E* 86: 046306 (in anisotropic porous media).

Han, X. H., Wang, Q., Park, Y. G., T'Joen, C., Sommers, A., and Jacobi, A. 2012. A Review of Metal Foam and Metal Matrix Composites for Heat Exchangers and Heat Sinks. *Heat Transfer Engineering* 33(12): 991–1009.

Ingham, D., and I. Pop. 1998. *Transport Phenomena in Porous Media*. Pergamon, Turkey.

Kaviany, M. 1995. *Principles of Heat Transfer in Porous Media: Mechanical Engineering Series*. Springer, New York.

Koponen, A., M K. J. Timonen, and D. Kandhai. 1998. Simulations of Single-Fluid Flow in Porous Media. *International Journal of Modern Physics C* 9(8): 1505–1521.

Kuwahara, F., M. Shirota, and A. Nakayama. 2001. A Numerical Study of Interfacial Convective Heat Transfer Coefficient in Two-Energy Equation Model for Convection in Porous Media. *International Journal of Heat and Mass Transfer* 44: 1153–1159.

Mancin, S., C. Zilio, A. Cavallini, and L. Rossetto. 2010. Pressure Drop during Air Flow in Aluminum Foams. *International Journal of Heat and Mass Transfer* 53(15): 3121–3130.

Mancin, S., C. Zilio, L. Rossetto, and A. Cavallini. 2012. Foam Height Effects on Heat Transfer Performance of 20 PPI Aluminum Foams. *Applied Thermal Engineering* 49: 55–60.

Nakayama, A. 1995. *PC-Aided Numerical Heat Transfer and Convective Flow*. CRC Press, Boca Raton, FL.

Nakayama, A., A. Kenji, C. Yang, Y. Sano, F. Kuwahara, and J. Liu. 2009. A Study on Interstitial Heat Transfer in Consolidated and Unconsolidated Porous Media. *Heat Mass Transfer* 45: 1365–1372.

Nakayama, A., and F. Kuwahara. 1999. A Macroscopic Turbulence Model for Flow in a Porous Medium. *Journal of Fluids Engineering* 121(2): 427–433.

Nakayama, A., F. Kuwahara, T. Umemoto, and T. Hayashi. 2002. Heat and Fluid Flow within an Anisotropic Porous Medium. *Journal of Heat Transfer* 124(4): 746–753.

Nield, D. A., and A. Bejan. 2013. *Convection in Porous Media*. Springer Science Business Media New York.

Ozgumus, T., and M. Mobedi. 2015. Effect of Pore to Throat Size Ratio on Interfacial Heat Transfer Coefficient of Porous Media. *Journal of Heat Transfer* 137(1): 12602.

Ozgumus, T., and M. Mobedi. 2016. Effect of Pore to Throat Size Ratio on Thermal Dispersion in Porous Media. *International Journal of Thermal Sciences* 104: 135–45.

Pathak, M. G., T. I. Mulcahey, and S. M. Ghiaasiaan. 2013. Conjugate Heat Transfer during Oscillatory Laminar Flow in Porous Media. *International Journal of Heat and Mass Transfer* 66: 23–30.

Pedras, M. H. J., and M. J. S. de Lemos. 2008. Thermal Dispersion in Porous Media as a Function of the Solid–Fluid Conductivity Ratio. *International Journal of Heat and Mass Transfer* 51(21–22): 5359–5367.

Petrasch, J., F. Meier, H. Friess, and A. Steinfeld. 2008. Tomography Based Determination of Permeability, Dupuit–Forchheimer Coefficient, and Interfacial Heat Transfer Coefficient in Reticulate Porous Ceramics. *International Journal of Heat and Fluid Flow* 29(1): 315–326.

Ranut, P., E. Nobile, and L. Mancini. 2014. High Resolution Microtomography-Based CFD Simulation of Flow and Heat Transfer in Aluminum Metal Foams. *Applied Thermal Engineering* 69(1–2): 230–240.

Sabet, S., M. Mobedi, M. Barisik, and A. Nakayama. 2018. Numerical Determination of Interfacial Heat Transfer Coefficient for an Aligned Dual Scale Porous Medium. *International Journal of Numerical Methods for Heat & Fluid Flow*. doi:10.1108/HFF-03-2018-0097.

Sarzynski, M. D., S. Schaefer, and O. O. Ochoa. 2012. Microstructure-Based Models for Multi-Functional Material Systems. *Mechanics of Advanced Materials and Structures* 19(6): 421–430.

Ucar, E., M. Mobedi, G. Altintas, and E. Glatt. 2015. Effect of Voxel Size in Flow Direction on Permeability and Forchheimer Coefficients Determined by Using Micro-Tomography Images of a Porous Medium. *Progress in Computational Fluid Dynamics* 15(5): 327–333.

Vafai, K. 2015. *Handbook of Porous Media*. CRC Press, Boca Raton, FL.

Vu, T. L., G. Lauriat, and O. Manca. 2014. Forced Convection of Air Through Networks of Square Rods or Cylinders Embedded in Microchannels. *Microfluidics and Nanofluidics* 16(1): 287–304.

Wang, M., J. Wang, N. Pan, and S. Chen. 2007. Mesoscopic Predictions of the Effective Thermal Conductivity for Microscale Random Porous Media. *Physical Review E* 75(3): 36702.

Xiao, F., and X. Yin. 2016. Geometry Models of Porous Media Based on Voronoi Tessellations and Their Porosity–Permeability Relations. *Computers & Mathematics with Applications* 72(2): 328– 348.

Xuan, Y. M., K. Zhao, and Q. Li. 2010. Investigation on Mass Diffusion Process in Porous Media Based on Lattice Boltzmann Method. *International Journal of Heat and Mass Transfer* 46 (10): 1039–1051.

Yang, C., and A. Nakayama. 2010. A Synthesis of Tortuosity and Dispersion in Effective Thermal Conductivity of Porous Media. *International Journal of Heat and Mass Transfer* 53(15): 3222–3230.

Young, P. G., T. B. H. Beresford-West, S. R. L. Coward, B. Notarberardino, B. Walker, and A. Abdul-Aziz. 2008. An Efficient Approach to Converting Three-Dimensional Image Data into Highly Accurate Computational Models. *Philosophical Transactions of Royal Society A* 366: 3155–3173.

Zafari, M., M. Panjepour, M. D. Emami, and M. Meratian. 2015. Microtomography-Based Numerical Simulation of Fluid Flow and Heat Transfer in Open Cell Metal Foams. *Applied Thermal Engineering* 80: 347–354.

10

Lattice Boltzmann Method for Modeling Convective Heat Transfer in Porous Media

Gholamreza Imani and Kamel Hooman

CONTENTS

10.1 Introduction...173
10.2 Standard LBM...175
    10.2.1 Lattice BGK (LBGK)..175
        10.2.1.1 Hydrodynamics..175
        10.2.1.2 Energy...176
    10.2.2 Multi-Relaxation Time LBM (MRT-LBM)..179
    10.2.3 Boundary and Interface Conditions..181
        10.2.3.1 No-Slip..181
        10.2.3.2 Conjugate Heat Transfer..183
    10.2.4 A Brief Review of the Applications 183
10.3 LBM Macroscale Models for Convection Heat Transfer in Porous Media 187
    10.3.1 Governing Equations...187
        10.3.1.1 Hydrodynamics..187
        10.3.1.2 Energy...189
    10.3.2 A Brief Review of the Applications 193
10.4 Conclusion 194
Nomenclature..195
References...196

10.1 Introduction

Convection in porous media has so many industrial applications, for example, in drying technology, energy storage systems, solar collectors, transpiration cooling, electronic cooling, metal foam heat exchangers, designed porous media, fuel cells, automobile industry, and biotechnology, to name a few (Nield and Bejan 2017; Vafai 2010, 2015). These problems are generally modeled based on either of the pore (micro) or macro scale [based on representative elementary volume (REV)-averaged] equations depending on the level of flow and heat transfer details one needs to obtain (see Figure 10.1). At the pore scale, the Navier–Stokes (N-S) and energy equations need to be solved, where, no-slip and conjugate heat transfer conditions are satisfied at the solid–fluid interfaces. Whereas, the volume-averaged equations, also called the generalized non-Darcy equation, supplement the N-S equations with the Darcy and Forchheimer drag terms to hydrodynamically account for the presence of the porous medium at the macroscale (Vafai 1981, 2015). The solid and fluid energy equations are also volume-averaged based on either of the local thermal equilibrium (LTE) or local thermal non-equilibrium (LTNE) assumption, depending on whether or not the local temperature difference between the solid and fluid phases can be ignored, respectively (Nield and Bejan 2017). The LTE assumption results in a single volume-averaged energy equation for an effective

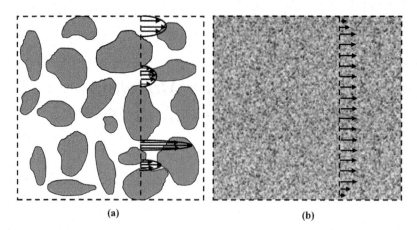

FIGURE 10.1 The schematic of different scales in porous media: (a) Pore; and (b) Macro (REV).

medium with effective thermophysical properties whereas, the LTNE assumption leads to two separate volume-averaged energy equations for the solid and fluid phases thermally communicating through an interfacial heat transfer source term (Nield and Bejan 2017). Volume-averaged equations are easier to solve, as the computational domain is simplified to the macroscale skipping over the microscale details. However, this simplification comes with inaccuracy, which is inherent to any averaging process. Hence, there are cases that pore-scale simulations are called for. However, pore-scale numerical simulation of the convection in porous media using the computational fluid dynamics (CFD) methods, such as finite difference (FD) and finite volume (FV), face formidable challenges on top of which is satisfying the boundary and interface conditions within the voids at the reasonable computational expense (Wang et al. 2007).

In 1988, McNamara and Zanetti (1988) introduced the lattice Boltzmann equation (LBE), also called the lattice Boltzmann method (LBM), as an alternative flow solver in an attempt to address the shortcomings of its predecessor, lattice gas automata (LGA). About a decade later, He and Luo (1997a, b) and Abe (1997) independently proved that the LBM could also be derived from the special discretization of the continuous Boltzmann transport equation (BTE) in both time and phase space. This discovery of the kinetic root of the LBM placed it on a rigorous theoretical foundation and made it very popular among the researchers. On top of the lower computational costs, the popularity of the LBM can also be attributed to the linear advection term, local collision operator, and easy calculation of the pressure from an equation of state (Chen and Doolen 1998). Another promising feature of the LBM is related to the capability of simulating fluid flows in different regimes (e.g., continuum, slip, and transition). For example, the governing equations of the continuum regime (i.e., N-S) can be recovered by performing the Chapman–Enskog analysis (Chapman et al. 1990) in the approximately incompressible limit of LBM (Chen and Doolen 1998). Moreover, the easy parallelization of the computer code due to the local collision operator makes the LBM an ideal choice for the pore-scale simulation of convection in porous media, a problem essentially associated with the high computational costs when CFD methods are used (Guo and Shu 2013).

LBM has now become a promising alternative numerical tool for simulating different transport phenomena with complex physics and geometries, such as transport phenomena in porous media (Chen and Doolen 1998; Succi 1989), which Succi (2001) classifies as the ones "ShouldUse" LBM. Following the continuum models, the LBM for modeling convection heat transfer in porous media has also been developed based on the pore and macroscales. LBM pore-scale models for convection heat transfer in porous media employ the standard LBM where the solid–fluid interface is modeled by applying the no-slip and conjugate heat transfer conditions at the pore scale. Therefore, over the past two decades, the pore-scale simulation of convection in porous media has been a drive for the development of the LBM no-slip and conjugate heat transfer schemes.

On the other hand, the LBM VAT-based (volume-averaged technique-based) models for fluid flow (Dardis and McCloskey 1998; Guo and Zhao 2002; Guo et al. 2009; Martys 2001; Spaid and Phelan 1997) and heat transfer (Chen et al. 2017a; Gao et al. 2014; Guo and Zhao 2005; Liu et al. 2014; Rong et al. 2010; Seta et al. 2006; Wang et al. 2016a, b) in porous media modify the standard LBM to hydrodynamically and thermally compensate for ignoring the details of the porous geometry. The work of Guo and Zhao (2002) can be considered as a central research in LBM macroscale modeling of the fluid flow through porous media. Guo and Zhao (2002) modified the standard LBM to recover the Brinkman–Forchheimer extended Darcy equation. Based on Guo and Zhao's (2002) model, different macroscale convection heat transfer models in porous media were developed assuming both LTE (Chen et al. 2017a; Guo and Zhao 2004, 2005; Liu et al. 2014; Rong et al. 2010; Seta et al. 2006a, b; Wang et al. 2016a) and LTNE (Gao et al. 2014; Wang et al. 2016b).

Currently, the LBM of fluid flow and heat transfer in porous media is a mature technique finding its way to more challenging problems, such as phase-change, multi-phase, multi-component, combustion, and so on. However, in this book chapter, in the interest of brevity, only the theory and applications of the available LBM pore-scale and macroscale models for a single-phase single-component convection in porous media will be discussed. As such, the essential elements of the LBM for pore-scale convection heat transfer in porous media, including the LBM no-slip and conjugate heat transfer schemes, will also be briefly reviewed.

10.2 Standard LBM

As discussed in the introduction section, the physics of pore-scale convection in porous media can be recovered by the standard LBM. With its root in the kinetic theory, LBM is considered to be a mesoscopic approach that fills the gap between the molecular dynamics (MD) and N-S, as simulation tools for the microscopic and macroscopic transport phenomena, respectively. That is, unlike MD, in which, one is interested in keeping track of individual particles, the LBM considers only the behavior of an ensemble of particles by introducing the single-particle distribution function $f(\mathbf{r},\xi,t)$, which is a key variable in the BTE. It is defined as the probability of finding a particular molecule with a given position $\mathbf{r}$ and a microscopic velocity ξ at a specified time t (Guo and Shu 2013; Mohammad 2011).

10.2.1 Lattice BGK (LBGK)

10.2.1.1 Hydrodynamics

The continuous BTE with the Bhatnagar–Gross–Krook (BGK) collision operator (Bhatnagar et al. 1954) describes the evolution of $f(\mathbf{r},\xi,t)$ as (Chapman et al. 1990):

$$\frac{\partial f(\mathbf{r},\xi,t)}{\partial t} + \xi.\nabla f(\mathbf{r},\xi,t) = -\frac{f(\mathbf{r},\xi,t) - f^{eq}(\mathbf{r},\xi,t)}{\gamma} \tag{10.1}$$

where γ is the relaxation time due to a collision that determines the rate at which $f(\mathbf{r},\xi,t)$ relaxes toward its local equilibrium. Here, $f^{eq}(\mathbf{r},\xi,t)$ is the Maxwell–Boltzmann distribution function represented by the following equation:

$$f^{eq}(\mathbf{r},\xi,t) = \frac{\rho}{(2\pi RT)^{D/2}} Exp\left(-\frac{(\xi - \mathbf{u})^2}{2RT}\right) \tag{10.2}$$

where ρ, $\mathbf{u}$, T, R, and D are the local macroscopic density, velocity, temperature, gas constant, and dimension of the space, respectively. The macroscopic variables ρ, $\mathbf{u}$, and T are the microscopic velocity moments of the $f(\mathbf{r},\xi,t)$ as follows (He and Luo 1997a, b):

$$\rho = \int f d\xi, \ \rho\mathbf{u} = \int \xi f d\xi, \ \rho\varepsilon = \frac{1}{2}\int (\xi - \mathbf{u})^2 f d\xi \tag{10.3}$$

where $\varepsilon = D_0 RT/2$ is the internal energy and D_0 is the number of degrees of freedom of molecules.

In 1997, He and Luo (1997a, b) proved that LBM can be derived from the special discretization of Equation 10.1 in both time t and phase space $(\mathbf{r}, \xi)$. Those authors derived the final discretized density evolution equation called the lattice BGK (LBGK) or single-relaxation time (SRT) LBM as presented in Equation 10.4. This density evolution equation is considered to be the "working horse" (Mohammad 2011) of the standard LBM.

$$f_i(\mathbf{r}+\mathbf{e}_i\delta t, t+\delta t) - f_i(\mathbf{r},t) = -\frac{f_i(\mathbf{r},t) - f_i^{eq}(\mathbf{r},t)}{\tau_v} \tag{10.4}$$

where $\tau_v = \gamma/\delta t$ is the non-dimensional velocity relaxation time, $\mathbf{e}_i$ are discrete microscopic velocities, and ω_i are the weight coefficients. The macroscopic variables can be calculated from the moments of the discrete density distribution function f_i as follows:

$$\rho = \sum_i f_i, \quad \rho\mathbf{u} = \sum_i \mathbf{e}_i f_i, \quad \rho\varepsilon = \frac{1}{2}\sum_i (\mathbf{e}_i - \mathbf{u})^2 f_i \tag{10.5}$$

As a part of the phase-space discretization, the lattice structure and equilibrium distribution function is determined, which results in different LBGK models generally referred to as $D_m Q_n$ (m-dimensional n-velocity) lattice models (Qian et al. 1992). The density equilibrium distribution function (EDF) in $D_m Q_n$ LBGK models can be expressed as follows:

$$f_i^{eq} = \omega_i \rho \left[1 + \frac{\mathbf{e}_i \cdot \mathbf{u}}{c_s^2} + \frac{(\mathbf{e}_i \cdot \mathbf{u})^2}{2c_s^4} - \frac{\mathbf{u} \cdot \mathbf{u}}{2c_s^2} \right] \tag{10.6}$$

where $c_s = \sqrt{RT}$ is the model sound speed. The most popular lattice models among the researchers are $D_2 Q_9$ model for 2D simulations and $D_3 Q_{15}$, and $D_3 Q_{19}$ models for 3D problems, as shown in Figure 10.2. In these models $c_s = c/\sqrt{3}$, where $c = \delta x/\delta t$ is the lattice velocity normally chosen equal to one, δx is the lattice spacing, and δt is the time step. The discrete velocity set $\mathbf{e}_i$ and the weight coefficients ω_i for $D_2 Q_9$, $D_3 Q_{15}$, and $D_3 Q_{19}$ models are presented in Table 10.1. The fluid pressure in the above-mentioned LBGK models is calculated by the equation of state as $p = \rho c_s^2$. Through performing the Chapman–Enskog expansion, the kinematic viscosity of the fluid for $D_2 Q_9$, $D_3 Q_{15}$, and $D_3 Q_{19}$ lattice models is derived as $\upsilon = (\tau_v - 0.5) c_s^2 \delta t$.

10.2.1.2 Energy

LBGK approaches for thermal flows are generally classified into the multi-speed (MS) and double distribution function (DDF) models (He et al. 1998). The MS models employ the higher order velocity terms in the density EDF in order to recover the macroscopic energy equation without using a separate distribution function for the temperature. Those models suffer from two shortcomings, namely, the severe numerical instability and narrow range of temperature variation (He et al. 1998). The DDF thermal models, on another hand, use a separate distribution function for the temperature (i.e., or internal energy), which enhances the above-mentioned shortcomings of the MS models. In early DDF models, called passive scalar (PS) models or *temperature-based* thermal LBM models, because the viscous heat dissipation and compression work were neglected, the temperature was treated as a passive scalar quantity advected by the flow field; therefore, a separate LB evolution equation similar to that of density was employed for the temperature field solely based on the conservation of the macroscopic energy equation, not the kinetic theory.

He et al. (1998) proposed a rigorous DDF thermal LBGK model based on the kinetic theory, which is an *internal energy-based* model, in which, the viscous heat dissipation and compression work were naturally incorporated. Those authors derived the continuous internal energy evolution equation from the BTE (refer to Equation 10.1) by replacing $f(\mathbf{r},\xi,t)$ with $g(\mathbf{r},\xi,t) = \left[(\xi - \mathbf{u})^2/2\right] f(\mathbf{r},\xi,t)$ as the internal energy distribution function and called the final internal energy transport equation the Boltzmann energy equation (BEE) presented as follows:

Lattice Boltzmann Method for Modeling Convective Heat Transfer in Porous Media

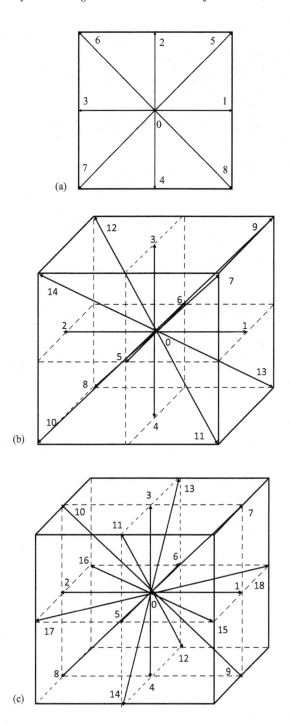

FIGURE 10.2 The schematic of different lattice models: (a) D_2Q_9; (b) D_3Q_{15}; and (c) D_3Q_{19}.

$$\frac{\partial g(\mathbf{r},\xi,t)}{\partial t}+\xi\cdot\nabla g(\mathbf{r},\xi,t)=\Omega(g)-f(\mathbf{r},\xi,t)\{(\xi-\mathbf{u})[\frac{\partial\mathbf{u}}{\partial t}+(\xi\cdot\nabla)\mathbf{u}]\} \quad (10.7)$$

where $\Omega(g)=-\left[f(\mathbf{r},\xi,t)-f^{eq}(\mathbf{r},\xi,t)\right](\xi-\mathbf{u})^2/2/\gamma$ is the internal energy collision operator. He et al. (1998) constructed the proposed DDF thermal model by properly discretizing Equation 10.7 in both time and phase space in a similar way that the LBM hydrodynamic model had been derived before

TABLE 10.1

Different Lattice Models with Model Specifications

Model	Discrete Velocity Set $\mathbf{e}_i$		Weights ω_i
D_2Q_9	$\mathbf{e}_i = \begin{cases} (0,0)c \\ (\pm 1,0)c,(0,\pm 1)c \\ (\pm 1,\pm 1)c \end{cases}$	$i=0$ $i=1-4$ $i=5-8$	$\omega_0 = 4/9$ $\omega_{1-4} = 1/9$ $\omega_{5-8} = 1/36$
D_3Q_{15}	$\mathbf{e}_i = \begin{cases} (0,0,0)c \\ (\pm 1,0,0)c,(0,\pm 1,0)c,(0,0,\pm 1)c \\ (\pm 1,\pm 1,\pm 1)c \end{cases}$	$i=0$ $i=1-6$ $i=7-14$	$\omega_0 = 2/9$ $\omega_{1-6} = 1/9$ $\omega_{7-14} = 1/72$
D_3Q_{19}	$\mathbf{e}_i = \begin{cases} (0,0,0)c \\ (\pm 1,0,0)c,(0,\pm 1,0)c,(0,0,\pm 1)c \\ (\pm 1,\pm 1,0)c,(\pm 1,0,\pm 1)c,(0,\pm 1,\pm 1)c \end{cases}$	$i=0$ $i=1-6$ $i=7-18$	$\omega_0 = 1/3$ $\omega_{1-6} = 1/18$ $\omega_{7-18} = 1/36$

from the BTE (He and Luo 1997a, b). However, the thermal model by He et al. (1998) involved a complicated source term (i.e., the second term on the right-hand side of Equation 10.7), which made it computationally expensive. Peng et al. (2003) and Shi et al. (2004) separately proposed two simplified forms of the model by He et al. (1998) widely used by the LBM community.

Peng et al. (2003) simplified the thermal LB model by He et al. (1998) for incompressible thermal flows by neglecting the viscous heat dissipation and compression work. In the model by Peng et al. (2003), the final discrete evolution equation for the internal energy distribution function is proposed as follows:

$$g_i\left(\mathbf{r}+\mathbf{e}_i\delta t,t+\delta t\right)-g_i\left(\mathbf{r},t\right)=-\frac{g_i\left(\mathbf{r},t\right)-g_i^{eq}\left(\mathbf{r},t\right)}{\tau_g} \tag{10.8}$$

where $g_i\left(\mathbf{r},t\right)$ is the discrete internal energy distribution function, and τ_g is the non-dimensional internal energy relaxation time. The macroscopic quantities, including density, velocity vector, internal energy $\varepsilon = D_0 RT/2$, and heat flux vector $\mathbf{q}$, at the solid surfaces can be expressed as:

$$\rho = \sum_i f_i, \ \mathbf{u} = \frac{1}{\rho}\left(\sum_i f_i\mathbf{e}_i\right), \ \rho\varepsilon = \sum_i g_i, \ \mathbf{q} = \left(\rho c_p\right)\frac{\tau_g - 0.5}{\tau_g}\sum_i g_i\mathbf{e}_i \tag{10.9}$$

for the D_2Q_9, D_3Q_{15}, and D_3Q_{19} lattice models (refer to Figure 10.2) the density EDF is calculated from Equation 10.6; however, (Peng et al. 2003, 2004) propose the following form for the internal energy EDF:

$$g_i^{eq} = \begin{cases} -\omega_i \rho\varepsilon \dfrac{\mathbf{u}\cdot\mathbf{u}}{2c_s^2} & i=0 \\[3mm] \omega_i \rho\varepsilon \left(a+b\dfrac{\mathbf{e}_i\cdot\mathbf{u}}{c_s^2}+\dfrac{\left(\mathbf{e}_i\cdot\mathbf{u}\right)^2}{2c_s^4}-\dfrac{\mathbf{u}\cdot\mathbf{u}}{2c_s^2}\right) & i=1-4:D_2Q_9;\ i=1-6:D_3Q_{15}\text{ and }D_3Q_{19} \\[3mm] \omega_i \rho\varepsilon \left(c+d\dfrac{\mathbf{e}_i\cdot\mathbf{u}}{c_s^2}+\dfrac{\left(\mathbf{e}_i\cdot\mathbf{u}\right)^2}{2c_s^4}-\dfrac{\mathbf{u}\cdot\mathbf{u}}{2c_s^2}\right) & i=5-8:D_2Q_9;\ i=7-14:D_3Q_{15};\ i=7-18:D_3Q_{19} \end{cases} \tag{10.10}$$

where a, b, c, d are, respectively, equal to $3/2, 1/2, 3, 2$ for the D_2Q_9 lattice, $1, 1/3, 3, 7/3$ for the D_3Q_{15} lattice, and $1, 1/3, 2, 4/3$ for the D_3Q_{19} lattice. Discrete velocity set $\mathbf{e}_i$ and weight coefficients ω_i are

Lattice Boltzmann Method for Modeling Convective Heat Transfer in Porous Media 179

available from Table 10.1. Performing the Chapman–Enskog expansion on the thermal LBGK model (Peng et al. 2003, 2004) for D_2Q_9, D_3Q_{15}, and D_3Q_{19} lattices, one can derive the following relations for the thermal diffusivity α of the fluid, for each case, respectively, as:

$$\alpha = 2(\tau_g - 0.5)c_s^2 \delta t, \; \alpha = \frac{5}{3}(\tau_g - 0.5)c_s^2 \delta t, \; \alpha = \frac{5}{3}(\tau_g - 0.5)c_s^2 \delta t \tag{10.11}$$

Shi et al. (2004) also proposed another DDF thermal LBGK model based on the original model by He et al. (1998). Those authors simplified the model (He et al. 1998) to incorporate only the viscous heat dissipation effects. Shi et al. (2004) further discussed that for the low-Prandtl or small-Eckert number thermal flows, their model could be simplified even more by neglecting the viscous heat dissipation term as well, which resulted in a model similar to the PS DDF thermal LBM. In the simplified model (Shi et al. 2004), the temperature evolution equation reads:

$$g_i(\mathbf{r} + \mathbf{e}_i \delta t, t + \delta t) - g_i(\mathbf{r}, t) = -\frac{g_i(\mathbf{r}, t) - g_i^{eq}(\mathbf{r}, t)}{\tau_g} \tag{10.12}$$

where the discrete velocity set $\mathbf{e}_i$ and the weight coefficients ω_i are available from Table 10.1 for the D_2Q_9, D_3Q_{15}, and D_3Q_{19} lattice models. In Equation 10.12, the temperature EDF $g_i^{eq}(\mathbf{r}, t)$ can be constructed as $g_i^{eq}(\mathbf{r}, t) = Tf_i^{eq}(\mathbf{r}, t)$, where $f_i^{eq}(\mathbf{r}, t)$ is available from Equation 10.6. In the model by Shi et al. (2004), the macroscopic quantities, as well as the fluid kinematic viscosity υ and thermal conductivity α, are available from the following equations, respectively:

$$\rho = \sum_i f_i, \; \mathbf{u} = \frac{1}{\rho}\left(\sum_i f_i \mathbf{e}_i\right), \; \rho T = \sum_i g_i, \; \mathbf{q} = (\rho c_p)\frac{\tau_g - 0.5}{\tau_g}\sum_i g_i \mathbf{e}_i \tag{10.13}$$

$$\upsilon = (\tau_v - 0.5)c_s^2 \delta t, \; \alpha = (\tau_g - 0.5)c_s^2 \delta t \tag{10.14}$$

It should be mentioned that in both DDF thermal models (Peng et al. 2003; Shi et al. 2004), the hydrodynamic part of the model are described by Eqs. 10.4–10.6. In thermal LBGK models, the final discrete evolution equations for the density and the internal energy (i.e., or temperature) are numerically solved through two main steps, namely, *propagation* (streaming) and *collision* as:

$$\text{Propagation:} \begin{cases} f_i(\mathbf{r} + \mathbf{e}_i \delta t, t + \delta t) = \hat{f}_i(\mathbf{r}, t) \\ g_i(\mathbf{r} + \mathbf{e}_i \delta t, t + \delta t) = \hat{g}_i(\mathbf{r}, t) \end{cases}; \; \text{Collision:} \begin{cases} \hat{f}_i(\mathbf{r}, t) = f_i(\mathbf{r}, t) - \dfrac{f_i(\mathbf{r}, t) - f_i^{eq}(\mathbf{r}, t)}{\tau_v} \\ \hat{g}_i(\mathbf{r}, t) = g_i(\mathbf{r}, t) - \dfrac{g_i(\mathbf{r}, t) - g_i^{eq}(\mathbf{r}, t)}{\tau_g} \end{cases} \tag{10.15}$$

where f_i and g_i are, respectively, the *pre-collision* density and internal energy (or temperature) distribution functions, and $\hat{f}_i$ and $\hat{g}_i$ are the corresponding *post-collision* populations, respectively. During the propagation step, the post-collision density and internal energy (or temperature) distribution functions from the previous time step propagate to the nearest neighboring lattices in each lattice direction $\mathbf{r} + \mathbf{e}_i \delta t$. The propagated distribution functions are then locally relaxed toward their equilibrium during the collision step (see Equation 10.15).

10.2.2 Multi-Relaxation Time LBM (MRT-LBM)

Thanks to its simplicity the LBGK (i.e., SRT-LBM) has been the most popular model among the LBM community (Guo and Shu 2013). However, this simplicity comes at the expense of reduced stability at low relaxation times while at higher relaxation times, accuracy is compromised (Lallemand and Luo 2000). This is because in SRT-LBM a single parameter (i.e., relaxation time) is responsible for relaxing all the conserved (i.e., density and mass fluxes in hydrodynamic model and temperature in the energy model) and nonconserved moments toward their equilibria. These deficiencies, as well as other well-known shortcomings of the SRT-LBM, were corrected by a multiple relaxation time (MRT) LBM proposed by d'Humieres (1999) around the same time that LBGK was proposed. In fact, the main difference between the MRT-LBM and SRT-LBM lies in the collision operator, that is, in the MRT-LBM instead of a single relaxation time there is a relaxation matrix (i.e., a diagonal matrix), where, each moment relaxes toward its equilibrium at a specific rate, which enhances the stability and accuracy of the model (d'Humieres 1992). In 2D MRT-LBM, normally D_2Q_9 and D_2Q_5 lattices are, respectively, used for the hydrodynamic and thermal modeling.

The discrete density and energy evolution equations for the $D_2Q_9 - D_2Q_5$ MRT-LBM are, respectively, available as follows (d'Humieres 1992; Mezrhab et al. 2010):

$$f_i\left(\mathbf{r}+\mathbf{e}_i\delta t,t+\delta t\right)-f_i\left(\mathbf{r},t\right)=-\mathbf{M}^{-1}\mathbf{S}\Big[\big|m(\mathbf{r},t)\big\rangle-\big|m^{eq}(\mathbf{r},t)\big\rangle\Big] \qquad (10.16)$$

$$g_i\left(\mathbf{r}+\mathbf{e}_i\delta t,t+\delta t\right)-g_i\left(\mathbf{r},t\right)=-\mathbf{N}^{-1}\mathbf{\Theta}\Big[\big|n(\mathbf{r},t)\big\rangle-\big|n^{eq}(\mathbf{r},t)\big\rangle\Big] \qquad (10.17)$$

where $\big|m(\mathbf{r},t)\big\rangle$ and $\big|n(\mathbf{r},t)\big\rangle$, are the microscopic velocity moment vector of the density, and temperature distribution functions (i.e., f and g), $\big|m^{eq}(\mathbf{r},t)\big\rangle$ and $\big|n^{eq}(\mathbf{r},t)\big\rangle$ are the equilibrium counterpart of the moment vectors that for the D_2Q_9 and D_2Q_5 lattice models with $c = 1$ are given by:

$$\big|m^{eq}(\mathbf{r},t)\big\rangle=\left(\rho,-2\rho+3|\mathbf{u}|^2, \rho-3|\mathbf{u}|^2, u_x,-u_x,u_y,-u_y,\left(u_x^2-u_y^2\right),u_xu_y\right) \qquad (10.18a)$$

$$\big|n^{eq}(\mathbf{r},t)\big\rangle=\left(T,u_xT,u_yT,aT,0\right) \qquad (10.18b)$$

where T is the macroscopic temperature, u_x and u_y are x and y components of the velocity, and a is a constant with a numerical value less than one for the stability of the D_2Q_5 thermal MRT-LBM to be ensured (Mezrhab et al. 2010).

$\mathbf{M}$ and $\mathbf{N}$ in Eqs. 10.16 and 10.17 are the hydrodynamic and energy transformation matrices (i.e., 9×9 and 5×5, respectively) that map the moment vector from the velocity space to the moment space (i.e., $\big|m(\mathbf{r},t)\big\rangle=\mathbf{M}\big|f(\mathbf{r},t)\big\rangle$ and $\big|n(\mathbf{r},t)\big\rangle=\mathbf{N}\big|g(\mathbf{r},t)\big\rangle$) or vice versa (i.e., $\big|f(\mathbf{r},t)\big\rangle=\mathbf{M}^{-1}\big|m(\mathbf{r},t)\big\rangle$ and $\big|g(\mathbf{r},t)\big\rangle=\mathbf{N}^{-1}\big|n(\mathbf{r},t)\big\rangle$). $\mathbf{M}$ and $\mathbf{N}$ for the D_2Q_9 and D_2Q_5 lattice models are available from d'Humieres (1992) and Mezrhab et al. (2010). In Eqs. 10.16 and 10.17, $\mathbf{S}$ and $\mathbf{\Theta}$ are the hydrodynamic and energy relaxation matrices presented as follows:

$$\mathbf{S}=diag\left(\zeta_\rho,\zeta_e,\zeta_\varphi,\zeta_j,\zeta_q,\zeta_j,\zeta_q,\zeta_\pi,\zeta_\pi\right); \ \mathbf{\Theta}=diag\left(\eta_T,\eta_\alpha,\eta_\alpha,\eta_v,\eta_v\right) \qquad (10.19)$$

where each element of $\mathbf{S}$ and $\mathbf{\Theta}$ is related to relaxing a specific physical quantity where $\zeta_\pi=1/\tau_v$ and $\eta_\alpha=1/\tau_g$. For further details on choosing other elements of the relaxation matrices $\mathbf{S}$ and $\mathbf{\Theta}$, the reader can consult (d'Humieres 1992; Mezrhab et al. 2010). Eqs. 10.16 and 10.17 are solved in two steps. First, the propagation step is performed in the velocity space similar to the SRT-LBM, then $\mathbf{M}$ and $\mathbf{N}$ are used to transform the propagated populations into the moment space. Next, the collision step is carried out in the moment space as:

$$\big|\hat{m}(\mathbf{r},t)\big\rangle=\big|m(\mathbf{r},t)\big\rangle-\mathbf{S}\Big[\big|m^{eq}(\mathbf{r},t)\big\rangle-\big|m(\mathbf{r},t)\big\rangle\Big];\big|\hat{n}(\mathbf{r},t)\big\rangle=\big|n(\mathbf{r},t)\big\rangle-\mathbf{\Theta}\Big[\big|n^{eq}(\mathbf{r},t)\big\rangle-\big|n(\mathbf{r},t)\big\rangle\Big] \quad (10.20)$$

where $\big|\hat{m}(\mathbf{r},t)\big\rangle$ and $\big|\hat{n}(\mathbf{r},t)\big\rangle$ are the post-collision moment vectors in the moment space. $\mathbf{M}^{-1}$ and $\mathbf{N}^{-1}$ are then used to map the post-collision moments back to the velocity space for the next propagation.

10.2.3 Boundary and Interface Conditions

Figure 10.3 illustrates the solid–fluid interface in an LBM pore-scale simulation of the convection heat transfer employing the D_2Q_9 lattice model. As seen in Figure 10.3 at nodes **A**, **B**, and **C**, for example, there are a number of unknown density f and internal energy g populations coming from the opposing medium (e.g., Ω_f or Ω_s) after the propagation step, which should be determined. The LBM no-slip and conjugate heat transfer schemes are formed based on finding those unknown incoming density and internal energy populations.

10.2.3.1 No-Slip

One particular feature of the LBM, that from the very beginning made it extremely attractive in dealing with pore-scale fluid flow simulations in the complex porous geometries, was the bounce back scheme Succi et al. (1989). The *standard bounce back* scheme assumes that the straight boundary is placed exactly on the lattice nodes as shown in Figure 10.4a, where an outgoing density population along a given lattice link is exactly reversed on the same link as follows:

$$\hat{f}_{\bar{i}}(\mathbf{r}_b, t) = \hat{f}_i(\mathbf{r}_f, t) \tag{10.21}$$

where $e_i = -e_{\bar{i}}$, $\hat{f}$ is the post-collision density distribution function, $\mathbf{r}_b$ is the boundary node, and $\mathbf{r}_f = \mathbf{r}_b - \mathbf{e}_i \delta t$ is the neighboring fluid node. The standard bounce back is of the first-order accuracy. Another bounce back scheme called the *halfway bounce back* finds the unknown density distribution populations at the boundary nodes from the same formulation as the standard bounce back scheme does (see Equation 10.21); however, the only difference is that the straight boundary is now placed at the middle of the lattice link connecting the interface solid and fluid neighboring nodes as illustrated in Figure 10.4b. The halfway bounce back scheme is considered to be of the second-order accuracy, which makes it extremely popular. In addition to the bounce back, other LBM schemes for satisfying the no-slip conditions were also presented, which are mostly of the second-order accuracy (Guo and Shu 2013). However, all the above-mentioned no-slip schemes are only applicable to the straight boundaries. Therefore, the researchers who employed the above-mentioned no-slip schemes in pore-scale fluid flow simulation in porous media, either considered simplified porous geometries such as an array of square or cubical blocks (Imani and Hooman 2017; Imani et al. 2013; Succi et al. 1989) or approximated the curved boundaries (e.g., a pack of sphere bed) with stair-like boundaries (Inamuro et al. 1999; Sheikh and Qiu 2018; van der Hoef et al. 2005) as shown in Figure 10.5. It should be stated that this geometry approximation brings about some errors. During the last two decades, a number of second-order schemes have also been presented for satisfying the no-slip conditions at the curved boundaries, in

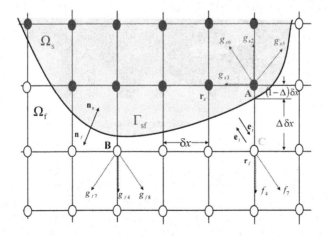

FIGURE 10.3 The schematic of a curved solid–fluid interface on a D_2Q_9 lattice model.

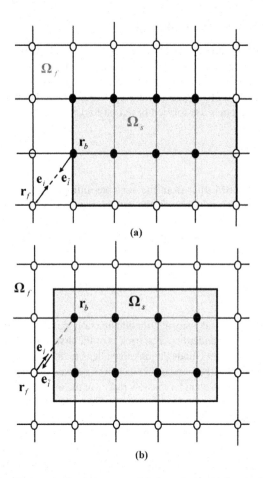

FIGURE 10.4 The schematic of bounce back schemes at straight boundaries: (a) Standard; and (b) Halfway.

FIGURE 10.5 A stair-like approximation of the pore space curved boundary in a metallic foam. (Reprinted with permission from Chiappini, D., *Int. J. Heat Mass. Transf.*, 117, 527–537, 2018.)

which, the accurate shape of the boundary is preserved. Among those schemes, the modified bounce back method of Filippova and Hänel (1997) and later Mei et al. (1999), the interpolation bounce back method of Bouzidi et al. (2001), and the non-equilibrium extrapolation method of Guo et al. (2002a) are widely implemented by others. Nonetheless, being less computationally intensive, the bounce back scheme with a stair-like approximation of the voids (see Figure 10.5) remained more popular than using the above-mentioned accurate curved boundary schemes (Inamuro et al. 1999; Sheikh and Qiu 2018; van der Hoef et al. 2005).

10.2.3.2 Conjugate Heat Transfer

Pore-scale simulation of convection heat transfer in porous media poses a solid–fluid conjugate heat transfer problem. The conjugate heat transfer conditions guarantee the continuity of the temperature and normal heat flux at the solid–fluid interface (i.e., Γ_{sf} in Figure 10.3) as follows:

$$T_f\big|_{\Gamma_{sf}} = T_s\big|_{\Gamma_{sf}}; \quad -k_f \frac{\partial T_f}{\partial \mathbf{n}}\Big|_{\Gamma_{sf}} = -k_s \frac{\partial T_s}{\partial \mathbf{n}}\Big|_{\Gamma_{sf}} \tag{10.22}$$

where Γ_{sf} denotes the solid–fluid interface, k_f and k_s are the fluid and solid thermal conductivities, respectively, and $\mathbf{n}$ is the normal unit vector of the interface.

The conjugate heat transfer is rather a new research topic in the LBM framework with a kickstart about a decade ago. The LBM conjugate heat transfer schemes can be generally categorized into three groups:

1. *The steady-state schemes*, in which, both solid and fluid phases are considered to have the same volumetric heat capacities equal to that of fluid. Hence, in those methods the conjugate heat transfer conditions are automatically satisfied at the interface through the propagation step without any special treatment (Karani and Huber 2015). From this category, the scheme by Wang et al. (2007) called the "*half-lattice division*" was the first attempt and the simplest method proposed in the LBM framework to simulate the conjugate heat transfer restricted to the straight interfaces in steady-state problems.

2. *The interpolation schemes*, where the unknown internal energy populations crossing the interface after the propagation step (see Figure 10.3) are corrected based on the continuity of the temperature and normal heat flux as well as employing the interpolation or extrapolation schemes. Meng et al. (2008) pioneered an LBM conjugate heat transfer scheme based on the counter-slip internal energy idea (D'Orazio and Succi 2004) for straight boundaries, which worked for both transient and steady problems. Those authors considered the unknown energy distribution functions crossing the interface to be the equilibrium populations of an unknown counter-slip internal energy, which could be determined by satisfying the conjugate heat transfer conditions (Imani et al. 2012a, b). From this category, Li et al. (2014) were the first group to present an LBM conjugate heat transfer scheme for the curved interfaces based on the bounce back idea and the interpolation method. To satisfy the conjugate heat transfer conditions, Li et al. (2014) considered the exact topology of the solid–fluid interface involved with finding the local unit vector normal to the interface, which made their formulation too complicated for pore-scale simulations in real porous media. Le et al. (2015) employed a second-order extrapolation of the temperature along the line normal to the interface in both solid and fluid media in an attempt to convert the conjugate heat transfer condition into a Dirichlet-type boundary condition. Mozafari-Shamsi et al. (2016b) presented a conjugate heat transfer scheme for curved boundaries based on the ghost-fluid thermal LBM (Khazaeli et al. 2013; Mozafari-Shamsi et al. 2016a).

3. *The modified thermal LBGK schemes*, in which, the thermal LBGK model is modified to take into account the spatial variation of the volumetric heat capacities within the heterogeneous media for temperature and normal heat flux to be automatically conserved at the conjugate interface, independent of the interface topology. Karani and Huber (2015) were pioneers in this category. Subsequent studies proposed different approaches for the inclusion of the spatial variation of the volumetric heat capacity into the standard thermal LBGK (Chen et al. 2017b; Hu et al. 2015; Lu et al. 2017; Wu et al. 2017).

10.2.4 A Brief Review of the Applications

LBM pore-scale simulation of the fluid flow in porous media is almost as old as the numerical method itself. Pore-scale simulation in porous media not only offers detailed information about the velocity and temperature distributions within the complicated pore space but also can serve as a numerical experiment platform for determining the macroscopic geometrical, thermophysical, and flow characteristics, including the permeability, effective thermal conductivity, pressure drop, and interstitial heat transfer coefficient from the microporous structure to be used in the macroscale models.

Succi et al. (1989) pioneered the use of LBGK with the bounce back scheme to simulate a low Reynolds number isothermal fluid flow through a randomly distributed three-dimensional (3D) blocks. Those authors successfully validated Darcy's law and calculated the permeability of the porous medium. Inamuro et al. (1999) employed the LBM to investigate the relation between the pressure drop and the Reynolds number considering a pore-scale simulation of the fluid flow across a sphere bead. Inamuro et al. (1999) found that the numerical results were in a good agreement with the Blake–Kozney and Ergun equations for the low and high Reynolds numbers, respectively. van der Hoef et al. (2005) used the LBGK model to simulate a low Reynolds number flow past mono and bidisperse random arrays of spheres. They validated their result for the permeability with the available experimental data and provided new relations for the permeability of the monodisperse and polydisperse porous media. Pan et al. (2006) evaluated the SRT and MRT-LBM in simulating the pore-scale fluid flow in porous media to conclude that the MRT-LBM was definitely superior to the SRT counterpart in terms of the accuracy and stability. Maier and Bernard (2010) investigated the accuracy of the different first-order and second-order no-slip schemes through LBGK pore-scale simulations of the fluid flow in porous media. Beugre et al. (2010) employed the MRT-LBM to simulate the fluid flow in a metallic foam. Those authors used the microtomography technique to reconstruct 3D images of the metallic foam as shown in Figure 10.6. Most recently, Sheikh and Qiu (2018) employed a GPU-based 3D MRT-LBM to investigate the fluid flow past a sphere bed in order to determine the velocity distribution in the pore space and drag force on individual particles for low and moderate Reynolds numbers. Those authors identified three velocity zones in the sphere bed, namely, the low velocity, high velocity, and recirculation zones as illustrated in Figure 10.7.

LBM pore-scale simulation of the conduction and convection heat transfer in porous media, on the other hand, is rather a new topic, as old as the first LBM conjugate heat transfer scheme proposed by Wang et al. (2007) about a decade ago. In a series of studies, the conjugate heat transfer scheme proposed by Wang et al. (2007) was used along with the thermal LBGK model (Peng et al. 2003) to predict the effective thermal conductivity of different saturated porous media (e.g., fibrous, open-cell random, and foam) Wang and Pan 2008; Wang et al. 2006, 2007a, b) through steady pore-scale conduction simulations. Zhao et al. (2010) employed the thermal LBGK model to investigate the pore-scale natural convection in an enclosure filled with arrays of the square and circular blocks. Imani et al. (2013) employed the simplified thermal LBGK model (Peng et al. 2003) together with the conjugate heat transfer scheme

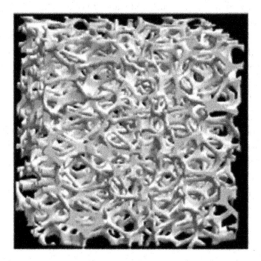

FIGURE 10.6 3D reconstructed metallic foam geometry via the computed tomography method. (Reprinted with permission from Beugre, D. et al. *J. Comput. Appl. Math.*, 234, 2128–2134, 2010.)

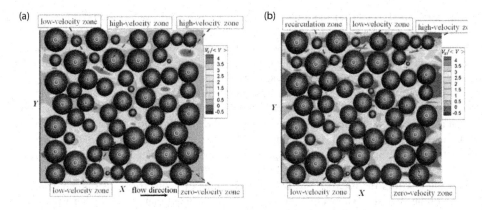

FIGURE 10.7 Velocity contours on the mid-plane of the spherical bed: (a) Re = 0.1; and (b) Re = 80 (Reprinted with permission from Bahman Sheikh and Tong Qiu "Pore-Scale Simulation and Statistical Investigation of Velocity and Drag Force Distribution of Flow through Randomly-Packed Porous Media under Low and Intermediate Reynolds Numbers" *Computers & fluids* 171, 15–28, 2018.)

(Meng et al. 2008) to investigate the pore-scale constant heat flux splitting between the solid and fluid phases at the boundary of a porous medium, through a forced convection simulation in a parallel plate channel filled with an array of conducting thermally connected square blocks. Those authors aimed to numerically evaluate the available macroscale constant heat flux split models at the porous boundary presented in (Alazmi et al. 2002) as models 1D, 1E, 1F, 2A, and 2B solely based on the geometrical and thermophysical parameters, where the flow parameters are absent. The first three models (1D, 1E, and 1F) consider that the constant heat flux splits between the solid and liquid phases based on their thermophysical and geometrical parameters; however, the last two models (2A and 2B) state that each phase receives a heat flux equal to the constant boundary heat flux regardless of thermophysical and geometrical parameters. Imani et al. (2013) based on their pore-scale results pointed out that the constant heat flux splitting at the wall of a porous medium is a complex process, which depends on the microscopic structure of the porous medium, porosity ϕ, solid–fluid thermal conductivity ratio λ, and the pore Peclet number Pe_p. As such, those authors compared their results with macroscale models 1D, 1E, 1F, 2A, and 2B (Alazmi et al. 2002) for $0.4 < \phi < 0.98$, $1 < \lambda < 1000$, and $7 < Pe_p < 350$ to determine the range of parameters over which those macroscale models fail to predict the details of heat flux splitting as shown in Table 10.2 by gray cells. As seen in Table 10.2, the models 2A and 2B deviate from the pore-scale results for $\lambda \gg 1$, because under this circumstance the resistance of the solid matrix to the heat flow considerably decreases, which violates the equality of heat fluxes assumed in models 2A and 2B. However, the most important

TABLE 10.2

The Range of Parameters over Which the Macroscale Models Fail to Predict Heat Flux Splitting at the Wall of the Porous Media (marked as gray cells)

Parameters \ Macro Scale Models	2A	2B	1D	1E	1F
$\phi \leq 0.7$, $\lambda = 1$					
$\phi > 0.7$, $\lambda = 1$					
$\phi \leq 0.7$, $\lambda \gg 1$	▓	▓			
$\phi > 0.7$, $\lambda \gg 1$	▓	▓			

Source: Data adapted from Imani, G. et al., *Transport Porous Med.*, 98, 631–649, 2013.

result from Table 10.2 is that for the higher values of porosity and solid-to-fluid thermal conductivity ratio, which are the case for most of the heat transfer applications of porous materials, available models fail to describe the constant heat flux splitting. That is because under these conditions, the effect of flow parameters on the heat flux splitting is pronounced, as a result, the macroscale models in which the flow parameters are absent cannot follow the pore-scale results.

In a subsequent work, Imani et al. (2017) used the thermal LBGK method with the conjugate heat transfer scheme (Meng et al. 2008) to investigate the pore-scale natural convection in an enclosure filled with a detached or attached bidisperse porous medium (BDPM) to assess the validity of the LTE assumption within the microporous media invoked by the available macroscale BDPM models (Nield and Kuznetsov 2006). Figure 10.8a and b, respectively, shows streamlines and isotherms within a BDPM with $\varepsilon_{mic} = 0.64$, $\varphi_{mac} = 0.85$, and $Ra = 10^7$. Imani et al. (2017) calculated the maximum percentage of non-equilibrium within the microporous over a wide range of geometrical, thermophysical, and flow parameters in the BDPM as shown in Table 10.3. They concluded that for the higher values of microporous porosity ε_{mic} and Rayleigh number, the LTE assumption within the microporous is not justified.

Liu and Wu (2016) employed the LBGK model to study the pore-scale forced convection in a 3D reconstructed porous medium with applying the constant temperature boundary condition at the solid–fluid interface instead of the conjugate heat transfer conditions. Chiappini (2018) employed a hybrid lattice Boltzmann method (i.e., an D_2Q_9 LBGK for the hydrodynamic and FV method for thermal simulations) to study natural convection in open-cell metal foams. The metal foam geometry was reconstructed via packing the elementary Kelvin unit cell. As shown in Figure 10.9, Chiappini (2018) compared the results of the Nusselt number from his simulation with those available from the experimental data.

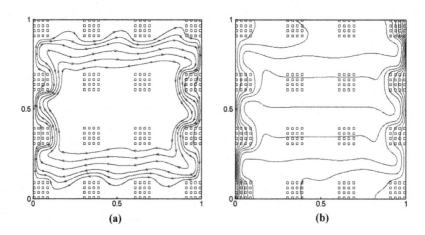

FIGURE 10.8 Natural convection in a BDPM with $\varepsilon_{mic} = 0.64$, $\varphi_{mac} = 0.85$, and $Ra = 10^7$: (a) Streamlines; and (b) Isotherms. (Reprinted with permission from Imani et al. "Lattice Boltzmann pore scale simulation of natural convection in a differentially heated enclosure filled with a detached or attached bidisperse porous medium" *Transport in Porous Media* 116, 91–113, 2017.)

TABLE 10.3

Max(%LTNE) Calculated within the Microporous for the Attached BDPM Geometry

	$\varepsilon_{mic} = 0.64$			$\varphi_{mac} = 0.64$		
Ra, λ	$\varphi_{mac} = 0.50$	$\varphi_{mac} = 0.64$	$\varphi_{mac} = 0.85$	$\varepsilon_{mic} = 0.50$	$\varepsilon_{mic} = 0.64$	$\varepsilon_{mic} = 0.85$
$10^6, 1$	1.4	1.0	0.6	0.6	1.0	4.2
$10^7, 1$	**8.8**	**5.4**	3.2	2.8	**5.4**	**10.0**
$10^6, 100$	2.5	0.5	0.5	1.0	0.5	3.4
$10^7, 100$	**7.0**	3.6	2.4	1.3	3.6	**16.0**

Source: Data from Imani, G. et al., *Transport Porous Med.*, 116, 91–113, 2017.

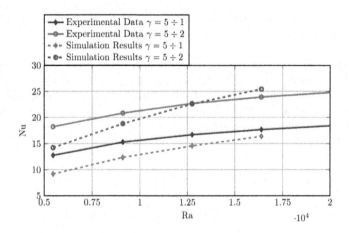

FIGURE 10.9 Comparing the results of average Nusselt number between the numerical and experimental data for two metal foam aspect ratio. (Reprinted with permission from Chiappini, D. *Int. J. Heat Mass. Transf.*, 117, 527–537, 2018.)

10.3 LBM Macroscale Models for Convection Heat Transfer in Porous Media

Unlike the pore-scale simulation, with macroscale models for convection in porous media details of the fluid flow and heat transfer within the pores cannot be recovered. Instead, the standard LBM is modified to hydrodynamically and thermally take into account the effective presence of the porous medium, generally by adding force and heat source terms, respectively, to the density and energy evolution equations, as well as modifying the density and temperature EDFs. As such, the LBM macroscale models for convection heat transfer in porous media strongly depend on the choice of LBM discrete forcing and heat source schemes; see (Guo et al. 2002b; McCracken et al. 2005) and (Cha and Zhao 2013; Chopard et al. 2009) for different LBM discrete forcing and heat source schemes.

10.3.1 Governing Equations

10.3.1.1 Hydrodynamics

The presented LBM macroscale models for simulating the fluid flow in porous media are based on modifying the LBGK model to take into account the momentum sink due to the presence of the porous medium. Spaid and Phelan (1997) were first to propose an LBM macroscale model (referred to as SP-LBM) for fluid flow simulation in porous media about two decades ago. Those authors adapted the density EDF in the standard LBM (i.e., LBGK) to recover the Brinkman-extended Darcy equation (i.e., $\nabla p = \mu \mathbf{u}/K + \mu_{eff} \nabla^2 \mathbf{u}$) in which $\mu_{eff} = \mu$ was assumed. Where μ_{eff} and μ, respectively, are the effective and fluid dynamic viscosities. In their work, the local macroscopic velocity, used in calculating the density EDF, was modified by incorporating the effect of the porous medium resistance $\mathbf{F}$ to the fluid flow, following the Shan–Chen LBM forcing method (Shan and Chen 1993) as $\mathbf{u}^{eq} = \mathbf{u} + \tau_v \mathbf{F}/\rho$, where $\mathbf{F} = -\mu \mathbf{u}/K$ is the Darcy's linear drag term, τ_v is the non-dimensional velocity relaxation time, with ρ and $\mathbf{u}$ being the volume-averaged fluid density and velocity calculated from Equation 10.5. Freed (1998) also used the same approach to incorporate the porous resistance force into the LBGK by modifying the equilibrium velocity.

Dardis and McCloskey (1998) modified the LBGK evolution equation (see Equation 10.4) to incorporate the hydrodynamic effects of the presence of the porous medium in order to recover the Brinkman-extended Darcy equation with $\mu_{eff} = \mu$ (which is not always the case). Those authors first introduced a continuous variable $n_s(\mathbf{r})$ as the solid scatterer density per node $\mathbf{r}$ where $0 < n_s(\mathbf{r}) < 1$. Then, to satisfy the no-slip condition due to the presence of the porous medium, the LBGK evolution equation was modified as follows:

$$f_i(\mathbf{r}+\mathbf{e}_i\delta t, t+\delta t) - f_i(\mathbf{r},t) = -\frac{f_i(\mathbf{r},t) - f_i^{eq}(\mathbf{r},t)}{\tau_v} + n_s(\mathbf{r}+\mathbf{e}_i\delta t)\left[f_{\bar{i}}(\mathbf{r}+\mathbf{e}_i\delta t, t) - f_i(\mathbf{r},t)\right] \quad (10.23)$$

where the second term on the right-hand side of Equation 10.23 is a partial bounce back term accounting for the presence of the porous medium. In their work, the effective kinematic viscosity was calculated as $\upsilon_{eff} = (\tau_v - 0.5)c_s^2 \delta t$.

Martys (2001) improved the LBM macroscale model proposed by Spaid and Phelan (1997) to recover the Brinkman-extended Darcy equation for fluid flow in porous media allowing $\mu_{eff} \neq \mu$ to be applied. He argued that the macroscale model proposed by Spaid and Phelan (1997) introduced some error terms of order $\tau_v^2 F^2$ into the Brinkman-extended Darcy equation where F was the body force. Martys (2001) proposed a modification to the evolution equation of the LBGK model to account for the presence of the porous medium as:

$$f_i\left(\mathbf{r}+\mathbf{e}_i \delta t, t+\delta t\right) - f_i\left(\mathbf{r},t\right) = -\frac{f_i\left(\mathbf{r},t\right) - f_i^{eq}\left(\mathbf{r},t\right)}{\tau_v} + \delta t F_i \tag{10.24}$$

where F_i is the discrete body force calculated from the following equation originally proposed by Luo (1993):

$$F_i = \omega_i \left[\frac{\left(\mathbf{e}_i - \mathbf{u}\right) \cdot \mathbf{F}}{c_s^2} + \frac{\mathbf{u}\mathbf{F}:\mathbf{e}_i\mathbf{e}_i}{c_s^4} \right] \tag{10.25}$$

Here, $\mathbf{F} = -\mu\mathbf{u}/K$ is Darcy's linear drag term and ω_i are the weight coefficients available from Table 10.1 for different lattice models. Although the proposed model by Martys (2001) improved the accuracy and stability of Spaid and Phelan's (1997) model, Guo et al. (2002b) proved that the discrete forcing method employed by Martys (2001) did not recover the correct hydrodynamics.

Guo and Zhao (2002) proposed a modified LBGK model called the generalized lattice Boltzmann equation (GLBE) that recovered the Brinkman–Forchheimer-extended Darcy equation. Those authors discussed that the previous LBM models based on the Brinkman-extended Darcy equation suffered from unrealistic velocity results due to ignoring the convective term in the momentum equation and were applicable only to low-speed flows. Guo and Zhao (2002) first introduced the effect of the porosity of the porous medium ϕ into the standard LBM density EDF (see Equation 10.6) as follows:

$$f_i^{eq} = \omega_i \rho \left[1 + \frac{\mathbf{e}_i \cdot \mathbf{u}}{c_s^2} + \frac{\left(\mathbf{e}_i \cdot \mathbf{u}\right)^2}{2\phi c_s^4} - \frac{\mathbf{u} \cdot \mathbf{u}}{2\phi c_s^2} \right] \tag{10.26}$$

Those authors also incorporated the total body force $\mathbf{F}$ due to the presence of the porous medium (i.e., Darcy and Forchheimer drag terms) as well as other external forces into the LBGK density evolution equation. To do so, an LBM discrete forcing scheme proposed by Guo et al. (2002b) was used to calculate the discrete body force in different lattice directions F_i, as shown in Equation 10.27. This discrete forcing method is proved to recover the correct hydrodynamic equation (Guo et al. 2002b):

$$F_i = \omega_i \left(1 - \frac{1}{2\tau_v}\right) \left[\frac{\mathbf{e}_i \cdot \mathbf{F}}{c_s^2} + \frac{\mathbf{u}\mathbf{F}:\mathbf{e}_i\mathbf{e}_i}{\phi c_s^4} - \frac{\mathbf{u} \cdot \mathbf{F}}{\phi c_s^2} \right] \tag{10.27}$$

where $\mathbf{F} = -\phi\upsilon\mathbf{u}/K - \left(\phi F_\varepsilon/\sqrt{K}\right) \mathbf{u}\left|\mathbf{u}\right| + \phi\mathbf{G}$ is the total body force, in which, the first and second terms are, respectively, Darcy and Forchheimer drag terms, and the last term accounts for the other body forces (e.g., buoyancy). F_ϕ is the geometrical function, and K is the permeability of the porous medium, which can be obtained from Ergun's experimental relation given for packed bed of spheres, respectively, as follows:

$$F_\phi = \frac{1.75}{\sqrt{150\phi^3}}, \quad K = \frac{\phi^3 d_p^2}{150\left(1-\phi\right)^2} \tag{10.28}$$

where d_p is the solid particle diameter in the packed bed. After calculating the microscopic discrete forces F_i, the LBGK evolution equation is modified as follows:

$$f_i(\mathbf{r}+\mathbf{e}_i\delta t,t+\delta t)-f_i(\mathbf{r},t)=-\frac{f_i(\mathbf{r},t)-f_i^{eq}(\mathbf{r},t)}{\tau_\nu}+\delta tF_i \qquad (10.29)$$

In the discrete forcing scheme (Guo et al. 2002b), the velocity needs to be shifted as $\mathbf{u}=1/\rho\left(\sum_i f_i\mathbf{e}_i+\mathbf{F}\delta t/2\right)$ to predict the correct hydrodynamic equation. Because $\mathbf{F}$ on the right-hand side of this shifted velocity formulation is a non-linear function of $\mathbf{u}$ itself, one cannot explicitly calculate $\mathbf{u}$ from this formulation. However, thanks to the quadrature nature of the velocity formulation, $\mathbf{u}$ can be explicitly derived as $\mathbf{u}=\mathbf{V}/\left(c_0+\sqrt{c_0^2+c_1|\mathbf{V}|}\right)$, where $\mathbf{V}=1/\rho(\sum_i f_i\mathbf{e}_i+\phi\mathbf{G}\delta t/2)$ is a temporarily defined velocity with $c_0=1/2[1+(\phi\delta t/2)\upsilon/K]$ and $c_1=(\phi\delta t/2)F_\varepsilon/\sqrt{K}$. In the GLBE model, the volume-averaged density and pressure are calculated from $\rho=\sum_i f_i$ and $p=\rho c_s^2/\phi$. After applying the Chapman–Enskog analysis to the GLBE model, Guo and Zhao (2002) proved that the Brinkman–Forchheimer-extended Darcy equation can be recovered:

$$\frac{\partial\mathbf{u}}{\partial t}+(\mathbf{u}.\nabla)\left(\frac{\mathbf{u}}{\phi}\right)=-\frac{1}{\rho}\nabla(\phi p)+\upsilon_{eff}\nabla^2\mathbf{u}+\mathbf{F} \qquad (10.30)$$

where the effective kinematic viscosity is recovered as $\upsilon_{eff}=c_s^2(\tau_\nu-0.5)\delta t$, which is not necessarily equal to the fluid kinematic viscosity υ. Those authors validated the proposed GLBE against the CFD results for three different benchmark problems, namely, the generalized Poiseuille flow, the plane Couette flow, and flow in a lid-driven cavity. As shown in Figure 10.10, the velocity distribution obtained from the GLBE for the Poiseuille flow agrees well with that of the FD method for different values of the Reynolds and Darcy numbers.

Guo et al. (2009) later modified the GLBE model (Guo and Zhao 2002) for incompressible fluids experiencing a large pressure gradient. This large pressure gradient can introduce the compressibility errors in the LBM models of flow in porous media based on the standard LBM, in which, an equation of state is used to calculate the pressure. Those authors proposed a pressure-based LBM, where, a pressure distribution function used instead of the density distribution function. However, in their model (Guo et al. 2009) an unwanted term appeared in the continuity equation when Chapman-Enskog expansion was performed.

10.3.1.2 Energy

10.3.1.2.1 LTE

Guo and Zhao (2005) proposed a thermal LBM macroscale model for convection heat transfer in porous media invoking the LTE assumption based on their GLBE flow model (Guo and Zhao 2002). Following the PS DDF thermal LBGK models, those authors employed a separate temperature distribution function to describe the temperature evolution equation:

$$T_i(\mathbf{r}+\mathbf{e}_i\delta t,t+\delta t)-T_i(\mathbf{r},t)=-\frac{T_i(\mathbf{r},t)-T_i^{eq}(\mathbf{r},t)}{\tau_g} \qquad (10.31)$$

where τ_g is the non-dimensional energy relaxation time, T_i is the discrete temperature distribution function, and T_i^{eq} is its equilibrium counterpart proposed as follows:

$$T_i^{eq}=\omega_i T\left(\sigma+\frac{\mathbf{u}.\mathbf{e}_i}{c_s^2}\right) \qquad (10.32)$$

Here, T is the volume-averaged fluid temperature defined as $\sigma T=\sum_i T_i$, in which, $\sigma=(\rho c_p)_{eff}/(\rho c_p)_f$ is the volumetric heat capacity ratio defined as the ratio of the effective volumetric heat capacity

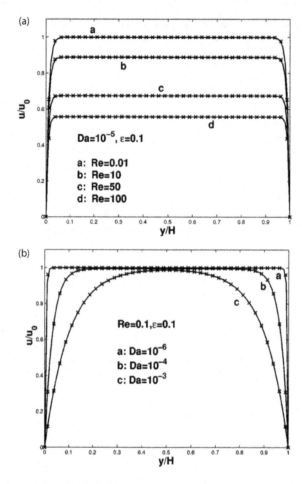

FIGURE 10.10 Comparison of the velocity distribution in a Poiseuille flow obtained from GLBE (symbols) and FD method (solid lines) for different (a) Reynolds; and (b) Darcy numbers. (Reprinted with permission from Guo, T. and Zhao, T., *Phys. Rev. E* 66, 3, 036304, 2002.)

$(\rho c_p)_{eff} = \phi(\rho c_p)_f + (1-\phi)(\rho c_p)_s$ to the fluid volumetric heat capacity $(\rho c_p)_f$. Through a Chapman–Enskog analysis, it can be shown that the presented model recovers the volume-averaged energy equation in porous media with LTE assumption, as presented in Equation 10.33, provided that the spatial and temporal variation of the volumetric heat capacity ratio σ could be ignored.

$$\sigma\frac{\partial T}{\partial t} + \mathbf{u}.\nabla T = \nabla.\left(\alpha_{eff}\nabla T\right) \quad (10.33)$$

In the model by Guo and Zhao (2005), the effective thermal diffusivity $\alpha_{eff} = \left[k_{eff} + k_d\right]/(\rho c_p)_f$ is derived to be $\alpha_{eff} = \sigma c_s^2(\tau_g - 0.5)\delta t$. Where $k_{eff} = \phi k_f + (1-\phi)k_s$ is the effective stagnant thermal conductivity of the porous medium, and k_d is thermal conductivity due to the thermal dispersion. Guo and Zhao (2005) validated their model through a number of natural convection problems with assuming $\sigma = 1$.

Based on the GLBE model for the macroscale fluid flow simulation in porous media (Guo and Zhao 2002), Seta et al. (2006a) employed the simplified DDF thermal LBGK model (Peng et al. 2003) to recover the LTE volume-averaged energy equation in the porous media with assuming $\sigma = 1$. Those authors applied their model to a natural convection problem in a porous enclosure and compared the computational time with that of the FD method to conclude the superiority of the LBM. Inspired by the GLBE model (Guo and Zhao 2002) and previous LBM axisymmetric thermal models, Rong et al. (2010) proposed a new thermal DDF model for axisymmetric convection heat transfer in porous media based on the LTE assumption. Wang et al. (2016a) modified the model by Guo and Zhao (2005) to enhance its stability at low values of the dynamic viscosity and effective thermal diffusivity. Liu and He (2017) extended Guo and Zhao's (2005) model to the MRT model. Those authors employed an D_2Q_9 MRT-LBM model for the hydrodynamic part, whereas, in their research, an D_2Q_5 MRT-LBM model was employed for modeling the thermal problem. Liu et al. (Guo and Zhao 2005) showed that their MRT-LBM model could enhance the accuracy and stability of the SRT (i.e., LBGK) model for low viscosity cases as shown in Figure 10.11. Recently, Chen et al. (2017a) argued that the models proposed by Guo and Zhao (2005) and Liu and He (2017) for convection heat transfer in porous media under the LTE assumption cannot recover the correct macroscopic energy equation in porous media with spatially varying volumetric heat capacity ratio. To address the issue, Chen et al. (2017a) introduced a constant reference volumetric heat capacity ratio σ_0 into the temperature EDF and proved that their model was capable of modeling the convection heat transfer in porous media with spatially varying volumetric heat capacity ratio. As shown in Figure 10.12, Chen et al. (2017a) compared the result of their model with that of Guo and Zhao (2005) through a natural convection heat transfer simulation in a bilayered porous medium to show the superiority of their model in dealing with spatially varying volumetric heat capacity ratio porous media.

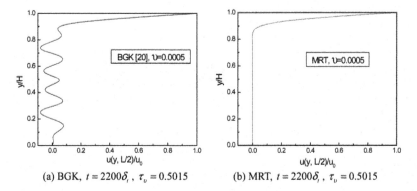

FIGURE 10.11 LBM result for the velocity distribution in a porous enclosure with low fluid viscosity ($\tau_\nu = 0.5015$): (a) SRT-LBM; and (b) MRT-LBM. (Reprinted with permission from Liu, Q. et al., *Int. J. Heat Mass. Transf.*, 73, 761–775, 2014.)

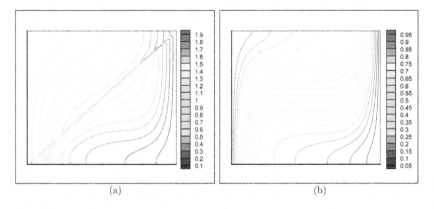

FIGURE 10.12 Isotherms for natural convection in a diagonally bilayered porous enclosure with $\sigma_2 = 2\sigma_1$: (a) Guo et al.'s model (2005); and (b) Chen et al.'s model. (Reprinted with permission from Chen, S. et al., *Int. J. Heat Mass. Transf.*, 111, 1019–1022, 2017.)

10.3.1.2.2 LTNE

In 2014, Gao et al. (2014) proposed a first-hand LBM macroscale model for simulating natural convection heat transfer in porous media under the LTNE condition. Those authors employed the GLBE model of Guo and Zhao (2002) for fluid flow modeling in porous media as well as the PS DDF thermal LBGK model for the thermal part. To recover the macroscopic volume-averaged energy equation in porous media under the LTNE condition, Gao et al. (2014) modified the thermal LBGK for the fluid and solid phases to incorporate the effects of the interfacial heat transfer between the two phases as well as the heat generation assumed within each phase, by adding a number of source terms (Guo et al. 2002b; McCracken et al. 2005) to the fluid and solid temperature evolution equations as follows:

$$g_{i,f}\left(\mathbf{r}+\mathbf{e}_i\delta t,t+\delta t\right)-g_{i,f}\left(\mathbf{r},t\right)=-\frac{g_{i,f}\left(\mathbf{r},t\right)-g_{i,f}^{eq}\left(\mathbf{r},t\right)}{\tau_{g,f}}+\delta t Sr_{i,f}+\frac{1}{2}\delta t^2\frac{\partial Sr_{i,f}}{\partial t}+\delta t Su_{i,f} \quad (10.34)$$

$$g_{i,s}\left(\mathbf{r}+\mathbf{e}_i\delta t,t+\delta t\right)-g_{i,s}\left(\mathbf{r},t\right)=-\frac{g_{i,s}\left(\mathbf{r},t\right)-g_{i,s}^{eq}\left(\mathbf{r},t\right)}{\tau_{g,s}}+\delta t Sr_{i,s}+\frac{1}{2}\delta t^2\frac{\partial Sr_{i,s}}{\partial t} \quad (10.35)$$

where $g_{i,f}$ and $g_{i,s}$ are, respectively, fluid and solid temperature distribution functions, $\tau_{g,f}$ and $\tau_{g,s}$ are, respectively, fluid and solid non-dimensional energy relaxation times, $Sr_{i,f}=\omega_i Sr_f$ is the fluid discrete source term, in which, $Sr_f=h_v\left(T_s-T_f\right)/\left[\phi\left(\rho c_p\right)_f\right]+q_f'''/\left(\rho c_p\right)_f$ is the fluid energy source term, where, the first and second terms in Sr_f, respectively, are related to the interfacial heat transfer (i.e., h_v is the volumetric heat transfer coefficient), and heat generation per unit volume that occurs within the fluid phase, $Sr_{i,s}=\omega_i Sr_f$ is the solid discrete source term, in which, $Sr_s=h_v\left(T_f-T_s\right)/\left[\left(1-\phi\right)\left(\rho c_p\right)_f\right]+q_s'''/\left(\rho c_p\right)$ is the solid energy source term, where, the first and second terms in Sr_s is related to, respectively, the interfacial heat transfer and heat generation per unit volume that occurs within the solid phase. T_s and T_f are the volume-averaged solid and fluid temperatures. In the model by Gao et al. (2014), $g_{i,f}^{eq}$, $g_{i,s}^{eq}$, and $Su_{i,f}$ are, respectively, the fluid and solid temperature EDF and a fluid discrete source term needed to recover the correct fluid energy equation calculated as follows:

$$\begin{cases} g_{i,f}^{eq}\left(\mathbf{r},t\right)=\omega_i T_f\left(1+\frac{\mathbf{e}_i.\mathbf{u}}{\phi c_s^2}\right) \\[3mm] g_{i,s}^{eq}\left(\mathbf{r},t\right)=\omega_i T_s \\[3mm] Su_{i,f}=\omega_i\frac{1}{\delta t}\left(1-\frac{1}{2\tau_{g,f}}\right)\left(\delta t\frac{\partial\mathbf{u}T_f.\mathbf{e}_i}{\partial t}-\frac{\delta t^2}{2}\frac{\partial^2\mathbf{u}T_f.\mathbf{e}_i}{\partial t^2}+O(\delta t^3)\right) \end{cases} \quad (10.36)$$

Gao et al. (2014) performed the Chapman–Enskog analysis and proved that their model could recover the volume-averaged fluid and solid energy equations in porous media with internal heat generations under the LTNE condition as follows:

$$\frac{\partial T_f}{\partial t}+\nabla.\left(\frac{\mathbf{u}T_f}{\phi}\right)=\frac{k_{e,f}}{\phi\left(\rho c_p\right)_f}\nabla^2 T_f+\frac{h_v\left(T_s-T_f\right)}{\phi\left(\rho c_p\right)_f}+\frac{q_f'''}{\left(\rho c_p\right)_f} \quad (10.37a)$$

$$\frac{\partial T_s}{\partial t}+=\frac{k_{e,s}}{\left(1-\phi\right)\left(\rho c_p\right)_s}\nabla^2 T_s+\frac{h_v\left(T_f-T_s\right)}{\left(1-\phi\right)\left(\rho c_p\right)_s}+\frac{q_s'''}{\left(\rho c_p\right)_s} \quad (10.37b)$$

Those authors verified their model against the available results from CFD methods through steady-state and transient natural convection problems in a porous enclosure. For the transient case, the results of

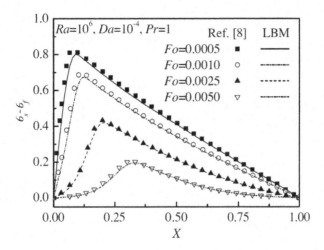

FIGURE 10.13 Solid-to-fluid temperature difference at the mid-height of a porous enclosure for different Fourier numbers. (Reprinted with permission from Gao, D. et al., *Int. J. Heat Mass. Tran.*, 70, 979–989, 2014.)

temperature non-equilibrium distribution at the mid-height of the enclosure are compared with those available from CFD methods for different Fourier numbers to show an excellent agreement as shown in Figure 10.13.

Later, Wang et al. (2016b) proposed a DDF LBGK model, based on the kinetic theory, for convection heat transfer in porous media under the LTNE condition, where, the viscous heat dissipation and compression work were incorporated naturally. In their model, the total energy distribution function was employed. Those authors claimed that the source terms added to the density and total energy evolution equations of their model to account for the presence of the porous medium were much less complicated than those of Gao et al. (2014).

10.3.2 A Brief Review of the Applications

In 2005, Guo and Zhao (2004) employed their thermal macroscale model (Guo and Zhao 2005) to simulate the natural convection in a cavity filled with a porous medium considering a temperature-dependent fluid viscosity. Wang and Afsharpoya (2006) employed the GLBE model to simulate the fluid flow through a channel partially filled with a porous medium in a fuel cell problem. Yan et al. (2006) successfully applied the thermal macroscale model (Guo and Zhao 2005) to study the natural convection problem in a cavity filled with a porous medium with a variable porosity. Shokouhmand et al. (2009) used the model (Guo and Zhao 2005) to investigate forced convection heat transfer in a parallel-plate channel with a centrally located porous insert under the LTE assumption. Those authors studied the effect of the porous insert blockage ratio Y_p and Darcy number Da on the average Nusselt number. As shown in Figure 10.14, Shokouhmand et al. (2009) reported that for lower Darcy numbers there was a blockage ratio at which the optimum Nusselt number could be achieved. The same authors (Shokouhmand et al. 2011) considered the fully developed convection heat transfer in a parallel-plate channel to investigate the effect of the porous insert position on the thermal performance of the channel. Mehrizi et al. (2013) employed the model by Seta et al. (2006a) to investigate the effect of porosity, Reynolds number, and Prandtl number on forced convection heat transfer in a porous heat exchanger. Rong et al. (2014) employed their axisymmetric thermal LBM model for convection heat transfer in porous media Rong et al. 2010) to investigate the heat transfer enhancement in a pipe partially filled with a porous medium with simulations performed on a GPU (graphics processing unit). Those authors investigated the effects of the porous layer thickness, Darcy number, and porosity on the performance evaluation criterion (PEC) of the pipe. Xu et al. (2017) used the thermal macroscale model proposed by Wang et al. (2016a) to study the double-diffusive natural

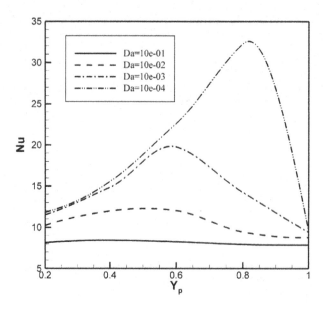

FIGURE 10.14 Nusselt number against the porous blockage ratio Y_p for different Darcy numbers. (Reprinted with permission from Shokouhmand, H. et al., *Int. Commun. Heat Mass Transf.*, 36, 378–384, 2009.)

convection around a heated cylinder in an enclosure filled with a porous medium. Chen et al. (2018) extended the thermal LBM model for convection heat transfer in porous media proposed by the same group (Chen et al. 2017a) to simulate the conjugate natural convection heat transfer in an open-ended cavity partially filled with a porous medium.

10.4 Conclusion

During the last three decades, LBM has become a promising numerical tool due to its explicit algorithm, easy coding, transient nature, linear advection term, local collision operator, and easy parallelization of the computer code, to name a few. The LBM has been proved to have a great potential in dealing with transport phenomena involved with complex interfaces, such as pore-scale fluid flow in porous media where it has shown great advantages over the conventional N-S solvers. As such, from the very beginning LBM was applied to pore-scale flow simulations in porous media mainly because of the extremely simple and computationally efficient no-slip method called the bounce back scheme, which was, however, only applicable to the straight boundaries (curved boundaries have been approximated by a stair-like approximation). On the other hand, LBM pore-scale convection heat transfer modeling in porous media, which is, in fact, a conjugate heat transfer problem, came to existence two decades later, which is still under development. So far, the researchers have invoked the simplified theoretical geometries to perform the pore-scale conjugate heat transfer in porous media to assess the macroscale models.

The main current challenges associated with LBM simulation of convection at the pore scale are how to efficiently implement the no-slip and conjugate heat transfer schemes along curved boundaries. Despite the excellent success achieved in this area, accurate no-slip and conjugate heat transfer schemes, which are consistent with the LBM in terms of the simplicity are yet to be developed. That is why the LBM pore-scale simulation of conjugate heat transfer in complex porous geometries, such as metallic foams, has not been performed yet. One possible approach for resolving this problem may be employing the non-uniform body-fitted meshes with LBM. Furthermore, the conformal mapping and coming up with a semi-LBM solver can be thought of as another ongoing line of development as well as the hybrid LBM methods.

Lattice Boltzmann Method for Modeling Convective Heat Transfer in Porous Media 195

During the past two decades, the macroscale models for fluid flow and heat transfer simulations in porous media have also been rapidly developing by modifying the standard LBM to recover the correct volume-averaged hydrodynamic and energy equations. In one hand, the solver wants to recover the pore-scale information, while on the other hand, VAT-based models have to rely on integrating over a REV. As such, developing simple yet accurate LBM discrete forcing and heat source schemes to precisely model the thermohydraulics of the porous medium is required. Despite the great success the macroscale models have achieved in fluid flow simulations in porous media, the thermal macroscale models still need to be further developed. In particular, more research into heat transfer in anisotropic porous media, bidispersed porous media, conjugate heat transfer in porous media, and porous-clear fluid interface, as well as turbulent heat transfer in porous media, are called for.

NOMENCLATURE

c	lattice velocity (m/s)
c_s	lattice sound speed (m/s)
$\mathbf{e}_i$	microscopic discrete velocity (m/s)
d_p	particle diameter in packed bed (m)
F_ϕ	geometrical factor (see Equation 10.28)
$\mathbf{F}$	body force (N)
F_i	discrete body force (see Equation 10.27)
f_i	density distribution function
f_i^{eq}	density EDF (see Equation 10.6)
g_i	internal energy distribution function
g_i^{eq}	internal energy EDF (see Equation 10.10)
h_v	volumetric heat transfer coefficient (W/m^3)
k_f	fluid thermal conductivity (W/mK)
k_s	solid thermal conductivity (W/mK)
K	permeability of porous media (m^2)
$\lvert m(\mathbf{r},t)\rangle$	density moment vector
$\mathbf{M}$	hydrodynamics transformation matrix
$\lvert n(\mathbf{r},t)\rangle$	energy moment vector
$\mathbf{N}$	energy transformation matrix
p	pressure (Pa)
q'''	volumetric heat generation (W/m^3)
S	hydrodynamics collision matrix
T_f	fluid phase temperature (K)
T_s	solid phase temperature (K)
$\mathbf{u} = (u_x, u_y)$	velocity vector (m/s)

Greek Letters

α	thermal diffusivity (m^2/s)
x, y	Cartesian coordinates (m)

Subscripts

f	fluid phase
s	solid phase
eff	effective
δx	lattice spacing (m)
δt	time step (s)
ε	internal energy

ε_{mic}	microporous porosity in BDPM
ε_{mac}	macroporous porosity in BDPM
θ	dimensionless temperature
Θ	energy collision matrix
λ	solid–fluid thermal conductivity ratio
μ	fluid dynamic viscosity $(\mathrm{N/m^2.s})$
υ	fluid kinematic viscosity $(\mathrm{m^2/s})$
ξ	continuous microscopic velocity $(\mathrm{m/s})$
ρ	fluid density $(\mathrm{kg/m^3})$
(ρc_p)	volumetric heat capacity $(\mathrm{J/Km^3})$
σ	volumetric heat capacity ratio
σ_0	reference volumetric heat capacity ratio
τ_v	non-dimensional velocity relaxation time
τ_g	non-dimensional energy relaxation time
ϕ	porosity of porous media
ω_i	weight coefficients

REFERENCES

Abe, T., Derivation of the lattice Boltzmann method by means of the discrete ordinate method for the Boltzmann equation. *Journal of Computational Physics*, 1997. **131**(1): p. 241–246.

Alazmi, B. and K. Vafai, Constant wall heat flux boundary conditions in porous media under local thermal non-equilibrium conditions. *International Journal of Heat and Mass Transfer*, 2002. **45**(15): p. 3071–3087.

Beugre, D. et al., Lattice Boltzmann 3D flow simulations on a metallic foam. *Journal of Computational and Applied Mathematics*, 2010. **234**(7): p. 2128–2134.

Bhatnagar, P.L., E.P. Gross, and M. Krook, A model for collision processes in gases. I. Small amplitude processes in charged and neutral one-component systems. *Physical Review*, 1954. **94**(3): p. 511.

Bouzidi, M.h., M. Firdaouss, and P. Lallemand, Momentum transfer of a Boltzmann-lattice fluid with boundaries. *Physics of Fluids*, 2001. **13**(11): p. 3452–3459.

Chai, Z. and T. Zhao, Lattice Boltzmann model for the convection-diffusion equation. *Physical Review E*, 2013. **87**(6): p. 063309.

Chapman, S., T.G. Cowling, and D. Burnett, *The Mathematical Theory of Non-Uniform Gases: An Account of the Kinetic Theory of Viscosity, Thermal Conduction and Diffusion in Gases*. 1990, London, UK: Cambridge university press.

Chen, S. and G.D. Doolen, Lattice Boltzmann method for fluid flows. *Annual Review of Fluid Mechanics*, 1998. **30**(1): p. 329–364.

Chen, S., B. Yang, and C. Zheng, A lattice Boltzmann model for heat transfer in porous media. *International Journal of Heat and Mass Transfer*, 2017a. **111**: p. 1019–1022.

Chen, S., Y. Yan, and W. Gong, A simple lattice Boltzmann model for conjugate heat transfer research. *International Journal of Heat and Mass Transfer*, 2017b. **107**: p. 862–870.

Chen, S., W. Gong, and Y. Yan, Conjugate natural convection heat transfer in an open-ended square cavity partially filled with porous media. *International Journal of Heat and Mass Transfer*, 2018. **124**: p. 368–380.

Chiappini, D., Numerical simulation of natural convection in open-cells metal foams. *International Journal of Heat and Mass Transfer*, 2018. **117**: p. 527–537.

Chopard, B., J. Falcone, and J. Latt, The lattice Boltzmann advection-diffusion model revisited. *The European Physical Journal Special Topics*, 2009. **171**(1): p. 245–249.

d'Humieres, D., Generalized lattice Boltzmann Equations, Rarefied Gas Dynamics: Theory and Simulations," *Progress in Astronautics and Aeronautics*, 1999. 159: p. 450–458.

D'Orazio, A. and S. Succi, Simulating two-dimensional thermal channel flows by means of a lattice Boltzmann method with new boundary conditions. *Future Generation Computer Systems*, 2004. **20**(6): p. 935–944.

Dardis, O. and J. McCloskey, Lattice Boltzmann scheme with real numbered solid density for the simulation of flow in porous media. *Physical Review E*, 1998. **57**(4): p. 4834.

Filippova, O. and D. Hänel, Lattice-Boltzmann simulation of gas-particle flow in filters. *Computers & Fluids*, 1997. **26**(7): p. 697–712.

Freed, D.M., Lattice-Boltzmann method for macroscopic porous media modeling. *International Journal of Modern Physics C*, 1998. **9**(08): p. 1491–1503.

Gao, D., Z. Chen, and L. Chen, A thermal lattice Boltzmann model for natural convection in porous media under local thermal non-equilibrium conditions. *International Journal of Heat and Mass Transfer*, 2014. **70**: p. 979–989.

Guo, Z. and C. Shu, Lattice Boltzmann method and its applications in engineering. In *Advances in Computational Fluid Dynamics*, Ed. C.-W. Shu and C. Shu. Vol. 3. 2013: World Scientific.

Guo, Z. and T. Zhao, Lattice Boltzmann model for incompressible flows through porous media. *Physical Review E*, 2002. **66**(3): p. 036304.

Guo, Z. and T. Zhao, Lattice Boltzmann simulation of natural convection with temperature-dependent viscosity in a porous cavity. *Progress in Computational Fluid Dynamics, an International Journal*, 2004. **5**(1–2): p. 110–117.

Guo, Z. and T. Zhao, A lattice Boltzmann model for convection heat transfer in porous media. *Numerical Heat Transfer, Part B*, 2005. **47**(2): p. 157–177.

Guo, Z., C. Zheng, and B. Shi, An extrapolation method for boundary conditions in lattice Boltzmann method. *Physics of Fluids*, 2002a. **14**(6): p. 2007–2010.

Guo, Z., C. Zheng, and B. Shi, Discrete lattice effects on the forcing term in the lattice Boltzmann method. *Physical Review E*, 2002b. **65**(4): p. 046308.

Guo, Z., C. Zheng, and B. Shi, Incompressible lattice Boltzmann model for porous flows with large pressure gradient. *Progress in Computational Fluid Dynamics, an International Journal*, 2009. **9**(3–5): p. 225–230.

He, X. and L.-S. Luo, A priori derivation of the lattice Boltzmann equation. *Physical Review E*, 1997a. **55**(6): p. R6333–R6336.

He, X. and L.-S. Luo, Theory of the lattice Boltzmann method: From the Boltzmann equation to the lattice Boltzmann equation. *Physical Review E*, 1997b. **56**(6): p. 6811.

He, X., S. Chen, and G.D. Doolen, A novel thermal model for the lattice Boltzmann method in incompressible limit. *Journal of Computational Physics*, 1998. **146**(1): p. 282–300.

Hu, Y. et al., Full Eulerian lattice Boltzmann model for conjugate heat transfer. *Physical Review E*, 2015. **92**(6): p. 063305.

Imani, G. and K. Hooman, Lattice Boltzmann pore scale simulation of natural convection in a differentially heated enclosure filled with a detached or attached bidisperse porous medium. *Transport in Porous Media*, 2017. **116**(1): p. 91–113.

Imani, G., M. Maerefat, and K. Hooman, Lattice Boltzmann simulation of conjugate heat transfer from multiple heated obstacles mounted in a walled parallel plate channel. *Numerical Heat Transfer, Part A: Applications*, 2012a. **62**(10): p. 798–821.

Imani, G. et al., Lattice Boltzmann method for simulating conjugate heat transfer from an obstacle mounted in a parallel-plate channel with the use of three different heat input methods. *Heat Transfer Research*, 2012b. **43**(6): p. 545–572.

Imani, G., M. Maerefat, and K. Hooman, Pore-scale numerical experiment on the effect of the pertinent parameters on heat flux splitting at the boundary of a porous medium. *Transport in Porous Media*, 2013. **98**(3): p. 631–649.

Inamuro, T., M. Yoshino, and F. Ogino, Lattice Boltzmann simulation of flows in a three-dimensional porous structure. *International Journal for Numerical Methods in Fluids*, 1999. **29**(7): p. 737–748.

Karani, H. and C. Huber, Lattice Boltzmann formulation for conjugate heat transfer in heterogeneous media. *Physical Review E*, 2015. **91**(2): p. 023304.

Khazaeli, R., S. Mortazavi, and M. Ashrafizaadeh, Application of a ghost fluid approach for a thermal lattice Boltzmann method. *Journal of Computational Physics*, 2013. **250**: p. 126–140.

Lallemand, P. and L.-S. Luo, Theory of the lattice Boltzmann method: Dispersion, dissipation, isotropy, Galilean invariance, and stability. *Physical Review E*, 2000. **61**(6): p. 6546.

Le, G., O. Oulaid, and J. Zhang, Counter-extrapolation method for conjugate interfaces in computational heat and mass transfer. *Physical Review E*, 2015. **91**(3): p. 033306.

Li, L. et al., Conjugate heat and mass transfer in the lattice Boltzmann equation method. *Physical Review E*, 2014. **89**(4): p. 043308.

Liu, Q. and Y.-L. He, Lattice Boltzmann simulations of convection heat transfer in porous media. *Physica A: Statistical Mechanics and its Applications*, 2017. **465**: p. 742–753.

Liu, Z. and H. Wu, Pore-scale study on flow and heat transfer in 3D reconstructed porous media using microtomography images. *Applied Thermal Engineering*, 2016. **100**: p. 602–610.

Liu, Q. et al., A multiple-relaxation-time lattice Boltzmann model for convection heat transfer in porous media. *International Journal of Heat and Mass Transfer*, 2014. **73**: p. 761–775.

Lu, J., H. Lei, and C. Dai, A lattice Boltzmann algorithm for simulating conjugate heat transfer through virtual heat capacity correction. *International Journal of Thermal Sciences*, 2017. **116**: p. 22–31.

Luo, L.-S., Lattice-gas automata and lattice Boltzmann equations for two-dimensional hydrodynamics, Ph.D. thesis, Georgia Institute of Technology, Atlanta, 1993.

Maier, R. and R. Bernard, Lattice-Boltzmann accuracy in pore-scale flow simulation. *Journal of Computational Physics*, 2010. **229**(2): p. 233–255.

Martys, N.S., Improved approximation of the Brinkman equation using a lattice Boltzmann method. *Physics of fluids*, 2001. **13**(6): p. 1807–1810.

McCracken, M.E. and J. Abraham, Multiple-relaxation-time lattice-Boltzmann model for multiphase flow. *Physical Review E*, 2005. **71**(3): p. 036701.

McNamara, G.R. and G. Zanetti, Use of the Boltzmann equation to simulate lattice-gas automata. *Physical Review Letters*, 1988. **61**(20): p. 2332–2335.

Mehrizi, A.A. et al., Lattice Boltzmann simulation of heat transfer enhancement in a cold plate using porous medium. *Journal of Heat Transfer*, 2013. **135**(11): p. 111006.

Mei, R., L.-S. Luo, and W. Shyy, An accurate curved boundary treatment in the lattice Boltzmann method. *Journal of Computational Physics*, 1999. **155**(2): p. 307–330.

Meng, F., M. Wang, and Z. Li, Lattice Boltzmann simulations of conjugate heat transfer in highfrequency oscillating flows. *International Journal of Heat and Fluid Flow*, 2008. **29**(4): p. 1203–1210.

Mezrhab, A. et al., Double MRT thermal lattice Boltzmann method for simulating convective flows. *Physics Letters A*, 2010. **374**(34): p. 3499–3507.

Mohammad, A.A., *Lattice Boltzmann Method: Fundamental and Engineering Applications with Computer Codes*. 2011, London, UK: Springer. 178.

Mozafari-Shamsi, M., M. Sefid, and G. Imani, Developing a ghost fluid lattice Boltzmann method for simulation of thermal Dirichlet and Neumann conditions at curved boundaries. *Numerical Heat Transfer, Part B: Fundamentals*, 2016a. **70**(3): p. 251–266.

Mozafari-Shamsi, M., M. Sefid, and G. Imani, New formulation for the simulation of the conjugate heat transfer at the curved interfaces based on the ghost fluid lattice Boltzmann method. *Numerical Heat Transfer Part B-Fundamentals*, 2016b. **70**(6): p. 559–576.

Nield, D.A. and A. Bejan, *Convection in Porous Media*. 2017, New York: Springer.

Nield, D. and A. Kuznetsov, The onset of convection in a bidisperse porous medium. *International Journal of Heat and Mass Transfer*, 2006. **49**(17–18): p. 3068–3074.

Pan, C., L.-S. Luo, and C.T. Miller, An evaluation of lattice Boltzmann schemes for porous medium flow simulation. *Computers & fluids*, 2006. **35**(8–9): p. 898–909.

Peng, Y., C. Shu, and Y. Chew, Simplified thermal lattice Boltzmann model for incompressible thermal flows. *Physical Review E*, 2003. **68**(2): p. 026701.

Peng, Y., C. Shu, and Y. Chew, A 3D incompressible thermal lattice Boltzmann model and its application to simulate natural convection in a cubic cavity. *Journal of Computational Physics*, 2004. **193**(1): p. 260–274.

Qian, Y., D. d'Humières, and P. Lallemand, Lattice BGK models for Navier–Stokes equation. *EPL (Europhysics Letters)*, 1992. **17**(6): p. 479.

Rong, F. et al., A lattice Boltzmann model for axisymmetric thermal flows through porous media. *International Journal of Heat and Mass Transfer*, 2010. **53**(23–24): p. 5519–5527.

Rong, F. et al., Numerical study of heat transfer enhancement in a pipe filled with porous media by axisymmetric TLB model based on GPU. *International Journal of Heat and Mass Transfer*, 2014. **70**: p. 1040–1049.

Seta, T., E. Takegoshi, and K. Okui, Lattice Boltzmann simulation of natural convection in porous media. *Mathematics and Computers in Simulation*, 2006a. **72**(2–6): p. 195–200.

Seta, T. et al., Thermal lattice Boltzmann model for incompressible flows through porous media. *Journal of Thermal Science and Technology*, 2006b. **1**(2): p. 90–100.

Shan, X. and H. Chen, Lattice Boltzmann model for simulating flows with multiple phases and components. *Physical Review E*, 1993. **47**(3): p. 1815.

Sheikh, B. and T. Qiu, Pore-scale simulation and statistical investigation of velocity and drag force distribution of flow through randomly-packed porous media under low and intermediate Reynolds numbers. *Computers & Fluids*, 2018. **171**: p. 15–28.

Shi, Y., T. Zhao, and Z. Guo, Thermal lattice Bhatnagar-Gross-Krook model for flows with viscous heat dissipation in the incompressible limit. *Physical Review E*, 2004. **70**(6): p. 066310.

Shi, B. et al., A new scheme for source term in LBGK model for convection–diffusion equation. *Computers & Mathematics with Applications*, 2008. **55**(7): p. 1568–1575.

Shokouhmand, H., F. Jam, and M. Salimpour, Simulation of laminar flow and convective heat transfer in conduits filled with porous media using Lattice Boltzmann Method. *International Communications in Heat and Mass Transfer*, 2009. **36**(4): p. 378–384.

Shokouhmand, H., F. Jam, and M. Salimpour, The effect of porous insert position on the enhanced heat transfer in partially filled channels. *International Communications in Heat and Mass Transfer*, 2011. **38**(8): p. 1162–1167.

Spaid, M.A. and F.R. Phelan Jr, Lattice Boltzmann methods for modeling microscale flow in fibrous porous media. *Physics of Fluids*, 1997. **9**(9): p. 2468–2474.

Succi, S., *The Lattice Boltzmann Equation for Fluid Dynamics and Beyond*. 2001, Oxford: Oxford University Press. 288.

Succi, S., E. Foti, and F. Higuera, Three-dimensional flows in complex geometries with the lattice Boltzmann method. *EPL (Europhysics Letters)*, 1989. **10**(5): p. 433.

Vafai, K., *Porous Media: Applications in Biological Systems and Biotechnology*. 2010: CRC Press.

Vafai, K., *Handbook of Porous Media*. 2015, Boca Raton, FL: CRC Press.

Vafai, K. and C. Tien, Boundary and inertia effects on flow and heat transfer in porous media. *International Journal of Heat and Mass Transfer*, 1981. **24**(2): p. 195–203.

van der Hoef, M.A., R. Beetstra, and J. Kuipers, Lattice-Boltzmann simulations of low-Reynolds-number flow past mono-and bidisperse arrays of spheres: Results for the permeability and drag force. *Journal of Fluid Mechanics*, 2005. **528**: p. 233–254.

Wang, L.-P. and B. Afsharpoya, Modeling fluid flow in fuel cells using the lattice-Boltzmann approach. *Mathematics and Computers in Simulation*, 2006. **72**(2–6): p. 242–248.

Wang, M. and N. Pan, Modeling and prediction of the effective thermal conductivity of random opencell porous foams. *International Journal of Heat and Mass Transfer*, 2008. **51**(5–6): p. 1325–1331.

Wang, J., M. Wang, and Z. Li, A lattice Boltzmann algorithm for fluid–solid conjugate heat transfer. *International Journal of Thermal Sciences*, 2007. **46**(3): p. 228–234.

Wang, L., J. Mi, and Z. Guo, A modified lattice Bhatnagar–Gross–Krook model for convection heat transfer in porous media. *International Journal of Heat and Mass Transfer*, 2016a. **94**: p. 269–291.

Wang, L. et al., A lattice Boltzmann model for thermal flows through porous media. *Applied Thermal Engineering*, 2016b. **108**: p. 66–75.

Wang, M. et al., Lattice Boltzmann modeling of the effective thermal conductivity for fibrous materials. 2007a.

Wang, M. et al., Mesoscopic predictions of the effective thermal conductivity for microscale random porous media. *Physical Review E*, 2007b. **75**(3): p. 036702.

Wang, M. et al., Three-dimensional effect on the effective thermal conductivity of porous media. *Journal of Physics D: Applied Physics*, 2006. **40**(1): p. 260.

Wu, W. et al., A lattice Boltzmann model for interphase conjugate heat transfer. *Numerical Heat Transfer, Part B: Fundamentals*, 2017. **72**(2): p. 130–151.

Xu, H. et al., Lattice Boltzmann simulation of the double diffusive natural convection and oscillation characteristics in an enclosure filled with porous medium. *International Communications in Heat and Mass Transfer*, 2017. **81**: p. 104–115.

Yan, W.-W. et al., Lattice Boltzmann simulation on natural convection heat transfer in a two-dimensional cavity filled with heterogeneously porous medium. *International Journal of Modern Physics C*, 2006. **17**(6): p. 771–783.

Zhao, C. et al., Numerical study of natural convection in porous media (metals) using Lattice Boltzmann Method (LBM). *International Journal of Heat and Fluid Flow*, 2010. **31**(5): p. 925–934.

Section III

Advanced Engineering Applications of Convection in Porous Media

11

Modeling Thermohydraulic Process in Enhanced Geothermal System Based on Two-Equation Thermal Model for Porous Media

Wenbo Huang, Wenjiong Cao, Guoling Wei, Yunlong Jin, and Fangming Jiang

CONTENTS

11.1 Introduction..203
11.2 EGS Reservoir Structure and Modeling Issues...204
11.3 Methodology 206
    11.3.1 Modeling Thermohydraulic Process in EGS 206
        11.3.1.1 Physical Model...206
        11.3.1.2 Governing Equations..206
        11.3.1.3 Non-Darcy Effect...207
        11.3.1.4 Heat Exchange between Fracture and Rock..208
        11.3.1.5 Thermophysical Properties of Working Fluid 209
    11.3.2 EGS Subsurface Geometry, Boundary Conditions, and Other Parameters 210
11.4 Results and Discussion 211
    11.4.1 Comparison of Water-EGS and SCCO$_2$-EGS...211
        11.4.1.1 Flow Behaviors 211
        11.4.1.2 Heat Extraction Performance 213
    11.4.2 Effects of Variable Thermophysical Properties of Fluid.....................................215
        11.4.2.1 Effects of Variable Density..216
        11.4.2.2 Effects of Variable Viscosity 216
        11.4.2.3 Effects of Variable Specific Heat Capacity 217
        11.4.2.4 Effects of Variable Thermal Conductivity...218
    11.4.3 Effects of Non-Darcy Flow...218
        11.4.3.1 EGS Performance with or without Non-Darcy Effects.......................218
        11.4.3.2 Criterion for Judging the Onset of Non-Darcy Flow 220
11.5 Summary...223
Acknowledgments...224
References..224

11.1 Introduction

Enhanced geothermal system (EGS) represents an advanced geothermal energy utilization technology whereas the idea itself is simple: for low-permeability rocks, a series of rock-fracturing procedures, such as hydraulic stimulation to create an artificially fractured reservoir, is performed. By circulating water through the stimulated region, heat can be continuously extracted from the rock, just like a natural hydrothermal system. Figure 11.1 illustrates the EGS concept and schematically shows its basic underground and earth-surface components. Heat transmission fluids are injected into the heat reservoir, and after being heated up by the hot rocks, they are extracted out for earth-surface power-generation and/or thermal utilizations.

203

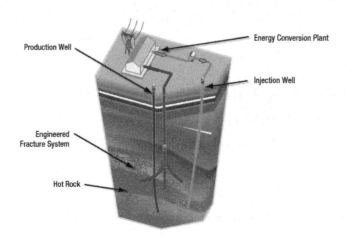

FIGURE 11.1 EGS cutaway diagram showing its basic underground and earth-surface components.

EGS has drawn wide interest across the world in the past 40 years. Until now, at least 31 EGS projects have been launched worldwide, but few of them are currently in real commercial operation (Breede et al. 2013). Significant financial risks are present for EGS projects, since the production potential of an EGS project is largely dependent on the quality of the fractures network in the EGS reservoir, which, at present, is still a major challenge to effective control. There exist at least two key technical issues for EGS power plant development. One is how to create an effective heat reservoir of sufficiently large volume with adequate permeability, the other is how to efficiently extract heat from the reservoir and at the same time keep the reservoir's service life as long as possible (Tester et al. 2006). Both issues are related with the subsurface thermohydraulic process. Numerical modeling may be a very effective and powerful tool to resolving these issues.

11.2 EGS Reservoir Structure and Modeling Issues

When the EGS concept was initially proposed by Robinson et al. (1971) of the Los Alamos National Laboratory (USA), the structure of engineered reservoirs was identified as essentially "penny-shaped" fractures connecting the injection and production wells, mainly based on the empirical hydraulic fracturing concepts of sedimentary rocks, which considered the induced fractures as planar and normal to the axis of the least principal stress (Lawn and Fuller 1975). Multiple parallel fracture structures were then suggested by Raleigh et al. (1974) in which the geothermal wells were drilled at an angle in a direction perpendicular to the expected orientation of the fractures to create a series of parallel cracks from a single well. The concept associated to single and multiple fractures assumed that the host rock behaved as a continuum; however, it was later found not being appropriate for the crystalline rock in most hot dry rock (HDR) cases, where the pre-existing joints were pervasive and much weaker than the host matrix (Baria et al. 1999). Further experiments dealing with field hydraulic fracturing indicated that the creation of new hydraulic fractures was not the dominant process; in fact, the shearing of natural joints and fractures favorably aligned with the principal directions of the local stress field was the prevailing process (Pine et al. 1984). The existing joints and fractures would prevent the extension of artificial fractures; meanwhile, they could be sheared and widened during the stimulation process thus causing a fracture network with increased permeability (Blanton 1982). Affected by the uneven distributed joints and faults in host rocks, the actual fractured HDR reservoir behaves like interconnected networks of flow paths with a few dominant routes acting as flow conduits (Batchelor 1986).

Since the heat reservoir of EGS contains a fracture network and is generally of strong heterogeneity, modeling fluid flow and heat transfer in industry-scale EGS reservoirs is still a challenging task (Lawn and Fuller 1975; Robinson et al. 1971). Simplification to the reservoir inner structure is a common strategy for numerical studies of EGS. The most prevalent simplification is the equivalent porous medium (EPM) model (Raleigh et al. 1974), which neglects the detailed inner structure of the heat reservoir and takes the reservoir as a continuum consisting of representative equivalent volumes (REVs) that have the same macroscopic properties as the fractured rocks. This treatment offers great convenience to model a large-scale heat reservoir since there is generally no detailed information about the fracture configuration in the reservoir available for use.

The existing EPM models can be roughly classified into two categories in terms of physical description to the fractured rock mass: the single porosity model and the dual porosity model. The single porosity model represents the property of fractures and rock matrix by a certain macroscopic approximation or average. This approach may be applicable to modeling flow and heat transfer processes in fractured media under near-thermodynamic-equilibrium conditions (Wu et al. 1999) which, however, is hard to accomplish when rapid flow and transport processes occur in fractured reservoirs (Jiang et al. 2014). The dual porosity model treats the heat reservoir as a porous medium consisting of two distinct porous regions: large-porosity fracture network and rock matrix of relatively smaller porosity (Warren and Root 1963). Global flow in the reservoir occurs mainly through the fracture network, which is described as an effective porous continuum. Rock matrix and fractures may exchange fluid (or heat) locally by means of "interporosity flow," which is driven by the difference in pressures (or temperatures) between matrix and fractures. The concept of dual porosity has been widely adopted in current geothermal models (Gelet et al. 2012; Sanyal and Butler 2005; Taron et al. 2009) though an accurate specification of model parameters may be very difficult, as a clear differentiation of these two porous regions is normally hard, even if not impossible.

Some recent researches considered another approach, called two-equation thermal model (also known as local thermal non-equilibrium model) (Gelet et al. 2013; Jiang et al. 2014; Shaik et al. 2011). The model accounts for the actual existence of temperature difference between the rock matrix and the fluid flowing in fractures and uses two energy equations to describe heat transfer in the rock matrix and in the fluid, respectively. Heat exchange between fluid and rock matrix is modeled by Newton's law of cooling, assuming the heat exchange rate is directly proportional to the fluid–rock temperature difference. The proportionality coefficient, termed as the equivalent solid–fluid heat transfer coefficient, can be measured experimentally or calculated from fine-scale models such as the fracture-scale model (Jiang et al. 2014; Gelet et al. 2013).

On the other hand, previous research work mostly assumed Darcy fluid flow in the porous medium model of EGS reservoirs (Brown 2000; Jiang et al. 2013; Pruess 2008), indicating a simple proportional relationship between the seepage velocity and the pressure gradient is assumed. However, the Darcy flow assumption is only valid when the pressure gradients and resultant flow velocities are sufficiently small, that is, when the Reynolds number is much less than the critical Reynolds number that is 1 for flow around a sphere. As the Reynolds number gradually increases, the non-linearity appears in the velocity-pressure gradient relationship, which is the so-called non-Darcy flow. Forchheimer (1901) deduced from experimental results, that an additional term in relation with the squared velocity is required to represent the non-Darcy flow effect, when the flow rates are high. In recent years, some literature work confirmed the existence of non-Darcy flow in EGS heat reservoirs, particularly within the region near the wellbore. For example, Kohl et al. (1997) investigated non-Darcy flow behavior in fractured rock, based on the field tests of the EGS site at Soultz-sous-Forêts, France. Zhang et al. (2012) employed the Forchheimer equation and a non-linear finite element model to study the non-Darcy effects in a geothermal reservoir. Results indicated that the Forchheimer flow in the near-well region led to very different fluid velocity and pressure scenarios in the wellbores.

Moreover, the large changes at temperature and pressure associated with EGS heat exploitation will lead to pronounced changes at the thermophysical properties of the heat transfer fluid, which will in turn affect the fluid flow and heat transport inside the EGS subsurface system. It is necessary to establish a variable thermophysical property EGS model to simulate the EGS heat extraction process and to predict the performance of EGS, including its lifetime and capacity.

This chapter reports a transient three-dimensional numerical model developed for the simulation of the heat extraction process in the EGS reservoir. The model treats the geothermal reservoir as a porous medium and adopts the two-equation thermal model to simulate the heat transfer process between the solid rock matrix and the fluid in fractures. The non-Darcy flow behavior of working fluid is studied based on the Forchheimer model. In addition, the model considers pressure- and temperature-dependent thermophysical properties of working fluid to investigate their effects on the heat extraction performance.

11.3 Methodology

11.3.1 Modeling Thermohydraulic Process in EGS

11.3.1.1 Physical Model

Physically, the EGS subsurface geometry consists of multiple domains: injection and production wells, heat reservoir, and rock enclosing the heat reservoir (e.g., the base and cap rock) (see Figure 11.2). The interfaces between different domains present difficulty and additional complexity when modeling the heat extraction process in EGS. In the present work, we treat the EGS subsurface geometry of interest as a single-domain consisting of multiple subregions: (1) Region 1 represents the heat reservoir; (2) Region 2 the rock enclosing the heat reservoir; and (3) Region 3 the injection and production wells, as schematically displayed by Figure 11.2. Different regions have distinct geophysical properties. The heat reservoir is looked as an equivalent porous medium, which is assumed to be isotropic and homogeneous in the present work, and is characterized by a single porosity ε and a finite permeability K. The rock enclosing the heat reservoir is impermeable to fluid, that is, $\varepsilon = 0$ and $K = 0$. The injection and production wells are looked as open channels, that is, $\varepsilon = 1$ and $K = \infty$. In this way, all the interfaces are taken as interior surfaces between two different subregions and will be handled automatically during the numerical modeling. This single-domain treatment circumvents typical difficulties about matching boundary conditions between domains in traditional multi-domain approaches and facilitates numerical implementation and simulation of the complete subsurface thermohydraulic process in EGS.

It is worth pointing out that the assumption of homogeneous porous medium to the heat reservoir can be relaxed for more practical applications, in which the heat reservoir may be heterogeneous and anisotropic. To account for fluid loss during EGS operation, the subregion 2 can be treated as fluid-permeable porous medium too.

11.3.1.2 Governing Equations

The model is focused on modeling and analyses of the subsurface thermohydraulic process in EGS. Owing to the single-domain treatment to the EGS subsurface geometry, this model can handle the

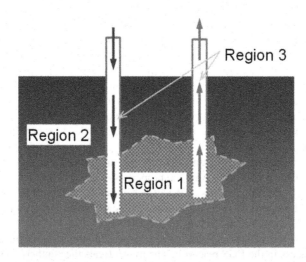

FIGURE 11.2 Schematic of the physical model of EGS subsurface geometry.

Two-Equation Thermal Model for Porous Media

involved heat transfer and fluid flow with ease. Major assumptions made in the derivation of model equations are summarized as follows:

1. The heat transmission fluid (i.e., water) is in liquid or liquid-like state and only single-phase flow is considered. In the reservoir, the fluid temperature considered in the present work is about 200°C, and the hydraulic pressure is high, >300 atm; therefore, it is reasonable to assume the water is in liquid or liquid-like state.
2. The porous heat reservoir is fluid saturated. That is to say, no immiscible gas is trapped in the rock fractures.
3. There do not exist any fluid–structure interactions, including chemical dissolution or deposition and physical pressing, etc.

The governing equations consist of a series of conservation equations. The single-phase transient incompressible flow is governed by the Navier–Stokes equations.

$$\frac{\partial(\varepsilon\rho)}{\partial t} + \nabla\cdot(\rho\mathbf{u}) = 0 \tag{11.1}$$

$$\frac{\partial(\varepsilon\rho\mathbf{u})}{\partial t} + \nabla\cdot(\rho\mathbf{u}\cdot\mathbf{u}) = -\nabla(\varepsilon P) + \nabla\cdot\mu^{eqv}\nabla\mathbf{u} - \frac{\mu}{K}\mathbf{u} + \varepsilon\rho\mathbf{g} + S_{ND} \tag{11.2}$$

Equations (11.1) and (11.2) describe the mass continuity of fluid and the momentum conservation, respectively. The application of the full-form momentum equation, Equation (11.2), enables a general treatment to the fluid flow in open-channel injection and production wells and in the porous heat reservoir. The third term on the right-hand side of Equation (11.2) represents the Darcy and non-Darcy resistance of seepage flow in the porous heat reservoir.

During heat extraction processes, the fluid is flowing in Region 1 and Region 3; heat stored in the rocks is convectively transferred to the fluid in the porous region, Region 3; heat transfer in Region 2 relies only on heat conduction. We consider local thermal non-equilibrium between the rock matrix and fluid flowing in the fractures of the porous heat reservoir, and thus employ two energy equations, Equations (11.3) and (11.4), to describe the heat conduction in HDR (or rock matrix) and the heat convection and advection in fluid, respectively.

$$\frac{\partial\left[\varepsilon(\rho c_p)_f T_f\right]}{\partial t} + \mathbf{u}\cdot\nabla\left[(\rho c_p)_f T_f\right] = \nabla\cdot\left(k_f^{eff}\nabla T_f\right) + ha(T_s - T_f) \tag{11.3}$$

$$\frac{\partial\left[(1-\varepsilon)(\rho c_p)_s T_s\right]}{\partial t} = \nabla\cdot\left(k_s^{eff}\nabla T_s\right) - ha(T_s - T_f) \tag{11.4}$$

Equation (11.3) describes the energy transport in the heat transmission fluid flowing in the fractures, and Equation (11.4) the heat transport in the rock matrix of the heat reservoir or in the surrounding impermeable rocks. In these two energy conservation equations, there is a term $\pm ha(T_s - T_f)$ introduced to model the heat exchange between solid rock matrix and fluid flowing in the fractures in the heat reservoir. Moreover, the Bruggeman correction with a correction factor of 1.5 is applied to determine the effective heat conductivity k^{eff}, that is, $k_s^{eff} = k_s(1-\varepsilon)^{1.5}$ and $k_f^{eff} = k_f\varepsilon^{1.5}$.

11.3.1.3 Non-Darcy Effect

The term S_{ND} in Equation (11.2) is the so-called non-Darcy term, which is commonly described by the Forchheimer model (Forchheimer 1901; Zhang and Xing 2012), that is,

$$S_{ND} = -\beta\rho_f|\mathbf{u}|\mathbf{u} \tag{11.5}$$

where β is the non-Darcy coefficient. The non-Darcy term, which formulates the flow resistance caused by the liquid–solid interactions, has been generally adopted for describing high-velocity flow behavior and validated experimentally and numerically by many researchers (Friedel and Voigt 2006; Janicek and Katz 1995; Kohl et al. 1997; Pascal and Quillian 1980; Zhang and Xing 2012). The non-Darcy flow coefficient β, which is associated with the intrinsic characteristics of porous medium, is a key parameter to predict flow behavior under non-Darcy flow regime. Generally, β is taken as a function of permeability and porosity or permeability only, and it is usually determined from laboratory measurements and/or practical well-tests.

11.3.1.4 Heat Exchange between Fracture and Rock

In terms of the present modeling framework, rock–fluid heat exchange in the heat reservoir of EGS is very important to the heat extraction. The heat exchange rate is calculated by the product of specific surface area (a) of fractures, the rock–fluid convective heat transfer coefficient (h), and the rock–fluid temperature difference ($T_s - T_f$), as expressed in Equations (11.3) or (11.4). The specific surface area of fractures is determined from the geometrical configuration of fractures; the rock–fluid convective heat transfer coefficient depends on the thermophysical properties of the fluid and the fluid flowing velocity. The product of h and a directly affects the rock–fluid heat exchange rate in the heat reservoir and dominates the temporal variation of fluid temperature at the outlet of production well. We can roughly estimate h and a based on the so-called parallel plate model, which geometrically approximates the complicated fracture network as N equidistant parallel fractures of spacing l (Figure 11.3). The reciprocal of h can be looked as the serial total thermal resistance from the heat convection in fluid and the heat conduction in rock. Simple derivation leads to Equations (11.6) and (11.7), where l is the spacing of fractures; h_w is convective heat transfer coefficient; A_{total} is the total heat transfer area of a REV; and V_{total} is the total volume of a REV.

$$h = \frac{1}{\dfrac{l}{2k_s} + \dfrac{1}{h_w}} = \frac{1}{\dfrac{l}{2k_s} + \dfrac{d}{Nuk_f}} = \frac{2\varepsilon Nuk_f k_s}{(Nuk_f + 2\varepsilon k_s)l} \tag{11.6}$$

$$a = \frac{A_{total}}{V_{total}} = \frac{N \cdot 2L}{N \cdot (L \cdot (d+l))} = \frac{2}{(1+\varepsilon)l} \tag{11.7}$$

Actual or more accurate values of h and a for practical EGS can be measured experimentally or calculated from fine-scale models.

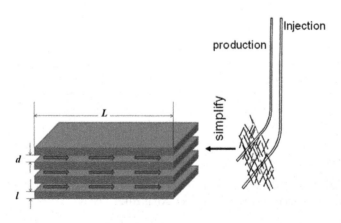

FIGURE 11.3 Schematic about the concept of parallel fracture simplification to the heat reservoir.

11.3.1.5 Thermophysical Properties of Working Fluid

Water is commonly used as the heat transfer fluid in EGS due to its high volumetric heat capacity and ease of obtaining. In 2000, Brown (2000) proposed a novel EGS concept that used supercritical CO_2 ($SCCO_2$) as the heat transfer fluid instead of water, which initiated the research topic about heat transfer fluid of EGS. $SCCO_2$ has liquid-like density and gas-like viscosity. It is the unique properties that make $SCCO_2$ a very attractive heat transfer fluid of EGS. Thermophysical properties of heat transfer fluid may experience significant changes during EGS operation. Especially for $SCCO_2$, its thermophysical properties show very strong dependence on temperature and pressure. Variable thermophysical properties of heat transfer fluid may lead to unusual thermal-hydraulic behaviors during EGS operation.

We model the thermophysical properties of water in terms of the International Association for the Properties of Water and Steam (IAPWS) data (Cooper and Dooley 2007, 2008; IAPWS 2011), and $SCCO_2$ in terms of data available in Heidaryan and Jarrahian (2013), Heidaryan et al. (2011), Jarrahian and Heidaryan (2012), and Span and Wagner (1996).

Figure 11.4 compares the thermophysical properties of water and $SCCO_2$ as functions of pressure and temperature. The thermophysical properties of water including density, viscosity, specific heat capacity, and thermal conductivity are generally larger than those of $SCCO_2$. The effect of pressure on the thermophysical properties of water is relatively weaker, whereas it can be very significant on the thermophysical properties of $SCCO_2$, particularly for the property, density. The density of $SCCO_2$ and the viscosity of water both exhibit strong temperature effects. For typical EGS applications, the density of water may vary within the range of 900 ~ 1000 kg/m^3, while that of $SCCO_2$ may vary within 500 ~ 900 kg/m^3. The stronger temperature-dependent density of $SCCO_2$ may arouse larger buoyancy along the reservoir/wellbore depth direction. The viscosity of $SCCO_2$ is notably smaller than that of water, leading to smaller flow resistance for the seepage flow of $SCCO_2$ in EGS geothermal reservoirs. However, the specific heat capacity and thermal conductivity of $SCCO_2$ are lower than those of water, which may diminish the heat extraction efficiency of $SCCO_2$-EGS.

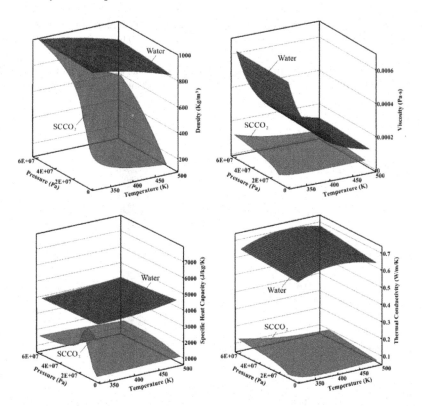

FIGURE 11.4 Thermophysical properties of water and $SCCO_2$ as functions of temperature and pressure.

11.3.2 EGS Subsurface Geometry, Boundary Conditions, and Other Parameters

The modeled EGS subsurface geometry consists of three subregions: (i) injection and production wells ($\varepsilon = 1$ and $K = \infty$); (ii) the geothermal reservoir (finite ε and K); and (iii) the rock enclosing the reservoir ($\varepsilon = 0$ and $K = 0$). With this treatment of the EGS subsurface geometry, the interfaces between the wells, the geothermal reservoir, and the surrounding rock are interior interfaces and are automatically handled in numerical modeling. Specially, the present work considers a five-spot well configuration EGS (one vertical injection well and four vertical production wells); the geometry and geometrical dimensions are schematically shown in Figure 11.5a. The simulated domain is a 2,000 × 6,000 × 2,000 m³ volume, where the horizontal plane at $z = 0$ indicates the earth surface. The reservoir, a 500 × 500 × 500 m³ porous cubic region, is located at a depth of about 4,000 m of the reservoir. The vertical injection well is positioned at the center of the horizontal plane, and the four production wells are arranged close to the four corners. There is a 50 m distance between the production well center and the nearby heat reservoir boundary. The injection and production wells are all 0.2 × 0.2 m² square-shaped in the xy plane. In the present work, only one-quarter of the EGS geometry is simulated because of the geometrical symmetry, as shown in Figure 11.5b. The boundary surfaces, including the xz-plane with the minimum y value and the yz-plane with the minimum x value, are set as symmetry boundaries; all the other external surfaces of the computational domain are specified as zero-flux boundaries for all primary variables, except for temperature, which has fixed values at these boundaries surfaces. The nonslip flow boundary condition is prescribed at the walls of the injection/production well boreholes. The fluid inlet and outlets have fixed pressure boundary conditions. The numerical mesh system is displayed in Figure 11.5b. The meshes are carefully designed to ensure sufficiently fine meshes in the injection and production wells and throughout the reservoir. Grid-independence tests performed indicate changes of less than 2% at the resultant pressure, velocity, and temperature distributions in the reservoir when the mesh system increases its numerical element number from ~150,000 to ~200,000. Further increasing the numerical element number to 270,000 generates negligible changes at the results. Therefore, the mesh system of 270,000 elements is adopted for the present numerical study.

The density, specific heat capacity, and thermal conductivity of the rock are set as $\rho_s = 2650$ kg/m³, $C_{ps} = 1000$ J/kg/K, $k_s = 2.4$ W/m/K, respectively (Jiang et al. 2013). The geothermal gradient is assumed to be 4 K/100 m, while the mean annual surface temperature is fixed at 300 K. Initially, the temperature of the subsurface working fluid is assumed to be locally the same as the rock temperature. During heat extraction, cold fluid, with a temperature of 343.15 K, is injected into the geothermal reservoir under a

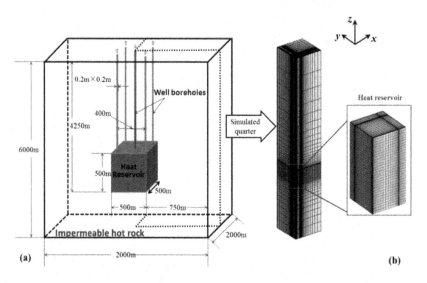

FIGURE 11.5 Model geometry and numerical mesh system: (a) geometry and geometrical dimensions of the considered five-spot well system; and (b) numerical mesh of the simulated domain.

prescribed 10 MPa pressure difference between the injection and the production wells. The pressure at the center of the geothermal reservoir is set at 40 MPa.

Reservoir permeability and porosity may be the two most important parameters, as they dominate the EGS heat extraction process. They dictate the flow distribution and the flow resistance (i.e., the needed external pump work) and thereby, they directly affect the EGS performance, including the heat extraction performance, geothermal reservoir lifetime, and economic performance. We assume that the EGS reservoir has been homogeneously fractured, being of constant and uniform porosity and permeability of 0.01 and 1.0×10^{-14} m^2 (Jiang et al. 2014), respectively.

The correlations of the non-Darcy coefficient β are taken from the literature. Friedel et al. (2006) summarized the experimental data from different sources and obtained a correlation for β. Janicek et al. (1955) established a correlation for β as a function of permeability and porosity for natural porous media. Pascal et al. (1980) described a method for predicting both the non-Darcy flow coefficient, β, and the vertical fracture length, based on a single-point, variable flow drawdown tests of shallow, low permeability gas reservoirs. We calculate the β values in terms of the three correlations of β, respectively, in Section 11.4.3. For the other cases in present work, the seepage flow of fluid in reservoirs is assumed to be Darcy flow, that is, $\beta = 0$.

In the present study, we take the product of h and a as a single parameter and assume it equal to 1.0 W/m^3/K. According to the parallel fracture simplification model (see Figure 11.3), $ha = 1.0$ W/m^3/K corresponds to convective heat transfer for fluid flowing in heat reservoir of parallel fractures with spacing $l \approx 3$ m.

11.4 Results and Discussion

11.4.1 Comparison of Water-EGS and SCCO$_2$-EGS

11.4.1.1 Flow Behaviors

Figure 11.6 compares the fluid mass flow rate of the water-EGS and the SCCO$_2$-EGS. The mass flow rate for both cases declines significantly during the very early period of EGS operation as the cold fluid (of higher viscosity) is being injected into the geothermal reservoir; later on, the fluid mass flow rate slightly declines as the rock is gradually cooled down by the heat transfer fluid. For most time during EGS operation, the fluid mass flow rate for the SCCO$_2$-EGS maintains at ~64 kg/s, more than three times of that for the water-EGS.

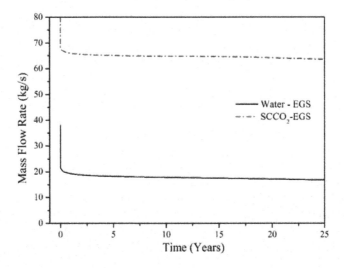

FIGURE 11.6 Mass flow rate of water-EGS and SCCO$_2$-EGS.

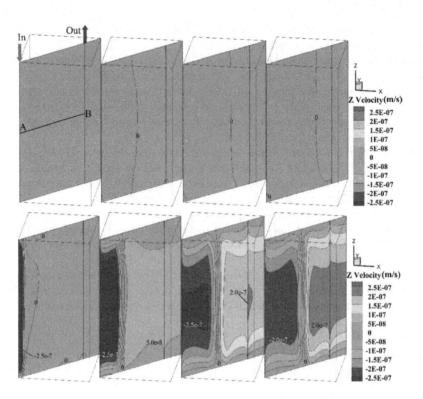

FIGURE 11.7 Z-velocity contour plots on a plane confined by the injection well and a production well. (Upper row: water-EGS; lower-row: SCCO$_2$-EGS. The plots from left to right correspond to time instants: 1, 5, 10, and 15 years, respectively.)

During EGS operation, heat exchange in the reservoir relies partially on natural convection of fluid. Figure 11.7 depicts Z-velocity contour plots on a plane confined by the injection well and a production well in the reservoir at four representative time instants: 1, 5, 10, and 15 years into the EGS operation for both the water-EGS and SCCO$_2$-EGS cases. For both cases, upon EGS operation, there forms first a negative Z-velocity region originates from the injection well and then expands toward the production well; at a relatively later instant into EGS operation (say, 5, 10, or 15 years), two disparate regions of negative and positive Z-velocity, respectively, coexist. Near the injection well, Z-velocity is negative as the cold fluid is injected downward; near the production well, it is positive as the hot fluid is mined upward. The Z-velocity magnitude for the SCCO$_2$-EGS case is significantly larger than that for the water-EGS case, indicating much stronger natural convective heat exchange for the former. The Z-velocity contour for the SCCO$_2$-EGS case also shows stronger time-dependence than that for the water-EGS case, as the thermophysical properties of the former are generally more sensitive to the variations of fluid temperature and pressure.

To shed more light on the Z-velocity distribution in the reservoir, we set a monitoring line, line AB, which connects the injection and production well centers on the mid-Z horizontal plane of the reservoir (see the first plot of Figure 11.7 for its position), and make a quantitative comparison of the Z-velocity profile along line AB between the water-EGS case and the SCCO$_2$-EGS case. To facilitate the comparison, we define a dimensionless parameter, which is the local Z-velocity divided by the velocity magnitude, that is, dimensionless Z-velocity $u_z/|u|$. Figure 11.8 shows the $u_z/|u|$ profiles along line AB for both the water-EGS and the SCCO$_2$-EGS cases at two time instants: 1 and 10 years into the EGS operation. It is worth pointing out that for both cases $u_z/|u|$ is about −1.0 in the injection well and 1.0 in the production well. However, in Figure 11.8, the vertical axis is truncated to keep the view window remaining within the range from −0.45 to 0.2 for better view. For the SCCO$_2$-EGS case, the $u_z/|u|$ profile is seen to have a peak negative value of about 0.05 at 1 year and 0.2 at 10 years, and a peak positive value of about 0.01 at 1 year and 0.08 at 10 years, which are evidently larger than the corresponding counterparts for

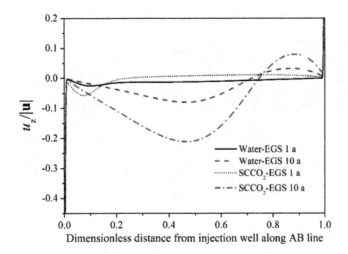

FIGURE 11.8 Dimensionless Z-velocity profile along the AB line in different time instant and working fluids.

the water-EGS. At 1 year, the $u_z/|\mathbf{u}|$ values are all slightly negative in the reservoir of the water-EGS. These findings corroborate again that much stronger natural convection flow exists in the SCCO$_2$-EGS reservoir.

11.4.1.2 Heat Extraction Performance

During EGS operation, the heat stored in the subsurface reservoir is extracted along with the seepage flow of heat transfer fluid. Evolution of rock temperature, which can be indicative of the heat extraction rate, is presented in Figure 11.9 for water- and SCCO$_2$-EGS both. The rock in the vicinity region surrounding the injection well is first cooled down by the injected cold fluid and a low-temperature region forms therein; this low-temperature region is seen to gradually expand toward the production wells. The expanding speed of low rock temperature region for the SCCO$_2$-EGS case (i.e., Case 2) is faster than that for the water-EGS case (i.e., Case 1), indicating the SCCO$_2$-EGS case has a higher heat extraction rate.

To evaluate the heat extraction performance of EGS, we define a parameter, heat extraction ratio, as

$$\theta(t) = \frac{\int (Q_{\text{out}}(t) h_{\text{out}}(t) - Q_{\text{in}}(t) h_{\text{in}}) dt}{\int_{V_h} \rho_s c_{ps} (T_{s,\text{ini}} - T_g) dv} \tag{11.8}$$

On the right-hand side of the above equation, the denominator represents the total heat stored in the geothermal reservoir in reference to the ground temperature, T_g; the numerator denotes the cumulative heat extracted by the heat transfer fluid, where Q denotes the fluid mass flow rate; h_{out} and h_{in} the specific enthalpy of the hot fluid at the production well outlet and of the cold fluid at the injection well inlet, respectively; the term $(Q_{\text{out}}(t) h_{\text{out}}(t) - Q_{\text{in}}(t) h_{\text{in}})$ represents the heat extraction rate; $T_{s,\text{ini}}$ represents the rock initial temperature; and V_h the volume of geothermal reservoir. Further, we can define the local heat extraction ratio in terms of the rock temperature, as

$$\theta_L(t) = \frac{T_{s,\text{ini}} - T_s(t)}{T_{s,\text{ini}} - T_g} \tag{11.9}$$

where $T_s(t)$ represents the rock temperature at time t.

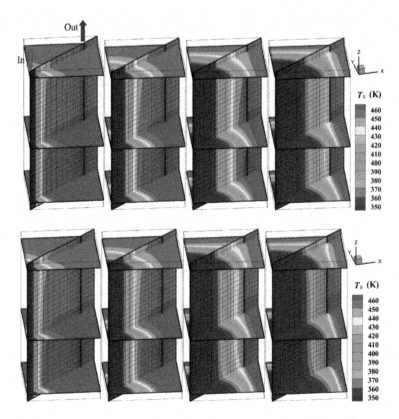

FIGURE 11.9 Rock temperature in the reservoir. (Upper row: water-EGS; lower-row: SCCO$_2$-EGS. The plots from left to right correspond to time instants: 1, 5, 10, and 15 years, respectively.)

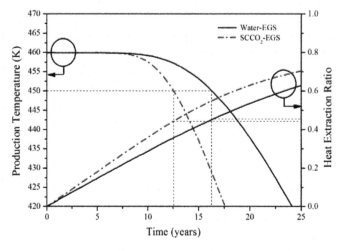

FIGURE 11.10 Production temperature and heat extraction ratio of water-EGS and SCCO$_2$-EGS.

Figure 11.10 makes a comparison of EGS production temperature and heat extraction ratio between the water-EGS and SCCO$_2$-EGS case, that is, Cases 1 and 2. The production temperature remains high, about 460 K, which is the initial average rock temperature in the reservoir, during the early stage of EGS operation, and then decreases for both cases, whereas the duration for the production temperature maintaining high is shorter and the subsequent temperature-drop speed is faster for the

Two-Equation Thermal Model for Porous Media 215

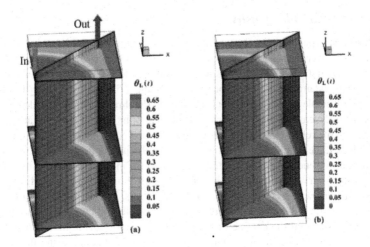

FIGURE 11.11 Local heat extraction ratio distribution in the reservoir at the end of EGS operation: (a): water-EGS at 16.3 years; and (b): SCCO$_2$-EGS at 12.5 years.

SCCO$_2$-EGS case. Taking the production temperature to be 10 K lower than the early-stage maximum production temperature as the EGS abandonment temperature (Tester et al. 2006), we can determine the EGS lifetimes are 12.5 and 16.3 years for the SCCO$_2$-EGS and water-EGS case, respectively. It is seen from Figure 11.10 that the heat extraction ratio for the SCCO$_2$-EGS case is higher than that for the water-EGS case. However, if we compare the heat extraction ratio values at the EGS ceasing-operation time, we see very small difference, about 0.44, for the SCCO$_2$-EGS case and 0.46 for the water-EGS case. The slightly larger heat extraction ratio that the water-EGS case has is caused by its longer lifetime and more thermal compensation the reservoir gets from the surrounding rocks (Chen and Jiang 2015).

Figure 11.11 displays the heat extraction ratio distribution in the reservoir at the end of EGS operation. However, careful inspection of Figure 11.11 finds easily that more heat has been extracted from the reservoir deep regions for the SCCO$_2$-EGS case, that is to say, the heat extraction for the SCCO$_2$-EGS case prefers to perform in the reservoir deep regions. This phenomenon is caused by the stronger natural convection of SCCO$_2$ fluid occurring in the reservoir, which has been discussed in detail in relation with Figures 11.7 and 11.8.

11.4.2 Effects of Variable Thermophysical Properties of Fluid

For real fluids, for example, water or SCCO$_2$, all the thermophysical properties are functions of temperature and pressure, thus are location dependent and vary with time during EGS heat extraction. This section gears to understand the effects of variable thermophysical properties of heat transfer fluid on EGS heat extraction performance. We evaluate the heat extraction performance of different cases by monitoring the real-time net electric power output W_e, which is defined as

$$W_e = \eta_H (Q_{out}(t)h_{out}(t) - Q_{in}(t)h_{in}) - \frac{\Delta p Q_{in}(t)}{\eta_P \rho_{in}(t)} \qquad (11.10)$$

where η_H denotes heat-to-electricity conversion efficiency, assumed to be constant, 0.14 (Tester et al. 2006) in the present work in terms only of the approximate production temperature range; and $(Q_{out}(t)h_{out}(t) - Q_{in}(t)h_{in})$ is the heat extraction rate, the same as in Equation (11.8). The second term on the right-hand side of Equation (11.10) represents the external pump work consumed to drive the heat transfer fluid flow through the geothermal reservoir; pressure head of the pump (Δp) is fixed at 10 MPa for all the cases; working efficiency of the pump (η_P) is assumed to be 0.9.

11.4.2.1 Effects of Variable Density

Figure 11.12 illustrates dependence of the net electric power output on fluid density. The temporal evolution of net electric power output show three distinct stages for both water- and SCCO$_2$-based cases. During the very early first stage, as short as a few months, the net electric power rapidly decreases. The second stage can last around 10 years for water-EGS and 7.5 years for SCCO$_2$-EGS, during which the EGS has relatively stable electric power output. During the third stage, the net electric power of EGS turns to decrease again, meaning the EGS is close to the end of its lifetime. Quantitatively, we define the time instant, at which the relative change of the net electric power output within one day is less than 5%, as the end of the first stage, and the time instant, at which the production temperature starts to decline from its high value, as the end of the second stage. Recalling the mass flow rate curves in Figure 11.6, the production temperature curves in Figure 11.10, and the results of rock temperature distribution presented in Figure 11.9, we deduce with ease the underlying mechanisms. The speedy drop of net electric power output during the first stage is associated with the fast mass flow rate decrease due to the significant increase of fluid viscosity, and the decline of the net power output during the third stage is associated with the so-called "thermal breakthrough" (Chen and Jiang 2015) of heat transfer fluid. For the two real fluid cases, the water density varies within 904.1–1000.0 kg/m^3 and the SCCO$_2$ density varies within 522.1–909.1 kg/m^3. Seen from Figure 11.12, the curve of net electric power output falls in-between the curves of 950 and 1000 kg/m^3 constant density cases for the real water case and 700 and 900 kg/m^3 constant density cases for the real SCCO$_2$ case.

To a great extent, the magnitude of net power output during the second stage represents the capacity of EGS power plant and the duration of this stage reflects the lifetime of this plant. Seen from Figure 11.12, the magnitude of net power output during the second stage is greatly affected by the density of fluid. For cases with water-based heat transfer fluid, the net power output during the second stage is 4.31 MW at 1000 kg/m^3 fluid density, 4.06 MW at 950 kg/m^3 fluid density, and 3.81 MW at 900 kg/m^3 fluid density; for cases with SCCO$_2$-based heat transfer fluid, the net power output during the second stage is 4.57 MW at 900 kg/m^3 fluid density, 2.77 MW at 700 kg/m^3 fluid density, and 0.95 MW at 500 kg/m^3 fluid density. The net power output is almost directly proportional to the density of fluid. Moreover, every 50 kg/m^3 decrease of fluid density leads to about 0.25 MW loss of the net power output for water-based cases, whereas about 0.45 MW loss for SCCO$_2$-based cases, indicating that the production performance of SCCO$_2$-EGS is more sensitive to the density variation of heat transfer fluid.

11.4.2.2 Effects of Variable Viscosity

The viscosity of fluid affects the heat extraction performance of EGS, as a lower viscosity would yield larger fluid velocities at a given pump work, and the net electric power output thus benefits from a larger mass flow rate of heat transfer fluid. Dependence of the net electric power output on fluid viscosity is

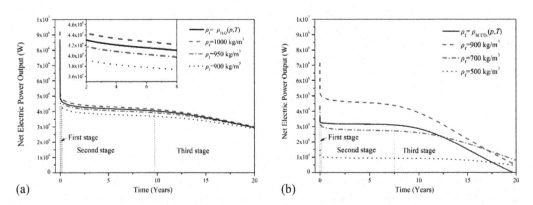

FIGURE 11.12 Net electric power output for cases with different fluid density scenarios: (a) water-based cases; and (b) SCCO$_2$-based cases.

Two-Equation Thermal Model for Porous Media

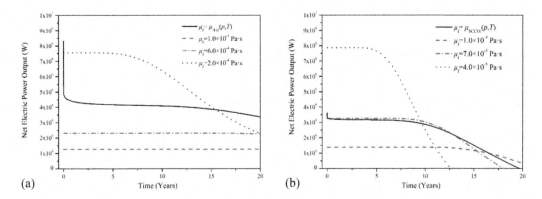

FIGURE 11.13 Net electric power output for cases with different fluid viscosity scenarios: (a) water-based cases; and (b) SCCO$_2$-based cases.

shown in Figure 11.13. Different from Figure 11.12, the first stage of net electric power temporal evolution is not seen for all the constant fluid viscosity cases as the early stage speedy drop of fluid mass flow rate shown in Figure 11.6 will not occur. Furthermore, the net electric power output prior to its last stage decline is approximately inversely proportional to the constant fluid viscosity assumed. For water-based cases, it is about 1.25 MW for the 1.0×10^{-3} Pa·s viscosity case, 2.30 MW for the 6.0×10^{-4} Pa·s viscosity case, and 7.55 MW for the 2.0×10^{-4} Pa·s viscosity case; for SCCO$_2$-based cases, it is about 1.38 MW for the 1.0×10^{-4} Pa·s viscosity case, 3.25 MW for the 7.0×10^{-5} Pa·s viscosity case, and 7.86 MW for the 4.0×10^{-5} Pa·s case.

For the two real fluid cases, the water viscosity varies within 1.6×10^{-4}–1.0×10^{-3} Pa·s and the SCCO$_2$ viscosity varies within 4.4×10^{-5}–9.4×10^{-5} Pa·s during EGS operation. Deduced from the results shown in Figure 11.13, simulations with constant fluid viscosity assumed may not be able to well predict the net electric power output of real EGS.

11.4.2.3 Effects of Variable Specific Heat Capacity

Figure 11.14 describes dependence of the net electric power output on fluid specific heat capacity. The specific heat capacity of fluid has positive effects on the net electric power output of EGS. For water-based cases, the net electric power output prior to its last stage decline is about 4.33 MW for the 4300 J/kg/K case, 4.21 MW for the 4200 J/kg/K case, and 4.09 MW for the 4100 J/kg/K; for SCCO$_2$-based cases, it

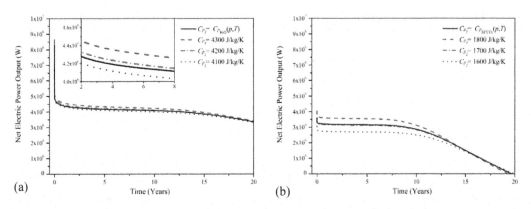

FIGURE 11.14 Net electric power output for cases with different fluid specific heat capacity scenarios: (a) water-based cases; and (b) SCCO$_2$-based cases.

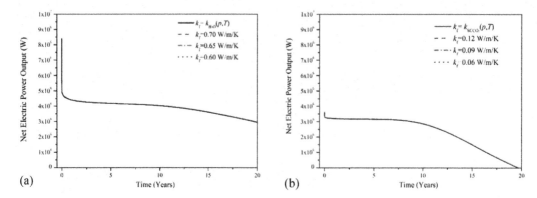

FIGURE 11.15 Net electric power output for cases with different fluid thermal conductivity scenarios: (a) water-based cases; and (b) SCCO$_2$-based cases.

is about 3.53 MW for the 1800 J/kg/K case, 3.12 MW for the 1700 J/kg/K case, and 2.69 MW for the 1600 J/kg/K case. Like the density effects discussed in subsection 3.2.1, the specific heat capacity of fluid also has directly proportional effects on the net electric power output, but at a much smaller proportion.

For the two real fluid cases, the specific heat capacity of water varies within 4093.0–4279.1 J/kg/K, and the specific heat capacity of SCCO$_2$ varies within 1593.9–1848.8 J/kg/K during EGS operation. Seen directly from Figure 11.14, a simulation with suitably assumed constant specific heat capacity of fluid is able to well predict the net electric power output of real EGS.

11.4.2.4 Effects of Variable Thermal Conductivity

Figure 11.15 displays dependence of the net electric power output on fluid thermal conductivity. For both water- and SCCO$_2$-based cases, the net electric power output curves are almost the same, indicating negligible effects of variable thermal conductivity of heat transfer fluid. We further deduce that heat extraction in EGS relies mainly on heat exchange between heat transfer fluid and hot rock.

From the above case studies, we find that the net electric power output of EGS has positive correlation to the density and specific heat capacity of heat transfer fluid, and negative correlation to the viscosity of fluid, whereas the thermal conductivity of fluid only has negligible effects on the net electric power output of EGS.

11.4.3 Effects of Non-Darcy Flow

To compare the Darcy and non-Darcy flow behaviors and further investigate the criterion for judging the onset of non-Darcy flow, we consider three correlations of β in the numerical simulations. With $K = 10.0$ mD and $\varepsilon = 0.01$, we calculate the β values in terms of the three correlations of β, respectively. The β correlations together with the calculated β values are listed in Table 11.1. Note that in Table 11.1, β has the unit of cm^{-1}, and K in mD. The β values calculated by different correlations show large differences.

We consider two groups of cases, as displayed in Table 11.2, one group for the study of non-Darcy effects on EGS heat extraction process, the other mainly for the derivation of a criterion that can be used to judge whether non-Darcy flow effects shall be considered or not when modeling EGS heat extraction processes.

11.4.3.1 EGS Performance with or without Non-Darcy Effects

During heat-mining processes in EGS, the heat transfer fluid is driven by the pressure gradient in the heat reservoir, extracting thermal energy from hot rocks. Figure 11.16 depicts the mass flow rate curves calculated for Cases 1–4. The non-Darcy effect has little impact on the mass flow rate in water-EGS but

TABLE 11.1

Correlations for the Non-Darcy Coefficient β

β Correlation	β Value (cm^{-1})
$\beta = \dfrac{4.1 \times 10^9}{K^{1.5}}$, (Jiang et al. 2013)	1.3×10^8
$\beta = \dfrac{1.82 \times 10^8}{K^{1.25} \varepsilon^{0.75}}$, (Forchheimer 1901)	3.2×10^8
$\beta = \dfrac{4.8 \times 10^{10}}{K^{1.176}}$, (Kohl et al. 1997)	3.2×10^9

TABLE 11.2
Simulated Cases

Group #	Case #	Working Fluid	β (cm^{-1})
1	Case 1	H$_2$O	Darcy
	Case 2	H$_2$O	1.3×10^8
	Case 3	SCCO$_2$	Darcy
	Case 4	SCCO$_2$	1.3×10^8
2	Case 5	H$_2$O	3.2×10^8
	Case 6	H$_2$O	3.2×10^9
	Case 7	SCCO$_2$	3.2×10^8
	Case 8	SCCO$_2$	3.2×10^9

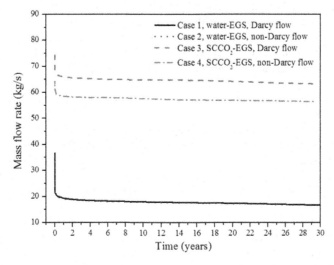

FIGURE 11.16 Mass flow rate as a function of EGS operation time in water- and SCCO$_2$-based EGS.

has a strong impact on that in SCCO$_2$-EGS. Under the assumption of prevailing Darcy flow, the mass flow rate of SCCO$_2$-EGS is about four times that of water-EGS, at the same given identical pressure differences between injection and production wells. However, when the fluid flow is non-Darcian, the mass flow rate of the SCCO$_2$-EGS is only about three times that of the water-EGS. This means that the advantage of SCCO$_2$-EGS, that is, significantly higher mass flow rates than in water-EGS, would be evidently weakened by the non-Darcy flow effect.

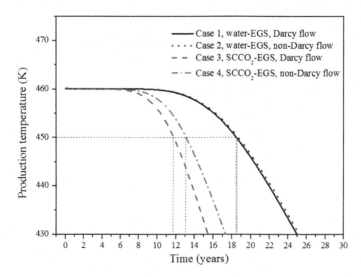

FIGURE 11.17 Non-Darcy effect on the production temperature curve for water- and SCCO$_2$-based EGSs.

Figure 11.17 shows the production temperature curves calculated for Cases 1–4. The displayed EGS production temperature is the volumetrically averaged fluid temperature at the outlet of production wells. At the beginning of fluid production, the production temperature is about 460 K for all water- and SCCO$_2$-based cases, which corresponds to the initial average rock temperature in the reservoir. Thereafter, the production temperature decreases with time. The duration for the production temperature, to remain at about 460 K, is longer for water-EGS cases compared with SCCO$_2$-EGS cases. Moreover, this duration is longer if non-Darcy flow is assumed, compared with the cases in which only Darcy flow is considered. It can be observed that the non-Darcy effects have a stronger impact on the SCCO$_2$-EGS production temperature, compared with the water-EGS. By defining the EGS abandonment temperature as the EGS production temperature equals to 10 K below its maximum value, we can determine the EGS lifetime from Figure 11.17 for the four cases. For the water-EGS, the lifetime is about 18.3 years under the Darcy flow assumption, while it is about 18.5 years when the non-Darcy flow is assumed. For the SCCO$_2$-EGS, the lifetime is about 11.8 years under the Darcy flow assumption and about 13.1 years for the non-Darcy case. The longer lifetime of the non-Darcy case is mainly caused by the reduced fluid mass flow rate, as shown in Figure 11.16.

Further, we calculated the real-time heat extraction rate, defined as the fluid enthalpy at the outlets of the production wells minus that at the inlet of the injection well, that is, $(Q_{out}(t)H_{out}(t) - Q_{inj}(t)H_{inj})$. Figure 11.18 shows the non-Darcy effects on the real-time heat extraction rate of EGS. The temporal evolution of the heat extraction rate shows three distinct stages for all four cases. During the first stage, ~1 year into the EGS operation, the heat extraction rate suffers from a sharp reduction, due to the rapid decrease in fluid mass flow rate (see in Figure 11.16). The second stage, with differing duration for each case, is characterized by a quasi-stable heat extraction rate. During the last stage, the heat extraction rate decreases at an accelerated rate, indicating that the EGS is close to the end of its lifetime. It is seen in Figure 11.18 that for water-EGS the non-Darcy effect is insignificant, whereas for SCCO$_2$-EGS the non-Darcy flow reduces the heat extraction rate, while, to some extent, also increases the duration of the stage, where a quasi-stable heat extraction rate can be observed.

11.4.3.2 Criterion for Judging the Onset of Non-Darcy Flow

There are two main dimensionless numbers that may be used to indicate the onset of non-Darcy flow, namely the Reynolds number, Re, and the Forchheimer number, Fo. The Reynolds number for fluid flow in porous media is commonly defined similar to that for fluid flow in pipes. However, it is difficult to determine the characteristic length due to the often complicated structure of porous heat reservoirs.

Two-Equation Thermal Model for Porous Media

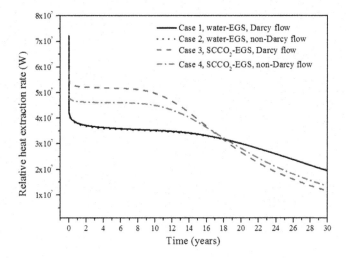

FIGURE 11.18 Non-Darcy effects on the real-time heat extraction rate versus EGS operation time for water- and SCCO$_2$-based EGSs.

To facilitate its use, Котяхов (1956) modified the definition of the Reynolds number for fluid flow in porous media as

$$\mathrm{Re} = 4\sqrt{2}\,\frac{\rho_\mathrm{f}|\mathbf{u}|\sqrt{K}}{\mu_\mathrm{f}\varepsilon^{3/2}} \qquad (11.11)$$

It was said that the seepage flow obeys Darcy's law when Re is less than a critical value, Re$_c$ = 0.2–0.3 (Котяхов 1956).

The Forchheimer number is defined as (Saboorian-Jooybari and Pourafshary 2015; Zhang and Xing 2012)

$$\mathrm{Fo} = \frac{\rho_\mathrm{f} K \beta |\mathbf{u}|}{\mu_\mathrm{f}} \qquad (11.12)$$

The Forchheimer number was also applied to identify whether the non-Darcy behavior must be considered or not (Saboorian-Jooybari and Pourafshary 2015; Zhang and Xing 2012). The critical value of Forchheimer number is in the range of 0.005–0.2 (Saboorian-Jooybari and Pourafshary 2015; Zhang and Xing 2012). For values of the Forchheimer number below the critical value, non-Darcy effects should be negligible. As shown by Equation (11.12), the Fo depends on three factors, namely: (1) mobility of the heat transfer fluid, $\rho_\mathrm{f}/\mu_\mathrm{f}$; (2) flow resistance properties, K and β, of the reservoir; and (3) the seepage velocity, $|\mathbf{u}|$. It is worth noticing that both the Reynolds number and Forchheimer number are directly proportional to the mobility of the fluid, if the porosity and the permeability of the reservoir as well as the seepage velocity in the reservoir are constant.

Intuitively, the non-Darcy effect will impose an additional resistance to the fluid flow and thus reduce the flow velocity. To quantify the non-Darcy effect, we define a new parameter, γ, namely, the relative velocity reduction, as

$$\gamma = \left|\frac{U - U_\mathrm{Darcy}}{U_\mathrm{Darcy}}\right| \times 100\% \qquad (11.13)$$

where U and U_Darcy denote the seepage velocity magnitude of the non-Darcy case and the corresponding Darcy case, respectively.

For the six cases (i.e., Cases 2, 4, 5–8, listed in Table 11.2) with non-Darcy flow considered, we calculate the γ values. Defining a monitoring line (i.e., Line AB) in the reservoir, we plot in Figure 11.19 the γ profile along this line for all the six cases at five years into the EGS operations. The position of Line

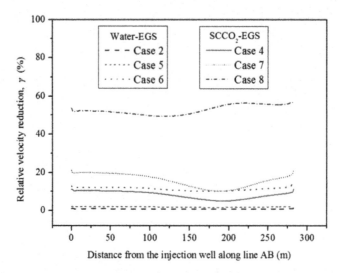

FIGURE 11.19 Relative velocity reduction in the heat reservoir due to non-Darcy flow after five years of EGS operation.

AB is indicated in Figure 11.7. It is located in the mid-*xy* plane of the heat reservoirs. Point A is positioned at the wall of the injection well borehole and Point B at the wall of the production well borehole. There is one more monitoring point, Point C, which is approximately at the mid-point of Line AB. The γ values are larger in the near-regions of the injection/production wells. The position of the largest γ value is always in the near-region of the production well. Except for Cases 2 and 5, where generally small γ values exist in the reservoirs, the other four cases all yield relatively large γ values (>10%), indicating more significant non-Darcy effects. The water-EGS cases do not show significant non-Darcy effects unless the non-Darcy flow coefficient, β, is sufficiently large. The SCCO$_2$-EGS cases all show significant non-Darcy effects.

To shed light on the correlation of non-Darcy effects and Re (or Fo), we tabulate the calculated γ, Re, and Fo values at the three monitoring points for the six cases (i.e., Cases 2, 4, 5–8) in Table 11.3. It indicates that increasing the non-Darcy coefficient, β, leads to more significant non-Darcy effects. At Point C, which is located at the wall of the production well borehole, γ takes on the largest value for each case. For water-EGS, γ is small, 1.1%–2.6% at Point C, if β takes the values of 1.3×10^8 cm^{-1} and 3.2×10^8 cm^{-1}, respectively. However, when a 3.2×10^9 cm^{-1} non-Darcy coefficient is given, the maximum γ value observed is 14.9% at Point C and even at Point B, the γ is 10.1%. The γ value is large for all the three SCCO$_2$-EGS cases considered, indicating significant non-Darcy effects.

Table 11.3 shows that the calculated Re values for water-EGS cases are small (<0.2), whereas for SCCO$_2$-EGS cases, they are large with a maximum of 2.25. There is no definite relation between the

TABLE 11.3

The Relative Velocity Reduction, Reynolds Number and Forchheimer Number at the Three Monitoring Points after 5 Years of EGS Operation

Case #	Heat Transfer Fluid	β (cm^{-1})	Fo$_{ch}$	γ Point A Point B	Point C (%)		Re Point A Point B	Point C		Fo Point A Point B	Point C	
2	H$_2$O	1.3×10^8	0.03	2.0	1.1	2.1	0.14	5.8×10^{-4}	0.13	0.03	1.3×10^{-4}	0.03
5	H$_2$O	3.2×10^8	0.07	2.5	1.7	2.6				0.08	3.3×10^{-4}	0.07
6	H$_2$O	3.2×10^9	0.74	13.1	10.1	14.9				0.78	3.2×10^{-3}	0.74
4	SCCO$_2$	1.3×10^8	0.35	11.0	7.2	11.1	2.25	5.4×10^{-3}	1.51	0.51	1.2×10^{-3}	0.35
7	SCCO$_2$	3.2×10^8	0.85	21.1	13.6	21.8				1.27	3.1×10^{-3}	0.85
8	SCCO$_2$	3.2×10^9	8.54	53.9	49.7	59.1				12.70	3.1×10^{-2}	8.54

Two-Equation Thermal Model for Porous Media 223

observed non-Darcy behavior and the Re values. Therefore, the Reynolds number is not a suitable parameter for judging the onset of non-Darcy flow in EGS heat reservoirs, which is an expected result, as the definition of the Reynolds number does not contain any information about the non-Darcy coefficient β.

Investigating the Fo values shown in Table 11.3, for Cases 4, 6, 7, and 8, which exhibit significant non-Darcy effects, the calculated Fo values are all large at Points A and B, and the larger the Fo, the more significant is the non-Darcy effect. For the other two cases (Cases 2 and 5) the calculated Fo values are small and no non-Darcy effects are observed. Therefore, the Forchheimer number shows a better correlation with γ and can thus be used as a criterion for judging the onset of non-Darcy flow.

The Forchheimer number evolves during EGS heat extraction operation and has strong location dependency. To calculate Fo, one needs to know the local fluid mobility and velocity. In the near-region of the injection well, the fluid experiences larger temperature changes and the fluid mobility may change significantly during EGS operations. Compared with Point A, determining the Fo at point B may be more robust to determine the onset of non-Darcy flow. Taking this into consideration, to facilitate practical uses, we propose to calculate the Forchheimer number at the wall of the production well in terms of the average fluid velocity and the fluid mobility at the initial average temperature in the reservoir. We name this Forchheimer number the characteristic Forchheimer number in EGS, Fo_{ch}, is defined it as

$$Fo_{ch} = \frac{QK\beta}{A\mu_{ref}} \tag{11.14}$$

Where Q denotes the mass flow rate; A is the cross-sectional surface area of the production well borehole, located within the heat reservoir; and μ_{ref} the reference fluid dynamic viscosity, determines in terms of the initial average reservoir temperature. The calculated Fo_{ch} values for the six cases are also shown in Table 11.3. Zhang and Xing. (2012) and Saboorian-Jooybari and Pourafshary (2015) suggest that the critical Forchheimer number is within 0.005–0.2. We suggest a critical Forchheimer number of 0.2, above which non-Darcy effects need to be considered.

It is worth pointing out that the Forchheimer model, Equation (11.5), is only one description, though commonly used, of non-Darcy flow. One other description is the Izbash model (Wen et al. 2008), which uses a high-order polynomial equation of fluid velocity to calculate the non-Darcy flow resistance. However, its theoretical foundation is not yet well established (Zhang and Nemcik 2013). Considering the macroscopic shearing effect between the fluid and the porous matrix, Brinkman (1949) added the second-order derivative of fluid velocity to the Darcy model, resulting in the Brinkman equation, which is suitable to describe non-Darcy flow in highly porous media. Using different non-Darcy flow models may lead to differing descriptions of non-Darcy flow behavior and regimes. However, because the Forchheimer model has some definite physical meaning and is more widely applicable, we recommend it as a proper model for describing non-Darcy flow in EGS heat reservoirs.

11.5 Summary

We presented with great details a three-dimensional transient model for a subsurface thermohydraulic process in EGS. This model has a couple of salient features. It treats the porous heat reservoir as an equivalent porous medium while it considers local thermal non-equilibrium between solid rock matrix and fluid flowing in the fractures and employs two energy conservation equations to describe heat transfer in the rock matrix and in the fractures, respectively. The variable properties of the water and supercritical carbon dioxide ($SCCO_2$) are considered in this model. The non-Darcy flow behavior of working fluid is also studied based on the Forchheimer model.

The heat extraction performances of water- and $SCCO_2$-EGS were compared. For the special quintuplet EGS (one injection well and four production wells) considered, at a given pressure drop between the injection well inlet and production well outlet: (1) the $SCCO_2$-EGS has much higher fluid mass flow rate than the water-EGS; (2) the $SCCO_2$-EGS has faster heat extraction rate, but with the same ceasing-operation criterion, the $SCCO_2$-EGS has shorter lifetime and the cumulative heat extraction amount at the end of operation for the $SCCO_2$-EGS is approximately the same as that for the water-EGS; and (3) stronger natural convection of fluid makes the heat extraction of $SCCO_2$-EGS more preferable to perform in the reservoir deep regions.

Effects of variable thermophysical properties of heat transfer fluid on EGS heat extraction performance were studied. For both water- and $SCCO_2$-EGS, the net electric power output was found to be positively related with the density and specific heat capacity of fluid, and negatively related with the viscosity of fluid, whereas the thermal conductivity of fluid shows little effect on the net electric power output. Nevertheless, the production performance of $SCCO_2$-EGS is generally more sensitive to the variation of fluid thermophysical properties. From viewpoint of pure numerical prediction, a simulation with suitably-assumed constant fluid density, specific heat capacity, and thermal conductivity may well predict the production performance of real EGS, whereas a simulation with constant fluid viscosity may not, especially during the very early stage operation of real EGS, the simulation gives completely wrong result.

The non-Darcy flow behavior was also discussed. It is found that non-Darcy effects decrease the mass flow rate of fluid injected and reduce the heat extraction rate of EGS, as a flow resistance in addition to the Darcy resistance is imposed to the seepage flow in EGS heat reservoirs. Compared with the water-EGS, the $SCCO_2$-EGS are more prone to experience much stronger non-Darcy flow due to the much larger mobility the $SCCO_2$ has. Non-Darcy flow in $SCCO_2$-EGSs may thus greatly reduce their heat extraction performance. Further, we analyze and propose a criterion to judge the onset of non-Darcy flow in EGS heat reservoirs. We take the fluid flow rate and the initial thermal state of the reservoir and calculate the characteristic Forchheimer number of an EGS. If the calculated Forchheimer number is larger than 0.2, the fluid flow in EGS heat reservoirs experiences non-negligible non-Darcy flow characteristic.

ACKNOWLEDGMENTS

Financial support received from the Strategic Priority Research Program of Chinese Academy of Sciences (XDA21060700), the China National Science Foundation and Guangdong-Province Joint Project (U1401232), the Key Scientific Development Project of Guangdong Province (2014A030308001), the Guangdong Key Laboratory of New and Renewable Energy Research and Development Foundation (Y709jf1001), and the China National Science Foundation (51406213) is gratefully acknowledged.

REFERENCES

Baria R., J. Baumgärtner, F. Rummel, R.J. Pine, Y. Sato, HDR/HWR reservoirs: Concepts, understanding and creation, *Geothermics*, 28(4) (1999) 533–552.

Batchelor A.S., Reservoir behaviour in a stimulated hot dry rock system, in: 11th Workshop on Geothermal Reservoir Engineering, Stanford University Calif, (1986).

Blanton T.L., An experimental study of interaction between hydraulically induced and pre-existing fractures, in: *SPE Unconventional Gas Recovery Symposium*, Society of Petroleum Engineers, (1982).

Breede K., K. Dzebisashvili, X. Liu, G. Falcone, A systematic review of enhanced (or engineered) geothermal systems: past, present and future, *Geothermal Energy*, 1(1) (2013) 1–27.

Brinkman H., A calculation of the viscous force exerted by a flowing fluid on a dense swarm of particles, *Flow, Turbulence and Combustion*, 1(1) (1949) 27.

Brown D.W., A hot dry rock geothermal energy concept utilizing supercritical CO2 instead of water, in: *Proceedings of the Twenty-Fifth Workshop on Geothermal Reservoir Engineering*, Stanford University, 2000, pp. 233–238.

Chen J., F. Jiang, Designing multi-well layout for enhanced geothermal system to better exploit hot dry rock geothermal energy, *Renewable Energy*, 74 (2015) 37–48.

Cooper J., R. Dooley, Revised Release on the IAPWS Industrial Formulation 1997 for the Thermodynamic Properties of Water and Steam/The International Association for the Properties of Water and Steam. (2007), There is no corresponding record for this reference.

Cooper J., R. Dooley, Release of the IAPWS formulation 2008 for the viscosity of ordinary water substance, in: *The International Association for the Properties of Water and Steam*, (2008).

Forchheimer P., Wasserbewegung durch boden, *Z. Ver. Deutsch, Ing.*, 45 (1901) 1782–1788.

Friedel T., H.-D. Voigt, Investigation of non-Darcy flow in tight-gas reservoirs with fractured wells, *Journal of Petroleum Science and Engineering*, 54(3–4) (2006) 112–128.

Gelet R., B. Loret, N. Khalili, A thermo-hydro-mechanical coupled model in local thermal nonequilibrium for fractured HDR reservoir with double porosity, *Journal of Geophysical Research-Solid Earth*, 117 (2012) B07205. doi:10.1029/2012JB009161.

Gelet R., B. Loret, N. Khalili, Thermal recovery from a fractured medium in local thermal nonequilibrium, *International Journal for Numerical and Analytical Methods in Geomechanics*, 37(15) (2013) 2471–2501.

Heidaryan E., A. Jarrahian, Modified Redlich Kwong equation of state for supercritical carbon dioxide, *The Journal of Supercritical Fluids*, 81 (2013) 92–98.

Heidaryan E., T. Hatami, M. Rahimi, J. Moghadasi, Viscosity of pure carbon dioxide at supercritical region: Measurement and correlation approach, *The Journal of Supercritical Fluids*, 56(2) (2011) 144–151.

IAPWS, Release on the IAPWS Formulation 2011 for the Thermal Conductivity of ordinary water substance, in: *International Association for the Properties of Water and Steam*, (2011).

Janicek J.D., D.L.V. Katz, Applications of unsteady state gas flow calculations, In: *Proc. U. of Michigan Research Conference*, (1955).

Jarrahian A., E. Heidaryan, A novel correlation approach to estimate thermal conductivity of pure carbon dioxide in the supercritical region, *The Journal of Supercritical Fluids*, 64 (2012) 39–45.

Jiang F., L. Luo, J. Chen, A novel three-dimensional transient model for subsurface heat exchange in enhanced geothermal systems, *International Communications in Heat and Mass Transfer* 41 (2013) 57–62.

Jiang F., J. Chen, W. Huang, L. Luo, A three-dimensional transient model for EGS subsurface thermohydraulic process, *Energy*, 72 (2014) 300–310.

Kohl T., K. Evans, R. Hopkirk, R. Jung, L. Rybach, Observation and simulation of non-Darcian flow transients in fractured rock, *Water Resources Research*, 33(3) (1997) 407–418.

Котяхов Ф., Основы физики нефтяного пласта, Гос. научнотехн. Изд-во нефтяной и горно-топливной лит-ры, (1956).

Lawn B.R., E. Fuller, Equilibrium penny-like cracks in indentation fracture, *Journal of Materials Science*, 10(12) (1975) 2016–2024.

Pascal H., R.G. Quillian, Analysis of vertical fracture length and non-Darcy flow coefficient using variable rate tests, in: *SPE Annual Technical Conference and Exhibition*, Society of Petroleum Engineers, (1980).

Pine R., A. Batchelor, Downward migration of shearing in jointed rock during hydraulic injections, *International Journal of Rock Mechanics and Mining Sciences & Geomechanics Abstracts*, Elsevier, 1984, pp. 249–263.

Pruess K., On production behavior of enhanced geothermal systems with CO2 as working fluid, *Energy Conversion and Management*, 49(6) (2008) 1446–1454.

Raleigh C., P. Witherspoon, A. Gringarten, Y. Ohnishi, Multiple hydraulic fracturing for the recovery of geothermal energy, *EOS, Transactions of American Geophysical Union*; (United States), 55(4) (1974).

Robinson E., R. Potter, B. McInteer, J. Rowley, D. Armstrong, R. Mills, *Preliminary Study of the Nuclear Subterrene*, Los Alamos Scientific Lab., New Mexico, 1971. doi:10.2172/4687637.

Saboorian-Jooybari H., P. Pourafshary, Significance of non-Darcy flow effect in fractured tight reservoirs, *Journal of Natural Gas Science and Engineering*, 24 (2015) 132–143.

Sanyal S.K., S.J. Butler, An analysis of power generation prospects from enhanced geothermal systems, *Geothermal Resources Council Transactions*, 29 (2005).

Shaik A.R., S.S. Rahman, N.H. Tran, T. Thanh, Numerical simulation of fluid-rock coupling heat transfer in naturally fractured geothermal system, *Applied Thermal Engineering*, 31(10) (2011) 1600–1606.

Span R., W. Wagner, A new equation of state for carbon dioxide covering the fluid region from the triple-point temperature to 1100 K at pressures up to 800 MPa, *Journal of Physical and Chemical Reference Data* 25(6) (1996) 1509–1596.

Taron J., D. Elsworth, K.-B. Min, Numerical simulation of thermal-hydrologic-mechanical-chemical processes in deformable, fractured porous media, *International Journal of Rock Mechanics and Mining Sciences*, 46(5) (2009) 842–854.

Tester J.W., B.J. Anderson, A. Batchelor, D. Blackwell, R. DiPippo, E. Drake, J. Garnish, B. Livesay, M. Moore, K. Nichols, The future of geothermal energy: Impact of enhanced geothermal systems (EGS) on the United States in the 21st Century, DOE Contract DE-AC07-05ID14517 Final Report, Massachusetts Institute of Technology, 2006.

Warren J., P.J. Root, The behavior of naturally fractured reservoirs, *Society of Petroleum Engineers Journal*, 3(03) (1963) 245–255.

Wen Z., G. Huang, H. Zhan, An analytical solution for non-Darcian flow in a confined aquifer using the power law function, *Advances in Water Resources*, 31(1) (2008) 44–55.

Wu Y.S., C. Haukwa, G.S. Bodvarsson, A site-scale model for fluid and heat flow in the unsaturated zone of Yucca Mountain, Nevada, *Journal of Contaminant Hydrology*, 38(1–3) (1999) 185–215.

Zhang J., H. Xing, Numerical modeling of non-Darcy flow in near-well region of a geothermal reservoir, *Geothermics*, 42 (2012) 78–86.

Zhang Z., J. Nemcik, Fluid flow regimes and nonlinear flow characteristics in deformable rock fractures, *Journal of Hydrology*, 477 (2013) 139–151.

12

Mixed Convection and Radiation Heat Transfer in Porous Media for Solar Thermal Applications

Simone Silvestri and Dirk Roekaerts

CONTENTS

12.1 Introduction...227
12.2 Modeling of Combined Convection and Radiation...228
12.3 Radiative Properties of Porous Media...233
      12.3.1 Experimental Approach..234
      12.3.2 Analytical Approach...235
      12.3.3 Statistical Approach..238
12.4 Solution Methods for the RTE...242
      12.4.1 Diffusion Approximation..242
      12.4.2 Spherical Harmonics (P_1)..243
      12.4.3 Finite Volume Method..244
      12.4.4 Monte Carlo Method..246
      12.4.5 Application of RTE Solution Methods to Porous Media...248
12.5 Effect of Radiative Heat Transfer on Convection in Porous Media....................................249
12.6 Volumetric Solar Absorbers..252
      12.6.1 Radiative Penetration Length versus Solid–Fluid Heat Flux..................................253
      12.6.2 Flow Instabilities and Hot Spots..256
12.7 Conclusion...259
References..259

12.1 Introduction

Facing the daunting problems of fossil fuel depletion and increased greenhouse gas emission, worldwide attention has turned toward the development of renewable energy systems. Among all the renewable sources available, solar energy has by far the highest theoretical potential. The earth intercepts an average of 89 PJ of solar radiation each second, which is three orders of magnitude larger than the current global energy demand Abbott (2010). To harvest part of this enormous solar power, an innovative concept consists of a transparent collector that allows direct solar absorption in the heat transfer fluid. This design is usually referred to as a *volumetric receiver* due to its capability of absorbing heat in a volume instead of on a surface (like the more classic parabolic trough receiver). The heat derived from solar concentration can then be converted into electricity through power cycles such as Rankine and Brayton cycles or chemical products such as CO- and H_2-based fuels through high temperature dissociation reactions (Lougou et al. 2018).

Avila-Marin (2011) and Avila-Marin et al. (2018) report a complete review on the status of volumetric solar receivers used in solar thermal power plants, comparing them to the already established parabolic trough collectors. They claim that, although not marketed yet, volumetric absorption might be the best alternative to tube receivers. With regard to solar thermochemical reactors many theoretical and experimental studies have been performed. In particular, Chueh et al. (2010) demonstrated the feasibility of a solar-driven thermocycle for the production of H_2 and CO_2 from H_2O and CO_2 dissociation using the reduction of nonstoichiometric ceria oxide arranged in a porous matrix. Many other chemical processes are objects of investigation, such as production of hydrogen from ferrous-metal-oxide-aided hydrolysis (Lougou et al. 2018; Villafan-Vidales et al. 2011) and solar methane reforming over catalytic porous media (Sang et al. 2012; Wang et al. 2014a).

Porous materials have long been recognized as good candidates for absorption materials in volumetric solar receivers because of three main reasons: (1) The effective radiative properties, dependent on the morphology of the material can be engineered to achieve a gradual absorption of sunlight through the receivers. This can efficiently minimize the radiative energy losses; (2) The large specific surface and, therefore, the large solid–fluid contact area, allows high heat and mass transfer performances, which reduces the detrimental temperature gradients in the absorbing phase; and (3) The absorption process occurs on a fixed matrix, which avoids problems related to solid transport, such as wall erosion or particle removal.

However, an effective design of high temperature porous media receivers poses many challenges regarding the interconnection of the multiple physical phenomena occurring. First of all, the sunlight absorption is highly dependent on the radiative properties of the solid matrix; hence, these need to be correctly characterized (Barreto et al. 2018; Du et al. 2007). Moreover, the high temperatures involved lead to a large impact of radiative emission, which causes a long-range heat transfer resulting in a nonlocal energy redistribution throughout the solid phase (Kaviany and Singh 1993). Finally, the temperature in the domain is largely dependent on the mass flow and transport properties of the fluid, which are affected by the radiative transfer process (Nimvari et al. 2018; Pitz-Paal et al. 1997. In addition to these questions, there are still many unresolved issues dealing with the continuous operation of such receivers (Lougou et al. 2018). One of the main issues is the occurrence of hot spots within the solid material. These high-temperature regions cause thermal stresses, which, if not addressed, lead to the sudden destruction of the receiver (Fend et al. 2004).

Due to these challenges, the risks associated with volumetric porous media receivers are still too high, and the efficiencies still too low, to allow commercially profitable options as opposed to traditional power production and standard thermochemical reactors. Nevertheless, the theoretical potential of volumetric absorption sparked a worlwide quest for improving design and efficiencies of the solar receiver by unveiling the underlying complex phenomena. Thanks to the increasing computational power, most of the ongoing research is being performed through the aid of computer simulation and numerical modeling, which involves the combination of convection, conduction, and radiation in irradiated porous media.

Therefore, within the framework of solar thermal absorbers, this chapter will provide the relevant tools and guidelines to simulate mixed convection and radiation in porous media. The structure will be organized as follows. First, a general introduction of modeling strategies for mixed convection and radiation will be presented. The main challenges will be addressed and the relevant solution methodologies will be reported. Second, different approaches for characterizing the radiative properties of a porous medium will be discussed, and the most common radiative transfer solution methods will be illustrated. Different numerical results regarding the effect of radiation on the heat transfer in a porous medium will be reported and, finally, the practical case of volumetric solar absorbers based on porous media will be briefly addressed.

12.2 Modeling of Combined Convection and Radiation

In general, radiative heat transfer manifests itself in the energy equation through the volumetric radiative heat source calculated as

$$Q^R = \nabla \cdot \boldsymbol{q}^R, \tag{12.1}$$

where $\boldsymbol{q}^R$ is the radiative heat flux in W/m^2. How to include this radiative source in the energy equation for the specific case of a porous medium has been subject of extensive study and, as a consequence, multiple approaches are available.

Mixed Convection and Radiation Heat Transfer in Porous Media for Solar Thermal Applications   229

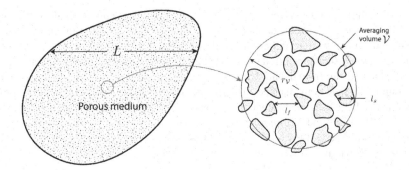

FIGURE 12.1 Length scales of porous media.

The most accurate approach consists of volume averaging the pore-scale governing equations within a reference volume $\mathcal{V}$. This requires that the length scales shown in Figure 12.1 obey the following constraints

$$l_s, l_f \ll r_\mathcal{V}, \quad r_\mathcal{V}^2 \ll L^2, \tag{12.2}$$

and that the solid phase can be considered uniformly distributed about the center of $\mathcal{V}$ (Whitaker 1999). The *superficial average*, for a quantity ψ in phase χ (where χ is either s or f), is then defined as

$$\langle \psi_\chi \rangle = \frac{1}{\mathcal{V}} \int_{\mathcal{V}_\chi} \psi_\chi dV \tag{12.3}$$

where $\mathcal{V}_\chi$ is the volume of phase χ contained in $\mathcal{V}$. On the other hand, better representation of the variable ψ_χ within volume $\mathcal{V}$ is given by the *intrinsic average*, defined as

$$\langle \psi_\chi \rangle^\chi = \frac{1}{\mathcal{V}_\chi} \int_{\mathcal{V}_\chi} \psi_\chi dV \tag{12.4}$$

These two averaging procedures are related by the value of the porosity defined as $\varepsilon_\chi = \mathcal{V}_\chi / \mathcal{V}$ such that

$$\langle \psi_\chi \rangle^\chi = \varepsilon_\chi \langle \psi_\chi \rangle \tag{12.5}$$

At a pore scale, without radiative heat transfer, the governing equations valid for each phase read

$$(\rho c_p)_s \frac{\partial T_s}{\partial t} = \nabla \cdot (k_s \nabla T_s) \qquad \text{in the solid phase} \tag{12.6a}$$

$$T_s = T_f \qquad \text{at the solid–fluid interface} \tag{12.6b}$$

$$-\mathbf{n}_{sf} \cdot k_s \nabla T_s = -\mathbf{n}_{sf} \cdot k_f \nabla T_f \qquad \text{at the solid–fluid interface} \tag{12.6c}$$

$$(\rho c_p)_f \left(\frac{\partial T_f}{\partial t} + \nabla \cdot (\mathbf{u}_f T_f) \right) = \nabla \cdot (k_f \nabla T_f) \qquad \text{in the fluid phase} \tag{12.6d}$$

The averaging of equations (12.6a–d) in $\mathcal{V}$ results in the appearance of several unclosed terms, which require an explicit expression. The closed form of evolution equations for the intrinsic average of T_s and T_f is obtained with the solution of the deviation problem related to the spatial deviation temperatures $\tilde{T}_s$ and $\tilde{T}_f$, where

$$\tilde{T}_\chi = T_\chi - \langle T_\chi \rangle^\chi \tag{12.7}$$

The full derivation of the problem is reported in Whitaker (1999) and Quintard et al. (1993). Based on the assumptions considered, it is possible to obtain the closure for a one- or a two-equation model. In the first case, the system is said to be in local thermal equilibrium (LTE), where $\langle T_s \rangle^s = \langle T_f \rangle^f = \langle T \rangle$ is assumed. In the second case, the system is referred to as local thermal non-equilibrium (LTNE), and the solid temperature is allowed to differ from the fluid temperature. This latter case is more suitable to treat radiative convective coupling in porous media since usually only the solid phase participates in the radiative heat transfer, leading to a large difference between $\langle T_s \rangle^s$ and $\langle T_f \rangle^f$ (Mahmoudi 2014). For a case of a volumetric solar receiver, the LTE model *cannot* be used since the heat transfer between the solid and the fluid phase is an important aspect of the receiver operations (Wang et al. 2013, 2014b). The closed averaged form of equations (12.6a–d), in a LTNE case, is given by (Quintard et al. 1997)

$$\varepsilon_s \left(\rho c_p \right)_s \frac{\partial \langle T_s \rangle^s}{\partial t} - \boldsymbol{u}_{sf} \cdot \nabla \langle T_f \rangle^f - \boldsymbol{u}_{ss} \cdot \nabla \langle T_s \rangle^s$$

$$= \nabla \cdot \left(\boldsymbol{K}_{sf} \cdot \nabla \langle T_f \rangle^f + \boldsymbol{K}_{ss} \cdot \nabla \langle T_s \rangle^s \right) - a_v h \left(\langle T_s \rangle^s - \langle T_f \rangle^f \right)$$

(12.8a)

$$\varepsilon_f \left(\rho c_p \right)_f \left(\frac{\partial \langle T_f \rangle^f}{\partial t} + \langle \boldsymbol{v}_f \rangle^f \nabla \langle T_f \rangle^f \right) - \boldsymbol{u}_{ff} \cdot \nabla \langle T_f \rangle^f - \boldsymbol{u}_{fs} \cdot \nabla \langle T_s \rangle^s$$

$$= \nabla \cdot \left(\boldsymbol{K}_{ff} \cdot \nabla \langle T_f \rangle^f + \boldsymbol{K}_{fs} \cdot \nabla \langle T_s \rangle^s \right) - a_v h \left(\langle T_f \rangle^f - \langle T_s \rangle^s \right)$$

(12.8b)

where $\boldsymbol{u}_{ss}, \boldsymbol{u}_{sf}, \boldsymbol{u}_{fs}, \boldsymbol{u}_{ff}, \boldsymbol{K}_{ss}, \boldsymbol{K}_{sf}, \boldsymbol{K}_{fs}, \boldsymbol{K}_{ff}$ and h are effective transport coefficients calculated with the solution of three different boundary problems. a_v is the specific surface area per unit volume. For a complete description of the problems involved, the reader is referred to Quintard et al. (1997).

The same averaging procedure has been performed by Leroy et al. (2013) with the inclusion of radiative heat transfer. In the specific case of an opaque solid phase and a transparent fluid phase, the radiative heat transfer is present only at the solid surface such that the heat flux boundary condition (12.6c) is modified to

$$-\boldsymbol{n}_{sf} \cdot k_s \nabla T_s - q_w^R = -\boldsymbol{n}_{sf} \cdot k_f \nabla T_f \qquad \text{at the solid-fluid interface.} \qquad (12.9)$$

q_w^R is the homogenized radiative flux impinging on the solid–fluid interface. This latter term can be calculated by assuming an equivalent semi-transparent medium with radiative heat source per unit volume Q^R, yielding

$$q_w^R = \frac{Q^R}{a_v}. \qquad (12.10)$$

This results in the addition of four closure variables $\left(\xi_s, \xi_f, \boldsymbol{p}_s, \boldsymbol{p}_f \right)$ to the right hand side (RHS) of Equations (12.8a and b)

$$\varepsilon_s \left(\rho c_p \right)_s \frac{\partial \langle T_s \rangle^s}{\partial t} = ... + \xi_s \langle Q^R \rangle + \nabla \cdot \left(\boldsymbol{p}_s \langle Q^R \rangle \right)$$

(12.11a)

$$\varepsilon_f \left(\rho c_p \right)_f \frac{\partial \langle T_f \rangle^f}{\partial t} = ... + \xi_f \langle Q^R \rangle + \nabla \cdot \left(\boldsymbol{p}_f \langle Q^R \rangle \right)$$

(12.11b)

This method requires the solution of an additional closure problem associated with the radiative closure variables. In particular,

Mixed Convection and Radiation Heat Transfer in Porous Media for Solar Thermal Applications 231

$$\xi_s = \frac{k_s}{a_v \mathcal{V}} \int_A \boldsymbol{n}_{sf} \cdot \nabla r_s d\mathcal{A} \,, \tag{12.12}$$

$$\xi_f = \frac{k_f}{a_v \mathcal{V}} \int_A \boldsymbol{n}_{fs} \cdot \nabla r_f d\mathcal{A} \,, \tag{12.13}$$

$$\boldsymbol{p}_s = \frac{k_s}{a_v \mathcal{V}} \int_A \boldsymbol{n}_{sf} r_s d\mathcal{A} \,, \tag{12.14}$$

$$\boldsymbol{p}_f = \frac{k_f}{a_v \mathcal{V}} \int_A \boldsymbol{n}_{fs} r_f d\mathcal{A} - (\rho c_p)_f \langle \tilde{v}_f r_f \rangle, \tag{12.15}$$

where $\mathcal{A}$ is the solid surface area in $\mathcal{V}$ and r_χ is the scaling factor between the deviation temperature of phase χ and the average radiative heat flux (i.e., $\tilde{T}_\chi = \ldots + r_\chi \langle q_w^R \rangle$). To eliminate the dependency of deviation temperature on deviation radiative heat flux, Leroy et al. (2013) assumed that the latter is related to the average radiative heat flux through a scaling factor α, such that $\tilde{q}_w^R = \alpha \langle q_w^R \rangle$. Furthermore, the radiative emission was related to the *radiative intrinsic average* temperature

$$\Theta_w = \frac{1}{\mathcal{V}^R} \int_{\mathcal{V}^R} T_w dV \,. \tag{12.16}$$

Θ_w required the definition of an additional averaging volume $\mathcal{V}^R$ based on a different length scale $r_\mathcal{V}^R \leq l_s, l_f$, introduced to track the variations of solid surface temperature within $\mathcal{V}$. For this reason, this method requires the knowledge of temperature at a pore scale. The solution of the averaged equations is no longer sufficient, and T_χ has to be calculated from Equation (12.7) after an assessment of the deviation temperature $\tilde{T}_\chi$. This approach, while complex and computationally expensive, is the most accurate available for coupling radiative heat transfer with convection and conduction in porous media.

On the other hand, most studies employ simplified versions of the volume-averaged energy equations, which are not strictly derived from the pore-scale transport Equation (12.42)

$$\varepsilon_s (\rho c_p)_s \frac{\partial \langle T_s \rangle^s}{\partial t} = \nabla \cdot \left(\boldsymbol{K}_s^* \cdot \nabla \langle T_s \rangle^s \right) - a_v h \left(\langle T_s \rangle^s - \langle T_f \rangle^f \right) \tag{12.17a}$$

$$\varepsilon_f (\rho c_p)_f \left(\frac{\partial \langle T_f \rangle^f}{\partial t} + \langle \mathbf{v}_f \rangle^f \nabla \langle T_f \rangle^f \right) = \nabla \cdot \left(\boldsymbol{K}_f^* \cdot \nabla \langle T_f \rangle^f \right) - a_v h \left(\langle T_f \rangle^f - \langle T_s \rangle^s \right) \tag{12.17b}$$

The thermal dispersion tensors $\boldsymbol{K}_s^*, \boldsymbol{K}_f^*$ and the volumetric heat transfer coefficient $a_v h$ are, in this case, usually determined empirically or by employing boundary layer correlations. By assuming isotropic medium, it is possible to further simplify $\boldsymbol{K}_s^*$ and $\boldsymbol{K}_f^*$ in scalar thermal dispersions k_s^* and k_f^*. Several expression for these terms, based on porosity and cell diameter, can be found in Vafai (2005). Coupling radiative heat transfer with Equations (12.17a, b) cannot follow a rigorous derivation based on pore-scale governing equation as done by Leroy et al. (2013) and is generally done by assuming that Q^R has no direct impact on the averaged fluid temperature $\langle T_f \rangle^f$ and is decoupled from the deviation temperatures $\tilde{T}_f$ and $\tilde{T}_s$. This simply results in the addition of Q^R on the right-hand side of Equation (12.17a) such that

$$\varepsilon_s (\rho c_p)_s \frac{\partial \langle T_s \rangle^s}{\partial t} = \nabla \cdot \left(\boldsymbol{K}_s^* \cdot \nabla \langle T_s \rangle^s \right) - a_v h \left(\langle T_s \rangle^s - \langle T_f \rangle^f \right) - Q^R \,. \tag{12.18}$$

This approach is widely used and proved to be effective in predicting radiative-convective coupling in a porous medium, especially in the framework of a solar thermal absorber (Martin et al. 1998; Pitz-Paal et al. 1997; Wu et al. 2011; Zaversky et al. 2018). Several additional assumptions are employed to calculate the value of Q^R.

1. The spatial deviation component of the radiative heat source is negligible compared to its superficial average (i.e., $Q^R = \langle Q^R \rangle$).
2. Q^R can be directly calculated from the intrinsic average solid temperature such that $Q^R = Q^R(\langle T_s \rangle^s)$. As a consequence the local variations of T_s within $\mathcal{V}$ are not accounted for. This assumption is akin to assuming that, with regards to radiation, $\tilde{T}_s$ is negligible compared to $\langle T_s \rangle^s$.
3. At scales equal and larger than $\mathcal{V}$, the porous medium can be identified with an equivalent semi-transparent medium with effective radiative properties. An example of this procedure for a combination of an opaque and solid phase is shown in Figure 12.2.

The radiative properties, considered here and explained in detail in the next section, are the absorption coefficient κ_λ, the scattering coefficient σ_λ, the extinction coefficient β_λ, the scattering albedo ω_λ, and the phase function Φ_λ where

$$\beta_\lambda = \kappa_\lambda + \sigma_\lambda, \quad \omega_\lambda = \frac{\sigma_\lambda}{\beta_\lambda}, \quad \int_{4\pi} \Phi_\lambda d\Omega = 4\pi \tag{12.19}$$

The radiative heat source can then be calculated as

$$Q^R = \nabla \cdot \int_0^\infty \underbrace{(\int_{4\pi} I_\lambda \mathbf{s} d\Omega)}_{\mathbf{q}_\lambda^R} d\lambda = \int_0^\infty \left(4\pi \kappa_\lambda I_{b,\lambda} - \kappa_\lambda G_\lambda\right) d\lambda \tag{12.20}$$

where the spectral incident radiation G_λ is given by

$$G_\lambda = \int_{4\pi} I_\lambda d\Omega \tag{12.21}$$

The spectral blackbody intensity $I_{b,\lambda}$ is a function of $\langle T_s \rangle^s$ while the spectral intensity I_λ obeys the Radiative Transfer Equation (RTE)

$$\mathbf{s} \cdot \nabla I_\lambda(\mathbf{s}) = \kappa_\lambda I_{b,\lambda} - \beta_\lambda I_\lambda(\mathbf{s}) + \frac{\sigma_{s,\lambda}}{4\pi} \int_{4\pi} I_\lambda(\mathbf{s}') \Phi_\lambda(\mathbf{s}' \cdot \mathbf{s}) d\Omega(\mathbf{s}') \tag{12.22}$$

The above equation tracks the evolution of radiation intensity along a propagation direction s. The first term on the RHS describes the emission from the solid matrix, the second is the extinction due to absorption and scattering, while the last one represents the augmentation due to the scattering

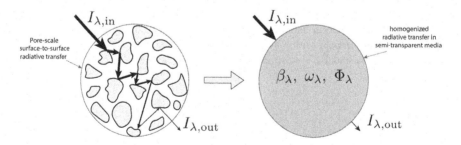

FIGURE 12.2 Visualization of the homogenization procedure: The left image illustrates the attenuation of spectral intensity due to surface-to-surface radiative transfer in a multiphase medium. The local temperature of the solid phase will increase due to surface absorption. The domain on the left is equated to a semi-transparent medium, characterized by a β_λ, ω_λ, and Φ_λ, which produces the same radiative field (right).

Mixed Convection and Radiation Heat Transfer in Porous Media for Solar Thermal Applications 233

in direction s. In the next sections, the solution methods for the radiative problem in porous media will be discussed (how to find radiative properties of the porous media, how to solve the radiative transfer equation).

12.3 Radiative Properties of Porous Media

Porous media are characterized by multiple phases arranged, in general, in a complex morphology. The challenge associated with radiative heat transfer in porous media is, therefore, to account for effective radiative properties, which are able to describe a homogeneous volume with an associated semi-transparent single phase. The porous medium can then be modeled in its entirety with a single extinction coefficient β_λ, scattering albedo ω_λ and phase function Φ_λ. In general, the use of effective properties is the most common approach since the necessary assumptions are valid for most cases, especially when dealing with porous media for solar thermal applications. The assumptions required to approximate the porous medium with a single semi-transparent phase are the following:

1. The size of the system is much larger than wavelength of radiation.
2. The pore diameters are small compared to the medium thickness.
3. The volume considered must be statistically homogeneous and isotropic. This assumption is generally verified if the volume considered is much larger than the pore diameter (i.e., if the previous assumption is valid).

   In addition to these assumptions, for solar thermal applications, stronger approximations are usually employed to simplify the description of radiative properties. It is commonly assumed that

4. The porous medium is composed of an opaque and a transparent phase.
5. The wavelength of radiation is much smaller than the pore diameter.

   In particular, assumption (5) leads to the possibility of calculating the radiative properties using the simple laws of geometrical optics, without resorting to the much more complicated Mie theory. Assumption (5) is usually verified since, in the considered applications, the dominant radiative heat transfer occurs in the visible part of the electromagnetic spectrum. Therefore, the characteristic dimension of the pores is much greater than the relevant wavelengths of radiative heat transfer, which are smaller than 5 μm.

The identification of effective radiative properties is specific to the investigated medium and different modeling procedures have been developed through the years to account for different levels of complexity. In particular, it is possible to distinguish between three main methodologies.

1. Identification through experimental measurements of transmissivity and reflectivity (experimental approach)
2. Direct calculation through idealized geometries (analytical approach)
3. Statistical identification through ray tracing in reconstructed 3D geometries (statistical approach)

While the first methodology employs simple calculations and requires no *a priori* knowledge of the material properties, the other two approaches are based on the knowledge of the local radiative properties of all the phases. If the phases involved are semi-transparent (ST), the refraction index, extinction coefficient, scattering coefficient, and phase function of the materials involved are needed at a pore scale $(n_\lambda, \beta_\lambda, \sigma_{s\lambda}$, and $\Phi_\lambda)$, while if the phase is opaque, the bi-directional reflectivity of the surface ρ_λ'' is sufficient. Due to the specific applications considered in this chapter, only the case of an opaque solid phase and transparent fluid phase will be considered.

12.3.1 Experimental Approach

This methodology involves the experimental assessment of spectral directional-hemispherical transmissivity (DHT_λ) and directional-hemispherical reflectivity (DHR_λ) of a sample. These are defined, as in Celzard et al. (2012), as the ratio of the spectral transmitted and reflected intensity to the incoming intensity $I_{\lambda 0}$ as

$$DHR_\lambda = -2\pi \frac{\int_{-1}^{0} I_\lambda(0,\mu)\mu d\mu}{I_{\lambda 0} d\Omega_0}, \quad DHT_\lambda = 2\pi \frac{\int_{0}^{1} I_\lambda(L,\mu)\mu d\mu}{I_{\lambda 0} d\Omega_0}, \quad (12.23)$$

where 0 and L are the front and the back of the sample, respectively. These two quantities are calculated by shining monochromatic light toward the sample and collecting the total outgoing intensity within the positive and negative hemispheres. A schematic of the experimental setup used by Celzard et al. (2012) to retrieve spectral measurement of tannin-based, glasslike, carbon foams is shown in Figure 12.3. A first approach to identify the radiative properties of the medium, as used by Mey et al. (2014), consists of employing Beer–Lambert's law to assess the extinction coefficient as

$$\beta_\lambda = -\frac{\ln(DHT_\lambda)}{h}, \quad (12.24)$$

where h is the length of the sample. The scattering albedo can be defined as

$$\omega_\lambda = DHT_\lambda + DHR_\lambda, \quad (12.25)$$

while the definition of the absorption and scattering coefficient follows

$$\kappa_\lambda = (1 - DHT_\lambda - DHR_\lambda)\beta_\lambda, \quad \sigma_\lambda = (DHT_\lambda + DHR_\lambda)\beta_\lambda, \quad (12.26)$$

A more refined identification method based on experimental measurements involves the solution of the inverse radiation problem. An example of the application of this approach can be found in Baillis et al. (2002), Celzard et al. (2012), and Loretz et al. (2008a). The inverse radiation problem applied to radiative properties identification consists of an iteration of the solution of the one-dimensional RTE within the porous structure. Between each iteration of the RTE solution, the values of β_λ, ω_λ, and Φ_λ are adjusted by minimizing an objective function. When the experimentally measured values of transmissivity and reflectivity are met, the final identification of the radiative properties is achieved. Different minimization techniques can be employed based on the available measurements.

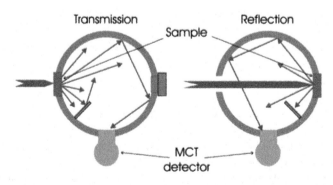

FIGURE 12.3 Schematic of the experimental apparatus used to calculate DHT_λ and DHR_λ. (From Celzard, A. et al., *Carbon*, 50, 4102–4113, 2012.)

As an example, Celzard et al. (2012) minimized the following objective function F_λ

$$F_\lambda = \left(\frac{\text{DHR}_{\lambda,\text{num}} - \text{DHR}_{\lambda,\text{exp}}}{\text{DHR}_{\lambda,\text{exp}}}\right)^2 + \left(\frac{\text{DHT}_{\lambda,\text{num}} - \text{DHT}_{\lambda,\text{exp}}}{\text{DHT}_{\lambda,\text{exp}}}\right)^2. \qquad (12.27)$$

Since they lacked bi-directional measurements, it was not possible to identify the phase function Φ_λ which was, therefore, assumed to be isotropic. Loretz et al. (2008a) used a different identification approach. In addition to the previous mentioned DHT$_\lambda$ and DHR$_\lambda$, their experimental apparatus measured the bi-directional transmissivity from the normal incoming to the normal outgoing direction DNT$_\lambda$. This measure was included in the objective function such that the identification could include an estimate of the scattering phase function. They approximated anisotropic scattering with a combination of diffuse and specular reflection within the medium based on an index of specularity $p_{s\lambda}$ such that

$$\Phi_\lambda(\Theta) = (1 - p_{s\lambda})\frac{8}{3\pi}(\sin\Theta - \Theta\cos\Theta) + p_{s\lambda}. \qquad (12.28)$$

when $p_{s\lambda} = 0$ (diffuse reflection) the phase function takes the form of scattering from a diffusely reflecting sphere, while $p_{s\lambda} = 1$ corresponds to purely specular reflection, for which the effective scattering from the homogenized semi-transparent media coincides with isotropic scattering. This last parameter was added to the variables to be determined. The increment of β_λ, ω_λ, and $p_{s\lambda}$ between the RTE iterations were connected to the difference between numerical and experimental values of DNT$_\lambda$, (DHT$_\lambda$ + DHR$_\lambda$) and DHR$_\lambda$/DHT$_\lambda$. Baillis et al. (2002) calculated the radiative properties of a polyurethane foam based on different measurements and assumptions to compare the results of different strategies. These involved the use of directional-hemispherical measurement or bi-directional measurement or a combination of the two. Also different assumptions for the phase functions were tested (i.e., isotropic Φ_λ or Henyey–Greenstein-type phase function). They showed that the value of the extinction coefficient does not vary significantly with the used measurements, while the scattering albedo is largely dependent on the strategy employed. Their conclusion is that the best approach consists of the usage of a combination of bi-directional and hemispherical transmissivities such that minimal *a priori* assumptions on the phase function are required.

12.3.2 Analytical Approach

Many studies have tried to calculate directly the volumetric radiative properties of porous media (in particular foams) based on idealized geometries. For such an analysis to take place, it is necessary to employ assumptions on the foam's morphology. The most popular approach consists of approximating the foam with a structure composed of regular polyhedra. Examples of idealized cell structures, as reported by Baillis and Sacadura (2000), are shown in Figure 12.4. In these structures, it is possible

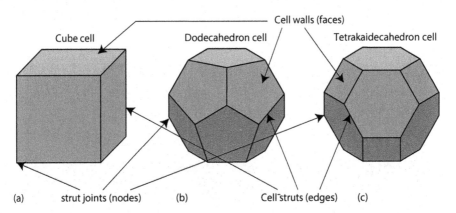

FIGURE 12.4 (a, b, c) Different cell models with cell structures as in Baillis et al. (From Baillis, D. et al., *Spec. Top. Rev. Porous Media*, 4, 111–136, 2013.)

TABLE 12.1

Geometrical Features of the Polyhedra in Figure 12.4 as Shown by Baillis and Sacadura

	Diameter	Volume	Struts/Volume	Mean Wall Area	Walls/Volume
Cube	l_i	l_i^3	$4/l_i^3$	D_{cell}^2	$3/D_{\text{cell}}^3$
Dodecahedron	$2.62 l_i$	$7.663 l_i$	$1.305/l_i^3$	$0.251 D_{\text{cell}}^2$	$14.081/D_{\text{cell}}^3$
Tetrakaidecahedron	$3 l_i$	$11.314 l_i$	$1.061/l_i^3$	$0.2391 D_{\text{cell}}^2$	$14/D_{\text{cell}}^3$

Source: Baillis, D. and Sacadura, J., *J. Quant. Spectros. Rad. Transfer*, 67, 327–363, 2000.

to distinguish walls, struts, and junctures (or joints), which vary based on the geometry considered. The geometrical characteristic of such cells, in terms of diameter, volume, struts per volume, mean wall area, and walls per volume are summarized in Table 12.1. With the knowledge of the material's local radiative properties (ρ_λ'') and the geometry of the foams, it is possible to derive analytical expressions for the extinction coefficient, the scattering coefficient, and the phase function.

Baillis and Sacadura (2000) provide an expression valid for open cell foams, where only the struts and the joints need to be considered (see Figure 12.5). If $\overline{G}_i$ (with $i = 1$ for struts and $i = 2$ for junctures) is the average geometrical cross section of the two different elements and N_i is their occurrence per unit volume, then

$$\beta = N_1 \overline{G}_1 + N_2 \overline{G}_2, \quad \sigma_\lambda = \rho_\lambda (N_1 \overline{G}_1 + N_2 \overline{G}_2), \quad \kappa_\lambda = (1 - \rho_\lambda)(N_1 \overline{G}_1 + N_2 \overline{G}_2). \quad (12.29)$$

The above equations assume that the solid phase is opaque such that no transmission occurs within the foam elements. Furthermore, it is possible to consider diffuse reflection, specular reflection, or a combination of both. In particular, while Baillis et al. (2002) and Baillis and Sacadura (2000) assumed diffuse reflection, Loretz et al. (2008a) based their model on the parameter of specularity $p_{s\lambda}$, which includes both effects in the volumetric phase function.

Another example of analytical modeling can be found in Coquard et al. (2009) and Placido et al. (2005). In particular, the latter approximate closed cell EPS and XPS foams with a pentagonal dodehadron cell structure composed of walls considered as thin slabs and struts represented by elongated cylinders. They use this idealized geometrical setup to calculate the spectral extinction coefficient corrected in the presence of anisotropic scattering

$$\beta_\lambda^* = \kappa_\lambda + \sigma_\lambda (1 - g_\lambda), \quad (12.30)$$

where g_λ is the anisotropy factor calculated as $\int_{-1}^{1} \Phi_\lambda(\mu) \mu d\mu$. Their approach consists of separating the calculation of β_λ^* into two different foams, one only composed of walls and the other one only

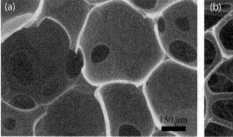
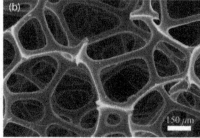

FIGURE 12.5 Difference between closed cells: (a) comprised of struts, walls, and junctures and open cells; and (b) foams that lack the presence of walls. (From Lee, Y. and Yoon, K., *Microporous Mesoporous Mat.*, 88, 176–186, 2006.)

comprising struts. The final result is obtained by averaging the expressions into one single weighted β_λ^* value. For the expression of the extinction coefficient in the struct foam, due to the small diameter of the cylinders, the usage of Mie theory was necessary. The two different β_λ^*'s take the form of

$$1\beta_{\text{struts},\lambda}^* = \frac{4\rho_{\text{foam}}}{\pi d_s \rho_{\text{bulk}}} \int_0^{\pi/2} \left(Q_{e,\lambda} - Q_{s,\lambda}\sin^2\phi - g_{\text{cyl}}Q_{e,\lambda}\cos^2\phi \right)\cos\phi d\phi, \quad (12.31a)$$

$$\beta_{\text{walls},\lambda}^* = \frac{4\rho_{\text{foam}}}{\pi d_w \rho_{\text{bulk}}} \int_0^{\pi/2} \left(R_\lambda \cos(2\phi) + 1 - T_\lambda \right)\sin(2\phi) d\phi, \quad (12.31b)$$

where ρ_{foam} and ρ_{bulk} are the foam and the bulk density, respectively. Moreover, $Q_{e,\lambda}, Q_{s,\lambda}$, and g_{cyl} are the extinction efficiency, the scattering efficiency, and the anisotropy factor, respectively, of an elongated cylinder of diameter d_s tilted at an angle ϕ, while R_λ and T_λ are the reflectivity and transmissivity of a thin slab of thickness d_w, tilted by an angle ϕ. The final β_λ^* is calculated under the independent scattering hypothesis by averaging with the volumetric fraction of struts in the volume occupied by the solid phase (f_s)

$$\beta_\lambda^* = (1 - f_s)\beta_{\text{walls},\lambda}^* + f_s \beta_{\text{struct},\lambda}^* \quad (12.32)$$

They finally apply their methodology to two different EPS foams with the same density (14.93 kg/m^3) and different cell diameter and show an excellent agreement between the calculated β_λ^* and the β_λ^* estimated from the experimental measurement of hemispherical reflectivity and transmissivity (see Figure 12.6). Several analytical relations for the extinction coefficient are given in (Coquard et al. 2012; Loretz et al. 2008a) based on the cell type and the inclusion of struct and junctures. All these relations are functions of the cell diameter and the porosity. As an example, one of the simplest analytical expression for β was developed by Glicksman et al. (1987) considering opaque solid, transparent fluid, independent scattering, and dodecahedron cells

$$\beta = 4.09\sqrt{\frac{1-\varepsilon}{D_{\text{cell}}^2}} \quad (12.33)$$

On the other hand, Loretz et al. (2008b) remove the dependency on porosity by relating β and ε to the geometrical features of the cell itself. They claim that a good model for the geometrical structure will be able to predict simultaneously the extinction coefficient and the porosity of the foam. The expression in (12.33) is, therefore divided into

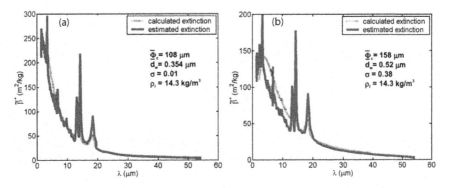

FIGURE 12.6 Comparison between analytically computed and experimentally assessed spectral extinction coefficient for two foams with smaller (a) and larger (b) cell diameter (From Placido, E. et al., *Infrared Phys. Technol.*, 46, 219–231, 2005.)

$$\beta = \frac{3}{4}1.305\frac{b}{a^2} \quad \text{and} \quad \varepsilon = 1 - 0.377\frac{b^2}{a^2}, \tag{12.34}$$

where a and b are the average struct length and thickness.

12.3.3 Statistical Approach

By employing a statistical approach, it is possible to model more accurately the properties of the porous medium. These methodologies are based on a ray-tracing Monte Carlo (MC) method and require a 3D reconstruction of the geometry of the porous medium at a tomographic scale $\left(l_t \ll l_s, l_f\right)$.

Tancrez and Taine (2004) developed the concept of Radiative Distribution Function Identification (RDFI). With this method, it is possible to identify the homogenized absorption coefficient, extinction coefficient, and phase function of a multiphase volume by calculating cumulative distribution functions, which describe the radiative transfer processes. Since a combination of an opaque solid and a transparent fluid is considered, the extinction occurs only at the interface between phases, and it is not wavelength dependent. If $F_e(s)$ is the probability of extinction within a distance s, it is possible to define an extinction cumulative distribution function, which depends only on the geometry of the medium, as

$$G_e(s) = \int_0^s F_e(s')ds' = \int_0^s \frac{1}{V_f}\frac{1}{4\pi}\int_{V_f}\int_{4\pi}\delta(s'-s_0)d\Omega dV ds', \tag{12.35}$$

where δ is the Dirac delta function, V_f is the fluid volume, and s_0 is the distance of a point in the volume V_f to the solid–fluid interface within the solid angle $d\Omega$. $G_e(s)$ represents the possibility that an intensity ray traveling in the porous media is extinguished within a distance s. It can also be thought as the probability that a point in the fluid domain has a distance s or lower to the porous surface. It is easy to see how this extinction cumulative distribution function is connected to the transmissivity of the medium

$$G_e(s) = 1 - \tau(s). \tag{12.36}$$

If the homogenized medium obeys the Beer–Lambert law, extinction within a distance s can be described with an appropriate effective extinction coefficient β such that

$$g_e(s) = 1 - e^{-\beta s}, \tag{12.37}$$

where $g_e(s)$ is the cumulative extinction distribution function for the equivalent homogenized beerian medium.

The RDFI approach consists of identifying $g_e(s)$ with $G_e(s)$ by specifying an effective extinction coefficient β. In practice, this is achieved by a least square fit method, which minimizes the following error

$$\varepsilon(\beta) = \left\{\sum_{s_i=0}^{N_s}[G_e(s_i) - g_e(s_i)]^2 \Big/ \sum_{s_i=0}^{N_s}[1 - G_e(s_i)]^2\right\}^{1/2} \tag{12.38}$$

where the summation is performed over all the discrete points s_i available from the numerical estimation of G_e. Since this approach provides the error between G_e and g_e, it allows the immediate assessment of the validity range of the beerian assumption for the equivalent semi-transparent medium. The effective absorption and scattering coefficients can also be identified using appropriate cumulative distribution functions. In particular, for the absorption probability

$$G_{a,\lambda}(s) = \int_0^s F_{a,\lambda}(s')ds' = \int_0^s \frac{1}{V_f}\frac{1}{4\pi}\int_{V_f}\int_{4\pi}\alpha'_\lambda(\gamma)\delta(s'-s_0)d\Omega dV ds', \tag{12.39}$$

Mixed Convection and Radiation Heat Transfer in Porous Media for Solar Thermal Applications 239

where α'_λ is the spectral directional absorptivity of the surface depending on the incidence angle γ. Again, this cumulative distribution function in an homogenized, beerian medium takes the form

$$g_{a,\lambda}(s) = \left(1 - e^{-\beta s}\right)\frac{\kappa_\lambda}{\beta}, \tag{12.40}$$

where κ_λ is the spectral effective absorption coefficient of such a semi-transparent medium. If the surface of the opaque phase is a diffuse reflector α'_λ is not a function of Ω, and, from Equation (12.39),

$$G_{a,\lambda}(s) = \alpha_{\lambda,\text{diff}}G_e(s). \tag{12.41}$$

This leads to the simple formulation of the absorption coefficient as

$$\kappa_\lambda = \alpha_{\lambda,\text{diff}}\beta \tag{12.42}$$

The scattering cumulative probability distribution function can be retrieved with the knowledge of $G_e(s)$ and $G_a(s)$ since

$$G_e(s) = G_{a,\lambda}(s) + G_{s,\lambda}(s) \tag{12.43}$$

The only remaining property to identify is the phase function, which expresses the probability that, in a scattering event, the ray redirects from direction s into an elementary solid angle $d\Omega$ centered over direction s_r. In the above considered porous medium

$$\Phi(s,s_r)d\Omega(s_r) = \frac{\displaystyle\int_{V_f} \rho''_\lambda(s,s_r,n)s \cdot n dV d\Omega_r(s_r)}{\displaystyle\int_{4\pi}\int_{V_f} \rho''_\lambda(s,s'_r,n)s \cdot n dV d\Omega_r(s'_r)}, \tag{12.44}$$

where n is the normal to the surface. By approximating the porous medium as statistically isotropic, it is possible to express $\Phi(s,s_r)$ as a function of the cosine angles between the incident ray, the reflected ray, and the normal to the surface, defined as

$$\mu_i = s \cdot n, \quad \mu_r = s_r \cdot n, \quad \mu_s = s \cdot s_r. \tag{12.45}$$

μ_s can be expressed in terms of μ_i and μ_r as

$$\mu_s = -\sqrt{\left(1-\mu_i\right)^2\left(1-\mu_r\right)^2}\cos(\phi_i - \phi_r) - \mu_i\mu_r, \tag{12.46}$$

where ϕ_i and ϕ_r are the azimuth angles of the incident and the reflected direction, respectively. By assuming that ρ''_λ is dependent only on the relative difference of the azimuthal angles and is independent of spatial location, the expression of $\Phi(\mu_s)$ simplifies to (Petrasch et al. 2007)

$$\Phi(\mu_s) = \frac{\displaystyle\int_0^1\int_0^{2\pi}\int_0^1 \delta(s \cdot s_r - \mu_s)\rho''_\lambda F(\mu_i)\mu_r d\mu_r d(\phi_i - \phi_r)d\mu_i}{\displaystyle\int_0^1\int_0^{2\pi}\int_0^1 \rho''_\lambda F(\mu_i)\mu_r d\mu_r d(\phi_i - \phi_r)d\mu_i}. \tag{12.47}$$

$F(\mu_i)$ is the probability distribution function associated with the incident direction. In this case, it is assumed to be independent of the distance between the emission and the extinction locations and of the direction of the ray.

In practice, to calculate the values of $G_e(s)$, $G_{a,\lambda}(s)$, and $\Phi(\mu_s)$ an MC procedure is employed. This entails generating a certain number of rays (N_r) with random starting position (within the transparent fluid volume) and random propagation direction. These rays are traced until extinction, which, in the considered case, occurs only upon impact with a solid surface. The path length to extinction is then calculated for the ray r as the distance between emission location and impact with the interface $s_{0,r}$. The probability distribution function of extinction within a distance s is then statistically estimated as

$$F_e(s) \approx \frac{1}{N_r} \sum_{r=1}^{N_r} \delta(s - s_{0,r}) \tag{12.48}$$

It is important to note that, if the starting position and the propagation directions of the ray are sampled from appropriate probability distribution functions

$$\lim_{N_r \to \infty} \frac{1}{N_r} \sum_{r=1}^{N_r} \delta(s - s_{0,r}) = F_e(s) \tag{12.49}$$

so to estimate $F_e(s)$ from Equation (12.48) a substantial number of rays has to be generated and traced. $G_e(s)$ can then be estimated using Equation (12.35). At the solid surface, extinction occurs either by absorption or by reflection (scattering). The absorption is calculated upon impact as

$$F_{a,\lambda}(s) \approx \frac{1}{N_r} \sum_{r=1}^{N_r} \alpha'_\lambda \, \delta(s - s_{0,r}) \tag{12.50}$$

Note that α'_λ depends on the angle of incidence γ. Then, $G_{a,\lambda}(s)$ can be estimated from Equation (12.39). If the absorptivity is non-gray and the interface is not a diffuse surface, the calculation of the absorption cumulative distribution function requires one to sample also a wavelength as an additional property of the ray. If α'_λ is diffuse, the absorption coefficient κ_λ can be directly obtained by Equation (12.42). Finally, $F(\mu_i)$ can be estimated as (Parthasarathy et al. 2012)

$$F(\mu_i) \approx \frac{1}{N_r} \sum_{r=1}^{N_r} \delta(\mu_i - \mu_{i0,r}), \tag{12.51}$$

where $\mu_{i0,r}$ is the incident direction cosine calculated upon impact of ray r with the solid–fluid interface. Several examples of the application of this statistical approach can be found, demonstrating the feasibility of the concept (Parthasarathy et al. 2012; Petrasch et al. 2007; Tancrez and Taine 2004).

A different statistical approach, capable of accounting for a semi-transparent solid phase, was developed by Coquard et al. (2010) for polymeric closed-cell foams. The basic principle of their method is the computation of the mean free path of the photons. The definition of the extinction coefficient follows:

$$\beta_\lambda = \frac{1}{l_{\text{mean}}} \tag{12.52}$$

This method still relies on an MC simulation. In the specific case of an opaque solid phase and a transparent fluid phase, where extinction occurs only upon impact with a solid surface, a large number of rays are launched from random starting position on the solid–fluid interface with sampled direction pointing away from the solid phase. The distance from the emission point to the impact point with the surface $(s_{0,r})$ is recorded and averaged to calculate the mean photon path. Therefore,

$$\beta \approx \left(\frac{1}{N_r} \sum_{r=1}^{N_r} s_{0,r} \right)^{-1} \tag{12.53}$$

The scattering albedo is calculated by counting the number of absorbed rays over the emitted total as

$$\omega_\lambda = \left(\frac{N_r - N_{\text{abs},\lambda}}{N_r} \right) \tag{12.54}$$

Again, the spectral albedo requires the sampling of a wavelength as one of the ray's properties. Finally, the scattering phase function is estimated as

$$\Phi_\lambda(\mu_s) = \frac{2F(\mu_s)}{\sin(\mu_s)(\pi/360)}, \tag{12.55}$$

where $F(\mu_s)$ is the probability distribution of μ_s which is recorded during the MC run. While the mathematical formulations of these two method differ, it is shown by Cunsolo et al. (2016) that if the medium strictly obeys the Beer–Lambert law, the two expressions are equivalent. This latter methodology has been used independently by Cunsolo et al. (2017) and Li et al. (2019) to accurately calculate the radiative properties of irregular open-cell foams made of semi-transparent media. They both accounted for the deviation of the struts shape from an ideal cylinder (see Figure 12.7). In particular, they considered the deviation from constant thickness with the parameter $t = d_{min}/d_{max}$ (where $t = 1$ represents constant thickness) and the deviation from a circular cross section with the parameter $k = R/r$ (where $k = 1$ represents a circle). The results obtained from their MC simulations allowed the parametrization of the extinction coefficient with a generalization of Glicksman's formal relation (12.33). Their proposed fit takes the form of

$$\beta = \frac{2.62\sqrt{1-\varepsilon}[1-0.22(1-t)^2][1+0.22(1-k)^2]}{D_{cell}} \tag{12.56}$$

for Cunsolo et al. (18) and

$$\beta = \frac{2.71\sqrt{1-\varepsilon}[1-0.19(1-t)^2][1+0.21(1-k)^2]}{D_{cell}} \tag{12.57}$$

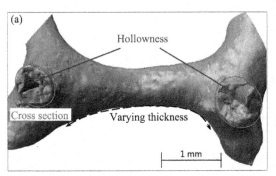 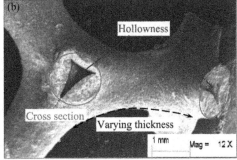

FIGURE 12.7 Real ceramic alumina strut as shown in Li et al. (a) 3D tomographic image. (b) 2D SEM image. The hollowness, varying thickness and cross-sectional shape have been taken into account in the analytical expression proposed. (From Li, Y. et al., *J. Quant. Spectros. Rad. Transfer*, 224, 325–342, 2019.)

242   *Convective Heat Transfer in Porous Media*

for Li et al. (2019). It is possible to notice that the two expressions are surprisingly similar despite the large statistical influence of the randomly generated 3D voronoi structures employed and, in addition, resemble quite closely the approximate Glicksman's model (12.33).

12.4 Solution Methods for the RTE

The solution of the RTE is known to be a challenging problem. The difficulties include the presence of different phenomena, such as emission, absorption, and scattering, the large wavelength dependence, and the propagation of intensity in the full spherical domain. Here we will briefly review the most successful RTE solution methods. A more extensive overview and guidelines for different applications are given in Coelho et al. (2016) and Modest (2013).

12.4.1 Diffusion Approximation

The diffusion approximation, also called the Rosseland approximation, is a simple formulation to treat optically thick media and avoid the cumbersome calculation of the full RTE. To derive Rosseland's formulation, we start by dividing the RTE by the extinction coefficient β_λ and normalizing the coordinate x with a geometrical length scale L, representative of the domain size over which radiative intensity changes significantly. In addition, the derivation of the diffusion approximation requires the assumption of isotropic scattering $\left(\Phi_\lambda = 1 \right)$

$$\frac{s_j}{\beta_\lambda L} \frac{\partial I_\lambda}{\partial x_j / L} = (1 - \omega_\lambda) I_{b,\lambda} - I_\lambda + \frac{\omega_\lambda}{4\pi} \int_{4\pi} I_\lambda d\Omega \tag{12.58}$$

The quantity $(\beta_\lambda L)$ represents the ratio between the newly defined length scale L and the radiative penetration depth $(1/\beta_\lambda)$. It can be also considered as the representative optical thickness τ_λ of the system. In the limit of large optical depths, $\tau_\lambda^{-1} \ll 1$. It is possible, then, to Taylor expand the intensity I_λ as a function of τ_λ^{-1} and truncate the series at the first-order term, such that

$$I_\lambda = I_\lambda^{(0)} + I_\lambda^{(1)} \tau_\lambda^{-1} + \dots \tag{12.59}$$

By substituting Equation (12.59) in Equation (12.58), it is possible to derive a closed form expressions for $I_\lambda^{(0)}$ and $I_\lambda^{(1)}$ under the large optical depth assumption. The final expression for intensity does not depend on propagation direction, but only on the local blackbody intensity and its spatial variation

$$I_\lambda \approx I_{b,\lambda} - \beta_\lambda^{-1} s \cdot \nabla I_{b,\lambda} . \tag{12.60}$$

leading to a simple expression for the spectral radiative source

$$Q_\lambda^R = \nabla \cdot \int_{4\pi} I_\lambda s d\Omega = -\nabla \cdot \left(\frac{4\pi}{3} \beta_\lambda^{-1} \nabla I_{b,\lambda} \right). \tag{12.61}$$

If integrated in wavelength, with the introduction of the Rosseland mean extinction coefficient

$$\beta_R^{-1} = \frac{1}{\int_0^\infty \frac{\partial I_{b,\lambda}}{\partial T} d\lambda} \int_0^\infty \beta_\lambda^{-1} \frac{\partial I_{b,\lambda}}{\partial T} d\lambda , \tag{12.62}$$

the final expression for Q^R yields

Mixed Convection and Radiation Heat Transfer in Porous Media for Solar Thermal Applications   243

$$Q^R = -\nabla \cdot (k_R \nabla T), \quad \text{with} \quad k_R = \frac{16\sigma_{SB}T^3}{3\beta_R},$$ (12.63)

where $\sigma_{SB} = 5.67 \cdot 10^{-8}$ is the Stefan–Boltzmann constant. Because of the formal similarity of Equation (12.63) with the conductive Fourier law, k_R is usually referred to as *radiative conductivity*.

The assumption of large optical thickness $(\tau_\lambda \gg 1)$ results in the following constraint on the zero- and first-order solution of radiative intensity

$$I_\lambda^{(0)} \gg I_\lambda^{(1)}\tau_\lambda^{-1},$$ (12.64)

which, after integration in propagation direction and wavelength, results in a validity condition depending on the temperaure field variation

$$\frac{1}{T}\frac{\partial T}{\partial x} \ll \beta_P(T)$$ (12.65)

where β_P is the Planck mean extinction coefficient. The diffusion approximation is derived by assuming isotropic scattering, but it can also be applied in cases with anisotropic scattering. If the scattering albedo is high enough, multiple scatterings increase drastically the path length to absorption increasing the isotropicity of radiative intensity. Gomart and Taine (2011) verified the accuracy of the diffusion approximation for an anisotropic scattering medium. They noticed that the diffusion approximation holds in a scattering case if β_P is substituted with an extinction coefficient, which accounts for anisotropic scattering $\beta^* = \kappa + \sigma(1-g)$ with g the anisotropy factor defined in the previous section. As a general guideline, the accuracy of the radiative conductive formulation is within 1% if at any scale and at any point in the medium, the temperature gradient satisfies (Gomart and Taine 2011)

$$\frac{1}{T}\frac{\partial T}{\partial x} < 0.03\beta^*(T,\omega,g)$$ (12.66)

The diffusion approximation fails close to the boundary of the porous medium where the optical thick assumption breaks. The distance from the boundary at which the diffusion approximation can be employed depends on the extinction coefficient: with larger optical thicknesses, the radiative Fourier law can be applied closer to the boundary. In practice, the diffusion approximation is valid if the distance from the boundary is larger than $5/\beta^*$ (Gomart and Taine 2011).

12.4.2 Spherical Harmonics (P_1)

The P_N method consists of a radiative intensity approximation with a truncated sum of angular-dependent spherical harmonics $Y_l^m(s)$ multiplied by spatial dependent coefficients $I_l^m(r)$.

$$I_\lambda^N = \sum_{l=0}^{N}\sum_{m=-l}^{m=l} I_l^m(r)Y_l^m(s), \quad \text{with} \quad \lim_{N\to\infty} I_\lambda^N = I_\lambda$$ (12.67)

In the P_1 solution method, the truncation of the above series occurs at the first order $(N = 1)$. Therefore

$$I_\lambda = I_0^0 Y_0^0 + I_1^{-1}Y_1^{-1} + I_1^0 Y_1^0 + I_1^1 Y_1^1$$ (12.68)

where the spherical harmonics take the value of (Modest 2013)

$$Y_0^0 = 1, \quad Y_1^{-1} = -\sin(\theta)\sin(\phi) = -s_y,$$ (12.69)

$$Y_1^0 = \cos(\theta) = s_z\ ,\quad Y_1^1 = -\sin(\theta)\cos(\phi) = -s_x. \tag{12.70}$$

Spectral intensity can then be expressed as

$$I_\lambda(r,s) = I_0^0(r) + c(r) \cdot s. \tag{12.71}$$

By substituting (12.71) into the RTE, it is easy to show that

$$I_0^0 = \frac{G_\lambda}{4\pi} \quad \text{and} \quad c(r) = \frac{3q_\lambda^R}{4\pi}. \tag{12.72}$$

An angular integration of the RTE multiplied by the propagation direction s leads to

$$\beta_\lambda^{-1}\nabla G_\lambda = -3q_1^R + \int_{4\pi}\left[\frac{\omega_\lambda}{4\pi}\int_{4\pi}\left(\frac{G_\lambda + 3q_\lambda^R \cdot s'}{4\pi}\right)\Phi_\lambda(s,s')d\Omega(s')\right]sd\Omega(s) \tag{12.73}$$

Making use of the fact that $\int_{4\pi}sd\Omega = 0$ and $\int_{4\pi}ssd\Omega = 4\pi/3$, it is straightforward to demonstrate that in case of isotropic scattering ($\Phi_\lambda = 1$)

$$\beta_\lambda^{-1}\nabla G_\lambda = -3q_\lambda^R \tag{12.74}$$

Combining the divergence of Equation (12.74) with (12.1), it is possible to obtain an elliptical equation for the spectral incident radiation

$$\frac{1}{3\kappa_\lambda}\nabla\frac{1}{\beta_\lambda}\nabla G_\lambda - G_\lambda = -4\pi I_{b,\lambda}. \tag{12.75}$$

The most common boundary conditions employed in literature are the Marshak's boundary conditions, which, in case of a diffuse boundary with emissivity $\varepsilon_{w,\lambda}$, read

$$-\frac{2 - \varepsilon_{w,\lambda}}{\varepsilon_{w,\lambda}}\frac{2}{3\beta_{w,\lambda}}n \cdot \nabla G_\lambda + G_\lambda = 4\pi I_{bw,\lambda} \tag{12.76}$$

where n is the unit direction vector normal to the surface. Equation (12.75) can only describe diffuse incident radiation and would fail in case of collimated intensity, as in the case of a directional flux in a volumetric solar absorber (Villafan-Vidales et al. 2011; Nimvari et al. 2018). This is because the truncation is performed only up to the first-order harmonic, while collimated radiation is well defined in angle and requires higher harmonics to be described. In such cases, the energy scattered from an additional collimated incident radiation $G_{c,\lambda}$, independent of the diffuse $G_{d,\lambda}$ is added to Equation (12.75).

$$\frac{1}{3}\nabla\frac{1}{\beta_\lambda}\nabla G_{d,\lambda} - G_{d,\lambda} = -4\kappa_\lambda\pi I_{b,\lambda} - \sigma_\lambda G_{c,\lambda}. \tag{12.77}$$

This formulation takes the name of modified differential approximation (MDA) (Equation 12.36). $G_{c,\lambda}$ can be calculated from exponential attenuation of collimated radiation in the porous medium. Finally, $G_\lambda = G_{d,\lambda} + G_{c,\lambda}$ and the radiative heat source is given by (12.20).

12.4.3 Finite Volume Method

The finite volume (FV) method is an alternative to the more common discrete ordinates (DO) method based on an angular discretization of the intensity propagation directions with an FV formulation. This method results in a fairly simple implementation, and it is particularly popular because of the easy coupling potential with CFD codes. The advantages of such a formulation when compared to DO is the full conservation of radiant energy and the reduced ray effects in optically thin media (Modest 2013). The starting point of the FV method is the integration of the RTE (12.22) over the control volume:

$$\int_V \nabla I_\lambda \cdot s dV = \int_V \kappa_\lambda I_{b\lambda} dV - \int_V \beta_\lambda I_\lambda dV + \int_V \frac{\sigma_s}{4\pi} \int_{4\pi} I_\lambda \Phi_\lambda(\Omega,s) d\Omega dV \qquad (12.78)$$

By making use of the divergence theorem

$$\int_V \nabla I_\lambda \cdot s dV = \int_A I_\lambda \boldsymbol{n} \cdot s dA \qquad (12.79)$$

where A represents the faces of the control volume and $\boldsymbol{n}$ the outward pointing unit normal vector. In order to represent the directional behavior in an FV fashion, the whole range of directions (a full sphere) is discretized in a set of control angles. By integrating over the control angle and changing order to the integrals

$$\int_A \int_{\Omega_i} I_\lambda \boldsymbol{n} \cdot s d\Omega dA = \int_V \int_{\Omega_i} \kappa_\lambda I_{b\lambda} dV - \int_V \int_{\Omega_i} \beta_\lambda I_\lambda d\Omega dV \qquad (12.80)$$

$$+ \int_V \int_{\Omega_i} \frac{\sigma_s}{4\pi} \int_{4\pi} I_\lambda \Phi_\lambda(\Omega',\Omega) d\Omega' d\Omega dV. \qquad (12.81)$$

The implementation of FV assumes constant intensity throughout the control angle Ω_i. The resulting discretized equation is then

$$\sum_k I_{ki}(\boldsymbol{n}_k \cdot \boldsymbol{s}_i) A_k = (S_{pi} - \beta_p I_{pi}) V \Omega_i, \qquad (12.82)$$

where:

$$S_{pi} = \kappa_p I_{bp} + \frac{\sigma_{sp}}{4\pi} \sum_{j=1}^{n} I_{pj} \Phi_p(\Omega_i,\Omega_j) \Omega_j, \qquad (12.83)$$

and

$$\boldsymbol{s}_i = \int_{\Omega_i} s d\Omega, \quad \boldsymbol{n}_k = \int_{A_k} n dA. \qquad (12.84)$$

The subscripts k and p identify variables on the faces and in the nodal point of the control volume ($k = n,s,w,e,t,b$), respectively, while the subscript i denotes the propagation direction. The wavelength dependency has been omitted for simplicity. Boundary conditions are obtained by integration over the surface area and control angle, hence, for diffuse boundaries $\left(\rho^s = 0\right)$

$$I_{w,out} = \varepsilon_w I_{bw} + (1-\varepsilon_w) \frac{\sum_{n_w \cdot s_i < 0} I_{w,in} |\boldsymbol{n}_w \cdot \boldsymbol{s}_i|}{\sum_{n_w \cdot s_i > 0} (\boldsymbol{n}_w \cdot \boldsymbol{s}_i)} \qquad (12.85)$$

The only remaining unknown is the facial intensity I_{ki} that must be derived from the nodal intensity I_{pi} by means of a spatial discretization schemes.

Several schemes were developed to discretize the facial intensity from the nodal intensity, characterized by different accuracy and computational requirements. The simplest discretization scheme is the STEP scheme, which assumes the facial intensity to be equal to the upstream nodal intensity. Therefore

$I_{ki} = I_{pi}$ if $\mathbf{n}_k \cdot \mathbf{s}_i > 0$ and $I_{ki} = I_{Ki}$ if $\mathbf{n}_k \cdot \mathbf{s}_i < 0$, where K represents the neighboring cells corresponding to the surface area k. The result of the implementation of this scheme is a fully explicit formulation of Equation (12.82)

$$I_{pi} = \frac{S_{pi} V \Omega_i - \sum_{\mathbf{n}_k \cdot \mathbf{s}_i < 0} I_{Ki} (\mathbf{n}_k \cdot \mathbf{s}_i) A_k}{\beta_p V \Omega_i + \sum_{\mathbf{n}_k \cdot \mathbf{s}_i > 0} (\mathbf{n}_k \cdot \mathbf{s}_i) A_k} \tag{12.86}$$

Because I_{Ki} is required for those faces for which $\mathbf{n}_k \cdot \mathbf{s}_i < 0$, it is sufficient to implement a marching scheme to obtain I_{pi}. If $\sigma_s \neq 0$ or $\rho_w \neq 0$, the propagation directions are coupled, which requires the iteration of Equation (12.86) until the desired tolerance is met. The STEP scheme results in a simple formulation and, akin to a fully implicit finite difference, avoids non-physical results. The drawback of using this method is the occurrence of a large truncation error and, therefore, a low degree of accuracy. Despite the drawbacks, due to the stability and the simple implementation, the STEP scheme is currently the most used spatial discretization scheme.

In order to reduce the truncation error associated with the use of the simple STEP scheme, numerous alternatives have been proposed. The most popular scheme is the diamond scheme, where the facial intensity is calculated as arithmetical average of the two neighboring nodal intensities.

$$I_{ki} = 0.5(I_{Ki} + I_{pi}) \tag{12.87}$$

The diamond scheme is still a first-order discretization scheme but is proven to produce more accurate results when compared to the simple STEP scheme. A severe limitation of the diamond scheme is dictated by stability issues. In addition, different studies using the diamond scheme encounter non-physical intensities, either negative or higher than the sum of incoming plus emitted intensity (Modest 2013). A more accurate and stable (but computationally expensive) method is the CLAM scheme. The CLAM scheme is a second-order bounded scheme that discretizes the facial intensity with a three-point approximation, requiring the knowledge of the upstream and downstream nodal intensities, as well as the nodal intensity from the second upstream cell. The facial intensity is then calculated as:

$$I_k = \begin{cases} I_u + \psi(I_d - I_u) & \text{if } 0 \leq \psi \leq 1 \\ I_u & \text{otherwise} \end{cases}, \tag{12.88}$$

$$\psi = \frac{I_u - I_{uu}}{I_d - I_{uu}} \tag{12.89}$$

where subscript u and uu denote the first and second node in the upstream direction, respectively, while subscript d denotes the first node in the downstream direction. The CLAM scheme results in a complicated and coupled formulation, since the set of equations become implicit and the intensity can no longer be calculated in a single sweep. In order to avoid an implicit formulation of Equation (12.82), the solution is first guessed from the STEP scheme, subsequently correcting with a source term, S_{clam}, which follows from an iterative procedure. CLAM scheme provides much more accurate solution to the RTE, specially in those limits in which first-order solutions fail, as for high optical thickness.

12.4.4 Monte Carlo Method

If a spectral description of radiative heat transfer is to be included, the state of art involves the use of an MC method. Compared to the above-mentioned RTE solution methods, the MC method can be considered the most accurate and flexible. Compared to all the other solution methods, the MC solution time grows much slower with problem complexity, allowing a detailed spectral description or the simulations of complex geometries, which are unfeasible with other methods such as P_1 or FV. An MC photon

Mixed Convection and Radiation Heat Transfer in Porous Media for Solar Thermal Applications   247

transport simulation consists of launching a number of photon beams starting from location in which a high-energy density is encountered and tracing them until fully depleted. MC methods can achieve a high degree of accuracy, being able to resolve exactly the RTE to an extent that is controllable by the number of statistical samples drawn and propagated. A drawback of these methods is the high computational cost, having to trace several rays (up to millions) in order to obtain a statistically significant result. Within a domain containing a non-gray absorbing and emitting medium, the radiative power emitted by cell i and absorbed within cell j is expressed by

$$Q_{i \to j}^R = \int_0^\infty \kappa_\lambda(T_i) I_{b\lambda}(T_i) \int_{V_i} \int_{4\pi} t_\lambda(i \to j)_m a_{\lambda j,m} d\Omega dV d\lambda. \tag{12.90}$$

t_λ is the spectral transmissivity between cell i and cell j following path m, and $a_{\lambda j,m}$ is the absorbed fraction in cell j along path m, which is given by

$$a_{\lambda j,m} = 1 - e^{-\kappa_\lambda(T_j) l_{j,m}}, \tag{12.91}$$

where $l_{j,m}$ is the distance traveled by the ray in cell j along path m. By introducing probability density functions

$$f_V = \frac{1}{V_i}, \quad f_\theta = \frac{\sin \theta}{2}, \quad f_\phi = \frac{1}{2\pi}, \quad f_\lambda = \frac{\kappa_\lambda I_{b\lambda}}{\int_0^\infty \kappa_\lambda I_{b\lambda}}, \tag{12.92}$$

it is possible to rewrite Equation (12.90) as

$$Q_{i \to j}^R = Q^{R,e}(T_i) \int_0^\infty f_\lambda \int_{V_i} f_V \int_0^{2\pi} f_\phi \int_0^\pi f_\theta A_{\lambda,m,i \to j} d\theta d\phi dV_i d\lambda \tag{12.93}$$

where $Q^{R,e}(T_i)$ is the power emitted by cell i.

$$A_{\lambda,m,i \to j} = t_\lambda(i \to j)_m a_{\lambda j,m}. \tag{12.94}$$

The MC method consists of a statistical estimation of the integrals in Equation (12.93) by sampling the properties of the statistical bundles from the PDFs in Equation (12.92). The resulting discretized equation then has the form

$$Q_{i \to j}^R \approx \frac{Q^{R,e}(T_i)}{N_r} \sum_{r=1}^{N_r} A_{r,i \to j} \tag{12.95}$$

The subscript r indicates a ray, characterized by its wavelength λ, and direction angles θ and ϕ (defining the path variable m), which are calculated inverting the following relations

$$R_\lambda = \int_0^\lambda f_{\lambda'} d\lambda', \quad R_\theta = \frac{1 - \cos \theta}{2}, \quad R_\phi = \frac{\phi}{2\pi}. \tag{12.96}$$

In the end, the radiative source for cell j can be calculated as

$$Q_j^R = Q_j^{R,e} - \sum_i Q_{i \to j}^R \tag{12.97}$$

12.4.5 Application of RTE Solution Methods to Porous Media

All the methods described above have been applied to the case of a solar volumetric receiver. The diffusion approximation is the approach that requires the least amount of computational resources and is, therefore, one of the most popular. As already discussed, this method requires the assumption of large optical thickness and isotropic radiation. For this reason, the diffusion approximation cannot handle collimated irradiation typical of a solar absorber.

Opposite to the diffusion aproximation is the MC method. The latter has a large flexibility to account for challenging problems, such as anisotropic scattering, complex geometries, spectral radiation, and collimated radiative flux. The accuracy of the MC method is only limited by the number of statistical samples used. In addition, the implementation is fairly simple and straightforward. However, the computational requirements of this method are high also for simple problems and, for this reason, an MC solver is rarely found coupled with computational fluid dynamics (CFD) in commercial packages. However, thanks to its capability of adapting to complex geometries, the MC method is generally employed in porous media to compute the solution of radiative heat transfer at a pore-scale resolution, which can then be used as reference for cheaper models.

On the other hand P_1 and FV models have a good balance in terms of computational requirements and accuracy, making them the favorite methods in combination with CFD solvers. These two methods are different in terms of accuracy and range of applicability. The FV is the most accurate differential method available, but it requires an angular discretization in addition to the CFD spatial discretization (Coelho et al. 2016). The angular mesh has to be tailored to the optical thickness of the system and to the spatial mesh to avoid ray effects. In particular, a large optical thickness domain requires a finer angular discretization to account for the high intensity gradients. For this reason, the FV is usually not the preferred choice in porous media, unless the dimensionality of the problem can be reduced to one dimension (Zaversky et al. 2018). The P_1 method is, generally, less accurate than the FV but also computationally cheaper. While it is not suitable for highly anisotropic or optically thin cases, these constraints are usually not restrictive in a porous medium, which is usually characterized by large extinction coefficients and isotropic radiation. For this reason, the P_1 method is by far the most popular solution method for radiative heat transfer in porous media.

Wang et al. (2014c) investigated the effect of different radiative models on the temperature of a solar volumetric receiver. In particular, they tested a P_1 and a diffusion approximation. The considered solar receiver had a total length of 50 mm and a total optical thickness of 75. They report a maximum difference in the calculated temperature among the two case of only 5% concluding that, in these conditions, the diffusion approximation is a suitable methodology to solve the RTE. On the other hand, they reported that the difference between the two cases increased with a decrease in fluid velocity or in optical thickness. While it is clear why the P_1 outperforms the diffusion approximation at lower τ values, the increase in method's difference with a decrease in fluid velocity can be linked to a larger relevance of radiative heat transfer when compared to convection.

In contrast, Hischier et al. (2009), performed the same comparison with opposite results. They computed the radiative source in a volumetric solar receiver benchmarking the results with an accurate MC simulation. Their receiver had a significantly lower optical thickness ($\tau = 3$) than the one of Wang et al. (2014c) and, therefore, the diffusion approximation was not able to reproduce the MC results. On the other hand, the P_1 agreed reasonably well with the MC benchmark.

Zaversky et al. (2018) performed a comparison between the discrete ordinates method (assimilable to the FV) and the diffusion approximation in a porous solar receiver. In the second case, to account for collimated radiation, they added an exponentially decaying source due to the incoming solar flux. Contrary to Wang et al. (2014c), they noticed a large deviation in the calculated temperatures among the two methods, both for the solid and for the fluid phase. This difference was related to the computation of radiative thermal losses at the front of the receiver. The diffusion approximation is not able to account for radiative penetration length; therefore, the losses are calculated using only the surface temperature. Contrarily, the DO method computes the losses emitted from the whole finite zone below the surface, which might have different temperatures than the front of the receiver. The results of the investigations from Wang et al. (2014c) and Zaversky et al. (2018) are shown in Figure 12.8.

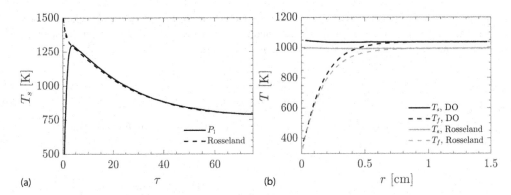

FIGURE 12.8 Comparison between different RTE solution method in the framework of volumetric solar receivers: (a) solid-phase temperature against optical thickness of the receiver in the case of Wang et al. (From Wang, F. et al., *Int. J. Heat Mass Trans.*, 78, 7–16, 2014c.); and (b) solid and fluid temperatures along the axial coordinate reported by Zaversky et al. (From Zaversky, F. et al., *Appl. Energ.*, 210, 351–375, 2018.)

In conclusion, although the diffusion approximation is suitable for describing the radiative heat transfer in the bulk of an optically thick porous medium, for the accurate description of the effects at the boundary an FV or MC method are recommended. In addition P_1 has been proven to be a viable cost-effective solution method in gray porous media for a wide range of optical thickness.

12.5 Effect of Radiative Heat Transfer on Convection in Porous Media

The effect of radiation on the heat transfer mechanism in porous media has been investigated numerically with different assumptions and methodologies. Theoretical studies that aim at identifying the role of radiation are usually concerned with the heat transfer in canonical idealized systems to isolate and parameterize the results. Some examples of numerical investigations will be reported and discussed in this section. The parameters of interest in this investigation are the Nusselt number (Nu), the Rayleigh number (Ra), and the conduction-radiation parameter (N), related to the Planck number (Pl), defined as

$$Nu = \frac{hL_{\text{ref}}}{k_f}, \quad Ra = \frac{Kg\alpha_f \Delta T_{\text{ref}} L_{\text{ref}}}{(\mu c_p / k)_f}, \quad N = \beta Pl = \frac{\beta k_s^*}{L_{\text{ref}} \sigma_{SB} T_{\text{ref}}^3} \quad (12.98)$$

with K the permeability of the porous medium, α_f the coefficient of volumetric expansion of the fluid, and h the convective heat transfer coefficient. While Ra and N are input parameters, the Nusselt number is an output quantity and changes in its value reflect the modification of the convective heat flux. The radiative heat transfer strength is connected to the conduction-radiation parameter: when $N \to \infty$ radiation is absent, while for $N \to 0$ the system reaches radiative equilibrium. All the studies have been performed while assuming gray medium. In a porous medium, any spectral effect will be smoothed, that is, washed out by the irregular orientation of the pore surfaces. It would require a very regular porous medium to see any net spectral effect. Therefore, the wavelength dependency of the solid matrix is not foreseen to largely influence the porous medium at the macroscopic scale, and the spectral dependency of the radiative properties will not be considered here. The influence of porosity and geometrical features, as explained in the previous section, is much more impactful.

In case of buoyancy-driven flows, the velocity is a result of a temperature differential. The inclusion of radiation modifies the temperature field, often non-locally, causing a change in flow pattern and heat transfer. Yih (1999) performed the first detailed study concerning the effects of radiative heat transfer on natural convection in porous media. He investigated the simplified case of an isothermal vertical cylinder embedded in a porous medium. He assumed LTE and treated the radiative heat transport with a diffusion formulation. The boundary conditions employed were

$$\text{at: } r = r_w : \quad v = 0, \quad T = T_w, \tag{12.99}$$

$$\text{at: } r \to \infty : \quad u = 0, \quad T = T_\infty, \tag{12.100}$$

where the w subscript stands for values at the cylinder wall. The Nusselt number along the height of the cylinder was defined as

$$Nu_x = \frac{q_w x}{k(T_w - T_\infty)}, \quad \text{where} \quad q_w = \left[(k + k_R) \frac{\partial T}{\partial r} \right]_{r=r_w}. \tag{12.101}$$

Figure 12.9 shows the variation of the Nusselt number with the conduction-radiation parameter (Figure 12.9a) and with the surface temperature excess ratio T_w/T_∞ (Figure 12.9b). It is possible to notice that radiation causes an increase in overall heat transfer (Nu). However, the effect of radiation on convection itself was not noticeable since radiative heat transfer was included in the definition of the Nusselt number (Equation 12.101). An extension to cylindrical annulus embedded in porous media was performed by Badruddin et al. (2006) with the same results as Yih (1999) (increase in total Nu with radiative heat transfer).

Chen et al. (2018) investigated the natural convection of a 2D saturated porous cavity bounded by a hot and a cold wall at the sides (with $T = T_h$ and T_c, respectively) and two adiabatic walls at the top and at the bottom. They used a more accurate approach consisting of an LTNE model for the energy equation and a discrete ordinate method for the radiative transport, which included isotropic scattering. Contrary to the previously mentioned studies, they visualized the direct effect of radiation by separating the Nusselt number into two different components—a convective and a radiative Nusselt number—defined as

$$Nu^C = -\frac{d}{dx}\left(\frac{T_f}{\Delta T}\right)_{x=0}, \quad Nu^R = \frac{\varepsilon_w}{4Pl\delta}\left[\left(\frac{T_s - T_c}{T_\text{ave}}\right)^4 - \int_{4\pi} \frac{I}{\sigma T_\text{ave}} d\Omega\right]_{x=0}, \tag{12.102}$$

where $\Delta T = T_h - T_c$, $T_\text{ave} = 0.5(T_h + T_c)$, and x is the horizontal spatial coordinate. Additionally, they separated the conduction-radiation parameter into the Planck number and extinction coefficient, representative of the optical thickness, and measured their impact individually. The results of their investigation is shown in Figure 12.10. By separating the convective and radiative Nusselt number, it is possible to conclude that the overall Nu increase because of radiation (Badruddin et al. 2006; Yih 1999) and is

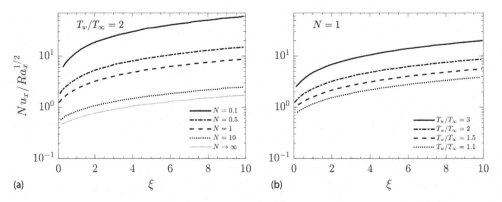

FIGURE 12.9 Effect of (a) conduction-radiation parameter and (b) surface temperature excess ratio on Nusselt number in a vertical cylinder embedded in a porous media. (From Badruddin, I. et al., *Int. Commun. HeatMass Transfer*, 33, 500–507, 2006.) ξ is the transverse curvature parameter defined as $\xi = 2x/(r_w Ra_x^{1/2})$, and $Ra_x = Kg\beta_f(T_w - T_\infty)x/(\mu c_p/k)_f$ is the Rayleigh number over the height of the cylinder.

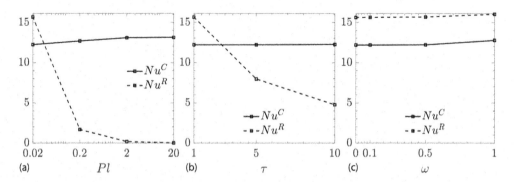

FIGURE 12.10 Effect of (a) Planck number, (b) optical thickness, and (c) scattering albedo on the conductive and radiative Nusselt numbers at the hot wall of a porous saturated cavity. (From Chen, Y. et al., *Int. Commun. Heat Mass Transfer*, 95, 80–91, 2018.)

due to the addition of a radiative heat transfer mode and not because of the increase in convective heat transfer. Indeed, Nu^C, in the case of Chen et al. (2018), has only really slight, almost negligible, changes with the radiative parameters Pl, τ, and ω. On the other hand, Nu^R is largely dependent on Pl and τ. The scattering albedo doesn't seem to have a large impact in the overall heat transfer behavior.

The same increase in Nusselt number was noticed in forced convection cases (channel flow) (Mahmoudi 2014; Talukdar et al. 2004). In particular, Talukdar et al. (2004) performed the same investigation as Yih (1999) in a channel flow with isothermal walls kept at T_w. They used an LTE formulation for the energy equation and a discrete transfer method (Modest 2013) for the radiative transport. They focused on the effect of the parameter γ (the porous medium shape parameter) defined as

$$\gamma = \left(\frac{H^2 \varepsilon}{K}\right)^{1/2}, \tag{12.103}$$

where H is the height of the channel. The Nusselt number was defined in their channel flow case as

$$Nu = \underbrace{-\frac{2}{\theta_w - \theta_m}\left[\frac{\partial \theta}{\partial \eta_y}\right]_{\eta_y=0}}_{Nu^C} + \underbrace{\frac{2}{\theta_w - \theta_m}\frac{2\beta H}{4N}\left[\Psi^R\right]_{\eta_y=0}}_{Nu^R} \tag{12.104}$$

where $\theta = T/T_{\text{ref}}$, $\eta_y = y/H$, and Ψ^R is the non-dimensional radiative heat flux at the wall. Figure 12.11 shows the results they obtained in terms of effect of radiation on the Nusselt number. Again, as in the case of natural convection in a vertical cylinder and a porous cavity, the overall Nusselt number increases with radiative heat transfer. It is interesting to note that, contrary to the porous cavity of Chen et al. (2018), the Nusselt number in a forced convection case increases with β. Nevertheless, despite the LTE formulation, convective Nusselt number is found to be independent on the radiative parameters (Figure 12.11) confirming the results obtained by Chen et al. (2018).

Mahmoudi (2014) investigated the effect of radiative heat transfer on forced convection in a 2D porous channel flow. He used an LTNE formulation combined with a DO method for the solution of radiative transport. The porous medium was assumed to be gray, emitting, and isotropically scattering. A constant convective heat flux was prescribed on the walls of the channel. The results showed that the overall Nusselt number increases with the inclusion of radiation also in the investigated case. Furthermore, the accuracy of the LTE condition in radiative porous media was analyzed by calculating the differential between solid and fluid temperature. The conclusion was that with the inclusion of radiative heat transfer the difference between the solid and the fluid temperature decreased. However, the boundary conditions of the investigated system can provide an explanation. The heat entered the porous medium through conduction from

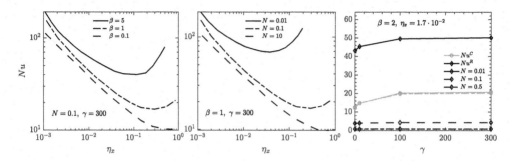

FIGURE 12.11 Effect of the (a) extinction coefficient and the (b) convection-radiation parameter on the total Nusselt number and (c) difference between convective and radiative Nusselt numbers in a porous channel flow with isothermal walls. (From Talukdar, P., et al., *Int. J. Heat Mass Transfer*, 47, 1001–1013, 2004.)

the walls and, in the case of radiative heat transfer, left through radiative emission at the outlet, where a relatively low temperature for radiative heat transfer was imposed $(T_{\text{out, surrounding}} = 300 \text{ K})$. This boundary condition resulted in a direct reduction of solid temperature when compared to a non-radiative case, except near the boundaries where additional heat from surface emission was available to the solid phase. The reduction of fluid temperature was only a consequence of a decrease in T_s, so for the fluid phase, radiation had an indirect, milder impact. Therefore, the lower temperature differential could be linked to the fact that radiative heat transfer introduced energy losses in the solid phase. If the non-radiative case would be characterized by $T_f > T_s$, such as combustion in porous media (Farzaneh et al. 2007; Talukdar et al. 2004), for the particular boundary condition chosen, the temperature differential would increase. In a case of a volumetric solar receiver, the solid phase receives a considerable amount of energy through radiation, larger than the energy lost through emission, causing a large temperature differential. For this reason, the LTE model is never recommended.

An example can be seen in Wang et al. (2014b), which investigated the effect of radiation in a 1D porous medium. They used an LTNE formulation and a diffuse approximation for radiative heat transfer. The solid phase was heated by an impinging radiative flux of 1 MW/m^2 on the inlet boundary. N does not influence the absorption at the surface due to the use of the diffusion approximation, yet, due to the large incoming radiative heat flux T_s was found to be always significantly larger than T_f. To evaluate the effect of radiation on heat transfer, they divided the Nusselt number into three contributions defined as

$$Nu_f = \frac{L}{\theta_{mf}-1}\frac{d\theta_f}{dX}, \quad Nu_{sc} = \frac{(1-R)L}{\theta_{ms}-1}\frac{d\theta_s}{dX}, \quad Nu_{sr} = \frac{RL}{\theta_{ms}-1}\frac{d\theta_s}{dX}, \qquad (12.105)$$

where θ is the non-dimensional temperature, $X = x/L$ is the non-dimensional coordinate, and $R = k_s^* / (k_s^* + k_R)$ is the ratio of thermal conductivity to the total conductivity (k_R is the radiative conductivity of the porous medium defined in equation (12.63)). The subscript m stands for an integral average over the lenght of the media (L). Nu_f represents the convection between solid and fluid phase while Nu_{sc} and Nu_{sr} conduction and radiation is through the solid phase, respectively. The computed value of Nu_f (not shown here) was independent of N, confirming that radiative heat transfer does not directly modify convection in a porous medium. The other contributions have opposite dependency on the conduction-radiation parameter N. Namely, with a decrease of N, Nu_{sc} decreases and Nu_{sr} increases. This result was expected since a decrease on N represents a larger influence of radiative heat transfer over conduction.

12.6 Volumetric Solar Absorbers

A schematic illustrating the operating principle of a porous medium volumetric solar absorber is shown in Figure 12.12. The incident solar flux impinges on the receiver and penetrates the porous medium. The solid phase absorbs the solar radiation and heats up. Finally, the heat is exchanged with the fluid phase. There are

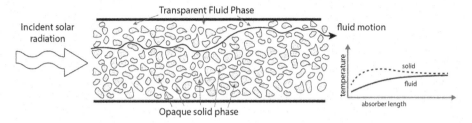

FIGURE 12.12 Schematic of the volumetric solar absorber concept. The plot qualitatively illustrates the desired temperature profiles.

three main requirements to consider when designing a volumetric solar absorber: (1) the amount of solid phase should be sufficient to completely absorb the incoming solar flux; (2) the temperature near the glass window should be minimized to reduce radiative losses; and (3) the heat transfer between solid and fluid phase should be large enough to allow an effective cooling of the absorber. This translates into two main objectives to achieve. First, to have a desired temperature profile, which requires a trade-off between the solar penetration length and a the solid–fluid heat transfer. Second, to avoid hot spots and have stable operating conditions. These two issues will be addressed in the following section.

12.6.1 Radiative Penetration Length versus Solid–Fluid Heat Flux

The porosity of the absorbing media is the most important parameter in terms of efficiency. It has a non-linear effect since it impacts both radiative heat transfer and convective heat transfer independently. As discussed in Section 3, the extinction coefficient is mostly defined by the morphology of the solid phase and, for this reason, a change in porosity strongly impacts the penetration depth of solar radiation in the receiver.

In terms of solar absorption, the porosity must be low enough and the receiver long enough to allow complete absorption. As an example, at an optical depth $\tau = 7$, 99.9% of the incoming solar flux is absorbed. However, an excessively low porosity affects the absorber efficiency negatively. First of all it causes a large absorption coefficient that might lead to rapid absorption at the front of the receiver. In this case, the energy accumulates near the glass window leading to high temperatures and large temperature gradients. This has two negative consequences: (1) the high temperature solid phase emits most of its energy outward leading to high losses; and (2) due to the large temperature gradient, thermal stresses deteriorate the material, which results in sudden breakage of the solid matrix. On the contrary, an excessively large porosity can mitigate the large maximum temperature and radiative losses but results in a small specific surface, which causes a decrease of the solid–fluid heat transfer.

Wang et al. (2013) investigated the role of porosity on the temperature distribution in a solar receiver with an LTNE formulation and a diffusion approximation for radiative transport. To address the concentrated solar radiation, they assumed the incoming radiative heat flux to be absorbed on the front boundary, claiming that, due to the large optical thickness, radiation can travel only short distances. They computed the distribution of this incoming flux with an MC ray tracing solver and used the result as a boundary condition for the porous media. As noticed by the calculated temperature distributions shown in Figure 12.13, the highest porosity value leads to the highest solid temperature. However, in their case, the effect of porosity is linked only to the change of the transport properties of the porous medium and the radiative redistribution throughout the solid phase. For this reason, the outlet temperature of the receiver is, although only slightly, lower for a higher porosity. Indeed, the penetration length of solar radiation, which would otherwise depend strongly on ε, is considered 0 in all the cases. This is reflected on the solid temperature profile, which shows an unwanted maximum at $L = 0$. This would cause large radiative losses that would drastically decrease the receiver's efficiency. They also concluded that a higher porosity leads to an increase in the thermal non-equilibrium region.

The effect of porosity on penetration depth is clearly shown in the work of Wu et al. (2011). Their heat transfer modeling consisted of a simplified LTNE formulation, Equations (12.17b and 12.18), combined

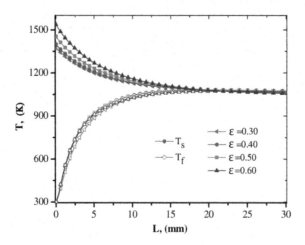

FIGURE 12.13 Solid and fluid temperature distributions in a 2D porous medium solar receiver with radially non-uniform solar heat flux at $x = 0$. (From ang, F. et al., *Int. J. Heat Mass Transfer*, 62, 247–254, 2013.)

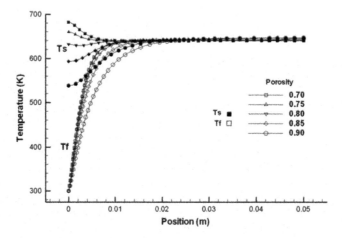

FIGURE 12.14 Solid and fluid temperature distributions in a 2D porous medium solar receiver as computed. (From Wu, Z. et al., *Solar Energ.*, 85, 2374–2385, 2011.)

with a P_1 approximation for radiative heat transfer modified to account for collimated solar irradiation (12.77). G_c was calculated assuming exponential attenuation with the receiver length $G_c = I_0 e^{-\beta z}$, where β was computed with the correlation in Vafai (2005) obtained in the geometrical optics limit. The results obtained in terms of temperature distribution within the receiver are shown in Figure 12.14. As opposed to Figure 12.13, while taking into account solar penetration depth, increasing ε results in a decrease of the maximum temperature for both the solid and the fluid phase. Moreover, this maximum is located deeper within the receiver. Nonetheless, the outlet temperature increases, again only slightly, with porosity leading to larger efficiencies. As a guideline for assessing the receiver's efficiency, Wu et al. (2011) state that for a receiver working at 1000 K, a difference of 100 K in the front surface temperature would lead in a 5% difference in efficiency. It is interesting to notice that, again, the non-thermal equilibrium region grows in size for an increased porosity.

To address the optimization of the penetration depth, several works investigated the innovative concept of *gradual porosity*. The objective is to engineer a solid mesh, which is characterized by a low porosity at the front of the receiver, which increases gradually with the depth. In this way, the penetration length of the solar flux can be increased with the largest absorption occurring at the end of the reactor.

This would lower the temperature gradient and maximum temperature while still ensuring a complete absorption of the incoming solar radiation. In addition, the low porosity at the back of the receiver would allow a high rate of solid to fluid heat exchange. Experimental tests of the gradual porosity concept have shown the benefits of this technology (Avila-Marin et al. 2018; Zaversky et al. 2018).

Avila-Marin et al. (2018) investigated 26 wire meshes with different arrangements. In particular, single-layered, double-layered, and triple-layered meshes were tested. All the configuration ensured total solar absorption at the end of the receiver but, while the single-layered meshes were characterized by a unique porosity value, the double- and triple-layered meshes were engineered to have lower porosity values on successive layers. They confirmed the trade-off in terms of single porosity materials. In particular, the best performing single layer mesh was the one with highest porosity combined with intermediate properties in terms of extinction coefficient and specific surface area. In their case, the high porosity resulted in a satisfactory penetration length but hindered the solid-to-fluid heat transfer due to the low specific surface area. The double-layer porosity, in general, was found to increase the receiver's efficiency thanks to the improved heat transfer in the second layer, but no further improvement was noticed for the triple-layer configurations. They conclude that the best performance is obtained with a double-layer solid mesh arranged in the following manner: the front layer should have the highest possible porosity while the second layer should maximize the solid–fluid interfacial area.

Zaversky et al. (2018) further investigate the gradual porosity concept by performing experimental and numerical investigation of single-, double-, and triple-layer configurations. They confirmed that the highest possible thermal efficiency is achieved by the largest porosity available, but, contrary to Avila-Marin et al. (2018), they noticed no influence of the second (or third) layer's properties in the thermal efficiency.

Du et al. (2007) optimized the porosity profile of a volumetric solar absorber in terms of reflectivity and absorptivity of the receiver. They wanted to achieve the maximum absorptivity with the lowest reflectivity possible. They performed MC ray tracing simulations to calculate the extinction coefficient of a porous material constructed through a 3D reconstruction process. Based on the radiative properties, the performance of the receiver was assessed and, through a genetic algorithm, porosity and cell diameter were modified. The MC simulation was iterated until an optimal converged configuration was reached. The final result is shown in Figure 12.15. The porosity gradually varies from a value of 0.95 at the front to 0.65 at the back of the receiver. This ensures an optimal solar absorption as shown in Figure 12.16.

Finally, Nimvari et al. (2018) approached the problem of large temperature gradient from a different angle. They computed the temperature profiles in a volumetric solar receiver using an LTNE model and P_1 approximation for radiative heat transfer. Instead of proposing a gradual porosity, they proposed a non-uniform velocity distribution, such that a larger convective heat transfer can be achieved in the hot regions of the receiver. In particular, a larger mass flow was directed toward the front of the receiver to increase the cooling of the solid matrix. They observed that a larger mass flow was effective in reducing the maximum temperature and concluded that this could represent a viable strategy to avoid mechanical failure of the absorber.

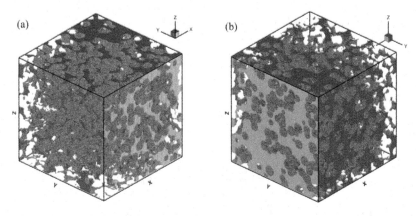

FIGURE 12.15 A gradual porosity mesh as developed by Du et al.: (a) view from the front; and (b) view from the back. (From Du, S. et al., *Appl. Energ.*, 207, 27–35, 2007.)

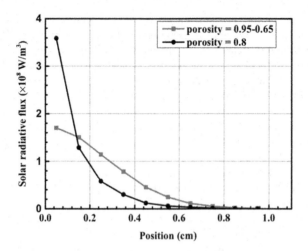

FIGURE 12.16 Solar radiative flux in gradual porosity solar receivers as calculated by Du et al. . As clearly demonstrated, with a gradual porosity it is possible to achieve an optimal solar flux profile. (From Du, S. et al., *Appl. Energ.*, 207, 27–35, 2007.)

12.6.2 Flow Instabilities and Hot Spots

Flow instabilities are a serious issue in volumetric solar receivers. The operations of the absorber are dependent on exterior weather conditions, which are hardly controllable. Therefore, the response of the system to a shift in energy input must be as mild as possible to achieve a constant efficiency. In addition, if flow instabilities propagate and grow, they can cause hot spots, which increase the local temperature and lead to material breakage. Therefore, it is important to understand the behavior of the system in different operational conditions.

The calculation of flow instabilities inside an irradiated porous media will be explored with a 1D absorber configuration and simplifying assumptions to illustrate the concepts. The derivation will follow closely the one performed in Becker et al. (2006). Other analyses with different assumptions, leading to the same conclusions, can be found in references Kribus et al. (1996) and Pitz-Paal et al. (1997). The pressure drop in a simplified 1D porous medium is described by the Darcy–Forchheimer law

$$\frac{dp}{dx} = -\frac{\mu_f}{K}v_f - \frac{\rho_f}{K_i}v_f^2 \qquad (12.106)$$

where K is the permeability, and K_i is known as the inertial permeability. By assuming an ideal gas law for the fluid phase, the integration of the above equation yields

$$\frac{p_i^2 - p_o^2}{2R} = \frac{\dot{m}_f}{K}\int_0^L \mu_f T_f dx + \frac{\dot{m}_f^2}{K_i}\int_0^L T_f dx \qquad (12.107)$$

where p_i, p_o, and $\dot{m}_f$ are the inlet pressure, the outlet pressure, and the mass flow in the receiver, respectively. Becker et al. (2006) consider the case of a constant radiative influx G_s. By considering a global energy balance within the absorber, neglecting radial heat transfer and assuming local thermal equilibrium

$$\underbrace{G_s}_{\text{solar input}} = \underbrace{(\dot{m}c_p)_f(T_o - T_i)}_{\text{enhalpy increase}} + \underbrace{\xi\sigma_{SB}T_o^4}_{\text{radiative losses}} \qquad (12.108)$$

where T_i and T_o are the inlet and outlet temperatures, respectively. ξ is a parameter that corrects for the thermal radiative losses. It is lower than one in the case of volumetric operations, while in the case of $\xi > 1$

the absorber is more similar to a parabolic trough. Note that the above equation approximates better a receiver in which the solar flux is perpendicular to the flow direction, instead of parallel. The mass flow $\dot{m}_f$ is then a result of the concurrent action of radiative absorption and radiative losses

$$\dot{m}_f = \frac{G_s - \xi \sigma_{SB} T_o^4}{c_{pf}(T_o - T_i)} \tag{12.109}$$

Given a inlet temperature and a power source G_s, for a certain absorber configuration (ξ known), the mass flow is uniquely defined by the outlet temperature. If all temperature changes are assumed to take place at the front of the receiver, such that the temperature profile can be roughly considered constant, Equation (12.107) yields

$$\frac{p_i^2 - p_o^2}{2RL} = \left(\frac{G_s - \xi \sigma_{SB} T_o^4}{c_{pf}(T_o - T_i)}\right)\left(\frac{1}{K}\mu_i \frac{T_o^{\varphi+1}}{T_i^{\varphi}} + \left(\frac{G_s - \xi \sigma_{SB} T_o^4}{c_{pf}(T_o - T_i)}\right)\frac{1}{K_i} T_o\right) \tag{12.110}$$

In the above equation, the viscosity is assumed to obey the following law

$$\mu = \mu_{\text{ref}}\left(\frac{T}{T_{\text{ref}}}\right)^{\varphi} \tag{12.111}$$

Figure 12.17 shows the profile of the left hand side (LHS) of Equation (12.110) plotted against outlet temperature for a porous medium, which has a linear relation between pressure derivative and velocity $(K_i \to \infty)$. One would predict that at higher pressure difference, the mass flow and the outlet temperature would decrease (Kribus et al. 1996). The system behaves as expected for low outlet temperatures. This is also strictly true for all outlet temperatures if viscosity is constant $(\varphi = 0)$. On the other hand, if viscosity increases with temperature $(\varphi > 0)$, which is the case for most common gasses (in air $\varphi \approx 0.7$), it is possible to identify a region in which outlet temperature increases with the pressure differential. This occurs because with a higher viscosity the flow resistance in the porous medium increases, thus reducing the mass flow within the system. With a lower mass flow, the convective heat transfer diminishes, and the flow is not able to effectively cool the irradiated solid phase. At a certain temperature, the influence of variable viscosity dominates and creates a feedback loop, which results in a continuous decrease of mass flow and increase of temperature. For this reason, the operation of the receiver in this regime is not stable.

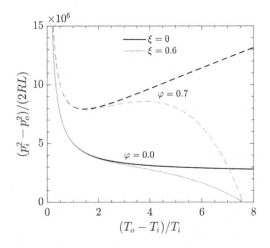

FIGURE 12.17 Pressure difference parameter against non-dimensional outlet temperature. Effect of viscosity exponent φ and radiative losses correction factor ξ.

Kribus et al. (1996) have defined a blow parameter B as

$$B = \frac{\gamma}{\gamma-1} \frac{K(p_i^2 - p_o^2)}{\mu_i G_s L^2} \quad (12.112)$$

They showed that (for $K_i \to \infty$ and $\xi = 0$) depending on φ there is a certain minimum value $B_{\min}$, which corresponds to an upper limit to the heat source a solar receiver can handle. This heat source has to be

$$G_s \leq \frac{\gamma}{\gamma-1} \frac{K(p_i^2 - p_o^2)}{\mu_i L^2 B_{\min}(\varphi)} \quad (12.113)$$

Above this solar concentration level, the feedback loop causes a choking of the receiver. As a consequence, no solution for the mass flow is found at lower B.

On the other hand, if radiative losses are considered $(\xi > 0)$ a higher temperature causes larger radiative emissions, which cool the receiver and restore the inverse relation between pressure differential and mass flow. In this case, there is a solution for the mass flow at all pressure difference values, and flow choking does not proceed to infinite temperatures but settles to a definite, small, mass flow value.

Figure 12.18 shows the influence of various parameters in the behavior of the system with a loss correction factor of $\xi = 0.6$. As expected, the instable behavior increases with the viscosity exponent φ (Figure 12.18a). The same occurs with the increase in the solar heat flux G_s (Figure 12.18b). It is noticeable that in certain conditions a single value of $p_i^2 - p_o^2$ corresponds to multiple solutions for the mass flow. An example is the curve for $G_s = 3000$ in Figure 12.18b, which has three distinct operation points for one single pressure difference value. Since the pressure difference must be the same for parallel sections in the receiver, it is possible that some regions stay at low temperature, while others achieve high temperatures, corresponding to the right-most operating point in Figure 12.18b. The occurrence of these hot spots would result in thermal stress and cause breakage of the porous material (Pitz-Paal et al. 1997). It is recommended, therefore, to operate in a regime in which $p_i^2 - p_o^2$ allows only one mass flow value. In addition, most of the curves are characterized by a small gradient region. In these conditions, a small pressure perturbation, which can be caused by a local variation in wind pressure or incident radiation, can cause a large change in outlet temperature and, therefore, efficiency Kribus et al. (1996). For this reason, and to avoid flow instabilities, Kribus et al. (1996) recommend the use of pressurized loops. At higher operating pressures, it is possible to shift the operations to the large gradient region and, consequently, have better control on the output of the receiver.

If the quadratic term in Equation (12.110) dominates (i.e., K_i is low enough), the detrimental effect of variable viscosity is reduced, and a monotonic decrease of mass flow with pressure difference is restored,

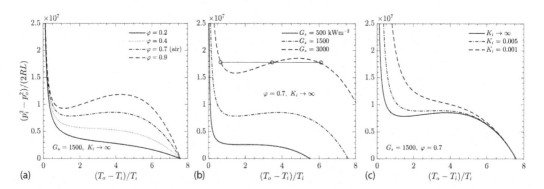

FIGURE 12.18 Effect of different parameter on the pressure characteristic with a loss factor of $\chi = 0.6$: (a) effect of viscosity exponent; (b) effect of solar irradiation; and (c) effect of inertial permeability.

leading to a stable and controllable operation of the receiver (Figure 12.18c). However, a quadratic characteristic of pressure loss is more common in low porosity materials due to the stagnation of the flow at the front end surfaces. Therefore, a low porosity reduces the risks of flow instability but works against the efficiency of the receiver (as detailed in the section above) (Pitz-Paal et al. 1997). For this reason, Pitz-Paal et al. (1997) suggest the use of orifices at the back of the volumetric absorber. These would dominate the pressure loss and lead to a stable pressure characteristic curve unaffected by local wind dishomogeneity. They also investigated the effect of radial heat transfer, which is neglected in the above derivation. They conclude that a high radial heat transfer reduces the instability caused by multiple mass flows occurring with the same pressure drop.

12.7 Conclusion

Different aspects of modeling radiative heat transfer in porous media, mainly in the framework of volumetric solar receivers, were presented.

First, the coupling of radiation with conduction and convection in the specific case of a porous medium was investigated. The most accurate approach involves a volume-averaging procedure with knowledge of interfacial temperature at a pore scale. On the other hand, most works assumes a direct dependency of the average radiative source on the average solid temperature. This approximation, while simplifying greatly the mathematical description, was found to be effective and accurate in the case of volumetric solar receivers.

Independently on the coupling approach, the radiative heat transfer process is always treated by assuming an equivalent semi-transparent medium with effective radiative properties. These can be identified with varying degrees of accuracy employing different strategies. Several of these strategies were presented in Section 12.3, ranging from experimental methods to the most accurate statistical approach. The calculated effective properties were found to be largely dependent on the morphology of the solid matrix (i.e., porosity and cell diameter) and mostly independent on wavelength. For this reason, the gray approximation is largely employed to describe the radiative heat transfer process in porous media.

The most common solution schemes for the RTE were presented in Section 12.4. It was shown that the diffusion approximation, which is the most simple approach in terms of implementation and computational expenses, is a viable choice for high temperature, optically thick systems, but it is unable to account for the radiative penetration length, which is required for a volumetric solar absorber. In this case, the P1 method provides a cheap and accurate alternative, and, for this reason, it has been successfully employed in many studies regarding porous receivers.

The effects of radiation on the heat transfer in a porous medium were reviewed in Section 12.5. In general, radiation was found to increase the overall heat transfer in the system. On the other hand, this increase was ascribed to the additional heat transfer process: convection and conduction appeared to be unaffected by radiation in all the investigated studies.

In conclusion, the challenges associated with the design of an efficient porous solar receiver were analyzed. The available literature showed that the physical phenomena determining the performance of porous volumetric solar receiver are quite well understood. Nevertheless, more in-depth studies on scale-up and robust continuous operation are still needed in order develop this device into an economic solar energy technology.

REFERENCES

Abbott D., Keeping the energy debate clean: How do we supply the world's energy needs? *Proceedings of the IEEE*, 98 (2010), pp. 42–66.

Avila-Marin A., Volumetric receivers in solar thermal power plants with central receiver system technology: A review, *Solar Energy*, 85 (2011), pp. 891–910.

Avila-Marin A., M. A. de Lara, and J. Fernandez-Reche, Experimental results of gradual porosity volumetric air receivers with wire meshes, *Renewable Energy*, 122 (2018), pp. 339–353.

Badruddin I., Z. Zainal, P. A. Narayana, and K. Seetharamu, Heat transfer by radiation and natural convection through a vertical annulus embedded in porous medium, *International Communation on Heat and Mass Transfer*, 33 (2006), pp. 500–507.

Baillis D., and J. Sacadura, Thermal radiation properties of dispersed media: Theoretical prediction and experimental characterization, *Journal of Quantitative Spectroscopy and Radiative Transfer*, 67 (2000), pp. 327–363.

Baillis D., M. Arduini-Schuster, and J. Sacadura, Identification of spectral radiative properties of polyurethane foam from hemispherical and bi-directional transmittance and reflectance measurements, *Journal of Quantitative Spectroscopy and Radiative Transfer*, 73 (2002), pp. 297–306.

Baillis D., R. Coquard, J. Randrianalisoa, L. Dombrovsky, and R. Viskanta, Thermal radiation properties of highly porous cellular foams, *Special Topics and Revies in Porous Media*, 4 (2013), pp. 111–136.

Barreto G., P. Canhoto, and M. Collares-Pereira, Three-dimensional modelling and analysis of solar radiation absorption in porous volumetric receivers, *Applied Energy*, 215 (2018), pp. 602–614.

Becker M., T. Fend, B. Hoffschmidt, R. Pitz-Paal, O. Reutter, V. Stamatov, M. Steven, and D. Trimis, Theoretical and numerical investigation of flow stability in porous materials applied as volumetric solar receivers, *Solar Energy*, 80 (2006), pp. 1241–1248.

Celzard F., G. Tondi, D. Lacroix, G. Jeandel, B. Monod, V. Fierro, and A. Pizzi, Radiative properties of tannin-based, glasslike, carbon foams, *Carbon*, 50 (2012), pp. 4102–4113.

Chen Y., B. Li, J. Zhang, and Z. Qian, Influences of radiative characteristics on free convection in a saturated porous cavity under thermal non-equilibrium condition, *International Communications in Heat and Mass Transfer*, 95 (2018), pp. 80–91.

Chueh M., C. Falter, M. Abbott, D. Scipio, P. Furler, S. Haile, and A. Steinfeld, High-flux solar-driven thermochemical dissociation of CO_2 and H_2O using nonstoichiometric ceria, *Science*, 330 (2010), pp. 1797–1801.

Coelho P., M. Mancini, and D. Roekaerts, Thermal radiation, in *ERCOFTAC Best Practice Guidlines, Computational Fluid Dynamics of Turbulent Combustion*, L. Vervisch and D. Roekaerts, eds., Ercoftac, Brussels, Belgium, 2016, ch. 3, pp. 77–127.

Coquard R., D. Baillis, and D. Quenard, Radiative properties of expanded polystyrene foams, *Journal of Heat Transfer*, 131 (2009), pp. 012702-1–012702-10.

Coquard R., D. Baillis, and E. Marie, Numerical investigation of the radiative properties of polymeric foams from tomographic images, *Journal of Thermophysics and Heat Transfer*, 24 (2010), pp. 647–658.

Coquard R., D. Rochais, and D. Baillis, Conductive and radiative heat transfer in ceramic and metal foams at fire temperaturess, *Fire Technology*, 48 (2012), pp. 699–732.

Cunsolo S., R. Coquard, D. Baillis, and N. Bianco, Radiative properties modeling of open cell solid foam: Review and new analytical law, *International Journal of Thermal Sciences*, 104 (2016), pp. 122–134.

Cunsolo S., R. Coquard, D. Baillis, W. Chiu, and N. Bianco, Radiative properties of irregular open cell solid foams, *International Journal of Thermal Sciences*, 117 (2017), pp. 77–89.

Du S., Q. Ren, and Y. He, Optical and radiative properties analysis and optimization study of the gradually-varied volumetric solar receiver, *Applied Energy*, 207 (2007), pp. 27–35.

Farzaneh M., R. Ebrahimi, M. Shams, and M. Shafiey, Two-dimensional numerical simulation of combustion and heat transfer in porous burners, *Engineering Letters*, 15 (2007), pp. 1–6.

Fend T., B. Hoffschmidt, R. Pitz-Paal, O. Reutter, and P. Rietbrock, Porous materials as open volumetric solar receivers: Experimental determination of thermophysical properties, *Energy*, 29 (2004), pp. 823–833.

Glicksman L., M. Schuetz, and M. Sinofsky, Radiation heat transfer in foam insulation, *International Journal of Heat and Mass Transfer*, 30 (1987), pp. 187–197.

Gomart H., and J. Taine, Validity criterion of the radiative Fourier law for an absorbing and scattering medium, *Physical Review E*, 83 (2011), pp. 021202-1–8.

Hischier I., D. Hess, W. Lipinski, M. Modest, and A. Steinfeld, Heat transfer analysis of a novel pressurized air receiver for concentrated solar power via combined cycles, *Journal of Thermal Science and Engineering Applications*, 1 (2009), pp. 041002-1–6.

Kaviany M., and B. Singh, Radiative heat transfer in porous media, *Advances in Heat Transfer*, 23 (1993), pp. 133–185.

Kribus A., H. Ries, and W. Spirkl, Inherent limitations of volumetric solar receivers, *Journal of Solar Energy Engineering*, 118 (1996), pp. 151–155.

Lee Y., and K. Yoon, Effect of composition of polyurethane foam template on the morphology of silicalite foam, *Microporous and Mesoporous Materials*, 88 (2006), pp. 176–186.

Leroy V., B. Goyeau, and J. Taine, Coupled upscaling approaches for conduction, convection, and radiation in porous media: Theoretical developments, *Transport in Porous Media*, 98 (2013), pp. 323–347.

Li Y., X. Xia, C. Sun, S. Zhang, and H. Tan, Volumetric radiative properties of irregular open-cell foams made from semitransparent absorbing-scattering media, *Journal of Quantitative Spectroscopy and Radiative Transfer*, 224 (2019), pp. 325–342.

Loretz M., M. Coquard, D. Baillis, and E. Maire, Metallic foams: Radiative properties/comparison between different models, *Journal of Quantitative Spectroscopy and Radiative Transfer*, 109 (2008a), pp. 16–27.

Loretz M., E. Maire, and D. Baillis, Analytical modelling of the radiative properties of metallic foams:contribution of x-ray tomography, *Advanced Engineering Materials*, 10 (2008b), pp. 352–360.

Lougou B., Y. Shuai, R. Pan, G. Chaffa, and H. Tan, Heat transfer and fluid flow analysis of porous medium solar thermochemical reactor with quartz glass cover, *International Journal of Heat and Mass Transfer*, 127 (2018), pp. 61–74.

Mahmoudi Y., Effect of thermal radiation on temperature differential in a porous medium under local thermal non-equilibrium condition, *International Journal of Heat and Mass Transfer*, 76 (2014), pp. 105–121.

Martin A., C. Saltiel, J. Chai, and W. Shyy, Convective and radiative internal heat transfer augmentation with fiber arrays, *International Journal of Heat and Mass Transfer*, 41 (1998), pp. 3431–3440.

Mey S., C. Caliot, G. Flamant, A. Kribus, and Y. Gray, Optimization of high temperature SiC volumetric solar absorber, *Energy Procedia*, 49 (2014), pp. 478–487.

Modest M., *Radiative Heat Transfer*, Academic Press, San Diego, CA, 2013.

Nimvari M. E., N. F. Jouybari, and Q. Esmaili, A new approach to mitigate intense temperature gradients in ceramic foam solar receivers, *Renewable Energy*, 122 (2018), pp. 2016–215.

Parthasarathy P., P. Habisreuther, and N. Zarzalis, Identification of radiative properties of reticulated ceramic porous inert media using ray tracing technique, *Journal of Quantitative Spectroscopy and Radiative Transfer*, 113 (2012), pp. 1961–1969.

Petrasch J., P. Wyss, and A. Steinfeld, Tomography-based monte carlo determination of radiative properties of reticulate porous ceramics, *Journal of Quantitative Spectroscopy and Radiative Transfer*, 105 (2007), pp. 180–197.

Pitz-Paal R., B. Hoffschmidt, M. Böher, and M. Becker, Experimental and numerical evaluation of the performance and flow stability of different types of volumetric absorbers under non-homogeneous irradiation, *Solar Energy*, 60 (1997), pp. 135–150.

Placido E., M. C. Arduini-Schuster, and J. Kuhn, Thermal properties predictive model for insulating foams, *Infrared Physics and Technology*, 46 (2005), pp. 219–231.

Quintard M., and S. Whitaker, One- and two-equation models for transient diffusion processes in twphase systems, *Advances in Heat transfer*, 23 (1993), pp. 77–94.

Quintard M., M. Kaviany, and S. Whitaker, Two-medium treatment of heat transfer in porous media: Numerical results for effective properties, *Advances in Water Resources*, 20 (1997), pp. 77–94.

Sang L., B. Sun, H. Tan, C. Du, Y. Wu, and C. Ma, Catalytic reforming of methane with CO_2 over metal foam based monolithic catalyst, *International Journal of Hydrogen Energy*, 37 (2012), pp. 13037–13043.

Talukdar P., S. Mishra, D. Trimis, and F. Durst, Combined radiation and convection heat transfer in a porous channel bounded by isothermal parallel plates, *International Journal of Heat and Mass Transfer*, 47 (2004a), pp. 1001–1013.

Talukdar, S. Mishra, D. Trimis, and F. Durst, Heat transfer characteristic of a porous radiant burner under the influence of a 2-d radiation field, *Journal of Quantitative Spectroscopy and Radiative Transfer*, 84 (2004b), pp. 527–537.

Tancrez M., and J. Taine, Direct identification of absorption and scattering coefficients and phase function of a porous medium by a monte carlo technique, *International Journal of Heat and Mass Transfer*, 47 (2004), pp. 373–383.

Vafai K., *Handbook of Porous Media*, Taylor & Francis Group, New York, 2005.

Villafan-Vidales H., S. Abanades, C. Caliot, and H. Romero-Paredes, Heat transfer simulation in a thermochemical solar reactor based on a volumetric porous receiver, *Applied Thermal Engineering*, 31 (2011), pp. 3377–3386.

Wang F., Y. Shui, H. Tan, and C. Yu, Thermal performance analysis of porous media receiver with concentrated solar irradiation, *International Journal of Heat and Mass Transfer*, 62 (2013), pp. 247–254.

Wang F., Y. Shuai, Z. Wang, and Y. Leng, Thermal and chemical reaction performance analyses of steam methane reforming in porous media solar thermochemical reactor, *International Journal of Hydrogen Energy*, 39 (2014a), pp. 718–730.

Wang P., K. Vafai, and D. Liu, Analysis of radiative effect under local thermal non-equilibrium conditions in porous media-application to a solar air receiver, *Numerical Heat Transfer, Part A*, 65 (2014b), pp. 931–947.

Wang F., T. Jianyu, S. Yong, T. Heping, and C. Shuangxia, Thermal performance analysis of porous media solar receiver with different irradiative transfer models, *International Journal of Heat and Mass Transfer*, 78 (2014c), pp. 7–16.

Whitaker S., *The Method of Volume Averaging*, Kluwer Academic Publishers, Dordrecht, the Netherlands, 1999.

Wu Z., C. Caliot, G. Flamant, and Z. Wang, Coupled radiation and flow modeling in ceramic foam volumetric solar air receivers, *Solar Energy*, 85 (2011), pp. 2374–2385.

Yih K., Radiation effect on natural convection over a vertical cylinder embedded in porous media, *International Communation on Heat and Mass Transfer*, 26 (1999), pp. 259–267.

Zaversky F., L. Aldaz, M. Sanchez, A. Avila-Marin, M. Roldan, J. Fernandez-Reche, A. Füssel, W. Beckert, and J. Adler, Numerical and experimental evaluation and optimization of ceramic foams as solar absorber: *Single-layer vs multi-layer configurations*, *Applied Energy*, 210 (2018), pp. 351–375.

13

Transpiration Cooling Using Porous Material for Hypersonic Applications

Adriano Cerminara, Ralf Deiterding, and Neil D. Sandham

CONTENTS

13.1 Introduction..263
    13.1.1 Hypersonic Environment 263
    13.1.2 Boundary-Layer Thermal Characteristics 264
13.2 Requirements of Thermal Protection Systems 265
13.3 Film Cooling and Transpiration Cooling Technique...266
13.4 Numerical Method 268
    13.4.1 Governing Equations...268
    13.4.2 Code Features..269
13.5 Results from DNS in Hypersonic Flow over Porous Surfaces 270
    13.5.1 Flat Plate with Periodic Regular Pores 270
        13.5.1.1 Code Verification and Grid Study 271
        13.5.1.2 Effect of Wall Cooling...273
        13.5.1.3 3D Results—Effect of Reynolds Number 274
    13.5.2 Flat Plate Domain with Injection through Localized Slots 276
    13.5.3 Injection through a Porous Layer...280
13.6 Conclusion..284
Acknowledgments..284
References...285

13.1 Introduction

13.1.1 Hypersonic Environment

Hypersonic flight is characterized by very high values of temperature and heat flux reached at the wall, which can compromise the structural integrity of the vehicle. This is due to the large amount of kinetic energy converted into thermal energy by the shock forming in front of the body and by viscous effects inside the boundary layer. This produces high temperature peaks inside the boundary layer at small normal distances from the wall, which results in a high wall heat flux. Aerodynamic heating due to viscous effects inside the boundary layer is the main source of wall heating over most of the surface of slender bodies (e.g., cruise vehicles) flying at hypersonic speeds. The leading-edge region near the stagnation point is, in contrast, linked to the high temperatures reached at the edge of the boundary layer due to the front shock. This is the primary form of heating in the case of blunt bodies, such as the reentry capsules. For a reentry vehicle flying at a Mach number of 36, for example, as in the case of the Apollo reentry capsule, the temperature in the shock layer in front of the nose can reach values as high as 11,000 K (Anderson Jr., 2006).

13.1.2 Boundary-Layer Thermal Characteristics

In order to understand the heat-exchange mechanism between the flow and the wall and to predict the temperature levels and heat load to which the vehicle surface is exposed, it is necessary to consider the main thermal characteristics of a hypersonic boundary layer.

We need to distinguish between two separate regions of the body, namely the stagnation-point region and the surface downstream of the leading edge. If we consider the case of an isothermal wall with a temperature T_w (e.g., the room temperature in a hypersonic wind tunnel), the wall heat flux in the stagnation region depends on the temperature difference between the maximum value at the boundary-layer edge (linked to the temperature jump across the shock), and the value at the wall. Moreover, the heat flux depends also on the boundary-layer thickness (δ), which is, in turn, dependent on the shock-layer thickness and hence on the stand-off distance. As the stand-off distance decreases with the nose radius, this results in the heat flux at the stagnation point being dependent on the nose radius (R), and in particular higher for a sharper nose, with the proportionality $q_w \propto 1/\sqrt{R}$ (Anderson Jr., 2006). Hence, the leading-edge heat flux is higher for bodies with a sharper nose. In the case of chemical reactions and vibrational nonequilibrium, the temperature changes within the shock layer due to these effects need to be taken into account, as they determine the value of the temperature at the boundary-layer edge.

On the surface of a slender body, in contrast, the aerodynamic heating is caused by the dissipation of a large amount of kinetic energy (related to the high streamwise speed u at the boundary-layer edge) inside the boundary layer due to viscosity. This represents the main mechanism acting on the surface section downstream of the vehicle leading edge. As a consequence of the viscous effects, the temperature increases continually with reducing velocity as the wall is approached, until it reaches a peak at a certain distance from the surface, and then rapidly decreases to the fixed (low) value at the wall (T_w). As the temperature peak is located near the wall, this results in a steep gradient on the vehicle surface, that is, a high heat flux. This is a peculiar feature of high-speed boundary layers, as compared to low-speed (subsonic) cases, which is due to the fact that in the high-speed case, the contribution of the kinetic-energy dissipation in the energy balance within the boundary layer is no longer negligible.

The wall heat flux can be expressed by Fourier's law as

$$q_w = -k \frac{\partial T}{\partial y}\bigg|_{y=0}. \tag{13.1}$$

By considering the equations for a two-dimensional (2D) compressible laminar boundary layer along a flat plate, in the case of high-speed flow where the viscous effects cannot be neglected, it is possible to express the wall heat flux as a function of the total temperature of the flow and the tangential velocity gradient at the wall through the relation

$$q_w = \frac{k(T_w - T_{0\infty})}{U_\infty} \left(\frac{\partial u}{\partial y} \right)\bigg|_{y=0}. \tag{13.2}$$

Equation 13.2 shows that the wall heat flux depends on the difference between the surface temperature and the fluid total temperature. The total temperature can be obtained, for an isentropic flow, by the relation

$$T_{0\infty} = T_\infty \left(1 + \frac{\gamma - 1}{2} M^2 \right), \tag{13.3}$$

which shows the dependence on the square of the Mach number; thus, it assumes very high values (typically on the order of thousands of Kelvin degrees) for flows at high Mach numbers. This means that, in a hypothetical case in which no mechanisms of heat dissipation are present in the material exposed to a hypersonic flow, the surface temperature would increase starting from its initial value T_w up to the value of the fluid total temperature. This is of course not sustainable by the material forming the vehicle surface in a real case and would melt the surface, resulting in failure of the vehicle structure. Also, it should be mentioned that the maximum temperature that the wall would reach in a hypothetical case, and hence the temperature at which the adiabatic condition is reached (i.e., $q_w = 0$), is not exactly the fluid total temperature, but a value slightly lower, which is called adiabatic wall temperature (T_{aw}), and is defined as

$$T_{aw} = T_\infty \left(1 + \frac{\gamma - 1}{2} r M^2 \right). \tag{13.4}$$

In the above relation, r is the recovery factor, which is $r = Pr^{1/2}$ for laminar flows and $r = Pr^{1/3}$ in the turbulent case. The latter equation determines the adiabatic wall temperature to be lower than the total temperature of the fluid due to the Prandtl number being different from 1 in a real case.

Finally, Equation 13.2 also shows that the wall heat flux depends directly on the wall shear stress, thus on the skin friction coefficient, which highlights the dramatic effect that transition to turbulence has, in general, on the surface heat flux in hypersonic flows.

On the basis of the above discussion, it is clear that aerodynamic heating represents a significant challenge in the design process of a hypersonic vehicle. In order to reduce the wall heating effects, hypersonic vehicles need an appropriate thermal protection system (TPS) capable of storing, or radiating/dissipating outward, the high heat load and keeping the temperature of the internal structure within tolerable values. In the following section, the main forms of surface cooling for a hypersonic vehicle are reviewed, and the principal parameters and physical properties that are required for the thermal protective layer on the vehicle surface are presented and discussed.

13.2 Requirements of Thermal Protection Systems

The cooling techniques considered for the thermal protection system of hypersonic vehicles can be divided into two main classes, namely, active cooling and passive cooling. The active cooling systems make use of a coolant fluid (either gas or liquid) coming in direct contact with the vehicle surface, either with the internal or the external face, in order to keep its temperature within a tolerable range. A passive cooling system, in contrast, makes use of the heat resistance and insulation capabilities of the material forming the TPS to store and radiate outward the heat received from the hot boundary layer. For example, for lifting bodies, reusable ceramic-matrix and carbon-carbon composite materials have been shown to be the most suitable materials for the TPS, providing high-heat-load storage and thermal resistance capabilities. When the capability of radiating the heat outward is coupled with the presence of a fluid flowing inside the TPS structure, such a system is defined as a semi-passive cooling system. For example, in the case of a heat pipe, the entering heat is carried by an internal working fluid to other sections of the surface where it is radiated outward. Ablation is also considered a form of semi-passive cooling and is the technique used in the most extreme hypersonic environments, as in the case of ballistic atmospheric reentry. Ablative systems are generally made of a polymeric-matrix (e.g., epoxy or phenolic resin) composite material that uses a thermochemical degradation process (pyrolysis) of the polymer to carry out the absorbed heat via blowing of the gaseous products of pyrolysis, which carry out a high enthalpy content. Fillers of a composite material can be of different types (e.g., glass microspheres, phenolic microspheres, silica fibers, nylon fibers, and carbon fibers) and their role, in general, is to provide (dependent on the type) thermal resistance and insulation properties, as well as structural integrity (e.g., abrasion resistance) of the TPS. One of the most advanced ablators designed for extremely high space reentry speeds is the so-called Phenolic Impregnated Carbon Ablator (PICA), which is made of a low-density substrate of carbon fibers (Carbon FiberForm) impregnated with phenolic resin (Tran et al., 1997). This material was used for the TPS of the Stardust capsule, which reentered the Earth atmosphere at a speed of about 12.8 Km/s (the highest ever reached for an Earth reentry) experiencing a peak surface temperature of about 3,200 K at an altitude of 65 km (Winter and Trumble, 2010).

Besides their capability to "actively" eject the heat through the ablation mechanism, ablative systems owe their efficiency as TPS also to the high insulation (hence, passive cooling) properties of the "char," that is, the residual carbonaceous layer formed on the surface of the heat shield as a result of the pyrolysis process. This is a particularly porous layer, whose solid structure is composed exclusively by the "heaviest" and most stable chemical elements that did not undergo gasification, along with the fillers of the composite material. Hence, the char layer combines both the properties of a low thermal conductivity (k) and a high thermal capacity (c_p), thus behaving as a perfect heat sink/insulator, capable of storing a large quantity of enthalpy content at the surface while not allowing the heat to be transferred more deeply into the structure. Although the ablative systems can resist the highest heat flux and temperature values encountered in reentry flight, their disadvantage is that they are not reusable, as they consume during the reentry.

If one considers more moderate conditions, for example those relative to a lifting reentry (Space Shuttle-like reentry) or, in general, a cruise flight (which ideally could cover a Mach number range from about Mach 5 to Mach 10), reusability becomes a realistic requirement. In such conditions, the ultra-high-temperature ceramics (UHTC) have been demonstrated to be among the best candidates for the TPS material, thanks to their thermal properties. UHTC materials have a high resistance to thermal shocks, melting points typically in the range of 2000–3000 K, as well as a high thermal conductivity and a high reradiation capability. They are considered suitable for the leading-edge regions of a hypersonic vehicle, which are exposed to high peaks of the heat flux. Instead of storing the heat as in the case of thermal insulation (which requires a low thermal conductivity along with a high thermal capacity), these materials allow the heat being transferred locally to the surface at a high rate (e.g., at the stagnation point) and to be easily conducted within the surface, so that it can be redistributed over a larger area of the surface where it is reradiated back into the atmosphere. This allows a significant reduction in the heat load absorbed by the leading edge of the vehicle.

Hence, we can summarize the main characteristics of a TPS for a generic hypersonic vehicle as follows: (1) high heat storage capability, that is, high thermal capacity (specific heat, c_p); (2) low thermal conductivity (k), thus high porosity (low density) of the internal structure; (3) high reradiation capability, that is, high surface emissivity coefficient (ε); (4) an internal mechanism that either convects the heat directly outside (e.g., ablation) or transfers it to different regions of the surface where it gets dissipated by another mechanism of inverse heat-transfer (e.g., re-radiation); (5) light structure, that is, low specific weight and limited thickness; and (6) durability in operating condition.

Note that among the above characteristics, reusability is not mentioned, as it depends on the system being of ablative or non-ablative type, which, in turn, depends on the flight conditions. An important challenge of the next-generation TPS systems for hypersonic flight will be then to push the limit of the TPS reusability up to the higher Mach number conditions, for example, for reentry flight or very high-speed cruise flight. This drives the design of the system toward a solution that combines the main features of passive or semi-passive cooling with a direct form of active cooling aimed to decrease the high external heat flux to more sustainable values. A system based on transpiration cooling has, in general, the potential to fill the gap between reusability and efficiency in very harsh environments.

13.3 Film Cooling and Transpiration Cooling Technique

The principle of active cooling systems is to make use of a coolant to directly decrease the local temperature and heat flux in the external flow and/or in the internal structure. The film cooling technique (Heufer and Olivier, 2008; Keller and Kloker, 2014, 2016), for example, consists of injecting coolant into the hot boundary layer to create a thin cold film of fluid near the wall, thus modifying the thermal as well as kinematic characteristics of the local boundary layer. This directly reduces the external wall heat flux by reducing the gas temperature. If the coolant is injected through discrete holes on the wall, this is known as effusion cooling. This technique is commonly used for turbine blades in the form of injection from single holes (Wittig et al., 1996; Baldauf et al., 2001). However, in this case, three-dimensional (3D) vortical structures are generated in the flow field, which can reduce the average cooling effectiveness (Heufer and Olivier, 2008). In contrast, injection through 2D slots leads to higher cooling effectiveness

Transpiration Cooling Using Porous Material for Hypersonic Applications

by providing a spanwise homogeneous flow field (Keller and Kloker, 2016). This is the form of injection mostly investigated for supersonic flow conditions.

The reduction of the heat flux is actually due to a dual mechanism, that is, a decrease of the local gas temperature near the wall and an increase of the boundary-layer thickness as a consequence of the coolant mass injection, which leads to a decrease in the velocity and temperature gradients at the wall (hence, of the skin friction and heat flux). Downstream of the opening, a laminar boundary layer will be formed, in general, by two separate layers with different profiles, that is, an upper region of hot fluid and a bottom layer of cold fluid. The streamwise extent of this separated flow region should be as long as possible to maintain the benefits of cooling. However, after a certain distance, the two fluid layers mix together until a hot boundary-layer profile is re-established. Thus, the holes should be adequately distributed over the surface, and the coolant blowing ratio ($F = \rho_c\, v_c/\rho_\infty u_\infty$) should be controlled to provide a target cooling efficiency. The performance of a film-cooling system is quantified by the cooling effectiveness, which is defined as (Heufer and Olivier, 2006)

$$\eta = \frac{T_{aw} - T_{aw,c}}{T_{aw} - T_{0,c}}. \tag{13.5}$$

In the above equation, the term $T_{aw,c}$ is the adiabatic wall temperature in the presence of cooling, and $T_{0,c}$ is the total temperature of the coolant gas, which is equal to the temperature inside the plenum chamber. The cooling effectiveness in this form requires that the system is given sufficient time to shift from a steady state of maximum temperature equal to the adiabatic wall temperature of the hot gas, to another steady state in which the maximum wall temperature reached (corresponding to zero heat flux) is lower due to the effect of cooling. In a wind-tunnel experiment, the test time is usually too small to allow for the adiabatic conditions to be reached. The wall temperature can be considered, in general, as fixed and equal to the room temperature. The plenum temperature can be assumed equal to the room temperature as well.

In the transpiration cooling technique, instead, cooling is achieved by a cold fluid transpiring from a porous material. The cold fluid traversing the thickness of the porous material reduces the temperature of the interior porous walls and brings the absorbed heat content outside (a form of reverse heat flux similar to an ablative system). In this case, the coolant injection into the boundary layer is not localized but distributed over the surface, and irregular due to the interior structure of the porous material. For the flow through porous media at very low Reynolds numbers (lower than 1), the flow rate per unit mass and area coming out from the surface (named also specific discharge or superficial velocity) is linked to the pressure gradient within the porous layer, the fluid viscosity, and the permeability of the material from the Darcy law:

$$v = -\frac{K_D}{\mu}\frac{\partial p}{\partial x}, \tag{13.6}$$

where K_D is the permeability coefficient (the Darcy coefficient in this case), and x is taken as the direction of the traversing flow within the material.

For higher Reynolds numbers, however, inertial effects are important and need to be added to the driving force related to the pressure gradient. Hence, the equation with the terms rearranged to express the pressure gradient takes the form of the Darcy–Forchheimer equation (Langener et al., 2011; Ifti et al., 2018),

$$-\frac{\partial p}{\partial x} = \frac{\mu}{K_D}v + \frac{\rho}{K_F}v^2, \tag{13.7}$$

in which K_F is the Forchheimer permeability coefficient. Hence, in a transpiration cooling system, the material permeability plays a fundamental role for the blowing ratio and the cooling effectiveness.

The permeability coefficient K_D can, in turn, be expressed as a function of two other important parameters, namely, the porosity (φ) and the hydraulic tortuosity (T), as given in (Matyka et al., 2008)

$$K_D = c_0 \frac{\phi^3}{T^2 S^2}, \tag{13.8}$$

where S is the specific surface area, that is, the ratio of the interstitial surface area of the pores to the bulk volume, and c_0 a constant (Kozeny constant), which depends on the internal geometry. The tortuosity is defined as the ratio between the average path length (λ) of the fluid going from the bottom surface to the upper surface of the porous layer and the thickness of the layer (l),

$$T = \frac{\lambda}{l}. \tag{13.9}$$

The capabilities of a transpiration cooling system can then be assessed and optimized once the parameters defining the material permeability, namely, the porosity and tortuosity, are known. However, the above-mentioned models for the permeability and tortuosity are not universal and may differ significantly with the flow conditions and the material, in general. Also, inhomogeneities of the internal material structure will affect the flow traversing the porous layer, leading to non-negligible differences in the blowing ratio as well as the thermodynamic properties of the coolant gas on the surface. Hence, direct numerical simulations capable of resolving the smaller scales within the porous structure can provide physical insights into the flow through a porous medium in hypersonic flow conditions, which can have important consequences on the efficiency of the thermal protection system.

13.4 Numerical Method

13.4.1 Governing Equations

We consider numerical solutions of the 3D Navier–Stokes equations for compressible flows, written in conservation form, under the assumption of a perfect gas. The set of non-dimensional conservation equations in Cartesian coordinates can be written as

$$\frac{\partial \rho}{\partial t} + \frac{\partial \rho u_j}{\partial x_j} = 0, \tag{13.10}$$

$$\frac{\partial \rho u_i}{\partial t} + \frac{\partial \rho u_i u_j}{\partial x_j} = -\frac{\partial p}{\partial x_i} + \frac{1}{Re}\frac{\partial \tau_{ij}}{\partial x_j}, \tag{13.11}$$

$$\frac{\partial \rho E}{\partial t} + \frac{\partial \rho E u_j}{\partial x_j} = -\frac{1}{(\gamma-1)RePrM^2}\frac{\partial}{\partial x_j}\left(\mu\frac{\partial T}{\partial x_j}\right) - \frac{\partial p u_j}{\partial x_j} + \frac{1}{Re}\frac{\partial \tau_{ij}u_i}{\partial x_j}. \tag{13.12}$$

The terms ρ, ρu, ρv, ρw, and ρE are the conservative variables of the system of equations, where ρ is the density, u, v, and w are the velocity components, respectively, in the x, y, and z directions; and E is the total energy per unit mass. In the flux vectors, the terms p, T, τ_{ij}, and μ are, respectively, the pressure, the temperature, the components of the viscous stress tensor, and the dynamic viscosity of the flow. The non-dimensional quantities are obtained through normalization of the dimensional variables with their freestream reference values, namely ρ^*_∞, U^*_∞ (freestream main velocity), U^{*2}_∞ (energy per unit mass and volume), $\rho^*_\infty U^{*2}_\infty$ (term linked to the dynamic pressure, used to normalize pressure and shear stress components), and T^*_∞ and μ^*_∞. Note that the superscript ($*$) is used to denote dimensional values. The characteristic length chosen to normalize the length scales is the boundary-layer displacement thickness (δ^*). Time scales are normalized with respect to the fluid dynamic characteristic time (δ^*/U^*_∞), based on the velocity of the undisturbed flow and on the characteristic length. The terms Re, Pr, M, and γ are, respectively, the Reynolds, Prandtl, and Mach numbers, and the ratio of specific heats ($\gamma = c_p^*/c_v^*$), that is, the dimensionless parameters of the flow. The Reynolds number is defined with respect to the boundary-layer displacement thickness of the similarity solution, as $Re = (\rho^*_\infty, U^*_\infty \delta^*)/\mu^*_\infty$; the Prandtl number is set to 0.72 for air, and γ is equal to 1.4, as we are considering a perfect gas model. The dynamic viscosity is, in turn, expressed in terms of temperature by Sutherland's law

$$\mu = T^{3/2} \frac{1+C}{T+C}, \tag{13.13}$$

where the constant C represents the ratio between the Sutherland's constant (set to 110.4 K) and the reference temperature in the freestream T^*_∞. The viscous stresses are defined in terms of the velocity derivatives, under the assumption of a Newtonian fluid, as

$$\tau_{ij} = \mu \left(\frac{\partial u_i}{\partial x_j} + \frac{\partial u_j}{\partial x_i} - \frac{2}{3} \delta_{ij} \frac{\partial u_k}{\partial x_k} \right). \tag{13.14}$$

We also need a relation linking the total energy to the temperature, which in non-dimensional form can be expressed as

$$E = \frac{T}{\gamma(\gamma-1)M^2} + \frac{1}{2}\left(u^2 + v^2 + w^2\right). \tag{13.15}$$

Finally, the system of equations is closed by the equation of state for a perfect gas:

$$p = \frac{1}{\gamma M^2} \rho T. \tag{13.16}$$

13.4.2 Code Features

The numerical simulations have been carried out using the Adaptive Mesh Refinement in Object-oriented C++ (AMROC) software framework (Deiterding, 2005a, 2005b, 2011), which uses a hybrid WENO-centered-difference (WENO-CD) scheme in conjunction with a structured adaptive mesh refinement (SAMR) approach. The implementation relative to the base central scheme is equipped with the option to be tuned (or optimized) for spectral resolution improvement, which denotes the scheme as WENO-tuned-centered difference (WENO-TCD), as described by Hill and Pullin (2004). The shock-capturing filter, namely, the WENO scheme, corresponds to the type of WENO-symmetric-order-optimized (WENO-SYMOO) scheme as shown by Martin et al. (2006). The schemes are integrated with a switching function that turns on/off the WENO method at discontinuities/smooth regions.

The base scheme (central scheme) has been proven to have a maximum formal order of accuracy of 6, corresponding to a seven-point stencil, with relative stencil coefficients as in Ziegler et al. (2011). The order of accuracy can be properly specified, which changes the stencil, and can be also reduced (keeping constant the stencil) for optimization of the bandwidth resolution capabilities. For example, considering the seven-point stencil, if the optimization option is specified, the corresponding scheme will be a fourth-order accurate scheme with optimized bandwidth properties (Hill and Pullin, 2004; Lombardini et al., 2014), thus has a minimum dispersion error of the modified wavenumber, suitable for large eddy simulations (LESs). Details of the WENO-CD scheme can be found in Cerminara et al. (2018b). The code has been tested to the sixth-order accuracy (for both CD and the WENO parts) for both the 2D and the 3D porous wall test cases, giving very good results (which will be shown later in this paper), thus showing to be suitable for direct numerical simulation (DNS) computations of the current problem.

Finally, the code works in conjunction with a SAMR algorithm (Deiterding, 2005a, 2011; Pantano et al., 2007), through which the uniform grid is dynamically refined with consecutive finer grid levels during the iteration cycles, following a patchwise refinement strategy. This applies in regions of the flow where numerical instabilities, related to low initial grid resolution, are detected by means of error indicators.

13.5 Results from DNS in Hypersonic Flow over Porous Surfaces

In this section, we will be showing numerical results of simulations resolving a hypersonic flow over porous surfaces, with and without the injection of cold fluid. Two porous surface configurations have been considered, namely, a simpler case of slots, which is representative of a case of film cooling by effusion and a case of a regular porous structure with distributed solid spherical elements, which mimics a realistic porous material, as in the case of transpiration cooling. The flow field characteristics in both cases are presented, as well as numerical aspects linked to the resolution of the porous structure, and the cooling performance is discussed in relation to the effects on transition.

13.5.1 Flat Plate with Periodic Regular Pores

The first configuration is a streamwise and spanwise periodic flat plate with square-shaped holes. The simulations have been run for both a 2D and a 3D configuration, thus in the 2D configuration the pores reduce to slots. Figure 13.1 shows the geometrical configurations considered in the 2D (left) and 3D (right) simulations. The Mach number is 6, the Reynolds number based on the initial boundary layer displacement thickness is 6,000, the freestream temperature is $T_\infty = 216.65$ K, and the wall temperature is fixed to the adiabatic wall temperature $T_w = 1522.4$ K.

The reference length is the displacement thickness of the initial laminar boundary layer, determined from a similarity solution, through which the flow field is initialized in our numerical simulations. The simple 2D geometry has been chosen for validation test cases, consisting of two pores with regular shape in the x-direction. The size of the computational domain is 3 times the initial boundary-layer displacement thickness in the x-direction, and 11 times in the y-direction. The pores have an x-wise length of 0.75 and a depth of 1. The large extension of the domain in the y-direction is due to the need to allocate enough vertical space for the propagation of disturbance waves generated by the pores.

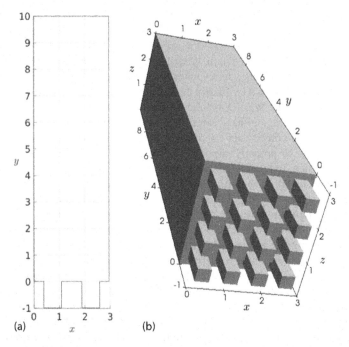

FIGURE 13.1 Computational domain geometries for the 2D (a) and 3D (b) simulation.

13.5.1.1 Code Verification and Grid Study

Preliminary 2D simulations have been carried out for the Reynolds number $Re = 6,000$ to validate the code and test its performances in comparison to shock/boundary-layer interaction (SBLI), a DNS code developed over a number of years at the University of Southampton, which has been extensively tested and validated in the past, for example by De Tullio et al. (2013).

Figure 13.2 shows solutions of the pressure field at the non-dimensional time $t = 10$ obtained with the codes AMROC and SBLI, respectively. The mesh resolutions out of the pores for the simulations run with AMROC and SBLI are 64×400 and 64×356 (in the x- and y-directions), respectively, whereas inside the pores 40 cells are used in the y-direction for the AMROC case and 131 points for the SBLI case. The difference in the number of cells/points in the y-direction outside and inside the pores between AMROC and SBLI is due to the fact that the grid used in AMROC is uniform in both directions, while the grid used in SBLI is stretched in the y-direction to increase the resolution toward the wall. The y-spacing in the inner-pore region is kept constant and equal to the spacing at the wall in SBLI.

As can be seen, very similar results are obtained with the two codes, which demonstrates the accuracy capabilities of our considered code (AMROC). Also, the figures show that pressure waves are formed at the pore edges, which propagate outward influencing the flow outside the boundary layer. It has been verified that the amplitude of the radiated pressure waves in the freestream, normalized with the freestream mean pressure, is about a factor of 1.4 higher than the amplitude of the associated normalized density waves, which characterizes acoustic disturbances.

Figure 13.3 shows a direct comparison between AMROC and SBLI for the density solution along x at $y = 1$. The result confirms the good agreement between the two different codes in accurately resolving the acoustic signal produced by the pores. A grid resolution study has been carried out for the case at $Re = 6,000$, based on the trend of the density along the streamwise direction at $y = 1$ (Figure 13.4) and the pressure along the wall-normal direction at $x = 1.5$ (Figure 13.5). Six different grid resolution levels (for the base grid level in the SAMR framework) have been analyzed to investigate the effect of the grid refinement in the x- and y-directions (both simultaneously and separately), namely: 32×440, 64×440, 128×440, 64×660, 128×660, 128×880 (indicating the number of cells in the x- and y-directions, respectively).

Both figures show a very good grid convergence. In particular, the solution for the oscillation peaks in Figure 13.4 is shown to be more sensitive to the grid resolution in the y-direction (when shifting from 440 to 660 cells), compared to the refinement in the x-direction. Figure 13.5 shows the pressure solution inside the boundary layer, which proves that at the considered grid resolution, that is, 64×440, without the need for adding dynamically more refinement levels (through the SAMR methodology), we already manage to resolve the boundary layer very well for the present configuration.

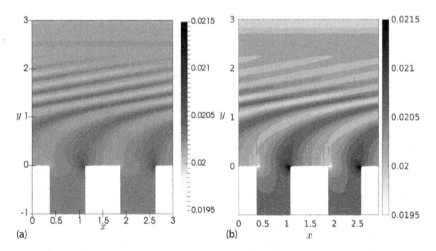

FIGURE 13.2 Pressure field at the time $t = 10$ obtained with AMROC (a) and SBLI (b).

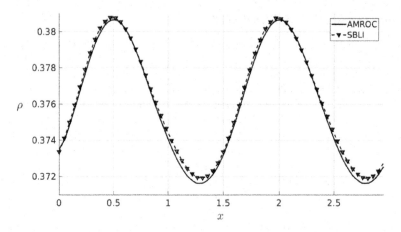

FIGURE 13.3 Comparison between AMROC and SBLI for the density trend along x at $y = 1$. Time $t = 10$.

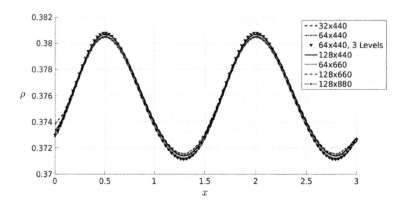

FIGURE 13.4 Grid resolution study for the density along x at $y = 1$, $Re = 6,000$.

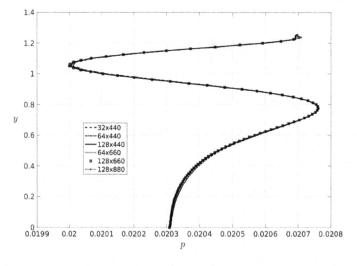

FIGURE 13.5 Grid resolution study for the pressure along y at $x = 1.5$, $Re = 6,000$.

13.5.1.2 Effect of Wall Cooling

The wall-cooling effect is analyzed in the 2D configuration for a Reynolds number $Re = 20,000$ by comparing results for the temperature and pressure fields obtained at time $t = 50$ for different wall-temperature conditions, namely: (1) $T_w = T_{ad}$ (i.e., adiabatic wall temperature, with $T_{ad} = 7.027 T_\infty$); (2) $T_w = 0.8 T_{ad}$; and (3) $T_w = 0.5 T_{ad}$. The two latter cases correspond to a wall-cooling effect with wall temperature decreased by 20% and 50% relative to the adiabatic value, respectively.

Figure 13.6 shows the temperature fields for the three above-mentioned wall-temperature conditions, while Figure 13.7 shows the corresponding result for the pressure field. As expected, with reference to Figure 13.6, wall-cooling produces an overall decrease of the temperature inside the boundary layer, which results in a lower boundary-layer thickness at the lower wall temperatures. This effect is related to the increase in density inside the boundary layer as a result of the cooling. The most remarkable effect of wall cooling is observed in the pressure field (shown in Figure 13.7), and in particular on the generated acoustic waves. It is evident that, as we move from the adiabatic wall-temperature case to the coldest case, a decrease of the wall temperature produces a decrease of the pressure inside the pores, which results in a stronger expansion at the left corner of each pore and in a stronger compression at the right corner. The pressure waves radiated by the pore edges into the boundary layer and the external field are then stronger in the colder wall cases.

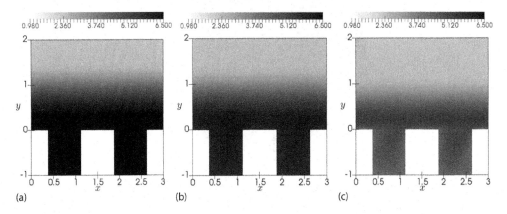

FIGURE 13.6 Temperature field in the near-pore region for the wall temperature fixed to the values $T_w = T_{ad}$ (a), $T_w = 0.8 T_{ad}$ (b), and $T_w = 0.5 T_{ad}$ (c). Time $t = 50$, $Re = 20,000$.

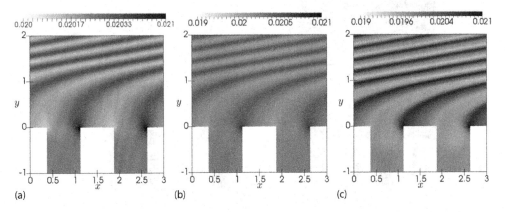

FIGURE 13.7 Pressure field in the near-pore region for the wall temperature fixed to the values $T_w = T_{ad}$ (a), $T_w = 0.8 T_{ad}$ (b), and $T_w = 0.5 T_{ad}$ (c). Time $t = 50$, $Re = 20,000$.

13.5.1.3 3D Results—Effect of Reynolds Number

The length of the 3D domain is 3 in both the x- and the z-directions, while the number of pores along each direction, compared to the 2D case, has been increased to 4, thus providing a 3D porous layer composed overall of 16 equally spaced square-shaped pores. The pores have a length of 0.375 in both the x- and the z-directions. This configuration corresponds to the configuration analyzed in the reference DNS study of De Tullio and Sandham (2010). High-resolution DNS simulations have been run for the 3D configuration with a mesh size $128 \times 660 \times 128$, using the sixth-order WENO-CD method, for both the Reynolds numbers ($Re = 6{,}000$ and $Re = 20{,}000$). The results of these simulations prove the high-resolution DNS capabilities of the considered hybrid method for the present case and show interesting physical features of a hypersonic flow over a porous coating.

Figure 13.8 shows a generic 3D view of the vertical velocity field, which gives a first impression of the 3D flow field and the orientation of the radiated waves outside the boundary layer along the streamwise and spanwise directions. Figure 13.9 shows a comparison of the pressure fields on a zy-slice at $x = 1.85$ for

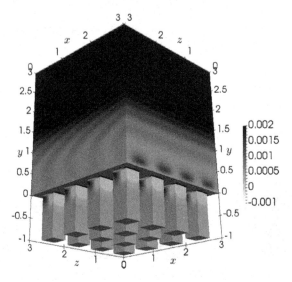

FIGURE 13.8 3D view of the vertical velocity field. Time $t = 50$, $Re = 20{,}000$. Note the inclination of the radiated waves with respect to the x-axis outside of the boundary layer, and the spanwise modulation of the velocity field in the immediate outer-pore region.

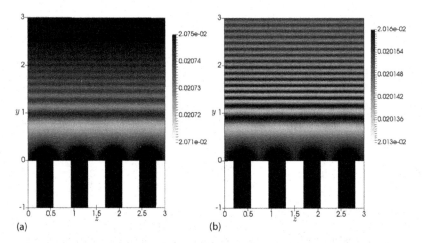

FIGURE 13.9 Pressure field on the zy-plane at $x = 1.85$ at $Re = 6{,}000$ (a) and $Re = 20{,}000$ (b). Time $t = 50$.

the two cases. Here, the color range has been restricted to a smaller band in order to better notice the differences between the two cases in the outer-pore region. We observe higher-amplitude radiated pressure waves and a higher pressure level within the pores for the lower Reynolds number case. A modulation of the pressure field along the spanwise direction in the immediate outer-pore zone is also evident, which is a result of the 3D porosity of the wall. A further analysis of the root-mean-square (RMS) levels of the pressure disturbances outside the boundary layer (conducted at the height $y = 1.5$) has shown RMS levels of the pressure fluctuations (normalized with the freestream pressure of the initial unperturbed flow) of 0.045 and 0.0154 for the lower and the higher Reynolds numbers, respectively. Thus, at the lower Reynolds number, the pressure waves radiated by the porous wall are approximately three times stronger than the waves radiated in the higher Reynolds number case.

The vertical velocity field in the pore region is analyzed both in an xy-plane at the spanwise position $z = 1.85$ (in Figure 13.10) and in an xz-plane at the vertical position $y = 0$ (in Figure 13.11), namely, at the surface plane. Figure 13.10 shows that higher absolute values of the inward (negative) and the outward (positive) vertical velocity are reached just above the pores at the lower Reynolds number, whereas more pronounced values are obtained for the higher Reynolds number inside the pores. A higher vertical velocity is observed outside of the boundary layer for the lower Reynolds number case, which is consistent with the flow being deflected by the presence of higher-amplitude radiated acoustic waves.

In Figure 13.11, the three-dimensionality of the inward and outward flow on the $y = 0$ surface is observed. Higher absolute values are seen for the lower Reynolds number case, consistent with the above discussion. However, in the lower Reynolds number case, the flow features shown in Figure 13.11 appear more two-dimensional than in the higher Reynolds number case. In fact, for the $Re = 6,000$ case, the

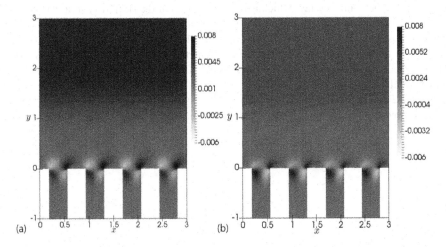

FIGURE 13.10 Vertical velocity field on the xy-plane at $z = 1.85$ at $Re = 6,000$ (a) and $Re = 20,000$ (b). Time $t = 50$.

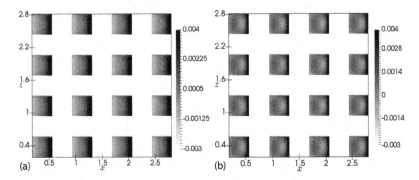

FIGURE 13.11 Vertical velocity field on the xz-plane at $y = 0$ at $Re = 6,000$ (a) and $Re = 20,000$ (b). Time $t = 50$.

maximum absolute values of the vertical velocity are localized in spanwise-elongated narrow strips adjacent to the pore left and right edges. At $Re = 20,000$, in contrast, the maximum of the inward velocity is detached from the left side and localized in a region between the pore center and the right side. This effect in turn produces a higher recirculation in the immediate inner-pore region, as seen from the higher internal absolute values shown in Figure 13.10 for the higher Reynolds number case.

13.5.2 Flat Plate Domain with Injection through Localized Slots

We consider in this section a long, flat plate domain with fluid injection through localized slots, with inlet and outlet boundary conditions. The present configuration has been considered for the study of the effects of fluid injection at different plenum pressures on transition, as well as on the cooling performance over the 3D surface. The flow conditions considered in the present study are related to ongoing experimental investigations in the field of film cooling in hypersonic flow and are as follows: Mach number $M = 5$, unit Reynolds number $Re_m = 12.6 \times 10^6$ 1/m, and freestream temperature $T_\infty = 81.7\text{K}$. The computational domain used for the case of a flat plate with slots and a plenum chamber is sketched in Figures 13.12 through 13.14. The size of the domain is $L_x \times L_y \times L_z = 200 \times 32 \times 50$, including the sum of the heights of slots and plenum chamber, equal to 2.5. Thus, the distance of the top boundary from the flat plate surface is 29.5. The slots have a height of 1.5, and the depth of the plenum chamber is 1. A close-up of the slot and the plenum chamber, along with the relative dimensions in an xy-plane, is presented in Figure 13.13. The initial condition is the laminar boundary layer from a similarity solution. The characteristic length is the boundary-layer displacement thickness of the similarity profile at the inflow boundary, that is, $\delta^* = 1$ mm. The Reynolds number based on the displacement thickness is then $Re = 12,600$. The boundary-layer thickness ($u_e = 0.99 U_\infty$) of the similarity inflow is $\delta^* = 1.25$ mm. The growth of the boundary layer in the streamwise direction for the initial state is then imposed following the method described in De Tullio et al. (2013). All the length scales in the present study are normalized with respect to δ^*. The inflow boundary in our numerical investigation corresponds a distance of 127 mm from the plate leading edge. Further details about the settings of the simulations can be found in Cerminara et al. (2018a).

At the inlet, a fixed inflow condition is imposed in the supersonic region of the boundary layer and in the uniform freestream, whereas an extrapolation condition is imposed in the subsonic region of the

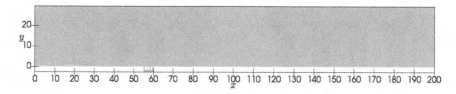

FIGURE 13.12 Computational domain, flat plate with slots, and plenum chamber located in the region $x = 55$–60.

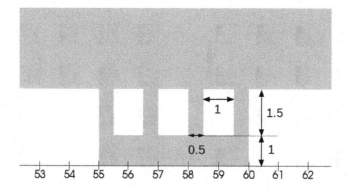

FIGURE 13.13 Close-up of the slot region.

boundary layer. The top boundary and the outlet are treated with a standard outflow boundary condition. A plenum boundary condition is used at the bottom boundary, through which the pressure and temperature are set equal to the corresponding stagnation values (p_0, T_0) in the plenum. Note that in the work of Keller and Kloker (2014), the same plenum boundary condition was set at the bottom boundary of each slot, as the presence of the plenum was not simulated. Our simulations, instead, consider the presence of a plenum chamber at the bottom of the slots. Figure 13.14 shows a top view on the surface of the 3D domain. The grid is uniform in each direction, with a size (for the base level in the AMR framework) of 800 × 384 × 100. Two and three overall grid levels are used in the AMR methodology, dependent on the particular case, which provides minimum cell sizes in x, y, and z of 0.125, 0.0415, and 0.25, for the two-level case, and 0.0625, 0.0207, and 0.125, for the three-level case. Simulations have been run for four different values of the plenum pressure, namely, $p_0 = 1.2p_\infty$, $1.5p_\infty$, $2p_\infty$, and $3p_\infty$, which provide gradually higher blowing ratios. Two grid levels have been used for the first three of the above-mentioned cases, whereas three levels have been used for the latter (i.e., the case with the highest plenum pressure), to better resolve the turbulent region. A constant plenum temperature of $T_0 = 3.67$, corresponding to a dimensional value of 300 K, is considered for all the configurations and for each case with different plenum pressures. Figure 13.15 shows results for streamlines colored with temperature in the injection

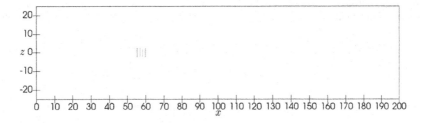

FIGURE 13.14 Top view of the domain in the xz-plane at $y = 0$.

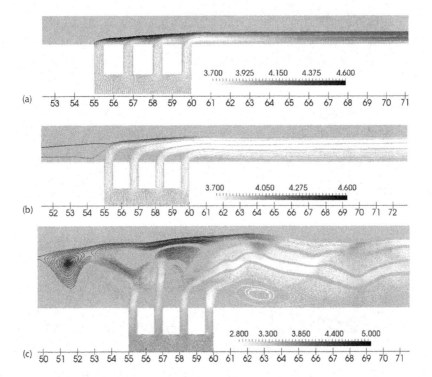

FIGURE 13.15 Temperature streamlines in the injection region: $p_0 = 1.2p_\infty$(a), $p_0 = 1.5p_\infty$(b), $p_0 = 3p_\infty$(c).

region at the different plenum pressures. These results provide an indication of how film cooling is active just downstream of the slots and the effects on the boundary layer of an increasing plenum pressure. As can be seen, for the lowest plenum pressure case, a very thin laminar layer of cold fluid is formed near the wall downstream of the slots and underneath the upper layer of hotter fluid coming from the upstream boundary layer. This is an ideal situation, as the boundary layer is only slightly disturbed by the coolant injection and transition is not triggered immediately downstream of the slots. Also, the thin coolant film is initially separated from the hot boundary layer (before mixing further downstream), which keeps the high temperatures of the hot boundary layer away from the wall. In the second case of $p_0 = 1.5p_\infty$, the coolant layer just downstream of the slots is still laminar, but thicker due to the higher blowing ratio. In this case, the hot boundary layer seems to be completely merged within the layer of cold fluid already at the slot location, and the cold layer is longer than in the previous case. These appear to be all positive aspects; however, the higher thickening of the boundary layer induced in the higher plenum pressure case needs to be taken into account in terms of the effect that this can have on transition. The flow separates upstream of the slots, and a small recirculation region is induced, as shown by the high-temperature streamlines on the left side of the figure. If this change in the flow enhances the boundary-layer unstable modes, or generates additional unstable modes, this may cause transition to happen at an earlier location than in the case without cooling. As we increase the plenum pressure the boundary layer gets more and more perturbed, and at the highest plenum pressure considered, it appears highly distorted and in a highly vortical state, which is likely to be a prelude to the occurrence of transition a short distance downstream.

Figure 13.16 shows the boundary layer state downstream of the slots through contours of the streamwise velocity at the height $y = 1$ for different plenum pressures. As we can see, at the plenum pressure

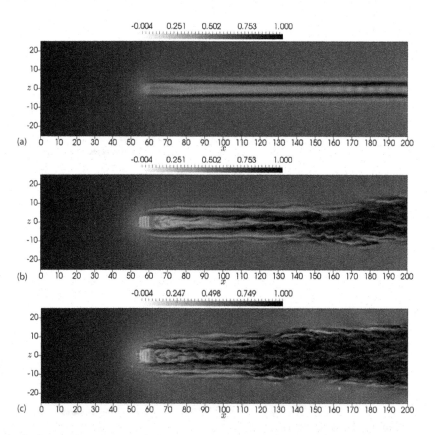

FIGURE 13.16 Streamwise velocity (u) inside the boundary layer at the height $y = 1 - p_0 = 1.5p_\infty$(a), $p_0 = 2p_\infty$(b), $p_0 = 3p_\infty$(c).

$p_0 = 1.5p_\infty$ the boundary layer is still laminar up to the end of the computational domain, but streamwise-elongated structures are generated, which modify the stability characteristics of the boundary layer further downstream. In particular, two high-velocity streaks are formed at the edges of the slots, which are positioned in between of two weaker low-velocity streaks at the sides and the larger low-velocity wake in the middle. Moreover, we notice the presence in the downstream region of 2D-wavy structures (normal to the streamwise direction) gradually growing along the wake, which interact with the streamwise streaks causing their progressive distortion. At the plenum pressure $p_0 = 2p_\infty$, the high-velocity streaks and the wake appear in a highly disturbed state just downstream of the slots, which causes the streaks to break down farther downstream and begin the transition process. At the highest plenum pressure case, a more violent transition process is triggered, with the streaks that appear partially fragmented and interact with the wake flow already in the early region downstream of the slots. In this case, the breakdown mechanism is much more rapid and leads to a turbulent state downstream of $x = 115$. The transition region is observed to spread downstream following a wedge-shaped structure.

Figure 13.17 shows the cooling effectiveness on the surface at the different plenum pressures. As can be seen, the higher values (close to 1) are localized around the slot region and along the wake. Whereas, values close to zero are reached along the main streamwise streaks and in the turbulent region, assuming locally negative values in high compression regions.

Figure 13.18 quantifies the cooling performance at different plenum pressures by showing the trend of the surface temperature along the midspan axis. As the wall is at adiabatic conditions, the temperature values represent the maximum temperature reached along the surface in the presence of cooling, which in turn gives an indication of the maximum temperature decrease induced by the coolant injection. As can be seen, for each case a strong temperature drop is reached at the slot location, which is then followed

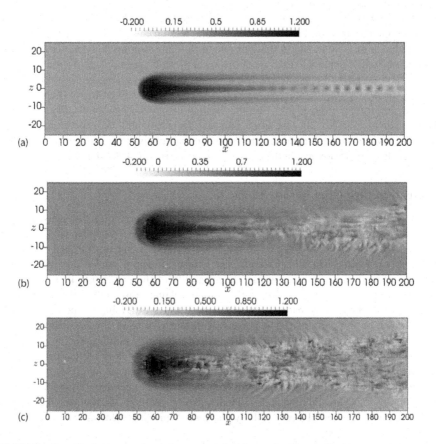

FIGURE 13.17 Cooling effectiveness on the surface: $p_0 = 1.5p_\infty$(a), $p_0 = 2p_\infty$(b), $p_0 = 3p_\infty$(c).

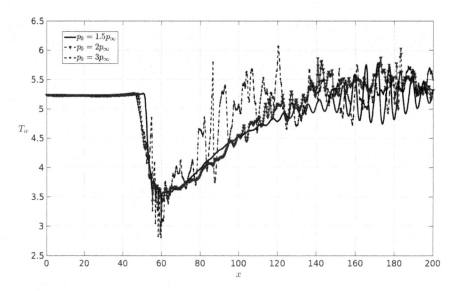

FIGURE 13.18 Surface temperature along the midspan axis ($z = 0$).

by a gradual increase downstream. In the case of plenum pressure $p_0 = 1.5p_\infty$ (laminar case), the temperature profile is smooth up to about $x = 100$, then pronounced oscillations, related to the 2D waves observed along the wake in Figure 13.16 (top), develop and grow downstream until reaching a peak at approximately $x = 170$. These oscillations may be representative of an unstable boundary-layer mode formed in this region of the flow. For the higher plenum pressure case, $p_0 = 2p_\infty$, the temperature profile presents irregular short-wavelength oscillations due to the transitioning flow, but the overall trend and absolute values are still comparable to those of the lower plenum pressure case over most of the domain length. At the highest plenum pressure case, in contrast, transition is triggered just downstream of the slot region, and the surface temperature shows a pronounced increase downstream of about $x = 80$. In the transitional region, from $x = 80$ to $x = 120$, the oscillations are stronger, with very high local peaks. Then (downstream of $x = 120$), after the turbulent state is reached, both their wavelength and amplitude reduce, and their average value is approximately equal to that relative to the other two cases. In this case, the cooling benefits are already lost in the early region downstream of the slots due to transition.

13.5.3 Injection through a Porous Layer

For a correct characterization of a transpiration cooling system, it is important to analyze the properties of the injected flow through a porous material. The Darcy–Forchheimer Equation (13.7) relates the pressure gradient across the thickness of the porous layer to the injected flow rate through the fluid viscosity, the inertial forces, and the permeability characteristics of the material. As we have seen in Section 3.1, the material permeability, or, in other words, the capability of the material to be traversed by a fluid flow, depends on its porosity, that is, the empty-to-total volume ratio, and on the tortuosity, which is in general defined as the average path length of the streamlines traversing the porous layer. The latter is hence dependent on the particular arrangement of the internal solid elements, for a given porosity. In the context of DNS simulations, a simplified model of an internal porous structure is represented by spherical particles of a certain radius, distributed within the porosity layer with a certain arrangement. The radius and number of the particles will determine, with respect to the total volume of the layer, the porosity as well as the internal length scale of the pores. In order to compare different tortuosity levels in a systematic way, it is more appropriate to compare different regular distributions of the spherical elements, instead of random ones. Hence, regularly distributed spherical elements can be used to carry out a parametric study of the flow features through a porous layer for different porosities and tortuosities. This can give important indications on the regime of validity of the Darcy–Forchheimer equation, and its modified versions, as well as providing insights in the physics of the flow through a porous medium by resolving the smaller porosity scales through direct numerical simulation.

In this section, we present preliminary results obtained considering the above-mentioned approach, which show some initial interesting physical aspects and provide a basis for future detailed studies aimed at the analysis of transpiration cooling systems. An example of a regular arrangement of spherical elements is shown in Figure 13.19, which represents the pressure field within a porous layer. In this case, 2D simulations have been run to carry out an initial grid study and an assessment of the porous structure model. Hence, the solid elements are to be considered as cylinders of an infinite length. The solid elements have been created by using an immersed boundary method (Deiterding et al., 2006; Deiterding, 2009). The same Cartesian method used for the other flow regions is then used to mesh the region within the porous layer. As can be seen, the element arrangement follows a face-centered cubic structure, where the xy-plane can be imagined as a face of a cube in a 3D space, with the same plane repeating periodically along the z-axis in a 3D case. Such an arrangement can be compared to a body-centered cubic type in a 3D simulation. In the present case, a porosity of 40% has been imposed, to mimic the behavior of realistic material samples currently investigated for transpiration cooling systems.

The computational domain is a local flat-plate domain with imposed periodicity in the x-direction. The domain dimensions are $Lx = 5$ and $Ly = 10.5$, the overall thickness of the region underneath the surface is 2, whereas the thickness of the porous layer is 1.156, including the diameter of the cylindrical elements at the edges. The freestream conditions are the same considered in Section 13.5.2 for the case of the flat plate with slots. The reference length is 1 mm, which corresponds to the similar boundary-layer displacement thickness of the initial condition. The cylinder elements have a radius of 0.078 mm, and the smallest distance between two adjacent elements is 0.02 mm. Hence, the minimum length scale of the internal porous structure is of the order of 20 µm in this case. The internal area defined by the centers of the cylindrical elements at the edges is 1×1 mm^2. A plenum boundary condition has been imposed at the bottom boundary, with $T_0 = 300$ K and $p_0 = 1.5 p_\infty$. As can be seen, a strong pressure gradient is formed in the lower sections of the porous layer, while the pressure adapts to the freestream value in the upper layer. Results refer to the simulation time $t = 20$.

The AMR methodology is very important for resolving the smaller scales within the porous layer. For the present case, a grid study has been carried out to find the minimum grid resolution, thus the

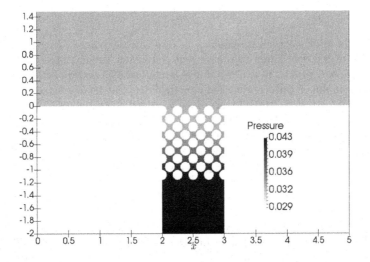

FIGURE 13.19 Pressure field.

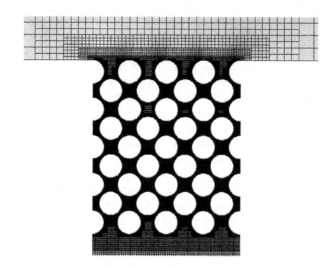

FIGURE 13.20 Grid with six AMR levels. Momentum in the y-direction plotted.

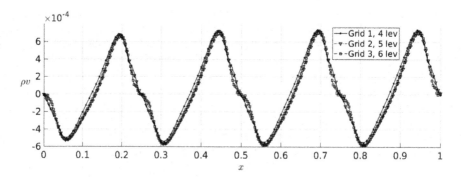

FIGURE 13.21 Blowing ratio along the x-direction on the surface—grid convergence study.

minimum number of refinement levels, required for resolving the porosity length scales with a negligible influence of the grid size. Three grids have been considered, namely, a base grid of 75 × 189 with four overall levels, a base grid of 50 × 126 with five levels, and a base grid of 50 × 126 with six levels. The corresponding minimum cell sizes in x and y are (in order from the first to the last of the grids listed above) $\Delta x_{min} = 0.0083, 0.00625, 0.003125$, and $\Delta y_{min} = 0.0069, 0.0052, 0.0026$. The grid resolution for the finest grid (with up to six levels) is shown in Figure 13.20. The results of the grid study are presented in Figure 13.21 for the blowing ratio along the x-direction on the surface ($y = 0$). As observed, the two more refined grids show, in general, a quite good agreement both at the surface and inside the porous layer. A local disagreement between these two grids is observed along the surface at the points where the blowing ratio profile crosses the zero value. At these points, in fact, there is an inflection of the profile, which is more pronounced for the highest grid as a result of a smoother surface of the cylindrical elements. Figure 13.22 shows the trend of the cooling effectiveness along the x-direction on the surface (for the finest grid), which reveals that in this case the cooling effectiveness is negative on the surface of the porous material. This is due to the fact that the imposed plenum pressure is not high enough to provide a sufficiently high blowing ratio to cool the wall of the porous material. In contrast to the case of the slots, the internal solid structure of the porous layer works as an obstacle to the injected flow. As a result, the flow undertakes a strong compression in the lower sections of the porous layer (as seen in Figure 13.19), which causes a further increase of the local temperature, that is, the temperature of the coolant. If the

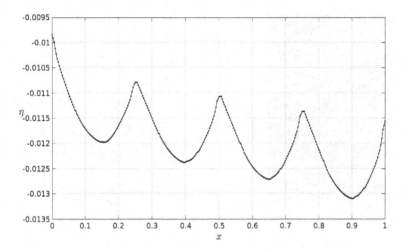

FIGURE 13.22 Cooling effectiveness along the *x*-direction on the surface.

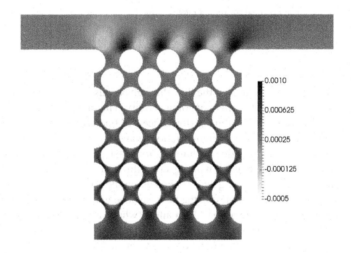

FIGURE 13.23 Field of the momentum in the *y*-direction (ρv). Close-up in the porous layer.

expansion in the upper sections of the porous layer is not high enough (i.e., the blowing ratio is relatively low), the injected fluid results to be a heated fluid from the internal flow compression, which does not perform a cooling effect, but a reverse effect instead. This proves that the thermodynamic effects taking place within the porous layer play a significant role on the cooling performance of transpiration cooling systems in hypersonic flow. Figure 13.23 shows a close-up of the blowing ratio (*y*-momentum) in the pore region and at the surface for the highest resolution grid. The figure highlights details of the blowing ratio, in particular the alternative negative and positive zones on the surface, as seen already in Figure 13.21, and the higher positive values reached in the points of minimum distance between two adjacent cylinders in the inner porous layer, correspond to the strong pressure gradient region observed in Figure 13.19. Figure 13.24 shows the streamwise and vertical velocity fields in the whole domain. These plots show that the boundary layer has not been significantly distorted by the blowing through the porous layer, as observed by the approximately constant boundary-layer thickness, but weak acoustic waves have been released outside of the boundary layer, which travel in the freestream at the Mach angle (11.5° for the present case).

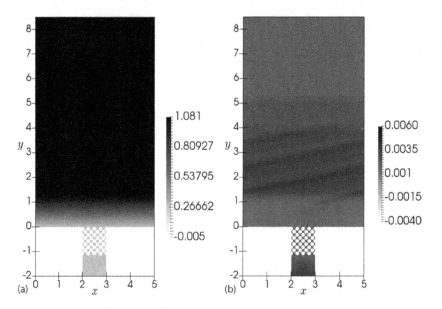

FIGURE 13.24 Streamwise (a) and vertical (b) velocity fields. Time $t = 20$.

13.6 Conclusion

The in-depth analysis of the main features of the flow through porous media in transpiration cooling systems for use on hypersonic vehicles is still an ongoing activity. The present results already show important physical aspects concerning wall-cooling performance and boundary-layer properties over a porous wall with and without injection. Different configurations have been considered and tested, namely, a periodic flat-plate domain with square-shaped holes, a long flat plate domain with injection through localized slots, and a local flat-plate domain with injection through a layer of distributed porosity. Flow regimes at Mach 5 and 6 have been investigated, dependent on the configuration.

Results indicate that the presence of holes on the body surface generates acoustic waves that are radiated into the freestream and propagate at the Mach angle, whose amplitude is dependent on the wall-temperature condition and the Reynolds number. For the slot injection case, perturbation induced by the local injection generates streaks inside the boundary layer, which interact with 2D waves growing along the wake farther downstream. At higher plenum pressures, transition occurs just beyond the slots, which is seen to cancel the benefits of wall cooling by promoting an earlier increase of the wall temperature with very high local peaks. Whereas, in the case of injection through a layer of distributed porosity, the obtained blowing ratio, for the same plenum pressure case, is much smaller, and the boundary layer only slightly distorted compared to the case of slot injection, as a consequence of the permeability of the porous layer.

Future 3D simulations including different porosities and different arrangements of the solid particles (i.e., different tortuosities) will assess the validity of the Darcy–Forchheimer equation for modeling the injected flow rate through a porous material at different hypersonic flow conditions, as well as giving further insights in the physics of the injected flow through porous media and its effects on the cooling performance and the transition mechanism in transpiration cooling applications.

ACKNOWLEDGMENTS

The authors would like to acknowledge support from the Engineering and Physical Sciences Research Council (EPSRC) under the Grant No. EP/P000878/1.

REFERENCES

Anderson Jr, J. D. (2006). *Hypersonic and High-Temperature Gas Dynamics*. Reston, VA: American Institute of Aeronautics and Astronautics.

Baldauf, S., Schulz, A., and Wittig, S. (1999). High resolution measurements of local heat transfer coefficients by discrete hole film cooling. *Journal of Turbomachinery*, 123(4):749–757.

Cerminara, A., Deiterding, R., and Sandham, N. (2018a). Direct numerical simulation of blowing in a hypersonic boundary layer on a flat plate with slots. In *2018 Fluid Dynamics Conference*, p. 3713.

Cerminara, A., Deiterding, R., and Sandham, N. (2018b). DNS of hypersonic flow over porous surfaces with a hybrid method. AIAA 2018-0600, AIAA Aerospace Sciences Meeting, Kissimmee, FL, 2018.

De Tullio, N. and Sandham, N. D. (2010). Direct numerical simulation of breakdown to turbulence in a Mach 6 boundary layer over a porous surface. *Physics of Fluids*, 22(9):094105.

De Tullio, N., Paredes, P., Sandham, N., and Theofilis, V. (2013). Laminar-turbulent transition induced by a discrete roughness element in a supersonic boundary layer. *Journal of Fluid Mechanics*, 735:613–646.

Deiterding, R. (2005a). Construction and application of an AMR algorithm for distributed memory computers. In *Adaptive Mesh Refinement-Theory and Applications*, pp. 361–372. Springer, Berlin, Heidelberg.

Deiterding, R. (2005b). Detonation structure simulation with AMROC. In *International Conference on High Performance Computing and Communications*, pp. 916–927, Springer.

Deiterding, R. (2009). A parallel adaptive method for simulating shock-induced combustion with detailed chemical kinetics in complex domains. *Computers & Structures*, 87:769–783.

Deiterding, R. (2011). Block-structured adaptive mesh refinement: Theory, implementation and application. *European Series in Applied and Industrial Mathematics: Proceedings*, 34:97–150.

Deiterding, R., Radovitzky, R., Mauch, S. P., Noels, L., Cummings, J. C., and Meiron, D. I. (2006). A virtual test facility for the efficient simulation of solid materials under high energy shock-wave loading. *Engineering with Computers*, 22(3–4):325–347.

Heufer, K. and Olivier, H. (2006). Film cooling for hypersonic flow conditions. In *Thermal Protection Systems and Hot Structures*, Proceedings of the 5th European Workshop held 17-19 May, 2006 at ESTEC, Noordwijk, The Netherlands, volume 631.

Heufer, K. and Olivier, H. (2008). Experimental and numerical study of cooling gas injection in laminar supersonic flow. *AIAA Journal*, 46(11):2741–2751.

Hill, D. J. and Pullin, D. I. (2004). Hybrid tuned center-difference-WENO method for large eddy simulations in the presence of strong shocks. *Journal of Computational Physics*, 194(2):435–450.

Ifti, H. S., Hermann, T., and McGilvray, M. (2018). Flow characterisation of transpiring porous media for hypersonic vehicles. In *22nd AIAA International Space Planes and Hypersonics Systems and Technologies Conference*, p. 5167.

Keller, M. A. and Kloker, M. J. (2014). Effusion cooling and flow tripping in laminar supersonic boundary-layer flow. *AIAA Journal*, 53(4):902–919.

Keller, M. A. and Kloker, M. J. (2016). Direct numerical simulation of foreign-gas film cooling in supersonic boundary-layer flow. *AIAA Journal*, 55(1):99–111.

Langener, T., Wolfersdorf, J. V., and Steelant, J. (2011). Experimental investigations on transpiration cooling for scramjet applications using different coolants. *AIAA Journal*, 49(7):1409–1419.

Lombardini, M., Pullin, D., and Meiron, D. (2014). Turbulent mixing driven by spherical implosions. part 1. Flow description and mixing-layer growth. *Journal of Fluid Mechanics*, 748:85–112.

Martin, M. P., Taylor, E. M., Wu, M., and Weirs, V. G. (2006). A bandwidth-optimized WENO scheme for the effective direct numerical simulation of compressible turbulence. *Journal of Computational Physics*, 220(1):270–289.

Matyka, M., Khalili, A., and Koza, Z. (2008). Tortuosity-porosity relation in porous media flow. *Physical Review E*, 78(2):026306.

Pantano, C., Deiterding, R., Hill, D. J., and Pullin, D. I. (2007). A low-numerical dissipation patch-based adaptive mesh refinement method for large-eddy simulation of compressible flmet. *Journal of Computational Physics*, 221(1):63–87.

Tran, H., Johnson, C., Rasky, D., Hui, F., Hsu, M., Chen, T., Chen, Y., Paragas, D., and Kobayashi, L. (1997). Phenolic impregnated carbon ablators pica as thermal protection system for discovery missions. NASA TM-110440, NASA, Washington, DC.

Winter, M. W. and Trumble, K. A. (2010). Spectroscopic observation of the stardust re-entry in the near UV with slit: Deduction of surface temperatures and plasma radiation. NASA Technical Report, ARC-E-DAA-TN1088, NASA Ames Research Center; Moffett Field, CA.

Wittig, S., Schulz, A., Gritsch, M., and Thole, K. (1996). Transonic film-cooling investigations: Effects of hole shapes and orientations. In ASME 1996 International Gas Turbine and Aeroengine Congress and Exhibition, p. V004T09A026. New York: American Society of Mechanical Engineers.

Ziegler, J. L., Deiterding, R., Shepherd, J. E., and Pullin, D. I. (2011). An adaptive high-order hybrid scheme for compressive, viscous flows with detailed chemistry. *Journal of Computational Physics*, 230(20):7598–7630.

14

Thermal Management and Heat Transfer Enhancement Using Porous Materials

Cong Qi, Kuo Huang, Jiaan Liu, Guohua Wang, and Yuying Yan

CONTENTS

14.1 Background 287
14.2 Porous Materials 288
    14.2.1 Concept of Porous Materials 288
    14.2.2 Preparation Methods 289
    14.2.3 Basic Characteristics 290
    14.2.4 Reconstruction Methods 291
        14.2.4.1 Experimental Reconstruction 291
        14.2.4.2 Numerical Reconstruction 291
14.3 Numerical Simulation Research Progress 292
    14.3.1 Two-Dimensional Structure 292
    14.3.2 Three-Dimensional Structure 293
14.4 Experimental Research Progress 296
    14.4.1 Heat Transfer Enhancement in Heat Exchanger System 296
    14.4.2 Heat Transfer Enhancement in Solar Power System 300
    14.4.3 Thermal Management in Battery Cooling System 301
14.5 Conclusions 304
Acknowledgments 305
References 305

14.1 Background

With the development of science and technology, in particular the rapid development of various power electronics and the applications in automotive, aerospace, numerous mobile devices, and data centers, there are increasing demands for heat exchangers to transfer high heat flux with increasingly smaller surface areas. In the field of industrial applications, the investment of heat-exchange equipment normally accounts for 30%–45% of the total. In addition to power electronics, automotive, and aerospace, heat-exchanges are traditionally widely applied to the power, petroleum, metallurgy, materials, chemical, food, and pharmaceutical industries. The effective utilizations of energy are largely dependent upon the effectiveness of heat exchanges. Shell-and-tube heat exchangers are the most widely used heat exchangers, but more than 80% of shell-and-tube heat exchangers still use smooth tube structures. Due to low heat-exchange efficiency, smooth tubes or conventional heat transfer surfaces often do not meet the requirement of increasing higher intensity and higher heat transfer efficiency of modern equipment. This has become more obvious for the cases such as cooling of high-temperature superconductors, thermal management of high-power electronic components or power batteries, and thermal control of spacecraft. This is due to the high-compact performance indexes (volume and weight) of heat exchangers;

common smooth surfaces cannot meet the requirement of high heat flux with limited heat transfer areas, and such smooth heat transfer surfaces must be replaced by other enhanced heat transfer surfaces.

The other important applications of porous structures to heat transfer are the recent development of heat pipes, including vapor chambers (Li et al. 2016a; Wang et al. 2014; Zhou et al. 2017). The innovative designs of new wicks or porous structures in heat pipes help improve the pumping effect by increasing capillary pressure and improving the heat transfer.

Porous materials, especially the metal foam materials, as excellent enhanced heat transfer surfaces, are widely used in the thermal management and heat transfer enhancement fields to improve the heat transfer performance of smooth surfaces.

14.2 Porous Materials

14.2.1 Concept of Porous Materials

Porous material was developed in the 20th century. It is a kind of material with a network structure formed by interconnected or closed pores (Figures 14.1 and 14.2 (Wang et al. 2018)). There are two kinds of structures: two-dimensional and three-dimensional. For the two-dimensional structure formed by the aggregation of polygonal pores on a flat surface, this kind of porous material can be called a honeycomb material (Figure 14.3). For the three-dimensional structure consisting of a large number of polyhedral shaped pores, the porous material can be called a foam material (Figure 14.4).

According to the pore size, porous materials can be divided into microporous materials (pore size less than 2 nm), mesoporous materials (pore size 2–50 nm), and macroporous materials (pore size greater than 50 nm).

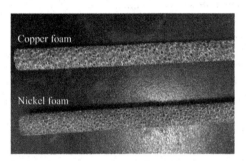

FIGURE 14.1 Metal foam materials. (From Wang, G. et al., *Therm. Sci.*, 22, S497–S505, 2018.)

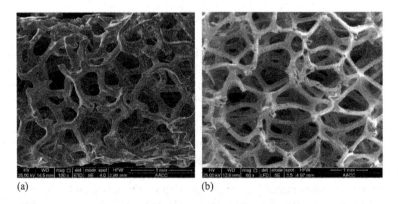

FIGURE 14.2 SEM photograph of metal foam materials: (a) copper; and (b) nickel. (From Wang, G. et al., *Therm. Sci.*, 22, S497–S505, 2018.)

FIGURE 14.3 Honeycomb material.

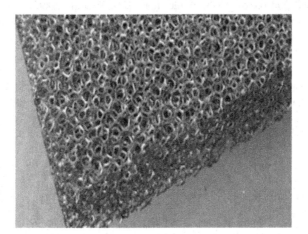

FIGURE 14.4 Foam material.

Porous materials generally have the advantages of low relative density, high specific strength, high specific surface area, light weight, sound insulation, heat insulation, and good permeability. This chapter takes foam metal material as an example to introduce its application in thermal management and heat transfer enhancement (Banhart 2001).

14.2.2 Preparation Methods

The preparation methods of porous metal foam mainly include four methods: (1) sintering method, (2) deposition method, (3) casting method, and (4) foaming method. The details of these preparation procedures are as follows:

1. Sintering method

    At high temperature, metal powders contact and interact with each other under the action of surface tension and capillary force. After cooling, the metal powders are connected and become metal foam. This method can greatly increase the porosity by adding water-soluble fillers, and it is suitable for the preparation of porous metal foam.

2. Deposition method

Chemical or physical methods are used to deposit the metallic material of the metal foam on the easily decomposed organic matter. This method is called the deposition method. The deposition method includes an electrodeposition method and a vapor deposition method. The electrodeposition method contains several steps: First, a polymer foam with open-cell structure is adopted, a strong oxidizer is used to corrode the foam under acidic conditions ($H_2Cr_2O_7/H_2SO_4/H_3PO_4$) to make its surface easy to be moistened by water and produce microscratch, and this step is called coarsening. Second, after coarsening, a Pd^{2+} in $PdCl_2$ solution is used to catalyze the surface, and this step is called activation. Third, a polymer foam is put into a plating bath, and electroless plating is carried out to obtain a metal layer uniformly attached in the conductive surface of the polymer. Afterward, the polymer foam is electroplated to obtain the desired metal/polymer composite foams. Finally, the metal foam is obtained after removing the polymer by heat treatment.

3. Casting method

The casting method can be divided into investment casting using polymer foam and infiltration casting around granular space holders. Investment casting: a polymer foam with open-cell structure is put into a container with a certain geometry, and the liquid refractory material is added around it. After the refractory material hardens, the polymer foam is heated to gasify, then the mold has the shape of the original polymer foam (Conde et al. 2006; Liu et al. 2017). Liquid metal is poured into the mold, and the refractory material is separated from metal after cooling. Lastly, the metal foam with the same shape of the original polymer foam can be obtained.

4. Foaming method

A foaming agent is added to the molten metal, and bubbles are generated by the separated gas. After cooling, metal foam with closed-cell structure can be obtained. This method is mainly used for the metal foam with low-melting point (such as Al, Sn) preparation. In order to overcome the inhomogeneity, high-speed stirring can be carried out to make the foaming agent dispersed throughout the melt, and solid particles with high-melting point can be added to increase the viscosity to avoid bubble escape.

14.2.3 Basic Characteristics

Basic characteristics of metal foam play an important role in the fields of thermal management and heat transfer enhancement. Basic characteristics of metal foam mainly contain porosity, equivalent aperture, pore density, relative density, and specific surface area. The details of these basic characteristics are as follows:

1. Porosity P

Porosity P is the proportion of the internal pores of a material to the total volume, and the corresponding equation is as follows:

$$P = \frac{V_0 - V}{V_0} \times 100\% \tag{14.1}$$

where P is the proportion, V_0 is the volume of a material in its natural state (apparent volume), and V is the absolute compact volume of a material.

2. Equivalent aperture D

Equivalent aperture D is the size of pore, which belongs to the description of porous materials from the "mesoscopic" level, and the corresponding equation is as follows:

$$D = \frac{4A}{P'} \tag{14.2}$$

where D is the equivalent aperture of pore, A is the cross-sectional area of pore, and P' is the wetted perimeter of pore.

3. Pore density pixels per inch (PPI)

Pore density PPI is the average number of pores per linear inch.

4. Relative density ρ

Relative density ρ is the relative value of density compared with base metal density, which belongs to the description of porous materials from the "macroscopic" level.

5. Specific surface area S

Specific surface area S is the total area of porous materials per unit volume.

14.2.4 Reconstruction Methods

The methods of porous media reconstruction can be divided into two categories: physical experiment and numerical reconstruction. Physical experiment method: the plane image of porous media is obtained by high-resolution instruments, such as the optical microscope and scanning electron microscope, and then the three-dimensional porous media model is obtained by image processing technology. Numerical reconstruction method: the porous media model is reconstructed by a mathematical method based on two-dimensional slice image analysis. The corresponding methods are as follows.

14.2.4.1 Experimental Reconstruction

Experimental reconstruction methods mainly include the CT scanning method and sequence slice grouping method. The details of these methods are as follows:

1. CT scanning method

The main operation steps of reconstruction of porous media by CT scanning are as follows: (1) CT scanning is carried out in samples to obtain projection data; (2) image processing technology is used to convert the scanned projection data into grayscale images; and (3) image binary segmentation technology is applied to obtain the relevant pore structure morphology of porous media.

2. Sequence slice grouping method

The basic operation steps of sequence section grouping method are as follows: (1) a layer of rock sample thin section parallel to the rock sample surface is cut off; (2) the rock sample surface is polished and a smooth rock sample surface is obtained; (3) scanning is carried out on the rock sample surface by scanning electron microscope, and other high-resolution instruments are used to obtain a two-dimensional slice image; and (4) the above operation steps are repeated, section images are combined, and finally, a three-dimensional porous media model is obtained.

14.2.4.2 Numerical Reconstruction

Numerical reconstruction methods mainly include a simulated annealing method, sequential indicator method, Gaussian field method, multiple-point statistics, and a process simulation method. The details of these methods are as follows:

1. Simulated annealing method

The simulated annealing method was developed by Hazlett (1997). This algorithm randomly generates a porous material with a certain porosity, and the system is constantly optimized until the random porous material meet the requirements of the research background of porous material model by adjusting the relative positions of pore and rock skeleton.

2. Sequential indicator method

The sequential indicator method takes the porosity and variation function of rock slice images as constraints and combines the sequential indicator simulation algorithm in statistics to reconstruct the porous materials.

3. Gaussian field method

   The Gaussian field method was first developed by Joshi (1974). Based on the statistical information of the two-dimensional core section, this method first generates a Gaussian field, which is composed of independent Gaussian variables, and then variables are correlated and further converted into digital models through transformation.

4. Multiple-point statistics

   Structural feature information is extracted from images of porous materials, and a feature structure database is formed. Then virtual and random porous media is generated from the extracted characteristic information.

5. Process simulation method

   Porous media is generated according to specific rules, and the morphology characteristics of porous materials can be changed by controlling the corresponding parameters.

14.3 Numerical Simulation Research Progress

Based on the reconstruction methods of porous materials, researchers have numerically simulated the thermal management and heat transfer enhancement of fluids in porous materials.

14.3.1 Two-Dimensional Structure

Sheikholeslami and Rokni (2018a) applied a control-volume-based finite-element method (CVFEM) to numerically simulate the heat transfer performance of Fe_3O_4-Ethylene glycol nanofluids in a porous-lid-driven cavity (Figure 14.5), and the effects of thermal radiation and external electric field were discussed. Sheikholeslami (2017a), Sheikholeslami and Shehzad (2017) and Sheikholeslami and Zeeshan (2017) also used CVFEM to numerically study the flow and heat transfer of nanofluids in a porous enclosure (Figure 14.6), a porous cavity with a hot cylinder (Figure 14.7), and a quarter round porous cavity (Figure 14.8).

The influence of Hartmann on heat transfer performance was discussed. Results indicated that the increasing Ha and decreasing Da, Ra are all beneficial to heat transfer enhancement.

Sheikholeslami and Rokni (2018b) numerically investigated the convection heat transfer of CuO-water nanofluids in a porous circular cavity (Figure 14.9). Results indicated that porosity

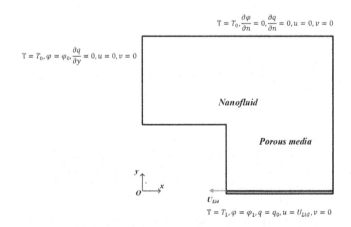

FIGURE 14.5 Porous-lid-driven cavity. (From Sheikholeslami, M. and Rokni, H.B., *Int. J. Heat Mass Transf.*, 118, 823–831, 2018a.)

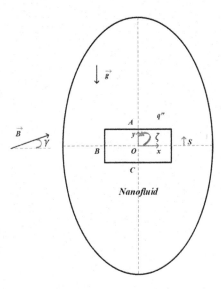

FIGURE 14.6 Porous enclosure. (From Sheikholeslami, M. and Zeeshan, A. *Comput. Methods Appl. Mech. Eng.*, 320, 68–81, 2017.)

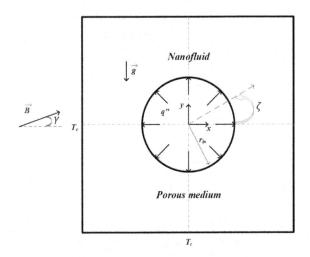

FIGURE 14.7 Porous cavity. (From Sheikholeslami, M. and Shehzad, S.A., *Int. J. Heat Mass Transf.*, 106, 1261–1269, 2017.)

has an opposite relationship with the temperature gradient. Sheikholeslami (2017b) numerically investigated the convection heat transfer of CuO-water nanofluids in a porous semicircular cavity (Figure 14.10). It was found that the temperature gradient increases with the increasing Darcy number and buoyancy forces.

14.3.2 Three-Dimensional Structure

Sheikholeslami et al. (2018a) also used CVFEM to numerically study the flow and heat transfer of nanofluids in a three-dimensional porous cavity (Figure 14.11). The influence of Hartmann on heat transfer was investigated. Sheikholeslami et al. (2018b) numerically researched the forced convection

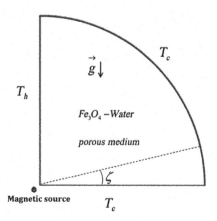

FIGURE 14.8 Porous curved cavity. (From Sheikholeslami, M. *Phys. Lett. A*, 381, 494–503, 2017a.)

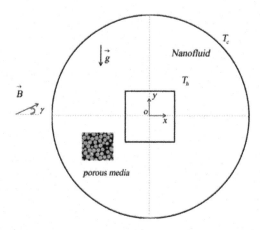

FIGURE 14.9 Porous circular cavity. (From Sheikholeslami, M. and Rokni, H.B. *Phys. Fluids*, 30, 012003, 2018b.)

heat transfer of Al_2O_3-water nanofluids in a porous circular cavity (Figure 14.12). Results showed that decreasing the Hartmann number and increasing the Darcy number are all advantageous to forced convection heat transfer.

In addition to the above studies on two- and three-dimensional porous cavities, other researchers have paid more attention to the heat transfer in porous media. Nazari and Toghraie (2017) numerically simulated the flow and heat transfer of CuO-water nanofluids in a sinusoidal channel with a porous medium. It was found that porous regions improve the heat transfer performance. Sheikholeslami et al. (2018c) numerically simulated the heat transfer of nanofluids in a porous energy storage system. Results presented that Lorentz forces can improve the discharging rate. Sheikholeslami (2018) numerically investigated the flow and heat transfer performance of $CuO-H_2O$ nanofluids flowing in a porous channel under a magnetic field. Results showed that the increasing Lorentz force reduces the velocity, and the increasing Hartmann number improves the heat transfer. In addition, many numerical researches have been carried out on thermal management and heat transfer enhancement using porous materials (Hatami et al. 2017; Sheikholeslami 2017c, d, e, f; Sheikholeslami and Rokni 2017; Sheikholeslami and Sadoughi 2017).

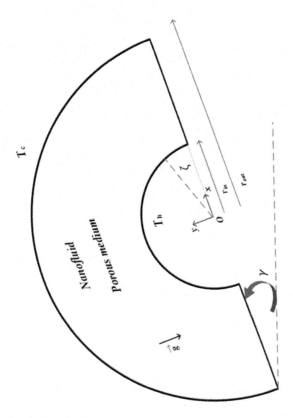

FIGURE 14.10 Porous semicircular cavity. (From Sheikholeslami, M. *Eng. Computation*, 34, 1939–1955, 2017b.)

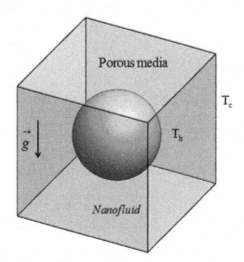

FIGURE 14.11 Three-dimensional porous cavity. (From Sheikholeslami, M. et al., *Int. J. Heat Mass Transf.*, 125, 375–386, 2018a.)

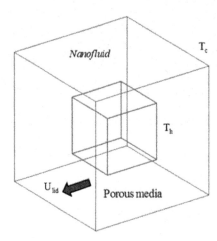

FIGURE 14.12 Porous cubic cavity. (From Sheikholeslami, M. et al., *Comput. Methods Appl. Mech. Eng.*, 338, 491–505, 2018b.)

14.4 Experimental Research Progress

14.4.1 Heat Transfer Enhancement in Heat Exchanger System

In addition to the numerical simulation researches, many experimental researches were carried out on the heat transfer enhancement of heat exchanger system with porous materials.

Wang et al. (2008) experimentally studied the flow and heat transfer characteristics of water in two kinds of metal foam tubes (Figures 14.1 and 14.13). It was found that copper foam tube shows higher heat transfer performance than nickel foam, and the comprehensive index increases with the Reynolds number.

Wan et al. (2018) experimentally investigated the thermohydraulic performance of TiO$_2$-water nanofluids flowing through a corrugated tube filled with copper foam (Figure 14.14). The effects of nanoparticle

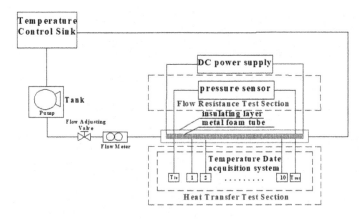

FIGURE 14.13 Experimental system. (From Wang, G. et al., *Therm. Sci.*, 22, S497–S505, 2018.)

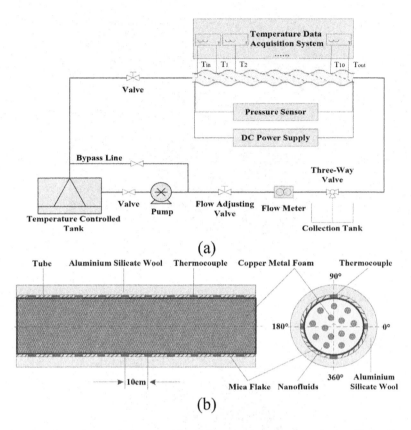

FIGURE 14.14 Experimental system: (a) schematic diagram; and (b) corrugated tube filled with copper foam. (From Wan, Y. et al., *Chin. J. Chem. Eng.*, 26, 2431–2440, 2018.)

mass fraction on the flow and heat transfer performance were studied. Results indicated that there is a critical Reynolds number ($Re = 2,400$) for the maximum comprehensive coefficient of performance, especially when $8,000 < Re < 12,000$, heat transfer performance can be improved dramatically with the increasing Reynolds number.

Sheikhnejad et al. (2017) experimentally investigated the flow and heat transfer characteristics of ferrofluid flowing through a horizontal tube partially filled with porous media (Figure 14.15), and the effects of a parallel magnet on heat transfer were discussed. Results indicated that porous media and magnetic fields are all beneficial to heat transfer enhancement.

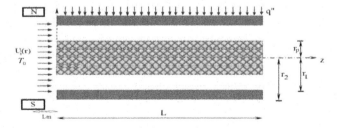

FIGURE 14.15 Horizontal tube partially filled porous media. (From Sheikhnejad, Y. et al., *J. Magn. Magn. Mater.*, 424, 16–25, 2017.)

FIGURE 14.16 Porous metal foam tube without magnetic field. (From Amani, M. et al., *Transport Porous Med.*, 116, 959–974, 2017a.)

FIGURE 14.17 Porous metal foam tube with magnetic field. (From Amani, M. et al., *Exp. Therm. Fluid Sci.*, 82, 439–449, 2017b.)

Amani et al. (2017a, b) experimentally studied the laminar forced convection heat transfer of Fe_3O_4-water nanofluids in a porous metal foam tube without and with magnetic field (Figures 14.16 and 14.17), respectively. The effects of nanoparticle concentration and Reynolds number on pressure drop and convection heat transfer were researched. Results showed that the increasing nanoparticle concentration and Reynolds number can cause an augment in Nusselt number.

Yan et al. (2015) carried out an experiment study on heat transfer enhancement in a liquid piston compressor/expander filled with porous media (Figure 14.18). Effects of five types of porous inserts on heat transfer performance were studied. Results presented that porous media can increase the efficiency/power density trade-off in a liquid piston compressor/expander.

Nazari et al. (2015) experimentally researched the convection heat transfer of Al_2O_3-water nanofluids flowing through a pipe filled with metal foam (Figure 14.19). The results were compared with water flowing through a same tube without metal foam. It was found that a significant improvement is shown in heat transfer using metal foam.

Thermal Management and Heat Transfer Enhancement Using Porous Materials

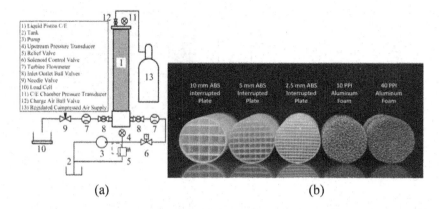

FIGURE 14.18 Experimental setup: (a) schematic diagram; and (b) five types of porous inserts. (From Yan, B. et al., *Appl. Energy*, 154, 40–50, 2015.)

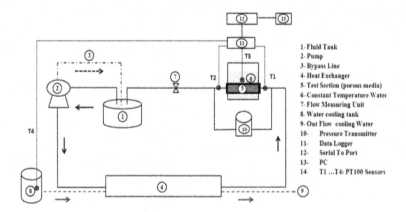

FIGURE 14.19 Experimental system. (From Nazari, M. et al., *Int. J. Therm. Sci.*, 88, 33–39, 2015.)

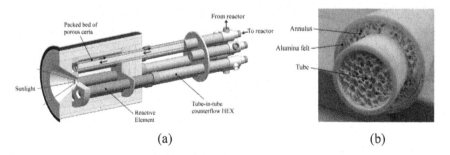

FIGURE 14.20 Experimental system: (a) heat exchanger; and (b) porous alumina. (From Banerjee, A. et al., *Appl. Therm. Eng.*, 75, 889–895, 2015.)

Banerjee et al. (2015) experimentally investigated the prototype counter-flow tube-in-tube heat exchanger (Figure 14.20). Results showed that heat transfer performance of heat exchanger with alumina reticulated porous ceramic is 9.5 times higher than that without alumina reticulated porous ceramic.

Mancin et al. (2013) experimentally researched the forced convection heat transfer of air flowing through metal foams (Figure 14.21). Effects of metal foam kind, pores per linear inch, porosity, and height on heat transfer performance were discussed. Results presented that copper foams show higher

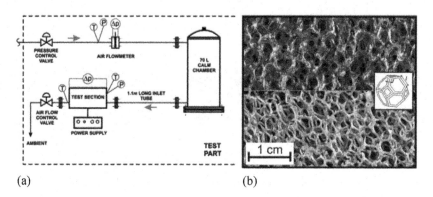

FIGURE 14.21 Experimental system: (a) schematic diagram; and (b) aluminum and copper foam. (From Mancin, S. et al., *Int. J. Heat Mass Transf.*, 62, 112–123, 2013.)

heat transfer performance than aluminum foams, and foam finned surface efficiency is the most important factor which affects the optimization of these enhanced surfaces.

From the above research, it can be obtained that porous materials are suitable to be used in heat exchanger system, and they play a crucial role in heat transfer enhancement.

14.4.2 Heat Transfer Enhancement in Solar Power System

Many experimental researches are also carried out on heat transfer enhancement in solar power system with porous materials.

Jouybari et al. (2017) experimentally studied the heat transfer performance of SiO_2-water nanofluids flowing through a solar collector with copper porous foam (Figure 14.22). Effects of nanoparticle concentration and porous media on heat transfer were discussed. Results showed that using the porous media and nanofluids can increase the thermal efficiency by 6%–8%.

Jamal-Abad et al. (2017a, b) experimentally investigated the efficiencies of a solar collector system with full and partially copper foam, respectively (Figures 14.23 and 14.24). The absorber is filled with copper foam to improve the heat transfer. Results presented that the overall loss coefficient can be decreased by 45% when the absorber is filled with copper foam; in addition, copper foam can also improve the collector efficiency.

Saedodin et al. (2017) experimentally studied the heat transfer performance of a flat-plate solar collector filled with porous copper foam (Figure 14.25). Results showed that the porous media can increase the Nusselt number and the maximum thermal efficiency by 82% and 18.5%, respectively.

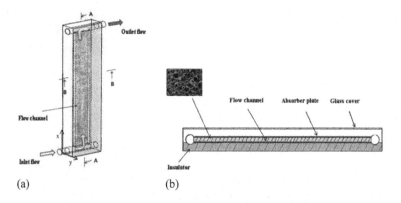

FIGURE 14.22 Experimental system: (a) schematic diagram; and (b) porous channel flat plate solar collector. (From Jouybari, H.J. et al., *Renew. Energ.*, 14, 1407–1418, 2017.)

Thermal Management and Heat Transfer Enhancement Using Porous Materials 301

FIGURE 14.23 Solar collector system: (a) schematic diagram; and (b) copper foam. (From Jamal-Abad, M.T. et al., *Renew. Energ.*, 107, 156–163, 2017a.)

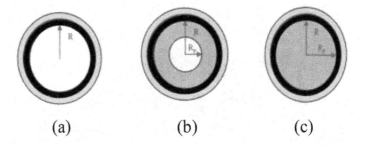

FIGURE 14.24 Three different configuration of absorber: (a) free porous media; (b) partially porous media; and (c) full porous media. (From Jamal-Abad, M.T. et al., *Int. J. Eng. Transactions B, Applications*, 30, 281–287, 2017b.)

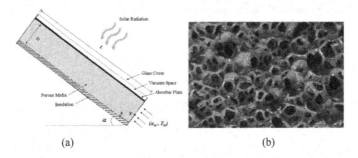

FIGURE 14.25 Solar collector system: (a) schematic diagram; and (b) copper foam. (From Saedodin, S. et al., *Energy Convers. Manag.*, 153, 278–287, 2017.)

Nima and Ali (2017) experimentally investigated the effects of metal foam on thermal performance of solar collector with different water flow rate (Figure 14.26). Results showed that using metal foam block in the riser of the solar collector can effectively improve the heat transfer performance by 100%.

From the above studies, it can be found that porous materials are widely used in the solar power system and are advantageous to the heat transfer enhancement.

14.4.3 Thermal Management in Battery Cooling System

Li et al. (2014) experimentally investigated the thermal management in lithium ion batteries using phase-change materials and porous metal foam (Figure 14.27). It was compared with pure-phase change materials and air cooling; results showed that air cooling cannot take away the all heat emission from lithium ion batteries and cause a high surface temperature. Phase-change materials can absorb most of heat emission and reduce the surface temperature, but the surface temperature distribution is not uniform.

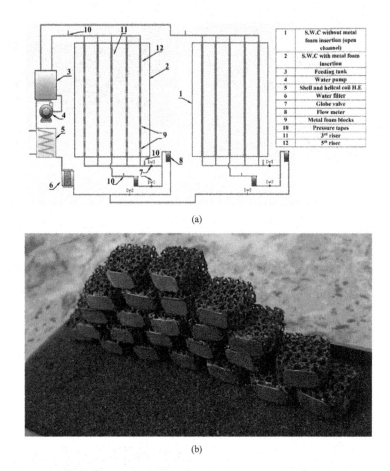

FIGURE 14.26 Solar collector system: (a) schematic diagram; and (b) copper foam. (From Nima, M.A. and Ali, A.M., *Arab. J. Sci. Eng.*, 42(11), 4863–4884, 2017.)

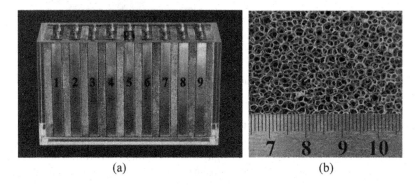

FIGURE 14.27 Battery cooling system: (a) battery pack design using copper foam-paraffin composite; and (b) copper foam. (From Li, W.Q. et al., *J. Power Sources*, 255, 9–15, 2014.)

Foam-paraffin composite not only can further reduce the surface temperature but also can improve the uniformity of the temperature distribution. The increasing pore density can increase the battery surface temperature.

Heat pipe are widely applied into battery cooling, and many researchers have paid more attention to the studies of heat pipe with porous materials.

Li et al. (2016b) experimentally studied the thermal performance of heat pipes with copper powder sintered-grooved composite structure (Figure 14.28). It was compared with pure grooved, sintered, and grooved with half sintered length, respectively. Results indicated that the porous wick structure (sintered and sintered-grooved) can remove the influence of centrifugal accelerations to a large extent. It was also found that porous sintered-grooved composite structure shows the best thermal performance.

Li et al. (2016c) experimentally researched the thermal performance of vapor chambers with two wick structures (copper foam and copper powder) (Figure 14.29). Effects of filling rates and particle size were discussed. Results showed that copper-foam-based vapor chambers show good temperature uniformity

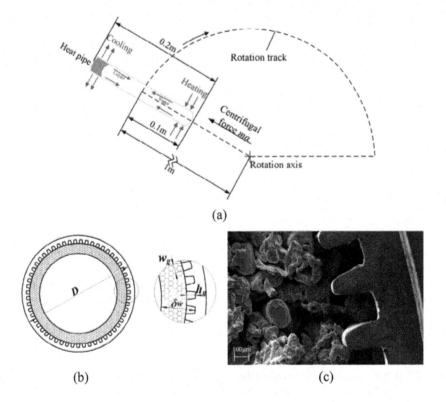

FIGURE 14.28 Heat pipes: (a) schematic of centrifugal experiments; (b) simplified cross section of SG: and (c) SEM of sintered-grooved composite. (From Li, Y. et al., *Appl. Therm. Eng.*, 96, 352–363, 2016b.)

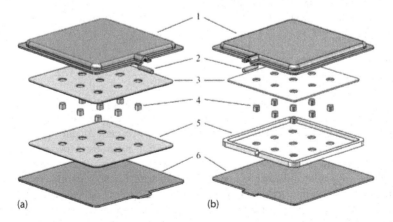

FIGURE 14.29 Structure of (a) CFVC; and (b) CPVC (1: top plates; 2: charging tubes; 3: condenser wicks; 4: copper solid columns; 5: evaporator wicks; 6: bottom plates). (From Li, Y. et al., *Exp. Therm. Fluid. Sci.*, 77, 132–143, 2016c.)

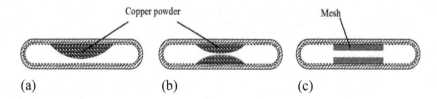

FIGURE 14.30 Schematic of three composite wick structures: (a) SSGW; (b) BSGW; and (c) MGW. (From Li, Y. et al., *Appl. Therm. Eng.*, 102, 487–499, 2016d.)

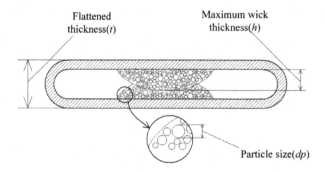

FIGURE 14.31 Heat pipe with porous sintered-wick structure. (From Li, Y. et al., *Appl. Therm. Eng.*, 86, 106–118, 2015.)

and copper-powder-sintered-based vapor chambers show low thermal resistance. A filling rate of 90% shows the shortest response time, and the particle size mainly affects the thermal resistance. Copper foam and copper powder are all advantageous to the improvement of thermal performance.

Li et al. (2016d) experimentally studied the thermal performance of heat pipe with various structures (single arch-shaped copper powder, bilateral arch-shaped copper powder, mesh sintered) (Figure 14.30). Results indicated that the heat pipe with single arch-shaped copper powder shows the smallest evaporation thermal resistance, and the heat pipe with mesh sintered shows the smallest condensation thermal resistance.

Li et al. (2015) experimentally investigated the thermal performance of a heat pipe with a porous arch-shaped sintered wick (Figure 14.31). The influence of different processing parameters on the thermal performance was discussed. Results presented that the heat transport capability decreases and thermal resistance increases with the decreasing flattened thickness.

From the above research, it can be found that porous materials can regulate and control the thermal performance of a battery cooling system and are beneficial to the thermal performance.

14.5 Conclusions

This chapter investigated the background of using porous materials; introduced the concepts of porous materials; four kinds of preparation methods (sintering method, deposition method, casting method, and foaming method); five main basic characteristics (porosity, equivalent aperture, pore density, relative density, and specific surface area); two reconstruction methods: experimental reconstruction (CT scanning method and sequence slice grouping method) and numerical reconstruction (simulated annealing method, sequential indicator method, Gaussian field method, multiple-point statistics, and process simulation method); numerical simulation research progress; and experimental research progress (heat transfer enhancement in heat exchanger system and solar power system, and thermal management in a battery cooling system).

Some suggestions on the future work are as follows:

1. New numerical models for flow and heat transfer of porous materials need to be developed in the future.
2. Flow and heat transfer performances of fluid flowing through porous materials under strong turbulence need to be investigated in the future.
3. Effects of geometries of porous materials on the flow and heat transfer characteristics of fluid need to be studied further in the future.
4. In order to reduce the flow resistance, the porous structure and heat transfer medium should be optimized.
5. New comprehensive evaluations for thermohydraulic performance of porous materials should be investigated from two points of thermal and energy efficiencies.

ACKNOWLEDGMENTS

This work is financially supported by the EU ThermaSMART project, H2020-MSCA-RISE (778104)-Smart thermal management of high power microprocessors using phase-change (ThermaSMART) and the Ningbo Science and Technology Bureau Technology Innovation Team (Grant No. 2016B10010).

REFERENCES

Amani M., M. Ameri, A. Kasaeian. Investigating the convection heat transfer of Fe3O4 nanofluid in a porous metal foam tube under constant magnetic field. *Experimental Thermal and Fluid Science*, 2017a, 82: 439–449.

Amani M., M. Ameri, A. Kasaeian. The experimental study of convection heat transfer characteristics and pressure drop of magnetite nanofluid in a porous metal foam tube. *Transport in Porous Media*, 2017b, 116(2): 959–974.

Banerjee A., R.B. Chandran, J.H. Davidson. Experimental investigation of a reticulated porous alumina heat exchanger for high temperature gas heat recovery. *Applied Thermal Engineering*, 2015, 75: 889–895.

Banhart J. Manufacture, characterisation and application of cellular metals and metal foams. *Progress in Materials Science*, 2001, 46(6): 559–632.

Conde Y., J.F. Despois, R. Goodall, A. Marmottant, L. Salvo, C. San Marchi, A. Mortensen. Replication processing of highly porous materials. *Advanced Engineering Materials*, 2006, 8(9): 795–803.

Hatami M., J. Zhou, J. Geng, D. Song, D. Jing. Optimization of a lid-driven T-shaped porous cavity to improve the nanofluids mixed convection heat transfer. *Journal of Molecular Liquids*, 2017, 231: 620–631.

Hazlett R.D. Statistical characterization and stochastic modeling of pore networks in relation to fluid flow. *Mathematical Geology*, 1997, 29(6): 801–822.

Jamal-Abad M.T., S. Saedodin, M. Aminy. Experimental investigation on a solar parabolic trough collector for absorber tube filled with porous media. *Renewable Energy*, 2017a, 107: 156–163.

Jamal-Abad M.T., S. Saedodin, M. Aminy. Experimental investigation on the effect of partially metal foam inside the absorber of parabolic trough solar collector. *International Journal of Engineering-Transactions B: Applications*, 2017b, 30(2): 281–287.

Joshi M. A class of stochastic models for porous media. PhD thesis, University of Kansas, 1974.

Jouybari H.J., S. Saedodin, A. Zamzamian, M.E. Nimvari, S. Wongwises. Effects of porous material and nanoparticles on the thermal performance of a flat plate solar collector: An experimental study. *Renewable Energy*, 2017, 114: 1407–1418.

Li W.Q., Z.G. Qu, Y.L. He, Y.B. Tao. Experimental study of a passive thermal management system for high-powered lithium ion batteries using porous metal foam saturated with phase change materials. *Journal of Power Sources*, 2014, 255: 9–15.

Li Y., J. He, H. He, Y. Yan, Z. Zeng, B. Li. Investigation of ultra-thin flattened heat pipes with sintered wick structure. *Applied Thermal Engineering*, 2015, 86: 106–118.

Li Y., S. Chen, B. He, Y. Yan, B. Li. Effects of vacuuming process parameters on the thermal performance of composite heat pipes. *Applied Thermal Engineering*, 2016a, 99: 32–41.

Li Y., Z. Li, C. Chen, Y. Yan, Z. Zeng, B. Li. Thermal responses of heat pipes with different wick structures under variable centrifugal accelerations. *Applied Thermal Engineering*, 2016b, 96: 352–363.

Li Y., Z. Li, W. Zhou, Z. Zeng, Y. Yan, B. Li. Experimental investigation of vapor chambers with different wick structures at various parameters. *Experimental Thermal and Fluid Science*, 2016c, 77: 132–143.

Li Y., W. Zhou, J. He, Y. Yan, B. Li, Z. Zeng. Thermal performance of ultra-thin flattened heat pipes with composite wick structure. *Applied Thermal Engineering*, 2016d, 102: 487–499.

Liu J., S. Shi, Z. Zheng, K. Huang, Y. Yan. Characterization and compressive properties of Ni/Mg hybrid foams. *Materials Science and Engineering: A*, 2017, 708: 329–335.

Mancin S., C. Zilio, A. Diani, L. Rossetto. Air forced convection through metal foams: Experimental results and modeling. *International Journal of Heat and Mass Transfer*, 2013, 62: 112–123.

Nazari S., D. Toghraie. Numerical simulation of heat transfer and fluid flow of Water-CuO Nanofluid in a sinusoidal channel with a porous medium. *Physica E: Low-dimensional Systems and Nanostructures*, 2017, 87: 134–140.

Nazari M., M. Ashouri, M.H. Kayhani, A. Tamayol. Experimental study of convective heat transfer of a nanofluid through a pipe filled with metal foam. *International Journal of Thermal Sciences*, 2015, 88: 33–39.

Nima M.A., A.M. Ali. Effect of metal foam insertion on thermal performance of flat-plate water solar collector under Iraqi climate conditions. *Arabian Journal for Science and Engineering*, 2017, 42(11): 4863–4884.

Saedodin S., S.A.H. Zamzamian, M.E. Nimvari, S. Wongwises, H.J. Jouybari. Performance evaluation of a flat-plate solar collector filled with porous metal foam: Experimental and numerical analysis. *Energy Conversion and Management*, 2017, 153: 278–287.

Sheikhnejad Y., R. Hosseini, M.S. Avval. Experimental study on heat transfer enhancement of laminar ferrofluid flow in horizontal tube partially filled porous media under fixed parallel magnet bars. *Journal of Magnetism and Magnetic Materials*, 2017, 424: 16–25.

Sheikholeslami M. Numerical simulation of magnetic nanofluid natural convection in porous media. *Physics Letters A*, 2017a, 381(5): 494–503.

Sheikholeslami M. Numerical investigation of MHD nanofluid free convective heat transfer in a porous tilted enclosure. *Engineering Computations*, 2017b, 34(6): 1939–1955.

Sheikholeslami M. Influence of magnetic field on nanofluid free convection in an open porous cavity by means of Lattice Boltzmann method. *Journal of Molecular Liquids*, 2017c, 234: 364–374.

Sheikholeslami M. CuO-water nanofluid free convection in a porous cavity considering Darcy law. *The European Physical Journal Plus*, 2017d, 132(1): 55.

Sheikholeslami M. Influence of Lorentz forces on nanofluid flow in a porous cylinder considering Darcy model. *Journal of Molecular Liquids*, 2017e, 225: 903–912.

Sheikholeslami M. Magnetohydrodynamic nanofluid forced convection in a porous lid driven cubic cavity using Lattice Boltzmann method. *Journal of Molecular Liquids*, 2017f, 231: 555–565.

Sheikholeslami M. Numerical investigation for CuO-H2O nanofluid flow in a porous channel with magnetic field using mesoscopic method. *Journal of Molecular Liquids*, 2018, 249: 739–746.

Sheikholeslami M., H.B. Rokni. Nanofluid convective heat transfer intensification in a porous circular cylinder. *Chemical Engineering and Processing: Process Intensification*, 2017, 120: 93–104.

Sheikholeslami M., H.B. Rokni. Numerical simulation for impact of Coulomb force on nanofluid heat transfer in a porous enclosure in presence of thermal radiation. *International Journal of Heat and Mass Transfer*, 2018a, 118: 823–831.

Sheikholeslami M., H.B. Rokni. Magnetic nanofluid flow and convective heat transfer in a porous cavity considering Brownian motion effects. *Physics of Fluids*, 2018b, 30(1): 012003.

Sheikholeslami M., M. Sadoughi. Mesoscopic method for MHD nanofluid flow inside a porous cavity considering various shapes of nanoparticles. *International Journal of Heat and Mass Transfer*, 2017, 113: 106–114.

Sheikholeslami M., S.A. Shehzad. Magnetohydrodynamic nanofluid convection in a porous enclosure considering heat flux boundary condition. *International Journal of Heat and Mass Transfer*, 2017, 106: 1261–1269.

Sheikholeslami M., S.A. Shehzad, Z. Li. Water based nanofluid free convection heat transfer in a three dimensional porous cavity with hot sphere obstacle in existence of Lorenz forces. *International Journal of Heat and Mass Transfer*, 2018a, 125: 375–386.

Sheikholeslami M., S.A. Shehzad, F.M. Abbasi, Z. Li. Nanofluid flow and forced convection heat transfer due to Lorentz forces in a porous lid driven cubic enclosure with hot obstacle. *Computer Methods in Applied Mechanics and Engineering*, 2018b, 338: 491–505.

Sheikholeslami M., Z. Li, A. Shafee. Lorentz forces effect on NEPCM heat transfer during solidification in a porous energy storage system. *International Journal of Heat and Mass Transfer*, 2018, 127: 665–674.

Sheikholeslami M., A. Zeeshan. Analysis of flow and heat transfer in water based nanofluid due to magnetic field in a porous enclosure with constant heat flux using CVFEM. *Computer Methods in Applied Mechanics and Engineering*, 2017, 320: 68–81.

Wan Y., R. Wu, C. Qi, G. Duan, R. Yang. Experimental study on thermo-hydraulic performances of nanofluids flowing through a corrugated tube filled with copper foam in heat exchange systems. *Chinese Journal of Chemical Engineering*, 2018, 26: 2431–2440.

Wang Q., J. Hong, Y. Yan. Biomimetic capillary inspired heat pipe wicks. *Journal of Bionic Engineering*, 2014, 11(3): 469–480.

Wang G., C. Qi, Y. Pan, C. Li. Experimental study on heat transfer and flow characteristics of two kinds of porous metal foam tubes filled with water. *Thermal Science*, 2018, 22(2): S497–S505.

Yan B., J. Wieberdink, F. Shirazi, P.Y. Li, T.W. Simon, J.D. Van de Ven. Experimental study of heat transfer enhancement in a liquid piston compressor/expander using porous media inserts. *Applied Energy*, 2015, 154: 40–50.

Zhou W., P. Xie, Y. Li, Y. Yan, B. Li. Thermal performance of ultra-thin flattened heat pipes. *Applied Thermal Engineering*, 2017, 117: 773–781.

15

Metal Foam Heat Exchangers

Simone Mancin

CONTENTS

15.1 Introduction...309
15.2 Air (Gas) Heat Exchangers..312
    15.2.1 General Overview...312
    15.2.2 Foamed Compact Heat Exchangers..312
        15.2.2.1 HVAC&R Heat Exchangers 312
        15.2.2.2 Air-Cooled Condenser 317
        15.2.2.3 Vehicles Applications..318
        15.2.2.4 Waste Energy Recovery 319
        15.2.2.5 Fuel Cell Application 321
    15.2.3 Foamed-Wrapped Tubes Heat Exchangers...321
    15.2.4 On the Foam-Finned Surface Performance 324
    15.2.5 Key Issues and Challenges in Metal Foam Heat Exchangers.............................327
        15.2.5.1 Thermal Contact Resistance..327
        15.2.5.2 Fouling Issue 328
15.3 Liquid and Two-Phase Heat Exchangers 331
15.4 Conclusion 333
Nomenclature..334
References..334

15.1 Introduction

The use of open-cell metal foams as enhanced surfaces on advanced heat exchangers has attracted the attention of both the scientific and industrial community for more than three decades so far. There are no doubts about the interesting properties exhibited by the open cell metal foams for heat transfer applications in many different fields, including harsh environments (Ashby et al. 2000; Liu et al. 2001a; Tuchinskiy 2005; Han et al. 2012b), among those:

- Low weight due to their high porosity (i.e., commonly greater than 90%).
- High specific surface area (roughly from 1000 to 3000 m^2 m^{-3} but up to 10,000 m^2 m^{-3} for compressed metal foams [Zhao 2012]).
- High gas permeability combined with a great tortuosity, thus promoting flow mixing and turbulence, and thus excellent heat transfer performance (Mancin et al. 2013).
- High material thermal conductivity.
- Resistance to thermal shock, wear, high temperature, humidity, and thermal cycling.

309

- High strength, stiffness, and toughness, suitable for high-pressure conditions.
- Easy control over material morphology (pore size and distribution).
- Machinability and weldability, allowing for the formation of complex part, limiting the contact resistance.
- Excellent noise attenuation.

Despite the listed positive properties exhibited by the metal foams, they also present a remarkable flow resistance and a generally high price. However, these interesting properties make metal foams a versatile engineering material for multi-functional components (Liu et al. 2001b). In fact, several applications for metal foams have been proposed, such as structural elements for aerospace, automotive and building systems, thermal management systems, filters and catalyst carrier, and others (Ismagilov et al. 2001; Wilkinson and Paserin 2004; Kim et al. 2005; Spoerke et al. 2005; Azzi et al. 2007; Kang et al. 2008; Koltsakis et al. 2008; Losito 2008). To date, metal foams have been proposed and studied by both industries and researchers in several components: combustion chambers, flame arrestors, cryogenics, strain isolation, cladding on buildings, geothermal operations, catalytic reactors, petroleum reservoirs, compact heat sinks for power electronics, solar radiation receivers, fuel cells, and heat exchangers for air conditioning and refrigeration systems (Mancin et al. 2013).

Figure 15.1 shows a few examples of industrial metal foam heat exchanger products, while Figure 15.2 presents some examples of heat exchangers prototypes studied by different research groups.

Considering the large number of applications and fields where metal foam heat exchangers have been applied, several classifications can be considered, including:

1. Application: air conditioning and refrigeration, thermal management of electronics components (i.e., heat sinks), solar receivers, latent and hybrid thermal energy storages, fuel cells, geothermal systems, etc.
2. Fluid: air or gases, liquids (water, refrigerants, etc.), phase change materials (PCMs), etc.
3. Heat transfer mode: single- or two-phase heat transfer (either liquid–vapor or solid–liquid)

The present book presents dedicated chapters to some of the cited applications: geothermal systems, fuel cells, and thermal management; thus, those topics are not described in detail in the following paragraphs.

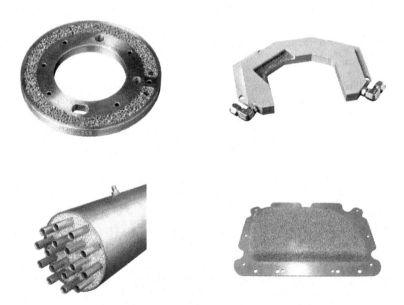

FIGURE 15.1 Examples of metal foam heat exchangers. http://ergaerospace.com/photo-gallery/.

Metal Foam Heat Exchangers

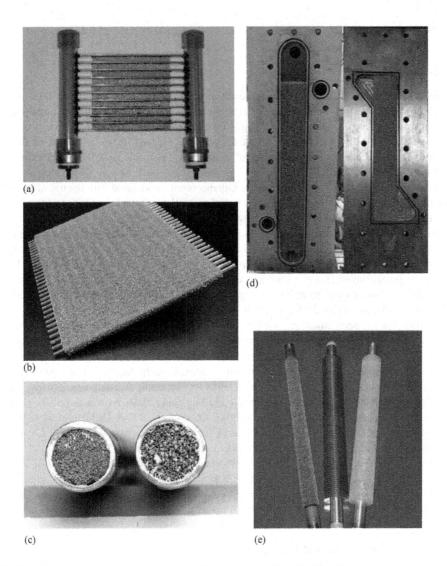

FIGURE 15.2 Examples of heat exchangers studied in the open literature: (a) air-cooled microchannel heat exchanger. (From Nawaz, K. et al., Experimental studies to evaluate the use of metal foams in highly compact air-cooling heat exchangers, *International Refrigeration and Air Conditioning Conference*, Paper 1150, 2010.) (b) FeCrAlY compact heat exchanger for high-temperature service. (From Zhao, C.Y. et al., *Thermal Transport Phenomena in Porvair Metal Foams and Sintered Beds, Final Report*, University of Cambridge, Cambridge, UK, 2001.) (c) metal foam-filled tubes using co-sintering technique. (From Lu, W. et al., *Int. J. Heat Mass Transf.*, 49, 2751–2761, 2006.) (d) compact heat exchanger. (From Kim, Y.D. and Kim, K.C., *Int. J. Heat Mass Transf.*, 130, 162–174, 2019.) and (e) foam wrapped and finned tubes. (From Chumpia, A. and Hooman, K., *Appl. Therm. Eng.*, 83, 121–130, 2015.)

Since the book refers to convection in metal foams, the latent and hybrid thermal energy storages are not here discussed. Moreover, the research on two-phase (e.g., flow boiling and condensation) heat transfer can still be considered at the pre-competitive stage. In fact, the available literature deals with the fundamental heat transfer during two-phase flow, among those: Diani et al. (2014), Pranoto and Leong (2014), Mancin et al. (2014a, b), Zhu et al. (2015a, b), Diani et al. (2015), Abadi et al. (2016a, b), Gao et al. (2018), and Li et al. (2019).

In order to congruently present the topic, the fluid is used to classify the metal foam heat exchangers; thus, the air (gases) heat exchangers are first presented, followed by the liquid and two-phase heat exchangers.

15.2 Air (Gas) Heat Exchangers

15.2.1 General Overview

At first sight, metal foams seem to be thought to act as extended surfaces in compact heat exchangers, and significant advantages can be achieved with the use of open-cell metal foam structures to manufacture highly effective, geometry-flexible, and multi-functional compact heat exchange devices. In fact, a heat exchanger is said to be compact if its surface area density exceeds 700 m^2 m^{-3}, which basically falls in the middle of the surface area density range of most of the open cell meal foam structures available on the market (Mancin et al. 2013).

Several mechanisms contribute to heat transfer enhancements associated with the use of metal foams, including interactions between the solid foam material and a through-flowing fluid, and the importance of achieving a quality metal-to-foam bond (brazing is preferred over epoxy-bonding) (Ashby and Lu 2003). Moreover, the foam structure itself plays a fundamental role in the final performance of the air foam-finned heat exchanger; in fact, the porosity and the pore density affect the shape and the thickness of the fibers and, consequently, they have a deep impact on both the heat transfer and fluid flow of the air through the porous layer. However, without the knowledge of the material and of the thickness of the foam layer, it is not possible to rationally compare different metal foam structures implemented as extended surfaces (Mancin et al. 2010a, b, 2013).

Over the past few decades, many research groups have studied the heat transfer and pressure-drop characteristics of porous media during either internal or external forced convective flow with the aim of understanding and, finally, modeling the heat transfer and fluid flow processes. These works cannot be directly referred to as applications of metal foams in heat exchangers. Most of the samples were electrically heated; thus, they might be classified as heat sinks for electronic thermal management. This is the case for instance of the works proposed by Calmidi and Mahajan (2000), Kim et al. (2001), Hwang et al. (2002), Callego and Klett (2003), Hsieh et al. (2004), Zhao et al. (2004), Dukhan et al. (2005), Giani et al. (2005), Straatmam et al. (2006), Dukhan and Chen (2007), Garriti et al. (2010), Cavallini et al. (2010), Mancin et al. (2010a, b, 2011, 2012a, b, 2013), Elayiaraja et al. (2010), Leong et al. (2011), De Schampheleire et al. (2013a), Abdi et al. (2014), Hooman (2014), Khashehchi et al. (2014, 2015, 2017), Hamadouche et al. (2018), Anuar et al. (2018a, b), and Orihuela et al. (2018). Besides, just a few have considered the application of the porous structure to real heat exchangers; in what follows, the description of the most recent attempts is reported and discussed.

15.2.2 Foamed Compact Heat Exchangers

15.2.2.1 HVAC&R Heat Exchangers

One of the first attempts to introduce and critically discuss the possible use of metal foams as extended surfaces in metal foam-finned heat exchangers was proposed by Kim et al. (2000). The authors experimentally compared the performance, in terms of modified Colburn factor j^* and friction factor f, of plate and fin heat exchangers implementing different metal foams and a reference-louvered fin. Figure 15.3 shows a schematic of the plate and fin heat exchanger studied and tested by Kim et al. (2000). Six aluminum-alloy metal foams were selected and tested to study the effects of both the porosity and of the pore density; in fact, three of those had similar porosity equal to 0.92 and different pore density, 10, 20, and 40 PPI, while the other three samples had 20 PPI with porosity of 0.89, 0.94, 0.96. The porous fin height was 9 mm.

The results are summarized in Figure 15.4, in which the modified Colburn factor j^* and the friction factor f are plotted as a function of the Reynolds number Re_H. The friction factor was much lower for low porosity foams, and the louvered fin exhibited slightly higher friction factor values than that of the porous fins at low Reynolds numbers. When the Reynolds number increased, however, the porous fins showed much higher friction factors as compared to the louvered fin. The modified j-factors of the porous fins decreased as the pore density increased or as the porosity decreased. Furthermore, the tested porous fins had a similar thermal performance as compared to the conventional louvered fin; however, the louvered fin showed a little better performance in terms of pressure drop.

Metal Foam Heat Exchangers

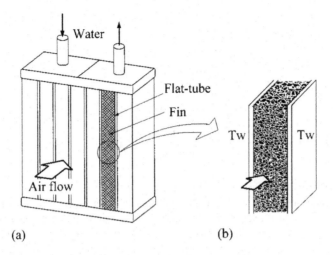

FIGURE 15.3 Schematic of the plate and fin heat exchanger studied and tested by Kim et al. (a) along with a drawing of one channel (b). (From Kim, S.Y. et al., *J. Heat Transf.*, 122, 572–578, 2000.)

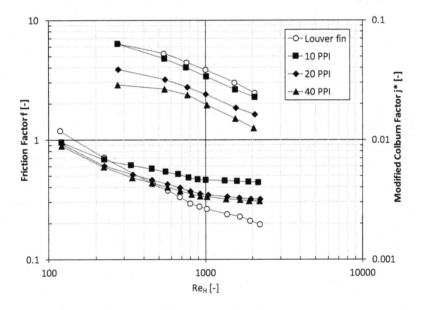

FIGURE 15.4 Comparison between the porous fins and the louvered fins in terms of modified Colburn factor j^* and friction factor. (Data from Kim, S.Y. et al., *J. Heat Transf.*, 122, 572–578, 2000.)

Kim et al. (2000) also proposed two simple correlations to estimate the modified Colburn factor and the friction factor, which can be applied to design plate-fin heat exchangers using metal foams. The modified Colburn factor j^* is defined and given by:

$$j^* = \frac{\eta^* \cdot h}{G \cdot c_p} \Pr^{2/3} = 13.73 \left(Re_H^{-0.489} \cdot Da^{0.451} \right) \tag{15.1}$$

where η^* is the overall surface efficiency, h is the heat transfer coefficient, G is the mass velocity referred to the frontal velocity, c_p is the heat capacity, while Pr is the Prandtl number. The Reynolds number Re_H is defined on the porous fin height and Equation (15.1) is valid $270 < Re_H < 2050$. Finally, the Darcy number $Da = K/H^2$, where K is the permeability of the metal foam.

The pressure drop Δp is given by:

$$p = \frac{L}{H}\left(\frac{1}{Re_H \cdot Da} + \frac{0.105}{Da^{0.5}}\right)\frac{G^2}{\rho} = f\frac{L}{H}\frac{G^2}{\rho} \qquad (15.2)$$

So, the friction factor f is given by:

$$f = \frac{1}{Re_H \cdot Da} + \frac{0.105}{Da^{0.5}} \qquad (15.3)$$

where Δp is the pressure drop, while L is the total length of the fin.

More than a decade later, Dai et al. (2012) conducted a theoretical study on the possible application of 40 PPI metal foam heat exchangers for heating, air conditioning, and refrigeration applications. The authors theoretically compared the performance of a flat tube, serpentine louver-fin heat exchanger, taken as representative of the state of the art for air-side heat exchanger design in HVAC&R (heating, ventilating, air conditioning, and refrigeration) systems with a foamed one.

Identical thermal-hydraulic requirements were imposed on the two heat exchangers and the volume, mass, and cost of the metal foam and louver fins were compared. The two heat exchangers achieved the requested performance, but the metal foam heat exchanger was smaller and lighter; however, the cost of this new component was remarkably higher as compared to the traditional louvered-fins one. At that time, a louvered stock could have been acquired at a cost of $7/kg while a metal foam could have been purchased at $466/kg. Considering the same costs for tubing ($5/kg) and for the brazing process $30, it was possible to estimate the cost of a metal foam heat exchanger, which, in the investigated range of operating conditions, varied with the foam thickness from $174 to $357, as compared to $38, which was the final cost of the louvered-fin heat exchanger (being the louvered fin fixed). The authors also stated that, in the case of 8-mm-thick foam layer, if the cost of the foam became $16/kg, the two heat exchangers would show an equal price. This work was carried out considering a 40 PPI aluminum foam, but the results can also be considered representative of other pore densities because the cost of the foam commonly does not depend on it. The same group experimentally explored the thermohydraulic performance of metal foam heat exchangers under dry operating conditions (Nawaz et al. 2017). Four different flat-tube heat exchangers with the same geometry (i.e., face area, flow depth, and fin dimensions) consisting of four different types of metal foams with 5, 10, 20, and 40 PPI and porosity greater than 0.93, were built and tested. An example of the tested heat exchanger is reported in Figure 15.2a; the foam fins were 15 mm high. Three different methods were used to join the metal foam to the tube, including thermal epoxy resin, thermal compound, and brazing. The authors stated that the brazing process resulted in a perfect contact (i.e., with a negligible thermal contact resistance) while the thermal epoxy resin exhibited a non-negligible thermal resistance, which was estimated to be around 0.55 m^2 K W^{-1}. The results showed that the pressure drops and the interstitial heat transfer coefficient increased with the pore density. The 5 PPI sample exhibited the lowest pressure drops and interstitial heat transfer coefficient while the 40 PPI sample exhibited the highest values.

Sertkaya et al. (2012) experimentally studied six air-cooled heat exchangers: three of them implemented aluminum foam as extended surface while the other three used convectional finned surface designs. The authors found that the metal foam heat exchangers presented lower heat transfer rates and higher pressure drops than conventional finned heat exchangers.

De Schampheleire et al. (2013b) compared two heat exchangers for an HVAC. As shown in Figure 15.5, one was a commercially available high-quality louvered-fin heat exchanger, while the second one was an in-house prototype made using 10 PPI open-cell aluminum foam. The comparison between both heat exchangers was done based on a performance evaluation criterion for foam defined as the ratio of the thermal conductance to friction factor. The results showed that the heat exchangers have a similar performance at low air velocity (1.1 m/s), while the louvered-fin heat exchanger performs slightly better at higher velocities. The authors also highlighted the influence of the contact resistance on the heat transfer penalization of the metal foam heat exchanger. The pressed-fit foam heat exchanger experienced very high contributions of the

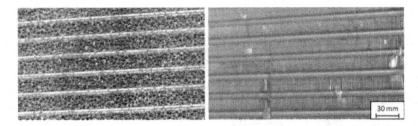

FIGURE 15.5 Photos of the two studied heat exchangers: foam (left) and louvered fins (right) by De Schampheleire et al. (2013b).

contact resistance (up to 70% of the overall thermal resistance), while for the louvered fin heat exchanger this contribution was much smaller (up to 11.1%), due to the presence of a fin collar. Thus, the authors stated that brazing can lower the overall thermal resistance by 44%. In this way, the hypothetical brazed foam heat exchanger would have the same performance at low velocities, while showing a 20% mass savings. More recently, Huisseune et al. (2014, 2015) proposed a comparison between metal foam heat exchangers and a finned one at low Reynolds number, typical of HVAC&R applications. The foamed heat exchangers showed up to six times higher heat transfer rate than the bare tube bundle at the same fan power. Differently, the louvered-fin heat exchanger outperformed all the simulated foamed heat exchangers for the same fan power. However, the authors stated that if the overall dimensions are not fixed, a metal foam heat exchanger having better performance than the louvered one can be designed by selecting the proper foam parameters, material, and dimensions.

Cicala et al. (2016) performed an experimental study to evaluate three aluminum-foamed heat exchangers, which showed better heat transfer behavior as compared to the bare tubes row. The authors suggested that the 20 PPI foam exhibited the best performance as compared to the 10 and 30 PPI ones.

In the application of HVAC&R systems, the air heat exchangers can also transfer latent heat in dehumidifying conditions. In these conditions, the condensed liquid retained in the extended surfaces has to be drained because it penalizes both the heat transfer and fluid flow. Unfortunately, for the metal foam heat exchangers under dehumidifying conditions, the interconnection of metal fibers makes the condensate more easily to accumulate in the structure, resulting in the decrement of the heat transfer coefficient and the increment of pressure drop (Costa et al. 2014; Nawaz 2014); hence, the liquid accumulated in the metal foam should be quickly drained out (Han et al. 2012a; Hu et al. 2016, 2017).

Han et al. (2012a) and Nawaz (2014) conducted dip tests to measure the water retention in grams per unit volume for four different aluminum foam samples. The 40 PPI sample with smaller-sized pores retains much more water than the 10 PPI sample does. Nawaz (2014) compared the water retention of a state-of-the-art louver fin against that of a 10 PPI metal foam during dip tests under identical test conditions. As reported in Figure 15.6b, the 10 PPI metal foam sample held much less water under both steady and transient conditions as compared to the louvered fin. Based on these data, one can anticipate that the metal foam heat exchanger will have a lower increase in pressure drop associated with wet-surface operation, compared to its dry-surface pressure drop, than a louvered-fin design. Furthermore, Nawaz (2014) and Nawaz and Jacoby (2018) measured the hydraulic performance of metal foam microchannel heat exchangers under dry, wet, and frosting conditions. As shown in Figure 15.6c and d, the authors found that the pressure drop increased quadratically as the face velocity increased, and this trend was observed under both dry- and wet-surface conditions. At a fixed-face velocity, for the 10 PPI sample, the pressure drops increased for wet-surface conditions as compared to dry-surface conditions; however, the increase was not as large as was manifested for many other compact heat exchangers, reinforcing the excellent condensate drainage behavior of metal foams. During the frosting tests, the air was blown at 4°C and 70% of relative humidity through a 10 PPI foam, while the coolant flowed at a temperature of −10°C. As shown in Figure 15.6e, the higher the face velocity, the larger the pressure drop at all times during the experiment. After some time, about the same amount of time at each face velocity, the pressure gradient becomes almost constant in time, indicating the heat exchanger has probably become fully loaded with frost (Figure 15.6f).

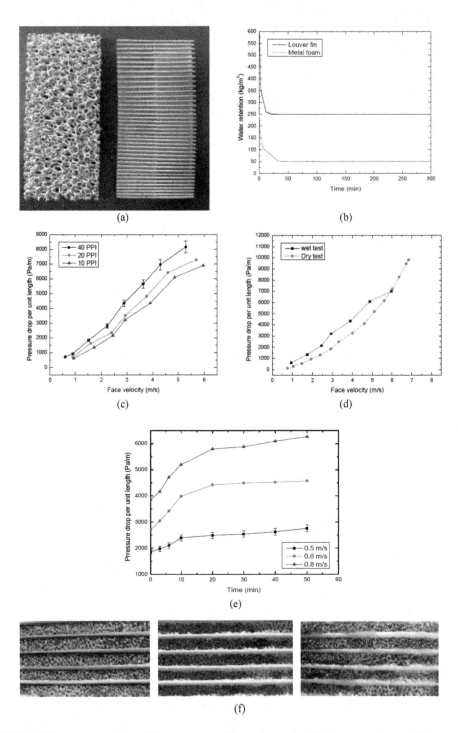

FIGURE 15.6 Water retention: (a) metal foam and state of the art louver fin; (b) dip tests results: metal foam vs. louver fins; (c) pressure gradients under wet conditions; (d) pressure gradients for 10 PPI metal foam: dry vs. wet conditions; (e) pressure gradients for 10 PPI metal foam under frosting conditions; and (f) frost growth for 10 PPI foam at different frontal velocities: from left to right, 0.5 m/s, 0.6 m/s, 0.8 m/s. (Data from Nawaz, K., *Aerogel Coated Metal Foams for Dehumidification Applications*, University of Illinois at Urbana-Champaign, IL, 2014.)

Hu et al. (2016) explored the performance of metal foams with pore density from 5 to 15 PPI under dehumidifying conditions by varying the frontal velocity from 0.5 to 2 m/s, air temperature from 27°C to 32°C, and relative humidity from 50% to 90%. Moreover, the authors compared the performance of the metal foams with a reference louver one; the results showed that, as the PPI of metal foam increased from 5 to 15, the total heat transfer decreased initially, and then increased due to the influence of condensing water. Compared to the louvered fin-and-tube heat exchanger with the same size, the heat transfer capacity of copper foam heat exchanger was enhanced by 69.2% ~ 127.2% under dehumidifying conditions, and the mass transfer rate was increased by 38% ~ 86%, meaning that the heat and mass transfer performance of metal foam is superior to that of louvered fin with the same volume. However, the wet air pressure drops were higher in the case of metal foams as compared to the louver case.

Hu et al. (2017, 2019) explored the effects of surface treatments on heat transfer and pressure drop of metal foams under dehumidifying conditions. In particular, Hu et al. (2019) studied three samples having 20 PPI, and the porosities 85%, 90%, and 95%. The authors concluded that the heat transfer coefficients in the hydrophobic and hydrophilic metal foams were larger by 4%–33% and 3%–21% than that in the uncoated metal foams, respectively, while the pressure drop in hydrophobic metal foams was increased by 3%–139% due to the large condensate droplets. In the case of hydrophilic metal foams, the pressure drop decreased by 1%–20% because of the condensate shape transformation from droplets to films. Finally, the performance of hydrophilic metal foam under dehumidifying conditions was found to be better than that of hydrophobic and uncoated metal foams. Hu et al. (2019) also suggested optimal porosities corresponding to the best comprehensive performance for hydrophobic and hydrophilic metal foams, which were 0.95 and 0.85, respectively.

15.2.2.2 Air-Cooled Condenser

Considering the air-cooled microchannel condensers for miniaturized refrigeration systems, Zilio et al. (2011) proposed using metal foams to enhance their air-side heat transfer performance for electronics cooling in aeronautical applications. The authors selected two 20-mm-thick aluminum foams with 5 and 10 PPI, and they compared their performance against a traditional finned surface by simulating the entire cycle, keeping constant the cooling capacity and the air pressure drops. In particular, Zilio et al. (2011) highlighted that the foamed air-cooled condenser achieved the same performance of the traditional finned one in terms of total power consumption of the cycle (i.e., compressor and air pumping power) but with remarkable mass and volume savings. The volume ratio and the mass saving were 0.44 and 79% in the case of 5 PPI foam and 0.44 and 77% in the case of 10 PPI one. Furthermore, the authors simulated the performance of the condenser of almost the same volume of the traditional one. This was obtained considering two slabs of foam, 20-mm-thick each, brazed on both sides of the microchannel condenser. (The configuration is similar to that studied by Ribeiro and Barbosa (2013) and depicted in Figure 15.7.)

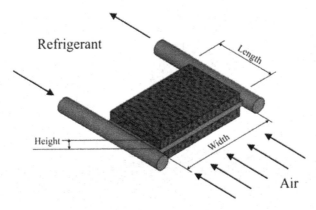

FIGURE 15.7 Air-cooled foamed microchannel condenser for miniaturized refrigeration unit. (Ribeiro, G.B. and Barbosa, Jr. J.R., *Appl. Therm. Eng.*, 51, 334–337, 2013.)

The results showed the foamed solutions still presented a remarkable mass saving, being 72% for 5 PPI and 68% for 10 PPI, due to the high porosity, and their total power consumptions were reduced by some 20%–25% thanks to the 7 K lower condensation saturation temperature that reduced the compressor pressure ratio (being fixed the evaporation temperature and the cooling capacity).

Ribeiro et al. (2012) investigated the use of copper foams for air-cooled microchannel condensers for miniaturized refrigeration system. The authors studied three different copper metal foam structures with different pore densities (10 and 20 PPI) and porosities (0.893 and 0.947). The results were compared against those obtained for a conventional condenser surface, with copper plain fins. The authors stated that for a fixed pumping power, the overall thermal conductances of the metal foam condensers were lower than that of a plain fin condenser with similar characteristics.

More recently, Ribeiro and Barbosa (2013) extended the analysis of Ribeiro et al. (2012) by comparing the performance results for the copper foam and copper plain-fin condensers with those obtained for aluminum louvered-fin condensers with similar overall dimensions. The authors concluded that in the range of the investigated operating conditions, aluminum louvered fins presented a better thermal-hydraulic performance than metal (copper) foams and, thus, can be considered as a more appropriate heat transfer enhancement medium for small-scale refrigeration cooling applications.

15.2.2.3 Vehicles Applications

Another interesting application for metal foam is in exhaust gas recirculation (EGR) systems where the foam can be used to cool the exhaust gas from an automobile engine. This will reduce the NOx emission and hence the compliance with the environmental regulation. Unfortunately, however, lowering the exhaust temperature increases the soot formation, which can lead to fouling of the foams.

Muley et al. (2012) developed, fabricated, and then experimentally assessed the thermohydraulic performance of a high-temperature metal foam heat exchanger for automotive exhaust gas recirculation system (Figure 15.8). The authors measured the heat transfer and pressure drop for stainless steel and Ni-Cr metal foam heat exchangers, and they compared the results with those collected for a state-of-the-art wavy plate-fin heat exchanger. The results showed that the foamed heat exchangers exhibited some heat transfer improvements, but the associated pressure drops were not compatible with EGR systems. The authors also stated that fouling and structural integrity of foams represent two issues to be addressed for the final deployment of this technology to industrial application. Lin et al. (2013) numerically investigated the possibility to reduce the air pressure drop during the flow through porous graphite foams heat exchangers in vehicles; the authors investigated four different configurations (baffle, pin-finned, corrugated, and wavy corrugated) of graphite foam fins. The results showed that the wavy corrugated foams presented high thermal performance and low pressure drop. Moreover, the wavy corrugated foam heat exchanger was compared against a conventional louvered aluminum-fin heat exchanger in terms of coefficient of performance (i.e., how much heat can be removed by a certain input of pumping power), power density (i.e., how much heat can be removed by a certain mass of fins), and compactness factor

FIGURE 15.8 Photo of the high-temperature foamed heat exchanger for EGR system tested by Muley et al. (From Muley, A. et al., *Heat Transf. Eng.*, 33, 42–51, 2012.)

(i.e., how much heat can be removed in a certain volume). The results demonstrated that, for a given frontal velocity, the louver fins always exhibit a greater coefficient of performance as compared to foamed fins. However, the wavy corrugated foam fins presented more than two times higher power density and more than 1.6 times higher compactness factor when compared to louver fins.

Fouling of heat exchangers is a complex phenomenon with some fundamental questions yet to be answered. The problem is even more difficult to understand when a metal foam is subject to fouling. As already stated, Muley et al. (2012) decided to completely discount the use of foams for gas coolers, as they are prone to fouling.

Interestingly, however, experimental data in Hooman and Malayeri (2016) showed that foam heat exchangers fit as EGR gas coolers much better than conventional heat exchangers. While a finned or louvered gas cooler is either expensive or impossible to clean (hence a replacement cost has to be budgeted for), a metal foam heat exchanger can be brush-cleaned and it is reusable. The heat transfer performance drop, compared to a clean heat exchanger, was only 17%. Experimental results in Hooman and Malayeri (2016) also pointed out that when a partly foam-filled channel is subject to particulate fouling, the deposition is not uniform, which had been the assumption made in some prior theoretical and numerical publications (Hooman et al. 2012; Odabaee et al. 2013a). A detailed discussion of the particulate fouling in metal foam heat exchangers is reported in Section 15.2.5.

15.2.2.4 Waste Energy Recovery

Worldwide, approximately 33% of the total consumed energy is directly discharged into the surroundings as waste heat (Hendricks and Choate 2006). Waste heat is abundantly available in power plants, chemical industries, petroleum industries, pulp and paper industries, internal combustion engines, and many other energy conversion processes (EIA 2015).

As analytically demonstrated by Lu et al. (2006) and Zhao et al. (2006), the heat transfer rate of the inside tube can be enhanced by more than 15 times when filling the tubes with metal foams (see Figure 15.2c); moreover, the authors also stated that metal foams can significantly enhance the heat transfer performance of tube-in-tube heat exchangers compared to that of conventional finned tube heat exchangers due to the high heat transfer area and strong flow mixing. Thus, metal foams can offer great possibility to design efficient heat exchangers for waste heat recovery systems.

Wang et al. (2016) implemented metal foam inserts to enhance the performance of a thermoelectric waste heat recovery system. Figure 15.9 shows the thermoelectric generator. The metal foam inserts were located in both the hot air and cold-water channels; three different pore densities were tested: 5, 10, and 20 PPI. The maximum power generation efficiency for the waste heat recovery system was

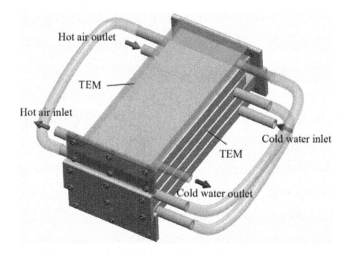

FIGURE 15.9 Schematic of the metal foam-filled thermoelectric generator tested by Wang et al. (From Wang, T. et al., *Energy Convers. Manag.*, 124, 13–19, 2016.)

around 2%, when the channels of the thermoelectric generator were filled with the 5 PPI foam inserts. This performance was around 30% higher than the value for the unfilled thermoelectric generator. It has to be pointed out that the inserts were just located inside the channel; thus, a great thermal contact resistance could have been present. As discussed later on par. 15.2.5, a better performance could have been achieved if the inserts had been brazed into the channels in order to minimize the thermal contact resistance.

Another application of waste heat recovery is that presented by Chen et al. (2019), who investigated compact exhaust heat exchangers using metal foams to improve the practicability of the waste heat recovery system for internal combustion engine. The authors compared six Ni metal foams heat exchangers with a conventional one. All the compact foamed heat exchangers performed better that the conventional one, while presenting only one-third of its volume and mass. The authors also studied the effect of the weight and pressure drop on power loss, demonstrating that the compact foamed heat exchangers had better overall performance as compared to the traditional one.

The waste heat thermal management and recovery call for efficient and compact heat exchangers that can withstand high temperature; Jazi et al. (2009) developed a technique to fabricate heat exchangers consisting of Nickel foam with thermal-sprayed skins of Inconel 625. Tsolas and Chandra (2012) and Hafeez et al. (2016) conducted preliminary tests on these spray-formed metal-foam heat exchangers, showing that they presented better performance as compared to the brazed foam heat exchanger because of the lower contact resistance between the foam strut and the external skin. More recently, Hafeez et al. (2017) compared the heat transfer during high-temperature gas flow through metal foam heat exchangers. 10 and 40 PPI foam structures were investigated, and the heat exchangers were made by either brazing Inconel sheets to foam or plasma spraying Inconel skins on the foam, as suggested by Jazi et al. (2009).

Figure 15.10 shows the two kinds of high-temperature heat exchangers. Again, thermal-sprayed heat exchangers were found to perform better as compared to the brazed ones; the 40 PPI heat exchangers presented better heat transfer but higher pressure drops as compared to the 10 PPI one. The authors also developed and validated a model to estimate the air temperature rise in the heat exchanger.

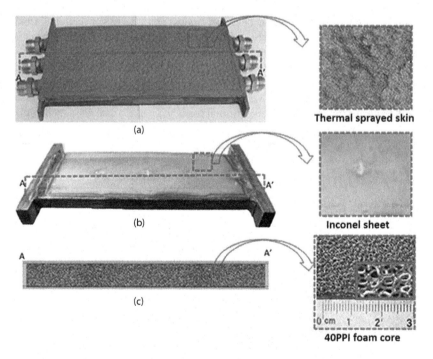

FIGURE 15.10 High temperature metal foam heat exchangers tested by Hafeez et al. (a) Thermally sprayed and (b) conventional brazed nickel foam heat exchangers and (c) foam core of heat exchanger only 40 PPI foam core is shown here. (a) Plasma spray heat exchanger with foam core; (b) conventional braze heat exchanger with foam core; and (c) section A-A foam core of the heat (40 PPI). (From Hafeez, P. et al., *ASME J. Heat Transf.*, 139, 121801-1–121801-11, 2017.)

15.2.2.5 Fuel Cell Application

Metal foams can also be used to reject high heat loads as in the case of fuel cell thermal management. Boyd and Hooman (2012) and Odabaee et al. (2013b) studied the application of metal foams in fuel cell thermal management (Figure 15.11), and they found that by keeping constant pressure drops, an air-cooled metal foam heat exchanger can be as efficient as a water-cooled one. This significantly reduces the size and cost of the fuel cell stack. The water-cooled system has to rely on an extra heat exchanger to cool the water (indirect cooling) while the air-cooled design directly dumps the heat to the ambient air. Practicality of such fuel cell designs were further improved in Fiedler et al. (2014) where the application of a compressive force to attach the foam to the graphite plate has been investigated. Fiedler et al. (2014) reported the thermal and electrical contact resistances for a foam-graphite plate compound. Their design reduces the assembly cost, by a factor of 5, compared to the alternative of brazing the foam to a graphite plate.

15.2.3 Foamed-Wrapped Tubes Heat Exchangers

In compact-foamed heat exchangers (i.e., internal configuration), the flow is bounded by solid walls. That is, the flow area available to the foam is the same as the total fluid area. Differently, in foamed-wrapped tube banks (i.e., external flow), only part of the flow path is covered by foam. Hence, part of the air flow can escape the foam and leave the heat exchanger without going through the foam. Flow-over tube banks covered with foam are similar to the partially blocked channels, since the approaching air may not go through the foam at all. It has to be pointed out that the optimum tube arrangement in the case of foamed-wrapped tube depends upon different parameters and, for a given application, definitely, differs from that it can be designed for aligned or staggered tube banks. It can be stated that filling all the available gaps between tubes with foam cannot be considered the optimal design. It incurs significant material cost, material that does not fully participate in the heat transfer process while still imposing a significant resistance to the flow. Moreover, because foams exhibit different thermohydraulic behavior compared to fins, their design should not follow the same path as that of fins. For instance, circular-finned tubes show better performance when they are set as staggered compared to the inline configuration. This is not the case with foams. One also has to be reminded that a multi-row bundle, say an N-row-finned tube bundle, does not necessarily lead to two times higher heat transfer if the number of rows are doubled, say

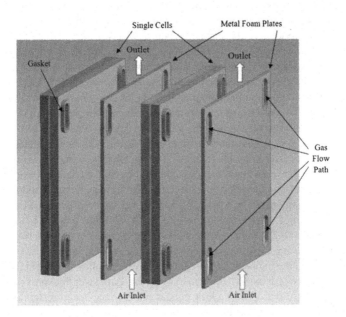

FIGURE 15.11 Odabaee et al. (2013) configuration of metal foam air-cooling system for fuel cells.

to 2N. This is despite the fact that the pressure drop grows linearly with the number of rows. That is, the per-row pressure drop increases pro rata while the per-row heat transfer does not show the same trend (it depends on a number of parameters including the Reynolds number). This is true for either finned or foam-wrapped bundles. Therefore, filling a box with foams and poking tubes through them is not the most efficient design for metal foam-wrapped tube bundles. It may be simpler to manufacture the bundle in this way, but the cost of adding foams where they cannot help the heat transfer process will outweigh the benefit (Hooman 2014).

T'Joen et al. (2010) first investigated foam-wrapped tubes foam by conducting wind tunnel experiments for a single row of aluminum tubes covered with thin layers (4–8 mm) of metal foam. The authors experimentally studied the impact of various parameters on the thermohydraulic performance of the foam-wrapped tubes, including the Reynolds number, the tube spacing, the foam height, and the type of foam. In fact, the comparison of the performance of foam-wrapped tubes with different foam height clearly showed that increasing the foam height reduces the exterior convective resistance, while at the same time increasing the pressure drop. It was also found that the air only penetrates the foam to a certain extent, which results in a decreasing performance as the foam height increases. Comparing the performance of heat exchangers with a different tube layout showed that although the smallest tube spacing resulted in the highest pressure drop, it also provided the largest heat transfer benefit, and thus has the best performance. According to those authors, a thin foam layer can lead to impressive overall performance when both the heat transfer and pressure drop are considered through a goodness factor. Moreover, T'Joen et al. (2010) also investigated the effects of the bonding resistance, demonstrating that the use of epoxy glue can lead to a remarkable penalization of the heat exchanger performance. The authors suggested that the research should focus into securing a cost-effective and efficient brazing process to connect metal foams to the tube surfaces. In particular, they stated that if an efficient bonding process (e.g., brazing) can be achieved, metal foam-covered tubes with a small tube spacing, small foam heights, and made of foam with a high specific surface area and thin struts potentially offer strong benefits at higher air velocities (>4 m/s) compared to helically finned tubes.

On the basis of this first experimental attempt, a few numerical and theoretical works have been conducted to first analyze the performance of the foamed-wrapped tubes in comparison with the finned tubes and conventional design for a given application. Among those, Odabaee and Hooman (2011) conducted an optimization study of metal foam heat exchangers aiming at replacing the finned tubes in air-cooled condensers of a geothermal power plant with foam-wrapped tubes. The authors demonstrated that the foam heat exchangers have two to six times higher performance factors compared with the conventional design while the pressure drop increase is within an acceptable range.

Later on, Odabaee and Hooman (2012) extended their previous work by numerically studying the heat transfer from a metal-foam wrapped tube bundle. The authors investigated the effects of key parameters, including the free stream velocity, longitudinal and transversal tube pitch, metal foam thickness, and characteristics of the foam (e.g., porosity, permeability, and form-drag coefficient) on heat transfer and pressure drop. The authors compared the numerical results for the foam-wrapped tube with the conventional finned-tube designs, from both heat transfer and pumping power point of view using the area goodness factor, $(j/f^{1/3})$ (Webb 1994; Shah and Sekulic 2003). The results showed that the area goodness factor of metal foam heat exchangers, at low range of stream velocities, can be five times higher than that of finned-tube heat exchangers. Moreover, the increasing of the tube pitch improved the area goodness factor while increasing the metal foam layer thickness did the reverse. In fact, Odabaee and Hooman (2012) stated that with a thin foam layer and large tube pitch, the heat transfer augmentation is more pronounced than the excess pressure drop caused by foams. For low air velocities through the foam, the heat transfer from the heated surface is pretty much conduction-dominated and flow penetrates to a certain depth, say three pores deep (T'Joen et al. 2010). Increasing the foam layer thickness or lowering the tube pitch does not improve the convection through the foam structure, as heat is extracted from the heated surface by conduction and removed by the high-velocity stream of air, which is flowing outside the foam. Hence, the net effect of blocking the flow passage by either increasing the foam layer thickness or decreasing the tube pitch is to increase the pressure drop with a little enhancement in the heat transfer rate.

Chumpia and Hooman (2014) compared heat transfer and pressure drop of a metal-wrapped tube to those of a finned tube. Figure 15.12 reports a photo of the tested samples. The results showed that,

Metal Foam Heat Exchangers

FIGURE 15.12 Specimens used by Chumpia and Hooman (2014). Based on either foam or fin height from left to right: foam-2 (5 mm), foam-1 (5 mm), foam-1 (12 mm), fin (15 mm), foam-1 (15 mm), and foam-1 (20 mm).

for thermal efficiency, overall thermal resistance decreased with the increase of foam layer thickness. However, this thermal advantage came at the expense of increasing pressure drop.

In a subsequent study, Chumpia and Hooman (2015) extended the work to a single-row air-cooled heat exchanger bundle in cross-flow and compared the results to those of industrial-finned tubes. Two sets of three tubular heat exchangers, constructed by wrapping aluminum foam of different thickness around cylindrical tubes, were tested for heat transfer performance and pressure drop characteristics. The authors could investigate the effects of the foam thickness and transversal pitch distance. In particular, two thicknesses were investigated, 5 mm (thin foam) and 15 mm (thick foam). The authors found that the performance gain in heat transfer was not increased in a direct proportion with the increase in foam layer thickness. Taking the heat transfer ratio between the "thick" and "thin" foams when both are arranged in a compact configuration reveals a gain by thick foam ranging from 1.8 at the smallest airflow to 1.5 at the largest airflow. Moreover, the authors stated that with good designs and sound technical strategies, the foam-wrapped heat exchangers with suitable foam thickness can give heat transfer benefit while keeping the pressure drop at the similar level as that caused by conventional finned tubes. It has been verified that one way of achieving this, if space is not a limiting factor, is to extend the pitch length such that the bundle acts as a collection of less resistive individual tubes.

More recently, Chumpia and Hooman (2018) conducted experiments on multi-row heat exchanger bundles composed of finned tubes or foam-wrapped ones. Also in this work, the authors tested two sets of aluminum foam-wrapped tubes, with foam layer thickness being 5 and 15 mm thick, respectively. Two-row and three-row bundles arrangements with fixed transversal and longitudinal pitch distances were tested; the air velocity was varied between 0.5 and 5.0 m s^{-1} under cross-flow. These tests permitted them to investigate the effects of foam layer thickness and the number of rows under staggered configuration. The results were compared against those obtained for an annular-finned tube bundle of 15 mm fin height of the same number of rows. Considering the foam bundles, it was observed that the second and third rows in the bundle had their heat transfer enhanced over the first row because of the turbulence generated by the preceding row(s). However, the pressure drops were still the major drawback in metal foam heat exchanger bundles; in fact, the friction factor of the foam bundle was found to be from three to six times larger than that of the finned bundle with similar dimensions.

A comparison of the heat transfer per tube for different bundles was made with single-tube results of Chumpia and Hooman (2014) to conclude that densely bundling the foam-wrapped tubes only increases the pressure drop while the heat transfer is not even higher than that of a single tube in isolation. Hence, if space is permitting, setting the foam-wrapped tubes sparsely will lead to better performance of both heat transfer and pressure drop, compared to current industry practice of densely bundling the tubes.

Alvandifar et al. (2018) performed a numerical work to study a novel design of partially metal foam-wrapped tubes. In particular, the authors studied the heat transfer and pressure drop for a tube bank with five rows of tubes wrapped by metal foam layers. The authors concluded that by covering bare tubes with the metal porous layers, the heat transfer rate can be augmented substantially in comparison to bare tube banks. Moreover, the numerical results confirmed that by increasing the porous layer thickness, the heat transfer rate remained almost constant while the pressure drop increased. In addition, by comparing the thermal performance of the porous-layer-wrapped tube bank to a finned tube bank of similar weight, it was shown that the thermal performance of the former was better than the latter, except at very low Re numbers or very high foam thicknesses. As shown in Figure 15.13, in order to mitigate the pressure drop issue, Alvandifar et al. (2018) proposed a novel arrangement in which the tubes are only partially covered by the foam layer. The numerical results showed that by means of the new arrangements, almost the same Nusselt number could have been achieved compared to the fully wrapped tube and, at the same time, the pressure drop, weight, and foam material usage in heat exchangers were reduced considerably. Furthermore, among all of the considered arrangements, arrangement A3 can be regarded as the best choice with 60% pressure drop reduction, 33% area goodness factor improvement, and 50% cut in the amount of the foam usage with respect to the fully wrapped foam tubes. Similar results were also found by Ghadikolaie et al. (2019) for the single partially foam-wrapped tube.

15.2.4 On the Foam-Finned Surface Performance

In the previous paragraphs, the internal and external flow in metal foam heat exchangers have been explored. Either in the case of internal or external flow, most of the authors compared the heat transfer performance of metal-foamed heat exchangers with that of a reference-finned surface. Different performance criteria have been selected and used, which could lead to different results. In general, according to Webb (1994) and Shah and Sekulic (2003), the area goodness factor can be considered a reliable way to compare the performance of largely different enhanced surfaces implemented in the heat exchangers. However, it has to be underlined that the use of goodness factors or performance evaluation factors should be conducted with care. One can define a goodness factor as the ratio of the overall heat transfer divided by the pressure drop. There are similar definitions using (j/f) instead of $(j/f^{1/3})$. All of the enhanced surfaces aim at getting high heat transfer at low pressure drops. One should be very careful in interpreting these factors and making comparisons if the aim is to go with a design comparison. Assuming that the "value" of added heat transfer is the same as the "penalty" incurred by excess pressure drop is not always appropriate. There are applications that the designer can accommodate extra pressure drop for instance

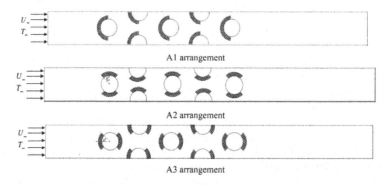

FIGURE 15.13 Novel partially wrapped metal foam tubes proposed by Alvandifar et al. (From Alvandifar, N. et al., *Int. J. Heat Mass Transf.*, 118, 171–181, 2018.)

even if heat transfer is only marginally improved. The same argument goes with the use of dimensionless heat transfer and pressure drop indicators (linear or with exponent). The main point here is that, while informative, such measures cannot be used to conclude that one design is better or worse than the other. They should only be used to provide extra information for the designer (Hooman 2014).

Moreover, the thermohydraulic performance of a foam-based heat exchanger depends upon several parameters, including the structural characteristics of the foam and the geometrical characteristics of the foam layer used as extended surface. In general, from the reviewed papers for compact foamed heat exchanger, it appeared that in most (if not all) of the cases, the comparison involved heat exchangers implementing metal foams with different porosity and/or pore density but with the same foam layer thickness. This implies that those heat exchangers might not have been the optimum design for the selected metal foams. This can be easily understood if considering the model proposed by Mancin et al. (2013) for the calculation of the heat transfer during internal flow in metal foams. The model permitted to estimate the heat transfer coefficient (h^*) as the product of the interstitial heat transfer coefficient (h) and of the foam-finned surface efficiency (η^*). For the sake of brevity, the equations are not reported here, but an example of the effect of the foam layer thickness on the heat transfer performance of aluminum foams with porosity equal to 0.93 and pore density ranging from 5 to 40 PPI is reported in Figures 15.14 through 15.16. Setting a frontal velocity of 2.5 m/s and considering air at ambient conditions (i.e., 25°C and ambient pressure), the values of the calculated interstitial heat transfer coefficients are reported in Figure 15.14 as a function of the pore density. It clearly appears that the interstitial heat transfer coefficient increases with the pore density. Figure 15.15 reports the effect of the foam layer thickness on the overall foam-finned surface efficiency as a function of the pore density, at the given operating conditions. The diagram clearly shows that the efficiency suddenly drops as the foam layer thickness increases, and the greater the pore density, the faster the heat transfer penalization. In particular, given a foam thickness of 5 mm (i.e., a 10-mm-thick layer for a microchannel heat exchanger), the overall foam-finned surface efficiency is equal to 0.93 for 5 PPI, 0.70 for 10 PPI, 0.41 for 20 PPI, and 0.24 for 40 PPI. Differently, if one would compare foams with the same overall foam-finned surface efficiency, for example $\eta^* = 0.5$, at the given operating conditions, foam layers with a thickness of 15, 7, 4, and around 2 mm for 5, 10, 20, and 40 PPI, respectively, should be considered. Thus, as reported in Figure 15.16, when multiplying the interstitial heat transfer coefficient with the overall foam-finned surface efficiency, the results permit to

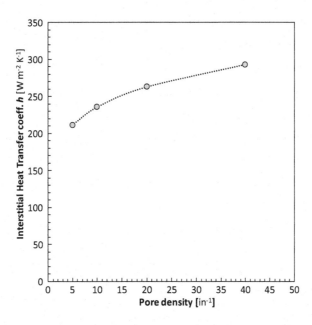

FIGURE 15.14 Interstitial heat transfer coefficient as a function of the pore density for aluminum foams with a porosity of 0.93. Frontal velocity 2.5 m/s. (Model by Mancin, S. et al., *Int. J. Heat Mass Transf.*, 62, 112–123, 2013.)

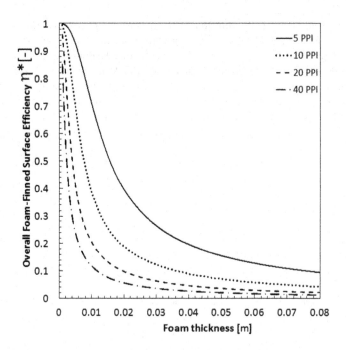

FIGURE 15.15 Overall foam-finned surface efficiency, η^* as a function of the foam layer thickness for aluminum foams with a porosity of 0.93. Air frontal velocity 2.5 m/s. (Model by Mancin, S. et al., *Int. J. Heat Mass Transf.*, 62, 112–123, 2013.)

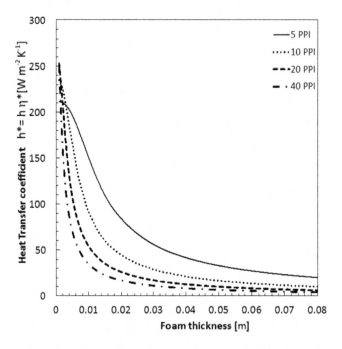

FIGURE 15.16 Heat transfer coefficient as a function of the foam layer thickness for aluminum foams with a porosity of 0.93. Air frontal velocity 2.5 m/s. (Model by Mancin, S. et al., *Int. J. Heat Mass Transf.*, 62, 112–123, 2013.)

Metal Foam Heat Exchangers 327

understand that it is not fair comparing different foams keeping constant the foam thickness. Thus, the foamed heat exchangers should be compared considering optimized solutions, and, for any given foam structure, the foam layer plays a fundamental role, and it should not be kept constant.

15.2.5 Key Issues and Challenges in Metal Foam Heat Exchangers

As clearly stated by Muley et al. (2012), key issues and challenges associated with the use of metal foams for heat-exchanger applications include precise measurements of their geometrical characteristic, reduction of excessive pressure drop, reduction of fouling, improved brazing/joining methods, and development of consistent test data reduction and design procedures. Among those, the authors underlined that significant research and development efforts should be devoted to addressing fouling and brazing/joining issues, and structural robustness before foams can be used effectively in heat-exchanger applications.

15.2.5.1 Thermal Contact Resistance

Considering the thermal contact resistance, the description should start from the analysis of the current technology: most of the air-cooled heat exchanger in HAVA&R as well as in many other applications use louver fins as extended surfaces. Elsherbini et al. (2003) reported an average thermal contact conductance of 9.44 kW m^{-2} K^{-1} (press-fit) for a louvered-fin heat exchanger with collars. As the fin collars completely overlap the tubes, the resulting contact resistance for the louvered heat exchanger became 6.1 10^{-4} K/W. For a varying airside mass flow rate, the relative contribution of the contact resistance to the overall thermal resistance ranges from 7% to 11%. In general, it can be stated that an average thermal contact resistance for the louver-fin heat exchanger can be up to 15% of the overall one.

In the case of foam applications, the contact resistance is hard to be estimated, and thus, it is commonly neglected (when possible) or lumped together with the external convective resistance.

However, this can lead to a great uncertainty in the evaluation of the performance of the heat exchanger prototypes, which use metal foams as extended surfaces. However, in most cases, the thermal contact resistance cannot be neglected, being very high (Odabaee and Hooman 2012); in fact, it can account for the most part of the overall heat exchanger contact resistance up to 70%, as described by De Schampheleire et al. (2013b). A few authors tried to minimize the thermal contact resistance of metal foams, using for instance a thermal grease to the interface (Jamin and Mohamad 2008). However, it has to be pointed out that bonding the foam to a substrate is a rather difficult task because its high porosity limits the contact points to many small spot-contacts instead of the typical line-contacts as in the case of fins. Sekulic et al. (2008) highlighted that poor contact between foam ligaments and substrate may have a sizeable influence on the overall performance. Besides, the authors suggested the most suitable brazing procedure for different foam types and materials.

De Jaeger et al. (2012) studied the contribution of the contact resistance by investigating four different bonding methods. The proposed estimation procedure permitted to estimate the contact resistance by minimizing the difference between the calculated heat transfer via a zeroth-order model and experimental data. The results of this work are very interesting because the authors demonstrated that by varying pore size, porosity, PPI value, aluminum alloy, foam height, air mass flow rate, air inlet temperature, and bonding method, only the bonding method had a large impact on the contact resistance. The values of the estimated contact resistance were: 0.70 10^{-3} m^2 K/W for brazing, 1.25 10^{-3} m^2 K/W for single-epoxy bonding, and 1.88 10^{-3} m^2 K/W for press-fit bonding. More recently, De Schampheleire et al. (2013b), based on the values of contact resistance estimated by De Jaeger et al. (2012), were able to evaluate the contact resistance of their foam heat exchanger being 0.010 K/W. This resulted in a contribution of the contact resistance to the overall resistance ranging from 49% to 70% for different air mass flow rate. Those results were comparable with those estimated by T'Joen et al. (2010) and Sadeghi et al. (2011). In fact, T'Joen et al. (2010) were able to estimate the contact resistance of the thin layer of epoxy glue used to fix the metal foam to the tube. Through destructive testing, the thickness of the epoxy layer could be measured to be 0.3 mm, and by modeling the epoxy layer as a cylinder annulus surrounding the tube, the contact resistance could have been calculated, resulting in contributions between 6% and 55% of the overall resistance. Sadeghi et al. (2011) experimentally measured the thermal contact resistance for a metal foam heat sink in a vacuum test chamber, by measuring the power of the heater and the temperatures of the heat sink. The authors tested four metal foams with different porosities (0.90–0.95) and PPI values (10 and 20), by varying the compression load on the metal foam. Relative contact resistance contributions up to 58% of the overall resistance were measured.

Fiedler et al. (2014) proposed a methodology to evaluate the thermal contact resistance of metal foams. The authors conducted numerical and experimental analyses on the thermal resistance of six copper foams. The results showed that in order to properly estimate the thermal contact resistance, a numerical and experimental approach is needed. Geometric analysis of computed tomography data indicates a linear dependence of these contact resistances and the conductive copper contact area within the considered range. The authors concluded that the application of the extrapolation technique for experimental data requires large specimens in order to achieve similar conductive contact areas and thus constant contact resistances.

In conclusion, this analysis reveals that the thermal contact resistance can hide the real performance of the metal foam heat exchanger leading to unfair comparisons between different technological solutions. So, an appropriate bonding method of the metal foam heat exchanger could reduce the thermal resistances, substantially (De Schampheleire et al. 2013b). The brazing method should be surely adopted as standard to lead to efficient thermal contact between the ligaments and the substrate.

15.2.5.2 Fouling Issue

Fouling of heat exchangers is a complex phenomenon with some fundamental questions yet to be answered. Besides a very limited published literature reports on fouling of metal foam heat exchangers, current studies on the porous structure metal foam could not explain the insight of fouling phenomenon due to its complex structure. The problem is even more difficult when one aims to understand when a metal foam is subject to fouling (Anuar et al. 2017).

As clearly stated by Hooman and Malayeri (2016), most of the past studies into the topic were conducted numerically (Odabaee et al. 2013a; Sauret et al. 2013, 2014; Sauret and Hooman 2014) with no experimental data available for validation. Hooman et al. (2012) offered a theoretical model that assumed uniform deposition in the pores and on the ligament surfaces; again, with no experimental data to compare the results with.

Furthermore, experiments show that there are favorable deposition locations depending on the foam geometry, particle size, heat exchanger configuration, and flow conditions. This has also been numerically demonstrated for external flow over tube bundles by Sauret and Hooman (2014). In particular, the authors showed that, as shown in Figures 15.17 and 15.18, depending on the particle size, keeping constant the air velocity at 3 m/s, there are favorable deposition locations. At constant particle diameter, the particle volume fraction increases with the particles diameter. Except for the smallest diameter (5 μm), the particles concentration is higher at the front of the tube than at the rear, leading to a preferential deposition area at the front. Moreover, the normal distributions of particles (Figures 15.17g and h) also show that for the distribution centered around the smallest diameter, a larger volume fraction of particles appears at the rear of the tube because of the entrainment of the smaller and lighter particles through the recirculation zone. One can note that the particles volume fraction for the normal distribution centered around /mean 10 μm is fairly similar to the constant diameter 10 μm.

However, this is not the case for 50 μm. Using a normal distribution of particles centered at a higher mean diameter tends to keep entrained all the particles together through the main jet. The deposition rate for the normal distribution centered around /mean 10 μm (Figure 15.18g) is similar to a constant diameter that can lay between 5 and 10 μm, while for the normal distribution centered around/mean 50 μm presented Figure 15.18h, the deposition rate is identical to a constant particle diameter lying between 50 and 100 μm.

Considering the permeability, Sauret and Hooman (2014) highlighted that the lower the permeability, the lower the velocity in the foam. This is especially relevant at the back of the tube where the region of really low velocities is increased. In addition, the traveling time of the particles decreases with the increase of the permeability. This leads to a higher likelihood of deposition at the rear of the tube for lower permeabilities. The region of deposition is also extended compared to higher permeabilities with an increase of approximately 50%, reaching a quarter of the foam at the rear of the tube.

Metal Foam Heat Exchangers

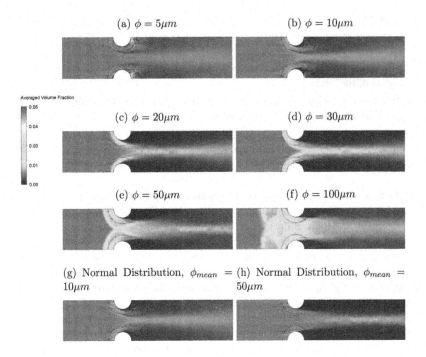

FIGURE 15.17 (a-h) Average volume fraction of particles for different constant particle sizes distributions at face velocity of 3 m/s in a tube bank, flow direction from left to right. (From Sauret, E. and Hooman, K., *Int. J. Heat Mass Transf.*, 79, 905–915, 2014.)

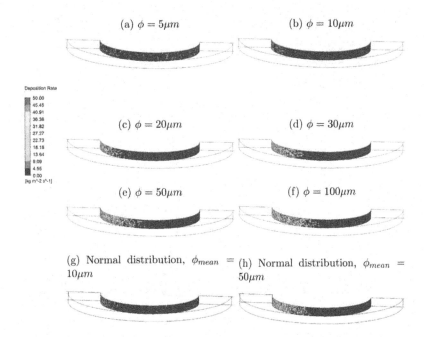

FIGURE 15.18 (a-h) Evolution of particle deposition rate with different particle sizes distributions at face velocity of 3 m/s, flow direction from left to right. (From Sauret, E. and Hooman, K., *Int. J. Heat Mass Transf.*, 79, 905–915, 2014.)

Finally, the authors suggested a likelihood deposition matrix as a method to understand the deposition phenomenon of a given particle; it is based on three indicators:

- Volume fraction of particles, which gives a clear information of the particle density in a particular region of the foam;
- The particle velocity, which indicates the likelihood of the particle to rapidly flow through the foam and so not deposit if the velocity is high; and
- Particle traveling time, which corresponds to the time that the particle has spent in the foam. If this time is high, the particle has been in the foam for a long time, which will potentially indicate a higher chance of deposition.

This can represent a useful tool to predict the regions more prone to deposition and thus to fouling. The results confirmed that for the smaller particles (less than 20 μm in diameter), the deposition will most likely occur inside the foam at the rear of the tube where the velocity is low due to the recirculation region. Moreover, the front of the tube is a potential region for deposition, as it has a high-volume fraction of particles with low velocities. The most unlikely regions of deposition appeared to be the halfway through the foam where the particle velocity is really high with short traveling time.

The EGR application highlights the fouling issue in a completely different geometry as compared to the case studied by Sauret and Hooman (2014).

Moreover, Hooman and Malayeri (2016) experimentally observed the fouling formation during the use of the gas coolers in the EGR system; in particular, the effect of the foam thickness and gas velocity was studied. Figure 15.19a and b show visual aspects of the same foams (20 PPI with 4 mm height) subject to

FIGURE 15.19 Visual observation of the foams with 20 PPI with 4 mm thickness after 6 h of fouling test: (a) 30 m/s of gas velocity; (b) 10 m/s of gas velocity; and (c) close-up view of the 20 PPI foam with 4 mm thickness at 10 m/s gas velocity after 6 h of fouling experiments. (From Hooman, K. and Malayeri, M.R., *Energy Convers. Manag.*, 117, 475–481, 2016.)

Metal Foam Heat Exchangers 331

fouling experiments at different gas flow rates. The foam subject to higher flow rate, that is, with 30 m/s as the gas velocity, showed a continuous smooth layer of soot formed both on inner and outer ligaments. However, at lower gas flow rates, that is, with 10 m/s, a discontinuous and rough fouling pattern was observed. For a more comprehensive understanding of the problem, Figure 15.19c is presented to show a close-up view of the foam subject to fouling test at 10 m/s gas speed: only the outer ligaments were fouled while the inner ones were still clear, and even the bare metal can be seen.

Hooman and Malayeri (2016) also investigated different cleaning methods to remove the formed fouling layer; the authors used brush cleaning with foams and tested the cleaned samples again to measure the deviation from the original (before fouling test) thermal resistance values. This is because brush-cleaning ability means to indicate the poor stickiness of deposited materials to the surface. The results showed that the 20 PPI samples exhibited a performance that was only lower than that of the clean (original) sample by about 17%. A subsequent tentative of using a combined cleaning approach with ultrasound and then brush achieved a performance within 11% of the clean foam. In the case of 40 PPI, the results were slightly worse, being 31% and 18% for brush cleaned only and ultrasound + brush cleaned case, respectively. Given the fact that the 20 PPI foam showed most of the fouling on the surface ligaments, effectiveness of a simple brush cleaning approach was not so surprising, but it clearly went against the perception that metal foam heat exchangers are impossible or very difficult/expensive to clean (Muley et al. 2012). In fact, compared to louvered or finned counterparts, the investigated metal foams were much easier to clean and reuse, which translated into significant savings and easier maintenance in the long run.

The combination of ultrasonic and mechanical (brush) cleaning seems to be the most suitable method for metal foam heat exchangers. Other methodologies, for instance those reviewed by Müller-Steinhagen et al. (2011) for other type of heat exchangers, can also be considered for metal foam ones, as long as the cleaning process does not deteriorate the foam microstructure-ligament layout (Anuar et al. 2017).

15.3 Liquid and Two-Phase Heat Exchangers

Boomsma et al. (2003) can be considered one of the first, if not the first, experimental study on liquid forced convection in metal foam heat exchangers for electronic cooling applications. The authors investigated the effects of the compression rate on the deionized water heat transfer and fluid flow; 40 PPI samples with nominal porosity of 0.92 and 0.95 were compressed at different compression rates to obtain a set of foams with lower porosities, then fashioned into heat exchangers. Boomsma et al. (2003) concluded that the compressed aluminum foams performed well not only in the heat transfer enhancement, but they also made a significant improvement in the efficiency over several commercially available heat exchangers, which operate under nearly identical conditions. The metal foam heat exchangers decreased the thermal resistance by nearly half when compared to currently used heat exchangers designed for the same applications.

Kim et al. (2008) experimentally investigated the single-phase and two-phase heat transfer characteristic of water and FC72 of three different metal foams, which had different pore size and porosity. The authors aimed at developing a compact and efficient cold plate for an LED package cooling. The investigated metal foams had 10 PPI with a porosity of 0.95, and 20 PPI with two different values of porosity 0.92 and 0.95.

The authors found that during single-phase heat transfer, a heat transfer coefficient of 10 kW m^{-2} K^{-1} was achieved with water flowing through the 20 PPI and 0.95 porosity, while a heat transfer coefficient of 2.85 kW m^{-2} K^{-1} was achieved with the FC-72 flowing in the same foam. As expected, for single-phase flow, the porous foam provided higher heat transfer rates than the open channel of the same dimensions, and with FC-72 the heat transfer capability reached the 35% of the values attained with water. However, the porous channel also provided greater pressure drop as compared to the empty channel. Considering the two-phase flow, the 10 PPI with 0.95 porosity sample exhibited the best behavior. The results showed that the porosity had a modest effect, up to 23% on the heat transfer coefficient at high heat fluxes. Alternately, at constant porosity, the pore density presented a greater effect. The larger the pore size, the greater the heat transfer.

Abadi et al. (2016c, d) experimentally investigated the single- and two-phase heat transfer of R245fa inside a plate heat exchanger with metal foam-filled channels. Figure 15.20 shows the metal foam

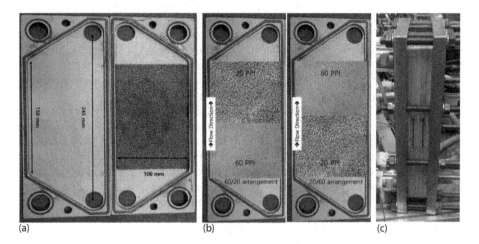

FIGURE 15.20 Metal foam arrangement: (a) channels with and without 20 PPI metal foam along with dimensions of one plate; (b) alternative metal foam arrangement, showing 20/60 and 60/20 configuration; and (c) heat exchanger. (From Abadi, G.B. et al., *Appl. Therm. Eng.*, 99, 790–801, 2016d.)

arrangements in the plate channels. Abadi et al. (2016c) investigated different metal foams with various pore density values of 20, 30, and 60 PPI. The channels were filled up with the foams with either uniform (Figure 15.20a) pore density or with two different pore densities along the flow direction (Figure 15.20b). The results showed that the maximum heat transfer enhancement was achieved by the 60 PPI foam, which increased the heat transfer coefficient by up to 5.6 times compared to the plate heat exchanger without metal foam inserts. However, as expected, the pressure drop penalization is also huge, and the 60 PPI foam inserts increased around 5.6 times the pressure drop of the empty channel.

When working as an evaporator for the Organic Rankine cycle (ORC) (Abadi et al. 2016d), the same plate heat exchanger achieved a heat duty around 10 kW. The authors stated that lower pore densities had a higher overall heat transfer coefficient, heat duty, and effectiveness. With 20 PPI metal foam inserts, the overall heat transfer coefficient of the unbrazed heat exchanger, the heat duty, effectiveness, and pressure drop increased by 130%, 100%, 120%, and 150%, respectively, as compared to the case without metal foams.

More recently, Abadi and Kim (2017a) experimentally investigated the use of metal foam inserts to improve the flow boiling heat transfer performance of plate heat exchanger evaporators with zeotropic refrigerant mixture. In particular, the authors studied a zeotropic mixture of R245fa and R134a (0.6/0.4 molar ratio), showing that the metal foam inserts can be used to mitigate the heat transfer degradation due to the mixture. In fact, the degraded heat transfer coefficient of the mixture compared to the pure refrigerants was recovered by the introduction of metal foams in the channels. In fact, the heat transfer coefficient increase compared to the empty channel evaporator was up to 2.3, 1.9, and 1.28 times for 20, 30, and 60 PPI foam evaporators, which was very similar to the enhancement factors for the pure R245fa refrigerant.

The same authors Abadi and Kim (2017b) experimentally investigated the single-phase heat transfer of R245fa in a 4-mm internal diameter tube filled up with foam. Samples of 20 and 30 PPI with porosity of 0.9 were used. The authors reported a great enhancement of the heat transfer coefficient and a comparable increment of the pressured drop. The results were similar to those collected by Mancin et al. (2014a).

Kim et al. (2016) proposed the application of metal foam heat exchangers for a high-performance liquefied natural gas regasification system. This new configuration was meant to design a system minimizing the mass while maximizing the thermal efficiency. The predicted performance of the new concept of foam-based braze plate heat exchanger was compared with that of a shell-and-tube heat exchanger. The results showed that the metal foam had the best performance at different channel heights according to the selected mass flow rate of the fluid. In the optimized configuration of both heat exchangers, the metal foam plate one had two times higher heat transfer rate and 20% lower pressure drop than the shell-and-tube one. The performance was increased as the mass flow rate of natural gas was increased.

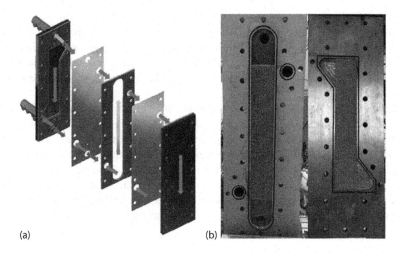

FIGURE 15.21 (a) Flow arrangement and (b) photo of a metal-foam-filled compact heat exchanger. (From Kim, Y.D. and Kim, K.C., *Int. J. Heat Mass Transf.*, 130, 162–174, 2019.)

More recently, Kim and Kim (2019) performed an experimental analysis of the use of nickel and copper foams to improve the efficiency of compact plate heat exchangers. Figure 15.21 shows the flow arrangement and a photo of the foam-filled plate heat exchanger. In particular, copper foams with 18 and 31 PPI and nickel foams with 24 and 33 PPI, with a porosity ranging between 0.87 and 0.93, were used. The experimental tests were run using R245fa as a working fluid. The authors considered the Colburn j-factor, the required pumping work per thermal resistance, and $(j/f^{1/3})$ area goodness factor to evaluate the overall thermal performance of the heat exchangers. The results showed that the Colburn j factor of the metal-foam-filled channel was up to 6.3 times higher than that of a plain channel. Consequently, the performance of the copper foam with 31 PPI, which exhibited the lowest required pumping power and overall thermal resistance, was significantly better.

The works by Nazari et al. (2015, 2017) proposed a combined synergetic use of nanofluid and metal foams to enhance the heat transfer in tubes. In particular, Nazari et al. (2015) studied the Al_2O_3/water nanofluid flow inside a 20 mm inner diameter tube, in which a 0.5 porosity metal foam structure was directly casted. It is worth underlining that in this case no thermal contact resistance was present because the foam was directly obtained by casting inside the tube. The use of the metal foam inside the tube led to a significant enhancement of heat transfer in comparison with the empty tube. Increasing the volume fraction of Alumina nanofluids led to an increase of the value of Nusselt number.

Nazari et al. (2017) also investigated the transient heat transfer from a liquid stored inside a closed reservoir. Different cooling methods, that is, the use of a metal foam-embedded tube, helical tube, and a straight tube, were used and compared for heat transfer from the fluid reservoir. The authors used a CuO/water nanofluid in various volume fractions as cooling fluid, and the results showed that the maximum thermal efficiency (in comparison with the straight tube) was 1.54 and 1.43 by using a metal foam tube and helical tube, respectively. In low flow rates ($Re < 3000$), the metal foam tube had better performance in comparison with the helical tube, while in large flow rates ($Re > 3000$), the helical tube was more effective.

15.4 Conclusion

This chapter presents a brief, yet comprehensive, description of the application of metal foams in heat exchangers. The literature has been critically reviewed to highlight the most interesting applications, by subdividing the topic in two main categories: gas (air) heat exchangers and liquid (including two-phase) heat exchangers. One can easily observe that most of the studies and applications uses air as working fluid and, in general, the gas heat exchangers have been largely studied. It can be stated that, taking into account the gas heat exchangers, the metal foams can be considered a promising enhanced surface. While for the liquid applications, most of the works studied the plate heat exchangers, in which the metal foam can be directly used as filler for the channels. A few works analyzed the tube filled up with foam. In general, it can be stated that the research activity on liquid heat transfer in metal foams is still at pre-competitive level.

The thermal contact resistance and the bonding process of metal foam still remain the main issues to be solved to achieve a successful deployment of this technology to the market. The brazing process seems to be the only method that assures a perfect contact between the ligaments and the substrate and, at the same time, induces a good structural robustness to the heat exchanger. The fouling might also be an issue for the metal foam heat exchangers, but the recent works on the topic showed that, surprisingly, they are much easier to clean and reuse as compared louver fins, which translates into significant savings and easier maintenance in the long run. On the basis of the results available in the open literature, it can be stated that the area goodness factor should be suggested as the best criterion to compare the performance of different heat exchangers.

NOMENCLATURE

c_p Specific heat capacity (J kg^{-1} K^{-1})
Da Darcy number (–)
f Friction factor (–)
G Mass velocity (kg m^{-2} s^{-1})
h Heat transfer coefficient (W m^{-2} K^{-1})
H Porous fin height (m)
j Colburn factor (–)
j^* Modified Colburn factor (–)
L Length (m)
Pr Prandtl number (–)
Re_H Reynolds number based on porous fin height [–]

Greek Letters

Δp Pressure drop (Pa)
η^* Overall foam-finned surface efficiency (–)
ρ Density (kg m^{-3})
ϕ Particle diameter (μm)

REFERENCES

Abadi, G.B., Moon, C., Kim, K.C., 2016a. Flow boiling visualization and heat transfer in metal-foam-filled mini tubes—Part I: Flow pattern map and experimental data, *International Journal of Heat and Mass Transfer* 98:857–867.

Abadi, G.B., Moon, C., Kim, K.C., 2016b. Flow boiling visualization and heat transfer in metal-foam-filled mini tubes—Part II: Developing predictive methods for heat transfer coefficient and pressure drop, *International Journal of Heat and Mass Transfer* 98:868–878.

Abadi, G.B., Moon, C., Kim, K.C., 2016c. Experimental study on single-phase heat transfer and pressure drop of refrigerants in a plate heat exchanger with metal-foam-filled channels, *Applied Thermal Engineering* 102:423–431.

Abadi, G.B., Kim, D.Y., Yoon, S.Y., Kim, K.C., 2016d. Thermal performance of a 10-kW phase-change plate heat exchanger with metal foam filled channels, *Applied Thermal Engineering* 99:790–801.

Abadi, G.B., Kim, K.C., 2017a. Enhancement of phase-change evaporators with zeotropic refrigerant mixture using metal foams, *International Journal of Heat and Mass Transfer* 106:908–919.

Abadi, G.B., Kim, K.C., 2017b. Experimental heat transfer and pressure drop in a metal foam filled channel, *Experimental Thermal and Fluid Science* 82:42–49.

Abdi, I.A., Hooman, K., Khashehchi, M., 2014. A comparison between the separated flow structures near the wake of a bare and a foam-covered circular cylinder, *Journal of Fluids Engineering*, 136(12):121203-1–121203-8.

Alvandifar, N., Saffar-Avval, M., Amani, E., 2018. Partially metal foam wrapped tube bundle as a novel generation of air-cooled heat exchangers, *International Journal of Heat and Mass Transfer* 118:171–181.

Anuar, F.S., Malayeri, M.R., Hooman, K., 2017. Particulate fouling and challenges of metal foam heat exchangers, *Heat Transfer Engineering* 38(7–8):730–742.

Anuar, F., Abdi, I.A., Hooman, K., 2018a. Flow visualization study of partially filled channel with aluminium foam block, *International Journal of Heat and Mass Transfer* 127:1197–1211.

Anuar, F.S., Abdi, I.A., Odabaee, M., Hooman, K., 2018b. Experimental study of fluid flow behaviour and pressure drop in channels partially filled with metal foams, *Experimental Thermal and Fluid Science* 99:117–128.

Ashby, M.F., Lu, T., 2003. Metal foams: A survey, *Science in China (Series B)* 46(6):521–532.

Ashby, M.F., Evans, A., Fleck, N.A., Gibson, L.J., Hutchinson, J.W., Wadley, H.N.G., 2000. *Metal Foams: A Design Guide*, Butterworth-Heinemann, Boston, MA.

Azzi, W., Roberts, W.L., and Rabiei, A., 2007. A study on pressure drop and heat transfer in open cell metal foams for jet engine applications, *Materials and Design* 28:569–574.

Boyd, B., Hooman K., 2012. Air-cooled micro-porous heat exchangers for thermal management of fuel cells, *International Communication Heat Mass Transfer* 39:363–367.

Boomsma, K., Poulikakos, D., Zwick, F., 2003. Metal foams as compact high performance heat exchangers, *Mechanics of Materials* 35:1161–1176.

Callego, N.C., Klett, J.W., 2003. Carbon foams for thermal management, *Carbon* 41:1461–1466.

Calmidi, V.V., Mahajan, R.L., 2000. Forced convection in high porosity metal foams, *ASME Journal Heat Transfer* 122:557–565.

Cavallini, A., Mancin, S., Rossetto, L., Zilio, C., 2010. Air flow in aluminum foam: Heat transfer and pressure drops measurements, *Experimental Heat Transfer* 23(1):94–105.

Chen, T., Shu, G., Tian, H., Ma, X., Wang, Y., Yang, H., 2019. Compact potential of exhaust heat exchangers for engine waste heat recovery using metal foams, *International Journal of Energy Research* 43:1–16.

Cicala, G., Cirillo, L., Diana, A., Manca, O., Nardini, S., 2016. Experimental Evaluation of Fluid Dynamic and Thermal Behaviors in Compact Heat Exchanger with Aluminum Foam, *Energy Procedia* 101: 1103–1110.

Chumpia, A., Hooman, K., 2014. Performance evaluation of single tubular aluminium foam heat exchangers, *Applied Thermal Engineering* 66:266–273.

Chumpia, A., Hooman, K., 2015. Performance evaluation of single tubular aluminium foam heat exchangers, *Applied Thermal Engineering* 83:121–130.

Costa, C., Miranda, V., Mantelli, M., Da Silva, A., Modenesi, C., Furlan, L., 2014. Experimental study of flexible, unstructured metal foams as condensation structures, *Experimental Thermal and Fluid Science* 57:102–110.

Dai, Z., Nawaz, K. Park, Y. Chen Q., Jacobi, A.M., 2012. A comparison of metal foam heat exchangers to compact multilouver designs for air-side heat transfer applications, *Heat Transfer Engineering* 33(1):21–30.

De Jaeger, P., T'Joen, C., Huisseune, H., Ameel, B., De Schampheleire, S., De Paepe, M., 2012. Assessing the influence of four cutting methods on the thermal contact resistance of open-cell aluminum foam, *International Journal of Heat and Mass Transfer* 55:6142–6151.

De Schampheleire, S., De Jaeger, P., Reynders, R., De Kerpel, K., Arneel, B., T'Joen, C., Huisseune, H., Lecompte, S., De Paepe. M., 2013a. Experimental study of buoyancy-driven flow in open-cell aluminium foam heat sinks, *Applied Thermal Engineering* 59:30–40.

De Schampheleire, S., De Jaeger, P., Huisseune, H., Ameel, B., T'Joen, C., De Kerpel, K., De Paepe. M., 2013b. Thermal hydraulic performance of 10 PPI aluminium foam as alternative for louvered fins in an HVAC heat exchanger, *Applied Thermal Engineering* 51:371–382.

Diani, A., Mancin, S., Rossetto, L., 2014. Experimental analysis of R134a flow boiling inside a 5 PPI copper foam, *Journal of Physics: Conference Series* 501:012017.

Diani, A., Mancin, S., Doretti, L., Rossetto, L., 2015. Low-GWP refrigerants flow boiling heat transfer in a 5 PPI copper foam, *International Journal of Multiphase Flow* 76:111–121.

Dukhan, N., Quiñoñes-Ramos, P., Cruz-Ruiz, E., Reyes, V.-M., Scott, E.P., 2005. One- dimensional heat transfer analysis in open-cell 10-ppi metal foam, *International Journal of Heat and Mass Transfer* 48:5112–5120.

Dukhan, N., Chen, K.-C., 2007. Heat transfer measurements in metal foam subjected to constant heat flux, *Experimental Thermal and Fluid Science* 32:624–631.

EIA, 2015. *Annual Energy Outlook 2015, With Projections to 2040*, Energy Information Administration, Washington, DC.

Elayiaraja, P., Harish, S., Wilson, L., Bensely, A., Lal, D.M., 2010. Experimental investigation on pressure drop and heat transfer characteristics of copper metal foam heat sink, *Experimental Heat Transfer* 23:185–195.

Elsherbini, A.I., Jacobi, A.M., Hrnjak, P.S., 2003. Experimental investigation of thermal contact resistances in plain fin-and-tube evaporators with collarless fins, *International Journal of Refrigeration* 26:527e536.

Fiedler, T., White, N., Dahari, M., Hooman, K., 2014. On the electrical and thermal contact resistance of metal foam, *International Journal of Heat and Mass Transfer* 72(8):565–571.

Garriti, P.T., Klausner J.F., Mei, R., 2010. Performance of aluminum and carbon foams for air side heat transfer augmentation, *ASME Journal of Heat Transfer* 132:121901-1–121901-9.

Ghadikolaie, M.M., Saffar-Avval, M., Mansoori, Z., Alvandifar, N., Rahmati, N., 2019. Heat transfer augmentation of a tube partially wrapped by metal porous layer as a potential novel tube for air cooled heat exchangers, *ASME Journal of Heat Transfer* 141:011802-1–011802-12.

Giani, L., Groppi, G., Tronconi, E., 2005. Heat transfer characterization of metallic foams, *Industrial & Engineering Chemistry Research* 44:9078–9085.

Gao, W., Xu, X., Liang, X., 2018. Flow boiling of R134a in an open-cell metal foam mini-channel evaporator, *International Journal of Heat and Mass Transfer* 126:103–115.

Hafeez, P., Yugeswaran, S., Chandra, S., Mostaghimi, J., Coyle, T.W., 2016. Fabrication of high-temperature heat exchangers by plasma spraying exterior skins on nickel foams, *Journal of Thermal Spray Technology* 25(5):1056–1067.

Hafeez, P., Chandra, S., Mostaghini, J., 2017. Heat transfer during high temperature gas flow through metal foam heat exchangers, *ASME Journal of Heat Transfer* 139:121801-1–121801-11.

Hamadouche, A., Azzi, A., Abboudi, S., Nebbali, R., 2018. Enhancement of heat exchanger thermal hydraulic performance using aluminum foams, *Experimental Thermal Fluid Science* 92:1–12.

Han, X., Kashif, N., Bock, J., Jacobi, A.M., 2012a. Open-cell metal foams for use in dehumidifying heat exchangers, *International Refrigeration and Air Conditioning Conference*. Paper 1312. West Lafayette, IN.

Han, X.H., Wang, Q., Park, Y.G., T'Joen, C., Sommers, A., Jacobi, A., 2012b. A review of metal foam and metal matrix composites for heat exchangers and heat sinks, *Heat Transfer Engineering* 33(12):991–1009.

Hendricks, T., Choate, W.T., 2006. *Engineering Scoping Study of Thermoelectric Generator Systems for Industrial Waste Heat Recovery*, Washington, DC, US Department of Energy.

Hooman, K., 2014. Thermohydraulics of porous heat exchangers: Full or partial blockage? In: K. Vafai, A. Bejan, A. Nakayama and O. Manca, *Proceedings of the 5th International Conference on Porous Media and Its Application in Science and Engineering*, Kona, Hawaii, HI.

Hooman, K., Tamayol, A., Malayeri, M.R., 2012. Impact of particulate deposition on the thermohydraulic performance of metal foam heat exchangers: A simplified theoretical model, *Journal of Heat Transfer* 134(9): 092601.1–092601.7.

Hooman, K., Malayeri, M.R., 2016. Metal foams as gas coolers for exhaust gas recirculation systems subjected to particulate fouling, *Energy Conversion and Management* 117:475–481.

Hsieh, W.H., Wu, J.Y., Shih, W.H., Chiu, W.C., 2004. Experimental investigation of heat- transfer characteristics of aluminum-foam heat sink, *International Journal of Heat and Mass Transfer* 47:5149–5157.

Hu, H., Weng, X., Zhuang, D., Ding, G., Lai, Z., Xu, X., 2016. Heat transfer and pressure drop characteristics of wet air flow in metal foam under dehumidifying conditions, *Applied Thermal Engineering* 93:1124–1134.

Hu, H., Lai, Z., Ding, G., 2017. Experimental investigation on water drainage characteristics of open-cell metal foams with different wettabilities, *International Journal of Refrigeration* 79:101–113.

Hu, H., Lai, Z., Ding, G., 2019. Influence of surface wettability on heat transfer and pressure drop characteristics of wet air in metal foam under dehumidifying conditions, *International Journal of Thermal Sciences* 135:331–343.

Huisseune, H., De Jaeger, P., De Schampheleire, S., Ameel, B., De Paepe, M., 2014. Simulation of an aluminum foam heat exchanger using the volume averaging technique, *Procedia Material Science* 4:353–358.

Huisseune, H., De Schampheleire, S., Ameel, B., De Paepe, M., 2015. Comparison of metal foam heat exchangers to a finned heat exchanger for low Reynolds number applications, *International Journal of Heat and Mass Transfer* 89:1–9.

Hwang, J.-J., Hwang, G.-J., Yeh, R.-H., Chao, C.-H., 2002. Measurement of interstitial convective heat transfer and frictional drag for flow across metal foams, *ASME Journal of Heat Transfer* 124:120–129.

Ismagilov, Z.R., Pushkarev, V.V., Podyacheva, O.Y., Koryabkina, N.A., Veringa, H., 2001. A catalytic heat exchanging tubular reactor for combining of high temperature exothermic and endothermic reactions, *Chemical Engineering Journal* 82:355–360.

Jamin, Y.L., Mohamad, A.A., 2008. Natural convection heat transfer enhancements from a cylinder using porous carbon foam: Experimental study, *ASME Journal of Heat Transfer* 130:122502-1–122502-6.

Jazi, H., Mostaghimi, J., Chandra, S., Pershin, L., Coyle, T., 2009. Spray-formed, metal-foam heat exchangers for high temperature applications, *ASME Journal of Thermal Science and Engineering Applications* 1(3):031008.

Kang, B., Kim, K., Lee, B., Noh, J., 2008. Heat transfer and acoustic properties of open cell aluminum foams, *Journal of Material Science and Technology* 24:1–6.

Khashehchi, M., Abdi, I.A., Hooman, K., Roesgen, T., 2014. A comparison between the wake behind finned and foamed circular cylinders in cross-flow, *Experimental Thermal and Fluid Science* 52:328–338.

Khashehchi, M., Abdi, I.A., Hooman, K., 2015. Characteristics of the wake behind a heated cylinder in relatively high Reynolds number, *International Journal of Heat and Mass Transfer* 86:589–599.

Khashehchi, M., Abdi, I.A., Hooman, K., 2017. A comparative analysis on the shed vortices from the wake of finned, foam-wrapped cylinders, *Fluid Dynamics Research* 49:4.

Kim, S.Y., Paek, J.W., Kang, B.H., 2000. Flow heat transfer correlations for porous fin in a plate-fin heat exchanger, *Journal of Heat Transfer* 122:572–578.

Kim, S.Y., Kang, B.H., Kim, J.-H., 2001. Forced convection from aluminium foam materials in an asymmetrically heated channel, *International Journal Heat and Mass Transfer* 44:1451–1454.

Kim, A., Hasan, M.A., Nahm, S.H., Cho, S.S., 2005. Evaluation of the compressive mechanical properties of Al-foams using electrical conductivity, *Composite Structures* 71:191–198.

Kim, D.W., Bar-Cohen, A., Han, B., 2008. Forced convection and flow boiling of a dielectric liquid in a foam filled channel, *IEEE Transaction* 86–94.

Kim, Y.D., Sung, T.H., Kim, K.C., 2016. Application of metal foam heat exchangers for a high performance liquefied natural gas regasification system, *Energy* 105:57–69.

Kim, Y.D., Kim, K.C., 2019. An experimental study on the thermal and hydraulic characteristics of open-cell nickel and copper foams for compact heat exchangers, *International Journal of Heat and Mass Transfer* 130:162–174.

Koltsakis, G., Katsaounis, D., Samaras, Z., Naumann, D., Saberi, S., Boehm, A., Markomanolakis, I., 2008. Development of metal foam based after treatment on a diesel passenger car, SAE paper.

Leong, K.C., Li, H.Y., Jin, L.W., Chai, J.C., 2011. Convective heat transfer in graphite foam heat sink with baffle and stagger structures, *ASME Journal Heat Transfer* 133:060902.

Li, H.W., Zhang, C.Z., Yang, D., Sun, B., Hong, W.P., 2019. Experimental investigation on flow boiling heat transfer characteristics of R141b refrigerant in parallel small channels filled with metal foam, *International Journal of Heat and Mass Transfer* 133:21–35.

Lin, W., Sundén, B., Yuan, J., 2013. A performance analysis of porous graphite foam heat exchangers in vehicles, *Applied Thermal Engineering* 50:1201–1210.

Liu, P.S., Liang, K.M., 2001a. Functional materials of porous metals made by P/M, electroplating and some other techniques, *Journal of Materials Science* 36:5059–5072.

Liu, P., Yu, B., Hu, A., Liang, K.M., and Gu, S.R., 2001b. Development in applications of porous metals, *Transaction of Nonferrous Metals Society of China* 11:630–638.

Losito, O., 2008. An Analytical characterization of metal foams for shielding applications, *Proceedings of the Progress in Electromagnetics Research Symposium*, Cambridge, New York, pp. 247–252.

Lu, W., Zhao, C.Y., Tassou, S.A., 2006. Thermal analysis on metal-foam filled heat exchangers. Part I: Metal-foam filled tubes, *International Journal of Heat and Mass Transfer* 49:2751–2761.

Mancin, S., Zilio, C., Cavallini, A., Rossetto, L., 2010a. Heat transfer during air flow in aluminum foams, *International Journal of Heat and Mass Transfer* 53(21–22):4976–4984.

Mancin, S., Zilio, C., Cavallini, A., Rossetto, L., 2010b. Pressure drop during air flow in aluminum foams, *International Journal of Heat and Mass Transfer* 53(15–16):3121–3130.

Mancin, S., Zilio, C., Rossetto, L., Cavallini, A., 2011. Heat transfer performance of aluminum foams, *ASME Journal of Heat Transfer* 133(6):060904.

Mancin, S., Zilio, C., Rossetto, L., Cavallini, A., 2012a. Foam height effects on heat transfer performance of 20 PPI aluminum foams, *Applied Thermal Engineering* (49):55–60.

Mancin, S., Zilio, C., Diani, A., Rossetto, L., 2012b. Experimental air heat transfer and pressure drop through copper foams, *Experimental Thermal and Fluid Science* (36):224–232.

Mancin, S., Zilio, C., Diani, A., Rossetto, L., 2013. Air forced convection through metal foams: Experimental results and modelling, *International Journal of Heat and Mass Transfer* 62:112–123.

Mancin, S., Diani, A., Doretti, L., Rossetto, L., 2014a. Liquid and flow boiling heat transfer inside a copper foam, *Procedia Materials Science* 4:346–351.

Mancin, S., Diani, A., Doretti, L., Rossetto, L., 2014b. R134a and R1234ze(E) liquid and flow boiling heat transfer in a high porosity copper foam, *International Journal of Heat and Mass Transfer* 74:77–87.

Muley, A., Kiser, C., Sundén, B., Shah, R.K., 2012. Foam heat exchangers: A technology assessment, *Heat Transfer Engineering* 33(1):42–51.

Müller-Steinhagen, H., Müller-Steinhagen, H., Malayeri, M.R., Watkinson, A.P., 2011. Heat exchanger fouling: Mitigation and cleaning strategies, *Heat Transfer Engineering* 32:189–196.

Nawaz, K., 2014. *Aerogel Coated Metal Foams for Dehumidification Applications*, University of Illinois, Urbana-Champaign, IL.

Nawaz, K., Bock, J., Dai, Z., Jacobi, A.M., 2010. Experimental studies to evaluate the use of metal foams in highly compact air-cooling heat exchangers, *International Refrigeration and Air Conditioning Conference*. Paper 1150. Purdue e-Pubs, Purdue University, West Lafayette, IN.

Nawaz, K., Bock, J., Jacobi, A.M., 2017. Thermal-hydraulic performance of metal foam heat exchangers under dry operating conditions, *Applied Thermal Engineering* 119:222–232.

Nawaz, K., Jacoby, A., 2018. Metal foams: Novel materials for air cooling and heating application- performance under dry, wet and frosting conditions, *Proceedings of 16th International Heat Transfer Conference*, IHTC-16, Beijing, China.

Nazari, M., Ashouri, M., Kayhani, M.H., Tamayol, A., 2015. Experimental study of convective heat transfer of a nanofluid through a pipe filled with metal foam, *International Journal of Thermal Sciences* 88:33–39.

Nazari, M., Baie, N.B., Ashouri, M., Shahmardan, M.M., Tamayol, A., 2017. Unsteady heat transfer from a reservoir fluid by employing metal foam tube, helically tube and straight tube: A comparative experimental study, *Applied Thermal Engineering* 111:39–48.

Odabaee, M., Hooman, K., 2011. Application of metal foams in air-cooled condensers for geothermal power plants: An optimization study, *International Communication in Heat and Mass Transfer* 38:838–843.

Odabaee, M., Hooman, K., 2012. Metal foam heat exchangers for heat transfer augmentation from a tube bank, *Applied Thermal Engineering* 36:456–463.

Odabaee, M., De Paepe, M., De Jaeger, P., T'Joen, C., Hooman, K., 2013a. Particle deposition effects on heat and fluid flow through a metal foam-wrapped tube bundle, *International Journal of Numerical Methods for Heat and Fluid Flow* 23(1):74–87.

Odabaee, M., Mancin, S., Hooman, K., 2013b. Metal foam heat exchangers for thermal management of fuel cell systems—An experimental study, *Experimental Thermal and Fluid Science* 51:214–219.

Orihuela, M.P., Anuar, F.S., Abdi, I.A., Odabaee, M., Hooman, K., 2018. Thermohydraulics of a metal foam-filled annulus, *International Journal of Heat and Mass Transfer* 117:95–106.

Pranoto, I., Leong, K.C., 2014. An experimental study of flow boiling heat transfer from porous foam structures in a channel, *Applied Thermal Engineering* 70:100–114.

Ribeiro, G.B., Barbosa, Jr. J.R., Prata, A.T., 2012. Performance of microchannel condensers with metal foams on the air-side: Application in small-scale refrigeration systems, *Applied Thermal Engineering* 36:152–160.

Ribeiro, G.B., Barbosa, Jr. J.R., 2013. Comparison of metal foam and louvered fins as air-side heat transfer enhancement media for miniaturized condensers, *Applied Thermal Engineering* 51:334–337.

Sadeghi, E., Hsieh, S., Bahrami, M., 2011. Thermal conductivity and contact resistance of metal foams, *Journal of Physics D: Applied Physics* 44:7.

Sauret, E., Abdi, I., Hooman, K., 2014. Fouling of waste heat recovery: Numerical and experimental results, *19th Australasian Fluid Mechanics Conference*, Melbourne, Australia.

Sauret, E., Hooman, K., 2014. Particle size distribution effects on preferential deposition areas in metal foam wrapped tube bundle, *International Journal of Heat and Mass Transfer* 79:905–915.

Sauret, E., Saha, S.C., Gu, Y., 2013. Numerical simulations of particle deposition in metal foam heat exchangers, *International Journal of Computation Materials Science and Engineering* 2:1350016.

Sekulic, D.P., Dakhoul, Y.M., Zhao, H., Liu, W., 2008. Aluminum foam compact heat exchanger: brazing technology development vs. thermal performance (CELLMET2008), *Proceedings of the International Symposium on Cellular Metals for Structural and Functional Applications*, Fraunhofer Center, Dresden, Germany, pp. 5e10.

Sertkaya, A.A., Altınısık, K., Dincer, K., 2012. Experimental investigation of thermal performance of aluminum finned heat exchangers and open-cell aluminum foam heat exchangers, *Experimental Thermal and Fluid Science* 36:86–92.

Shah, R.K., Sekulic, D.P., 2003. *Fundamentals of Heat Exchanger Design*, John Wiley & Sons, Hoboken, NJ.

Spoerke, E.D., Murray, N.G., Li, H., Brinson, L.C., Dunand, D.C., Stupp, S., 2005. A bioactive titanium foam scaffold for bone repair, *Acta Biomaterialia* 1:523–533.

Straatmam, A.G., Gallego N.C., Thompson B.E., Hangan H., 2006. Thermal characterization of porous carbon foam—Convection in parallel flow, *International Journal of Heat and Mass Transfer* 49:1991–1998.

T'Joen, C., De Jaeger, P., Huisseune, H., Van Herzeele, S., Vorst, N., De Paepe, M., 2010. Thermo-hydraulic study of a single row heat exchanger consisting of metal foam covered round tubes, *International Journal Heat and Mass Transfer* 53, 3262–3274.

Tsolas, N., Chandra, S., 2012, Forced convection heat transfer in spray formed copper and nickel foam heat exchanger tubes, *ASME Journal Heat Transfer* 134(6):062602.

Tuchinskiy, L., 2005. Novel fabrication technology for metal foam, *Journal of Advanced Materials* 37:60–65.

Wang, T., Luan, W., Liu, T., Tu, S.T., Yan, J., 2016. Performance enhancement of thermoelectric waste heat recovery system by using metal foam inserts, *Energy Conversion and Management* 124:13–19.

Webb, R.L., 1994. *Principles of Enhanced Heat Transfer*, John Wiley & Sons, New York.

Wilkinson, D.S., Paserin, V., 2004. Processing and properties of nickel foams for battery electrodes, *Proceedings of the 2nd Symposium on Advanced Materials for Energy Conversion*, Charlotte, NC.

Zhao, C.Y., Kim, T., Lu, T.J., 2001. *Thermal Transport Phenomena in Porvair Metal Foams and Sintered Beds, Final Report*, University of Cambridge, Cambridge, UK.

Zhao, C.Y., Kim, T., Lu, T.J., Hodson, H.P., 2004. Thermal transport in high porosity cellular foams, *Journal of Thermophysics and Heat Transfer* 18(3):309–317.

Zhao, C.Y., Lu, W., Tassou, S.A., 2006. Thermal analysis on metal-foam filled heat exchangers. Part II: Tube heat exchangers, *International Journal of Heat and Mass Transfer* 49:2762–2770.

Zhao, C.Y., 2012. Review on thermal transport in high porosity cellular metal foams with open cells, *International Journal of Heat and Mass Transfer* 55:3618–3632.

Zilio, C., Mancin, S., Diani, A., Rossetto, L. 2011. Aluminum foams as possible extended surfaces for air cooled condenser, *Proceedings of 23rd IIR International Congress of Refrigeration*, Prague, Czech Republic.

Zhu, Y., Hua, H., Sun S., Ding, G., 2015a. Flow boiling of refrigerant in horizontal metal-foam filled tubes: Part 1—Two-phase flow pattern visualization, *International Journal of Heat and Mass Transfer* 91:446–453.

Zhu, Y., Hua, H., Sun S., Ding, G., 2015b. Flow boiling of refrigerant in horizontal metal-foam filled tubes: Part 2—A flow-pattern based prediction method for heat transfer, *International Journal of Heat and Mass Transfer* 91:502–511.

16

Heat and Fluid Flow in Porous Media for Polymer-Electrolyte Fuel Cells

Prodip Kumar Das and Deepashree Thumbarathy

CONTENTS

16.1 Introduction 341
16.2 Operation Principle of Polymer-Electrolyte Fuel Cells 342
    16.2.1 Operating Temperature and Pressure 344
    16.2.2 Reactant Flow Rate and Relative Humidity 344
    16.2.3 Thermodynamic Efficiency 345
16.3 Structures of Porous Media in Polymer-Electrolyte Fuel Cells 345
    16.3.1 Catalyst Layer 345
    16.3.2 Gas Diffusion and Microporous Layers 346
16.4 Transport in Polymer-Electrolyte Fuel Cells 348
    16.4.1 Mass Diffusion 348
    16.4.2 Convection 349
    16.4.3 Conduction 349
    16.4.4 Migration and Proton Transport 350
    16.4.5 Electron Transport 350
16.5 Heat and Fluid Flow in Catalyst Layer and Gas-Diffusion Electrode 351
    16.5.1 Multi-Phase Volume-Averaging Method 351
    16.5.2 Multi-Phase Mixture and Two-Fluid Methods 353
        16.5.2.1 Transport of Gas Phase 354
        16.5.2.2 Transport of Liquid Phase 355
        16.5.2.3 Transport of Energy 356
16.6 Summary 356
Nomenclature 357
Acknowledgments 358
References 358

16.1 Introduction

Global warming and greenhouse gas (GHG) emissions are two critical issues currently addressed by scientists all over the world. Some GHGs (e.g., CO_2) are emitted to the atmosphere through natural processes and human activities. Among the human activities, internal combustion (IC) engine vehicles are significant producers of harmful greenhouse emissions. For instance, a quarter of domestic CO_2 and other GHG emissions in the UK come from the IC engine vehicles and drones (www.legislation.gov.uk, www.theccc.org.uk). The amount of GHG emissions that come from the combustion of fossil fuels is

341

about 90%. Recent research also shows that the amount of CO_2 produced from a small car can be reduced by as much as 72% when powered by a fuel cell running on hydrogen reformed from natural gas instead of a gasoline IC engine. In addition, the world's fossil fuel reserve is limited; hence, alternative and sustainable energy technologies are required for our next generation.

Zero-emission hydrogen fuel-cell vehicles and drones can reduce GHG emissions due to their ability to efficiently converting the chemical energy of hydrogen fuel to generate electricity without a combustion process. Fuel cells also have a broad range of other potential uses, from small-scale battery replacement applications to multi-megawatt distributed power generation. For vehicles and drones, polymer-electrolyte fuel cells (PEFCs) are considered a long-term solution for the decarburization of these sectors due to their capabilities of producing high-power densities under a rapid load change condition. Many large vehicle manufacturers and the drone industry are dedicated to hydrogen fuel cell technology, such as Toyota and Hyundai Motor.

In a broader sense, PEFCs can be divided into two categories: gas feed (hydrogen) and liquid feed (methanol). Hydrogen fuel cells can be further subdivided into proton-exchange-membrane fuel cells (PEMFCs) and anion-exchange-membrane fuel cells (AEMFCs). In PEMFCs and AEMFCs, hydrogen and oxygen (or air) are combined electrochemically across a polymer membrane generating electricity, water, and heat. Conversely, in methanol fuel cells, methanol and oxygen (or air) are combined electrochemically for generating electricity, water, and heat. The operation of these fuel cells involves a complex overlap of interrelated physicochemical processes, which include electrochemical reactions as well as transport of ions, electrons, energy, and species in gas and liquid phases across several porous layers. Though a significant improvement has been made over the past decades, transports of heat and fluid that flow through the porous players of PEFCs still need attention, as the cost and durability of PEFCs stifle their commercial utilization (Weber et al. 2014). As the interrelated physicochemical processes of these fuel cells are identical, the discussion in this chapter will focus on PEMFCs with some highlights of AEMFCs.

16.2 Operation Principle of Polymer-Electrolyte Fuel Cells

The images of a PEFC stack and single cell and the schematic of various cell components are shown in Figure 16.1. In a PEFC, the anode, electrolyte (membrane), and cathode are sandwiched together, known as a membrane electrode assembly (MEA). Both the anode and the cathode include a gas diffusion layer (GDL) and a catalyst layer (CL). Often a microporous layer (MPL) is sandwiched between GDL and CL for better mass transport. To achieve higher output voltages and higher power, cells can be combined with a large fuel cell stack (Figure 16.1).

Both the PEMFCs and the AEMFCs have identical cell components configurations. The key differences between them are the electrode reactions and the transports of protons and hydroxides ions, as shown in Figure 16.2. In PEMFCs, humidified H_2 gas is supplied into the anode gas flow channel.

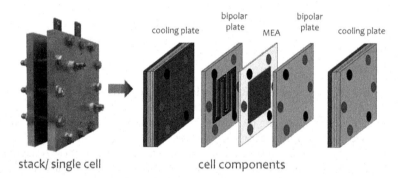

FIGURE 16.1 Images of a PEFC stack and single cell and the schematic of various cell components. (From www.fuelcellstore.com and EnSys research group, www.staff.ncl.ac.uk/prodip.das.)

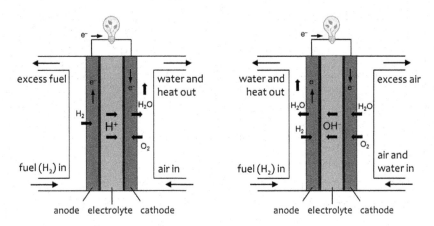

FIGURE 16.2 Schematic illustrations of a proton-exchange-membrane fuel cell (left) and an anion-exchange-membrane fuel cell (right).

In the presence of a catalyst, hydrogen molecules are stripped of their electrons to produce protons at the anode catalyst layer (aCL). Then protons transport through a proton-exchange membrane (PEM) to the cathode catalyst layer (cCL), and electrons travel through an external circuit to the cathode. At the cCL, protons and electrons recombine with oxygen to generate water. The electrode reactions for PEMFCs are:

$$\text{Anode: } 2H_2 \rightarrow 4H^+ + 4e^- \tag{16.1}$$

$$\text{Cathode: } 4H^+ + 4e^- + O_2 \rightarrow 2H_2O \tag{16.2}$$

Conversely, the oxygen reduction reaction at the cCL produces hydroxides ions (OH−) in AEMFCs and then transport through the anion-exchange membrane (AEM) to the anode. At the aCL, hydroxide ions combine with hydrogen to generate water and electrons. Electrons are then conducted through an external circuit to the cathode to complete the overall cell reaction. The electrode reactions for AEMFCs are:

$$\text{Anode: } 2H_2 + 4OH^- \rightarrow 4H_2O + 4e^- \tag{16.3}$$

$$\text{Cathode: } O_2 + 2H_2O + 4e^- \rightarrow 4OH^- \tag{16.4}$$

The aforementioned electrode reactions (Equations 16.1 through 16.4) are written in the single step form. However, multiple elementary reaction pathways are possible at each electrode (Li 2006), a discussion of which is beyond the scope of this chapter. For both PEMFCs and AEMFCs, the overall electrochemical reaction can be represented by the following:

$$2H_2 + O_2 \rightarrow 2H_2O + \text{Heat} + \text{Electrical Energy} \tag{16.5}$$

The electrical energy of a PEFC is estimated from the cell potential and current. While the maximum potential is dictated by the thermodynamics of the reactions, the current is limited by the kinetics and transport limitations within the cell and associated transport losses. Due to the losses (also known as overpotentials), the actual cell voltage is always lower than the equilibrium potential. A typical polarization (potential vs. current density) curve of a PEFC has three distinct regions of losses: activation loss, Ohmic loss, and mass transfer loss. Activation loss is due to the kinetics of charge transfer reactions in aCL and cCL, the Ohmic loss is due to the resistance of cell components, and mass transfer loss (concentration overpotential) is due to the limited rate of mass transfer. Figure 16.3 shows a typical

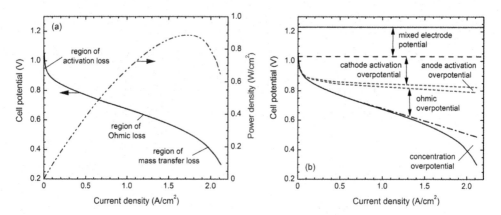

FIGURE 16.3 (a) Typical polarization and power density curves for PEMFCs with the region showing three main losses; and (b) a breakdown of various overpotentials.

performance curve of a PEMFC (also known as polarization curve) and a breakdown of various overpotentials (Das et al. 2007; Weber et al. 2014). An AEMFC will exhibit similar curves but the current density values will be lower (Machado et al. 2017, 2018).

16.2.1 Operating Temperature and Pressure

The most important operating parameters for a PEFC operation are temperature and pressure. A PEFC can be operated at room temperature as high as 80°C, while the operating pressure can be at ambient pressure or at a higher pressure. For the reaction that takes place at temperature, T and pressure p, the maximum potential can be obtained from a PEFC is given by the following expression (Li 2006)

$$E_r^0 = -\frac{\Delta g}{nF} = -\frac{\Delta h - T\Delta s}{nF} \tag{16.6}$$

where E_r^0 is the standard reversible cell potential, Δg is the change in the Gibbs free energy per mole of H_2, Δh is the enthalpy change per mole of H_2, Δs is the entropy change per mole of H_2, n is the number of electron transfer per mole of H_2, and F is the Faraday constant. For Equations (16.1) and (16.3), the value of n is 2. If pure H_2 and O_2 are used as fuel and reactant, the standard reversible cell potential (at 25°C and 1 atm) will be 1.229 and 1.185 V for the product water in liquid form and the vapor form, respectively. For higher temperatures, the reversible cell potential will be lower as it decreases with the temperature. On the other hand, the reversible cell potential will be higher at a higher pressure. Although higher operating temperature results in a lower reversible potential, the overall cell performance and the actual cell voltage can be higher at higher temperature due to better reaction kinetics.

16.2.2 Reactant Flow Rate and Relative Humidity

The reactants flow rates and relative humidity also play an important role in the overall performance of a PEFC. Both H_2 and O_2 (or air) are required to be humidified before supplying to the PEFCs for keeping the electrolyte membrane hydrated. Typical relative humidity is about 100% or less. The flow rate of H_2 and O_2 should always be equal or higher than the consumption rates of the fuel and oxidant in a PEFC. The consumption rates ($\dot{N}$) of H_2 and O_2 can be calculated based on Faraday's law using the following expressions:

$$\dot{N}_{H_2,\text{consumed}} = \frac{I}{2F} \text{ and } \dot{N}_{O_2,\text{consumed}} = \frac{I}{4F} \tag{16.7}$$

Heat and Fluid Flow in Porous Media for Polymer-Electrolyte Fuel Cells 345

where I is the cell current. Often the actual flow rates are higher than the consumption rates and represented by the parameter called stoichiometry, S_t:

$$S_{t,H_2} = \frac{\dot{N}_{H_2,\,in}}{\dot{N}_{H_2,\,consumed}} \text{ and } S_{t,O_2} = \frac{\dot{N}_{O_2,\,in}}{\dot{N}_{O_2,\,consumed}} \tag{16.8}$$

For typical PEFC operation, the stoichiometry ratios are about 1.1 ~ 1.2 for H_2 and 2 for O_2.

16.2.3 Thermodynamic Efficiency

The theoretical thermodynamic efficiency (also known as reversible efficiency) of a PEFC is defined as the ratio of maximum possible electrical work to the total chemical energy (Li 2006). Thus, we can write

$$\eta_{rev} = \frac{\Delta g}{\Delta h} = 1 - \frac{T \Delta s}{\Delta h} \tag{16.9}$$

For the standard operating pressure and temperature, the reversible efficiency of a PEFC is about 83%. However, the actual efficiency of a PEFC will be much lower than the reversible efficiency due to the losses illustrated in Figure 16.3.

16.3 Structures of Porous Media in Polymer-Electrolyte Fuel Cells

A rigorous description of PEFC operation requires the coupling of thermodynamics and kinetics of the reactions and conservation of species, energy, and momentum within a representative geometry. A discussion of which is beyond the scope of this chapter and can be found elsewhere (Weber et al. 2012). Thus, the focus is given on the conservation of species, energy, and momentum in this chapter.

To understand the transport of species, energy, and momentum, it is essential to understand the structures of PEFC components and the associated phases. In a PEFC, several phases co-exist together, including a liquid phase, a gas phase, and a solid phase. Further, the multi-phase mixture flows through narrow gas channels, porous gas diffusion layers, and porous catalyst layers. Therefore, the governing equations for PEFCs are influenced by the physical structure of the cell and co-existing phases.

The core part of a PEFC has nine layers. Four layers on the anode side, four layers on the cathode side, and a polymer electrolyte layer in the middle. The four layers on each side of the polymer electrolyte layer are a bipolar plate, a GDL, an MPL, and a CL. The flow channels are grooved on the bipolar plate. The average dimensions of the polymer electrolyte layer, CL, MPL, GDL, and flow channel are 50, 10, 40, 200, and 1000 μm, respectively. Each of these layers has several co-existing phases. A schematic illustration of the structures of these layers and various phases is shown in Figure 16.4. The porous media within a PEFC include the CLs and the diffusion media, which are often composed of multiple layers including GDLs and MPLs. In the following, an overview is given for these porous layers.

16.3.1 Catalyst Layer

PEFCs have two CLs where the electrochemical reactions take place. CLs are in direct contact with the polymer electrolyte and MPL, at both anode and cathode sides. These CLs have a complex structure of the membrane, supported carbon particles, catalyst particles, and void space. They must provide void spaces for the reactants to reach the catalyst surface and the products to be escaped. In addition, there should be a pathway for the electrons and protons to reach the reaction sites. The most widely used catalyst layer is the Pt/C CL, where platinum (Pt) nanoparticles dispersed onto the surfaces of larger carbon black particles. The typical pore size in Pt/C CL is about 0.05 μm, and the porosity is about 20% ~ 30%. The thickness of CLs varies between 10 and 50 μm. The Pt/C CL can have a Pt-loading of several mg per cm^2.

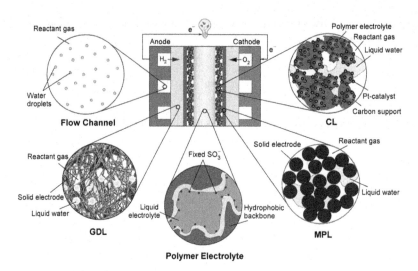

FIGURE 16.4 Physical structures of various layers of a PEFC.

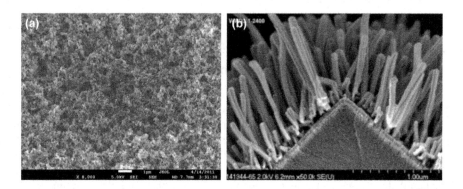

FIGURE 16.5 Scanning electron micrographs of (a) conventional Pt/C catalyst layer; and (b) NSTF (PtCoMn alloy) catalyst layer. (From Das, P.K. and Weber, A.Z., Water management in PEMFC with ultra-thin catalyst-layers, *Proceedings of the ASME 11th Fuel Cell Science, Engineering, and Technology Conference*, Paper No. FuelCell2013-18010:V001T01A002 2013, Debe, M.K., *Nature*, 486, 43–51, 2012.)

Due to the higher cost of Pt catalysts, there is a big push to reduce catalyst loading from the catalyst layers, which leads to the development of the nanostructured thin film (NSTF) catalyst layers. The key differences of NSTFs with conventional Pt/C layers are low Pt-loading, thinner, and no carbon support or ionomer, as the Pt catalyst is directly deposited to form electronically conductive and electrochemically active layer (Debe 2013; Holdcroft 2014). The NSTF CL is 20–30 times (typically 0.5 ~ 1 μm) thinner than the conventional Pt/C layer. Thus, it has significantly lower Pt-loading (in the order of 0.05–0.15 mg-Pt/cm²), which would be useful to reduce the cost of a PEFC system. The scanning electron micrographs of conventional Pt/C and NSTF catalyst layers are shown in Figure 16.5.

16.3.2 Gas Diffusion and Microporous Layers

The porous structure of a GDL is made by weaving carbon fibers into a carbon cloth or by pressing carbon fibers together into a carbon paper. Figure 16.6 shows the scanning electron micrographs of typical carbon paper GDL and carbon cloth GDL. The role of GDLs in PEFCs is to provide mechanical support while spreading the reactant gas and electrons over the electrode as well as allowing the

Heat and Fluid Flow in Porous Media for Polymer-Electrolyte Fuel Cells 347

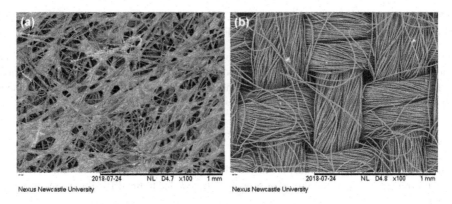

FIGURE 16.6 Scanning electron micrographs of typical (a) carbon paper GDL; and (b) carbon cloth GDL.

product water to transport from the catalyst layer to the gas flow channel. The GDLs are also rendered wet-proof by saturating the pores with hydrophobic poly-tetra-fluoro-ethylene (PTFE) emulsions, followed by drying and sintering to affix the PTFE particles to the carbon fiber to improve liquid transport, as visible in carbon paper GDL in Figure 16.6a. The typical pore size in GDL is about 10 μm, and the uncompressed porosity is about 70% ~ 80% (Nam et al. 2009). The thickness of GDLs varies between 100 and 300 μm.

Due to the higher porosity and pore size of GDLs compared with the CLs, most commercial GDLs often come with an MPL, which is a thin layer of carbon nanoparticles mixed with PTFE. For instance, the standard SIGRACET (C-type) MPL is based on 77 wt% carbon black and 23 wt% PTFE. The scanning electron micrograph of a typical MPL is shown in Figure 16.7. The MPL acts as a transition layer for the transports of gases and liquid between the macropores of GDL and the nanopores of adjacent CL. Moreover, MPLs act as a barrier for the liquid water and helps keep the water within the membrane from escaping and prevent drying out the membrane. The typical pore size in MPL varies between 0.1 and 0.3 μm and the porosity is about 50%. The thickness of MPLs varies between 10 and 100 μm. Often the combined layer of the GDL and MPL is referred to as a gas-diffusion electrode (GDE), as the transports through these two layers are fundamentally identical. Thus, the GDL and MPL will be referred to as GDE in the subsequent sections.

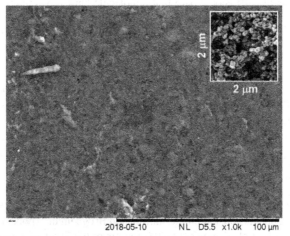

FIGURE 16.7 Scanning electron micrograph of a typical microporous layer.

16.4 Transport in Polymer-Electrolyte Fuel Cells

The transports of mass, energy, species, electron, and proton in PEFCs involve convection, diffusion, conduction, and migration. In this section, these processes are briefly described for GDE, CL, and membrane, while the transports in the flow channels and solid bipolar plates and the reaction kinetics are available in Ref. (Weber et al. 2012).

16.4.1 Mass Diffusion

The primary mechanism of mass transport inside PEFC porous media is the diffusion. Considering the dilute species transport theory, the flux of species i in phase k due to the diffusion is governed by Fick's first law

$$\mathbf{N}_{i,k} = -D_i \, \nabla c_{i,k} \tag{16.10}$$

where D_i is the diffusion coefficient of species i, and $c_{i,k}$ is the concentration species i in phase k. The diffusion coefficient can be predicted from the mean free path and average velocity for molecules in an ideal gas from the Maxwell–Boltzmann distribution, and it obeys the following expression

$$D \propto \frac{T^{\frac{3}{2}}}{p} \tag{16.11}$$

where T and p are the temperature and pressure, respectively. However, the actual diffusion coefficient in PEFC porous media can be different from the real diffusion coefficient. This is because the available cross section for diffusion is less than the free fluid, and the distance between one point and another in the porous medium. Thus, the diffusion coefficient should be corrected using the porosity and tortuosity of PEFC porous media, which is known as the effective diffusion coefficient.

$$D_i^{\text{eff}} = \frac{\varepsilon}{\tau} D_i \tag{16.12}$$

where D_i^{eff} is the effective diffusion coefficient of species i, ε is the porosity, and τ is the tortuosity. The tortuosity represents the diffusional path that a molecule must travel to cross a region of a certain thickness. The simplest approach to determine the tortuosity is the Bruggeman approximation, which gives

$$\tau = \varepsilon^{-0.5} \tag{16.13}$$

The Bruggeman approximation has widely been used for estimating the effective transport properties for CLs, GDLs, and MPLs. As the pores of these porous layers can be simultaneously filled with gas and liquid phases, the Bruggeman equation requires revision to account for the presence of the liquid phase. Considering the liquid saturation, the Bruggeman equation for two-phase scenarios can be revised as

$$D_i^{\text{eff}} = \varepsilon^m \left(1 - S_l\right)^n D_i \tag{16.14}$$

where S_l is the liquid saturation, m is the Bruggeman exponent (=1.5), and n is the saturation exponent. Often the saturation exponent is considered the same as the Bruggeman exponent. However, the value of the saturation exponent is strongly dependent on the amount of liquid saturation (Das et al. 2010; Hwang and Weber 2012). At low liquid saturation, the Bruggeman approximation is reasonable, but a higher-order correction is needed for the diffusivity at high liquid saturation. It is believed that the Bruggeman

Heat and Fluid Flow in Porous Media for Polymer-Electrolyte Fuel Cells 349

approximation underpredicts the diffusion inside the CLs, where the Knudsen diffusion can be significant. Thus, an appropriate measure will be required for the effective diffusion coefficient in CLs to account for the Knudsen diffusion as well (Nonoyama et al. 2011).

16.4.2 Convection

Convection is applicable to both mass and heat transfer. Due to the phase velocities, the fluid (gas or liquid) can be transported through the pores due to the advection. If the phase velocities are known, one can write the advection transport as

$$\left(\mathbf{N}_{i,k}\right)_{\text{advection}} = c_{i,k}\mathbf{v}_k \tag{16.15}$$

where $\mathbf{v}_k$ is the mass-averaged velocity of phase k. The mass-averaged velocity can be defined using Darcy's law as (Weber et al. 2014)

$$\mathbf{v}_k = -\frac{k_k}{\mu_k}\nabla p_k \tag{16.16}$$

where k_k is the effective permeability (a product of the absolute permeability and the relative permeability), μ_k is the viscosity, and p_k is the total pressure of phase k.

Similarly, the rate of change of thermal energy per unit volume of fluid due to the convection can be written as

$$\dot{\mathbf{q}}'''_{\text{convection}} = \rho_k C_{P_k}\mathbf{v}_k\nabla T_k \tag{16.17}$$

where ρ_k is the density; C_{P_k} is the heat capacity of phase k, which is a combination of the various components of that phase; and T_k is the temperature.

16.4.3 Conduction

Similar to mass diffusion, conduction plays a dominant role in PEFCs. The conduction is involved in the diffusion of thermal energy (as often known as heat conduction), conduction of electron via an external circuit, and conduction of proton via the electrolyte membrane. The conduction of protons and electrons is discussed later, while the heat conduction is described in this section.

Fourier's law of thermal conduction governs the heat conduction inside the PEFCs as well as in the solid bipolar plate, which can be written as:

$$\dot{\mathbf{q}}''_{\text{conduction}} = -k_{T,k}^{\text{eff}}\nabla T_k \tag{16.18}$$

where $\dot{\mathbf{q}}''_{\text{conduction}}$ is the rate of heat transfer by conduction, and $k_{T,k}^{\text{eff}}$ is the effective thermal conductivity of phase k.

The effective thermal conductivity of PEFC porous media is a function of porosity and thermal conductivity of the solid phase, the gas species, and the liquid water. The simplest approach to estimate the effective thermal conductivity is the Bruggeman approximation, as described earlier. There are many complex and more accurate formulations available in the open literature for effective thermal conductivity (Das et al. 2010; Zamel et al. 2010). Moreover, the thermal conductivity varies significantly within the porous media, which leads to a higher through-plane value than the in-plane value. Thus, careful attention will be required for estimating the effective thermal conductivity and taking care of the anisotropy of the porous media, as the anisotropic effective thermal conductivity can have a strong influence on the temperature profiles inside the GDEs (Pasaogullari et al. 2007).

16.4.4 Migration and Proton Transport

Migration takes place in the electrolyte membrane due to the potential difference across the membrane. This is analogous to the diffusion due to the concentration gradient or the heat transport due to the temperature gradient. The migration of the charged species in the electrolyte membrane can be written as (Mench 2008)

$$\left(\mathbf{N}_{i,k}\right)_{\text{migration}} = -z_i u_i F c_{i,k} \nabla \Phi \tag{16.19}$$

where z_i is the charge number for the ionic species, u_i is the mobility of the charged species, F is Faraday's constant, and $\nabla \Phi$ is the electrical field gradient. The mobility of species i is related to the diffusivity via the Nernst–Einstein relationship as

$$u_i = \frac{D_i}{RT} \tag{16.20}$$

where D_i is the diffusion coefficient, R is ideal gas constant (=8.31 J mole^{-1} K^{-1}), and T is the temperature. Along with the migration, the protons in the membrane can also be transported via diffusion and convection. Combining these three modes of proton transport, we have the Nernst–Planck equation governing the proton transport as (Mench 2008)

$$\mathbf{N}_{i,k} = -D_i \nabla c_{i,k} + c_{i,k} \mathbf{v}_k - z_i u_i F c_{i,k} \nabla \Phi \tag{16.21}$$

where the first two terms represent the diffusion and convection of protons through the membrane.

16.4.5 Electron Transport

The solid phase of a PEFC consists of a solid electrode and electrolyte membrane phase. The electrons are transported through the solid phase, while the electrolyte membrane acts as an insulator. The transport of electrons is governed by the current balance as

$$\nabla \cdot \left(\sigma_s^{\text{eff}} \nabla \Phi_s\right) = i_h^{rxn} \tag{16.22}$$

where Φ_s is the electronic potential, σ_s^{eff} is the effective electronic conductivity, and i_h^{rxn} is the overall reaction current density. The effective electronic conductivity takes into account the volume fraction of electronically conductive material in the CLs, GDLs, and MPLs. Similar to other effective properties, the effective electronic conductivity can also be estimated using the Bruggeman correlation, as it provides a good approximation of the effective electronic conductivity (Das et al. 2010). The overall reaction current density is given by

$$i_h^{rxn} = A_v \left(1 - S_l\right) i_h \tag{16.23}$$

where i_h is the transfer current density, and A_v is the specific interfacial area between the ionically and electronically conductive phases, which can be estimated from (Das et al. 2008)

$$A_v = \frac{A_s m_{\text{Pt}}}{\delta_{\text{CL}}} \tag{16.24}$$

where A_s is the catalyst surface area per unit mass of the catalyst, m_{Pt} is the catalyst mass loading per unit area, and δ_{CL} is the thickness of the catalyst layer. Often the catalyst mass loading is considered uniform throughout the thickness of the CL. However, the distributions of catalyst particles within the CL can be inhomogeneous (Xing et al. 2017, 2018).

Heat and Fluid Flow in Porous Media for Polymer-Electrolyte Fuel Cells 351

For PEFCs, the hydrogen oxidation reaction (HOR) and the oxygen reduction reaction (ORR) occur at the aCL and the cCL, respectively. The transfer current densities (i_h) for these reactions are governed by the Butler–Volmer kinetics, as

$$i_h^{\text{HOR}} = i_0^{\text{HOR}} \left[\frac{P_{\text{H}_2}}{P_{\text{H}_2}^{\text{ref}}} \exp\left(\frac{\alpha_a F \eta_{\text{HOR}}}{RT} \right) - \exp\left(\frac{-\alpha_c F \eta_{\text{HOR}}}{RT} \right) \right] \tag{16.25}$$

and

$$i_h^{\text{ORR}} = i_0^{\text{ORR}} \left[\exp\left(\frac{\alpha_a F \eta_{\text{ORR}}}{RT} \right) - \frac{P_{\text{O}_2}}{P_{\text{O}_2}^{\text{ref}}} \exp\left(\frac{-\alpha_c F \eta_{\text{ORR}}}{RT} \right) \right] \tag{16.26}$$

In these equations, i_0 is the exchange current density, α_a and α_c are the anodic and cathodic transfer coefficients, P_{H_2} and $P_{\text{H}_2}^{\text{ref}}$ are the partial pressure and reference partial pressure for H_2, and P_{O_2} and $P_{\text{O}_2}^{\text{ref}}$ are the partial pressure and reference partial pressure for O_2, respectively. The term η represents the overpotential, which can be determined from

$$\eta = \Phi_s - \Phi_m - U_h \tag{16.27}$$

where Φ_s is the electronic or solid-phase potential, and Φ_m is the ionic or membrane-phase potential. The standard potentials for the PEFC reactions are (Newman and Thomas-Alyea 2004)

$$U_h = 0 \text{ for HOR}$$
$$= 1.250 + 0.230\left(1 - \frac{T}{273.16} \right) \text{ for ORR} \tag{16.28}$$

The parameter values related to Pt/C and NSTF reaction kinetics are available in the literature (Balliet and Newman 2011a, b; Ahluwalia et al. 2012).

Apart from the aforementioned transport processes, radiation heat transfer can take place. However, the operating temperatures of PEFCs are in the range of room temperature to about 80°C. Thus, the magnitude of the radiation heat transfer will be relatively smaller compared to the convective heat transfer from the cell surface to the surrounding. A simple correction of the overall convective heat transfer coefficient can take care of the convective-radiative heat transfer together.

16.5 Heat and Fluid Flow in Catalyst Layer and Gas-Diffusion Electrode

As the multi-component reactant gases are present throughout the CLs and the GDEs, the governing equations for each phase are needed to derive using a volume-averaging method. The basic principle of volume-averaging procedure for multi-phase flow is to average the single-phase conservation equation over an elementary volume, where several phases coexist.

16.5.1 Multi-Phase Volume-Averaging Method

When two or more miscible fluids occupy the void space of a porous layer, they mix because of diffusive and dispersive effects, leading ultimately to a multi-component mixture. In order to capture the interactions between the phases, a volume-averaging procedure can be applied to the conservation equations for each species. The volume-averaging procedure integrates the conservation equation for each

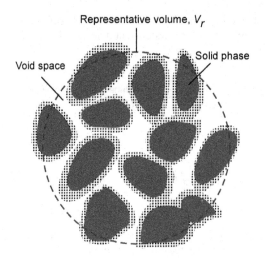

FIGURE 16.8 Representative volume for volume-averaging method.

species over a representative volume. For any quantity Ψ associated with phase k (can be liquid, gas, or solid phase) in PEFC porous media, the volume average of Ψ_k, which is also known as superficial average, can be defined as

$$\langle \Psi_k \rangle = \frac{1}{V_r} \int_{V_r} \Psi_k dV \tag{16.29}$$

where V_r is the representative volume, and Ψ_k is taken to be zero in other phases. The representative volume is illustrated in Figure 16.8.

In addition to the superficial average term, the phase-volume or intrinsic-average quantities can be defined as

$$\langle \Psi_k \rangle^k = \frac{1}{V_k} \int_{V_k} \Psi_k dV \tag{16.30}$$

where V_k is the representative volume of phase k. These two quantities (superficial average and intrinsic average) are related through

$$\frac{\langle \Psi_k \rangle}{\langle \Psi_k \rangle^k} = \frac{V_k}{V_r} = \varepsilon_k \tag{16.31}$$

in which ε_k is the volume fraction of phase k within the representative volume.

Using the volume-average method, the governing volume-averaged equations for multi-phase transport processes for phase k (liquid, gas, or solid phase) can be written as follows (Whitaker 1971; Gray 1975; Kaviany 1995; Baschuk and Li 2005; Das 2010):

Conservation of mass:

$$\frac{\partial}{\partial t}\left(\varepsilon_k \langle \rho_k \rangle^k\right) + \nabla \cdot \left(\varepsilon_k \langle \rho_k \rangle^k \langle \mathbf{v}_k \rangle^k\right) = \Gamma_{M,k} \tag{16.32}$$

Conservation of momentum:

$$\frac{\partial}{\partial t}\left(\varepsilon_k \langle \rho_k \rangle^k \langle \mathbf{v}_k \rangle^k\right) + \nabla \cdot \left(\varepsilon_k \langle \rho_k \rangle^k \langle \mathbf{v}_k \rangle^k \langle \mathbf{v}_k \rangle^k\right) + \nabla\left(\varepsilon_k \langle p_k \rangle^k\right)$$
$$- \langle p_k \rangle^k \nabla \varepsilon_k - \nabla \cdot \left(\varepsilon_k \langle \tau_k \rangle^k\right) - \varepsilon_k \langle \mathbf{b} \rangle^k = \Gamma_{F,k} \quad (16.33)$$

Conservation of species:

$$\frac{\partial}{\partial t}\left(\varepsilon_k \langle \rho_k \rangle^k \langle \omega_k^\alpha \rangle^k\right) + \nabla \cdot \left(\varepsilon_k \langle \rho_k \rangle^k \langle \omega_k^\alpha \rangle^k \langle \mathbf{v}_k \rangle^k + \varepsilon_k \langle J_k^\alpha \rangle^k\right) = \Gamma_{S,k}^\alpha \quad (16.34)$$

Conservation of energy:

$$\frac{\partial}{\partial t}\left(\sum_k \sum_{\alpha \neq \alpha^\pm} \varepsilon_k \langle \rho_k \rangle^k \langle \omega_k \rangle^k C_{p,k}^\alpha \langle T \rangle^k\right) + \nabla \cdot \left(\sum_k \sum_{\alpha \neq \alpha^\pm} \varepsilon_k \langle F_\alpha^k \rangle^k C_{p,k}^\alpha \langle T \rangle^k\right)$$
$$= -\nabla \cdot \langle q \rangle^k + Q_v + Q_{jle} + Q_{rxn} \quad (16.35)$$

where the subscript k denotes the phase, τ_k is the viscous stress, $\mathbf{b}$ is the body force, ω_k^α is the mass fraction of species α within phase k, and J_k^α is the mass flux of species α due to molecular diffusion. The term $\langle T \rangle^k$ represents the equilibrium temperature of all co-existing phases, and Q denotes the energy consumption or production (heating/cooling due to evaporation/condensation, Joule heating, and the heat of reaction). In the conservation of energy equation, the total mass flux of species α is defined as

$$F_\alpha^k = \rho_k \omega_k^\alpha \mathbf{v}_k + J_k^\alpha \quad (16.36)$$

In Equations (16.32) through (16.34), the right-hand terms represent the source terms, and each of these terms can vary based on each layer and phase. Moreover, these terms include the interfacial source terms (Baschuk and Li 2005).

16.5.2 Multi-Phase Mixture and Two-Fluid Methods

There are two approaches widely used for the porous layers of PEFCs, namely, multi-phase mixture and two-fluid methods. In the first approach, it is considered that the gas and liquid phases are non-continuous and form a pseudo-fluid or multi-phase mixture. Thus, all the transport equations can be divided into two groups. One will be applying for the multi-phase mixture phase and the other for the solid phase. The main advantage of using a multi-phase mixture approach is: it does not require to solve both liquid and gas phases and no approximation will be required for the interfacial source terms, as they will be canceled out, except the solid–fluid interfacial source term. However, the multi-phase mixture method requires a careful estimation of the properties of the mixture. Often the effective medium theory is used to estimate physicochemical properties of the multi-phase mixture, which is, in fact, an approximation and can lead to erroneous results. The detailed explanation of the multi-phase mixture, associated conservation equations, and mixture properties is available elsewhere (Baschuk and Li 2005).

In reality, both gas and liquid phases co-exist in continuous phase inside the porous layers of a PEFC. Thus, the two-fluid (gas and liquid water) method can provide a more realistic picture of transports inside the porous layers of a PEFC. In the two-fluid method, both the gas- and liquid-phase conservation equations are required to solve together with appropriate approximations. One key approximation is required for the interfacial interaction between the gas and liquid phase. This can be simplified by considering the phase-change of liquid water (as liquid phase has only liquid water) and assuming negligible interactions for the other species. In the following, a discussion is provided for the two-fluid method and related conservation equations for the GDEs and CLs.

16.5.2.1 Transport of Gas Phase

For the gas phase transport through the GDEs and CLs, we can recall Equations (16.32) through (16.34) and replace phase k with gas phase (g). Thus, the conservations of mass, momentum, and species can be written as

$$\frac{\partial}{\partial t}\left(\varepsilon_g \rho_g\right)+\nabla\cdot\left(\varepsilon_g \rho_g \mathbf{v}_g\right)=\Gamma_{M,g} \tag{16.37}$$

$$\frac{\partial}{\partial t}\left(\varepsilon_g \rho_g \mathbf{v}_g\right)+\nabla\cdot\left(\varepsilon_g \rho_g \mathbf{v}_g \mathbf{v}_g\right)+\nabla\left(\varepsilon_g p_g\right)-\nabla\cdot\left(\varepsilon_g \tau_g\right)=\Gamma_{F,g} \tag{16.38}$$

$$\frac{\partial}{\partial t}\left(\varepsilon_g \rho_g \omega_g^\alpha\right)+\nabla\cdot\left(\varepsilon_g \rho_g \omega_g^\alpha \mathbf{v}_g + \varepsilon_g \mathbf{J}_g^\alpha\right)=\Gamma_{S,g}^\alpha \tag{16.39}$$

For the momentum equation, the body force term is not applicable inside the PEFC porous media. For the species transport in the gas phase, an ideal three-component gas mixture (water vapor, O_2, and N_2) should be considered for the cathode side and an ideal two-component gas mixture (water vapor and H_2) should be considered for the anode side.

Assuming the production of water in the cathode catalyst layer is in liquid form, the source terms for the conservation of mass equation in the GDEs (anode and cathode), cCL, and aCL are

$$\Gamma_{M,g} = \begin{cases} A\left(p_{\text{sat}} - x_g^{H_2O} p_g\right) & \in \text{ GDEs} \\ M_{O_2} A_v \dot{\wp}_{O_2} + A\left(p_{\text{sat}} - x_g^{H_2O} p_g\right) & \in \text{ cCL} \\ M_{H_2} A_v \dot{\wp}_{H_2} + A\left(p_{\text{sat}} - x_g^{H_2O} p_g\right) & \in \text{ aCL} \end{cases} \tag{16.40}$$

where p_{sat} is the saturation pressure, $x_g^{H_2O}$ is the water vapor mole fraction, M is the molecular weight, and $\dot{\wp}$ is the consumption of species in the catalyst layers (Baschuk and Li 2005). The term A represents a constant that represents the interfacial mass-transfer rate of water between the gas and liquid phases (Das 2010).

The source terms for the conservation of momentum equation can be simplified as a generalized Darcy term for GDEs and CLs as

$$\Gamma_{F,g} = -\frac{\varepsilon_g \mu_g}{K k_{rg}} \mathbf{v}_g \tag{16.41}$$

where μ_g is the viscosity of the gas mixture, K is the permeability, and k_{rg} is the relative permeability of the gas phase. Conversely, the source terms for the conservation of species equation are

$$\Gamma_{s,g}^{H_2O} = A\left(p_{\text{sat}} - x_g^{H_2O} p_g\right) \tag{16.42}$$

$$\Gamma_{s,g}^{O_2} = \begin{cases} 0 & \notin \text{ cCL} \\ M_{O_2} A_v \dot{\wp}_{O_2} & \in \text{ cCL} \end{cases} \tag{16.43}$$

Heat and Fluid Flow in Porous Media for Polymer-Electrolyte Fuel Cells 355

$$\Gamma_{s,g}^{H_2} = \begin{cases} 0 & \notin aCL \\ M_{H_2} A_v \dot{\wp}_{H_2} & \in aCL \end{cases} \qquad (16.44)$$

16.5.2.2 Transport of Liquid Phase

The liquid phase in PEFCs consists of liquid water only. Thus, the conservation of species is not required for the liquid water, and the conservation of mass and momentum can be combined together for porous GDEs and CLs using Darcy's law, which gives

$$\frac{\partial}{\partial t}(\varepsilon_l \rho_l) + \nabla \cdot \left(-\frac{\rho_l K k_{rl}}{\mu_l}\nabla p_l\right) = \Gamma_{M,l} \qquad (16.45)$$

where k_{rl} is the relative permeability of liquid water. The liquid pressure inside the pores can be related to the gas pressure through the capillary pressure (p_c) as

$$p_c = p_l - p_g \qquad (16.46)$$

The capillary pressure is also a function of liquid saturation (S_l), which is governed by the following expression (Das et al. 2011)

$$p_c = \gamma \cos \theta_c \left(\frac{\varepsilon}{K}\right) F(S_l) \qquad (16.47)$$

where γ is the surface tension, θ_c is the contact angle, and $F(S_l)$ is the Leverett function (Leverett 1941). The Leverett function is mainly applicable to the packed beds and only includes the influence of porosity and permeability, but widely used for PEFCs. For GDEs and CLs, it is equally important to take account of wetting properties, such as breakthrough pressure and adhesion force (Das et al. 2012; Santamaria et al. 2014). Thus, the experimentally measured capillary pressure-saturation relationship should be considered for GDEs and CLs.

Using Equations (16.46) and (16.47) in Equation (16.45), the liquid water transport equation can be expressed in terms of the liquid water saturation as,

$$\frac{\partial}{\partial t}(\varepsilon_l \rho_l) + \nabla \cdot (-D_c \nabla S_l) = \Gamma_{S,l} \qquad (16.48)$$

where D_c is the capillary diffusivity for the liquid water, which can be written as

$$D_c = \frac{K k_{rl}}{\mu_l} \rho_l \frac{dp_c}{dS_l} \qquad (16.49)$$

The right-hand term of Equation (16.48) is

$$\Gamma_{S,l} = \begin{cases} -A\left(p_{sat} - x_g^{H_2O} p_g\right) + \nabla \cdot \left(\rho_l \frac{k_{rl}}{\mu_l k_{rg}}\mathbf{v}_g\right) & \notin cCL \\ M_{H_2O} A_v \dot{\wp}_{H_2O} - A\left(p_{sat} - x_g^{H_2O} p_g\right) + \nabla \cdot \left(\rho_l \frac{k_{rl}}{\mu_l k_{rg}}\mathbf{v}_g\right) & \in cCL \end{cases} \qquad (16.50)$$

The last term in the above equation represents the amount of liquid water transport hindered by the gas flow, which can be neglected for the GDEs if the gas pressure is uniform.

16.5.2.3 Transport of Energy

For the PEFCs porous media, it is realistic to assume that all phases will be in a local thermal equilibrium inside the pores, thus, the conservations of energy can be simplified as

$$\sum_{\alpha} \varepsilon_{\alpha} \rho_{\alpha} C_{p,\alpha} \left(\frac{\partial T}{\partial t} + \mathbf{v}_{\alpha} \cdot \nabla T \right) = \nabla \cdot \left(k_T^{\text{eff}} \nabla T \right) + Q_v + Q_{\text{jle}} + Q_{\text{rxn}} \tag{16.51}$$

where k_T^{eff} is the effective thermal conductivity, which is a combination of the bulk thermal conductivities for the various phases. The source terms for Equations (16.51) can be expressed as

$$Q_v = \begin{cases} -\Delta H_v R_v & \in \text{GDEs} \\ -\Delta H_v \left(R_v + R_{v,M} \right) & \in \text{CLs} \end{cases} \tag{16.52}$$

$$Q_{\text{jle}} = \begin{cases} \dfrac{\mathbf{i}_s \cdot \mathbf{i}_s}{\sigma_s^{\text{eff}}} & \in \text{GDEs} \\[3mm] \dfrac{\mathbf{i}_s \cdot \mathbf{i}_s}{\sigma_s^{\text{eff}}} + \dfrac{\mathbf{i}_m \cdot \mathbf{i}_m}{\sigma_m^{\text{eff}}} & \in \text{CLs} \end{cases} \tag{16.53}$$

$$Q_{\text{rxn}} = \begin{cases} 0 & \in \text{GDEs} \\ i_h^{rxn} \left(\eta_h + \Pi_h \right) & \in \text{CLs} \end{cases} \tag{16.54}$$

where ΔH_v is the heat of vaporization, R_v is the rate of evaporation, $R_{v,M}$ is the rate of evaporation from the membrane, i_h^{rxn} is the overall reaction rate for h, and Π_h is the Peltier coefficient (Weber and Newman 2006; Zenyuk et al. 2016).

The governing conservation equations presented in this section only provide the transport of gas and liquid phases inside the GDEs and CLs. However, these transports processes are closely linked with the transports inside the membrane and the gas flow channels as well as the transport of electron via the external circuit. Thus, the entire sets of transport equations are needed along with appropriate boundary conditions and related physicochemical properties to solve the governing conservation equations for the PEFCs. There are many studies available in the open literature that will provide the reader a clear idea of transports of mass, heat, electron, and protons in other layers of PEFCs and related boundary conditions (Weber et al. 2014).

16.6 Summary

This chapter summarizes, in brief, the transport processes involved in PEFC porous media (both GDEs and CLs). It provides insights to understand the structures of various components of PEFCs and the associated phases. The governing conservation equations provide an understanding of the transport conditions of gas and liquid phases inside the GDEs and CLs. The chapter highlights the vast complexities of transport within the polymer electrolyte fuel cells, and the governing equations and transport properties for PEFCs are influenced by the physical structure of the cell and the co-existing phases.

NOMENCLATURE
Abbreviation

aCL	anode catalyst layer
aGDL	anode gas diffusion layer
aMPL	anode microporous layer
cCL	cathode catalyst layer
cGDL	cathode gas diffusion layer
cMPL	cathode microporous layer
AEMFC	anion-exchange-membrane fuel cell
GDE	gas diffusion electrode
HOR	hydrogen-oxidation reaction
ORR	oxygen-reduction reaction
PEFC	polymer-electrolyte fuel cell
PEMFC	proton-exchange-membrane fuel cell

English

A_v	specific interfacial area, (1/m)
A_s	catalyst surface area per unit mass of the catalyst, (cm^2/mg)
$c_{i,k}$	concentration species i in phase k, $(mole/m^3)$
C_p	specific heat capacity, $(J/kg \cdot K)$
D_i	Fickian diffusion coefficient of species i, (m^2/s)
$D_{i,j}$	diffusion coefficient of i in j, (m^2/s)
F	Faraday's constant, 96487 (C/equiv)
i	current density, (A/m^2)
k_T	thermal conductivity, $(J/m^2 \cdot K)$
m_{Pt}	platinum loading, (mg/cm^2)
$\mathbf{N}_{i,k}$	superficial flux density of species i in phase k, $(mol/m^2 \cdot s)$
p_i	partial pressure of species i, (Pa)
p_c	capillary pressure, (Pa)
p	pressure, (Pa)
q	heat flux, $(J/m^2 \cdot s)$
Q	source term for energy equation, $(J/m^3 \cdot s)$
R	ideal gas constant, 8.31 $(J/mol \cdot K)$
S_l	liquid saturation
Δs	entropy of reaction, $(J/mol \cdot K)$
t	time, (s)
T	temperature, (K)
u_i	mobility of species i, $(m^2 \cdot mol/J \cdot s)$
U_h	reversible cell potential of reaction h, (V)
$\mathbf{v}$	velocity vector, (m/s)

Greek Letters

α_a	anodic transfer coefficient
α_c	cathodic transfer coefficient
δ	thickness, (m)
ε_k	volume fraction of phase k
ε	porosity

ρ_k	density of phase k, (kg/m³)
σ	conductivity, (S/m)
η	overpotential, (V)
μ	viscosity, (Pa·s)
τ	stress tensor, (Pa)
τ	tortuosity
ω_k	mass fraction of phase (k)
Φ	potential, (V)

Subscripts/Superscripts

CL	catalyst layer
eff	effective properties
g	gas phase
GDL	gas diffusion layer
i	arbitrary species
j	arbitrary species
k	arbitrary phase
l	liquid phase
m	membrane phase
ref	reference value
s	solid phase

ACKNOWLEDGMENTS

The authors would like to thank Dr. Yasser Mahmoudi Larimi, Dr. Kamel Hooman, and Prof. Kambiz Vafai for the invitation to write this chapter. The financial support from the Engineering and Physical Sciences Research Council (EPSRC) via a research grant (EP/P03098X/1) is gratefully acknowledged. The authors also acknowledge the in-kind support of SGL CARBON GmbH and for providing GDL materials.

REFERENCES

Ahluwalia, R. K., X. Wang, A. Lajunen, A. J. Steinbach, S. M. Hendricks, M. J. Kurkowski, and M. K. Debe. 2012. "Kinetics of oxygen reduction reaction on nanostructured thin-film platinum alloy catalyst." *Journal of Power Sources* 215:77–88.

Balliet, R. J., and J. Newman. 2011a. "Cold-start modeling of a polymer-electrolyte fuel cell containing an ultrathin cathode." *Journal of the Electrochemical Society* 158 (9):B1142–B1149.

Balliet, R. J., and J. Newman. 2011b. "Cold start of a polymer-electrolyte fuel cell i. Development of a two-dimensional model." *Journal of the Electrochemical Society* 158 (8):B927–B938.

Baschuk, J. J., and X. Li. 2005. "A general formulation for a mathematical PEM fuel cell model." *Journal of Power Sources* 142 (1–2):134–153. doi:10.1016/j.jpowsour.2004.09.027.

Das, P. K. 2010. "Transport phenomena in cathode catalyst layer of PEM fuel cells." PhD Thesis, University of Waterloo.

Das, P. K., A. Grippin, A. Kwong, and A. Z. Weber. 2012. "Liquid-water-droplet adhesion-force measurements on fresh and aged fuel-cell gas-diffusion layers." *Journal of the Electrochemical Society* 159 (5):B489–B496. doi:10.1149/2.052205jes.

Das, P. K., X. Li, and Z. S. Liu. 2007. "Analytical approach to polymer electrolyte membrane fuel cell performance and optimization." *Journal of Electroanalytical Chemistry* 604 (2):72–90. doi:10.1016/j.jelechem.2007.02.028.

Das, P. K., X. Li, and Z. S. Liu. 2008. "A three-dimensional agglomerate model for the cathode catalyst layer of PEM fuel cells." *Journal of Power Sources* 179:186–199. doi:10.1016/j.jpowsour.2007.12.085.

Das, P. K., X. Li, and Z. S. Liu. 2010. "Effective transport coefficients in PEM fuel cell catalyst and gas diffusion layers: Beyond Bruggeman approximation." *Applied Energy* 87 (9):2785–2796. doi:10.1016/j.apenergy.2009.05.006.

Das, P. K., X. Li, Z. Xie, and Z. S. Liu. 2011. "Effects of catalyst layer structure and wettability on liquid water transport in polymer electrolyte membrane fuel cell." *International Journal of Energy Research* 35 (15):1325–1339. doi:10.1002/er.1873.

Das, P. K., and A. Z. Weber. 2013. "Water management in PEMFC with ultra-thin catalyst-layers." *Proceedings of the ASME 11th Fuel Cell Science, Engineering, and Technology Conference*, July 14–19, 2013, Minneapolis, MN: The American Society of Mechanical Engineers. doi:10.1115/FuelCell2013-18010.

Debe, M. K. 2012. "Electrocatalyst approaches and challenges for automotive fuel cells." *Nature* 486 (7401):43–51. doi:10.1038/nature11115.

Debe, M. K. 2013. "Tutorial on the fundamental characteristics and practical properties of nanostructured thin film (NSTF) catalysts." *Journal of the Electrochemical Society* 160 (6):F522–F534. doi:10.1149/2.049306jes.

Gray, W. G. 1975. "A derivation of the equations for multiphase transport." *Chemical Engineering Science* 30 (2):229–233.

Holdcroft, S. 2014. "Fuel cell catalyst layers: A polymer science perspective." *Chemistry of Materials* 26 (1):381–393. doi:10.1021/cm401445h.

Hwang, G. S., and A. Z. Weber. 2012. "Effective-diffusivity measurement of partially-saturated fuel-cell gas-diffusion layers." *Journal of the Electrochemical Society* 159 (11):F683–F692. doi:10.1149/2.024211jes.

Kaviany, M. 1995. *Principles of Heat Transfer in Porous Media*. New York: Springer.

Leverett, M. C. 1941. "Capillary behavior in porous solids." *Petroleum Division Transactions of the American Institute of Mining and Metallurgical Engineers* 142:152–169.

Li, X. 2006. *Principles of Fuel Cells*. New York: Taylor & Francis Group.

Machado, B. S., N. Chakraborty, and P. K. Das. 2017. "Influences of flow direction, temperature and relative humidity on the performance of a representative anion exchange membrane fuel cell: A computational analysis." *International Journal of Hydrogen Energy* 42 (9):6310–6323. doi:10.1016/j.ijhydene.2016.12.003.

Machado, B. S., N. Chakraborty, M. Mamlouk, and P. K. Das. 2018. "A three-dimensional agglomerate model of an anion exchange membrane fuel cell." *Journal of Electrochemical Energy Conversion and Storage* 15:011004-1–011004-12. doi:10.1115/1.4037942.

Mench, M. M. 2008. *Fuel cell engines*. Hoboken, NJ: John Wiley & Sons.

Nam, J. H., K. J. Lee, G. S. Hwang, C. J. Kim, and M. Kaviany. 2009. "Microporous layer for water morphology control in PEMFC." *International Journal of Heat and Mass Transfer* 52 (11–12):2779–2791. doi:10.1016/j.ijheatmasstransfer.2009.01.002.

Newman, J., and K. E. Thomas-Alyea. 2004. *Electrochemical Systems*. 3rd ed. New York: John Wiley & Sons.

Nonoyama, N., S. Okazaki, A. Z. Weber, Y. Ikogi, and T. Yoshida. 2011. "Analysis of oxygen-transport diffusion resistance in proton-exchange-membrane fuel cells." *Journal of The Electrochemical Society* 158 (4):B416–B423.

Pasaogullari, U., P. P. Mukherjee, C. Y. Wang, and K. S. Chen. 2007. "Anisotropic heat and water transport in a PEFC cathode gas diffusion layer." *Journal of the Electrochemical Society* 154 (8):B823–B834. doi:10.1149/1.2745714.

Santamaria, A. D., P. K. Das, J. C. MacDonald, and A. Z. Weber. 2014. "Liquid-water interactions with gas-diffusion-layer surfaces." *Journal of the Electrochemical Society* 161 (12):F1184–F1193. doi:10.1149/2.0321412jes.

Weber, A. Z., S. Balasubramanian, and P. K. Das. 2012. Proton exchange membrane fuel cells. Edited by K. Sundmacher. Vol. 41, *Advances in Chemical Engineering*. San Diego: Academic Press. doi:10.1016/B978-0-12-386874-9.00003-8

Weber, A. Z., R. L. Borup, R. M. Darling, P. K. Das, T. J. Dursch, W. B. Gu, D. Harvey et al. 2014. "A critical review of modeling transport phenomena in polymer-electrolyte fuel cells." *Journal of the Electrochemical Society* 161 (12):F1254–F1299. doi:10.1149/2.0751412jes.

Weber, A. Z., and J. Newman. 2006. "Coupled thermal and water management in polymer electrolyte fuel cells." *Journal of the Electrochemical Society* 153 (12):A2205–A2214.

Whitaker, S. 1971. "The transport equations for multi-phase systems." *Chemical Engineering Science* 28 (1):139–147.

www.legislation.gov.uk. 2008. "Climate change act." accessed September 2018. http://www.legislation.gov uk.

www.theccc.org.uk. 2012. "Committee on climate change technical report." accessed September 2018. https://www.theccc.org.uk.

Xing, L., W. D. Shi, P. K. Das, and K. Scott. 2017. "Inhomogeneous distribution of platinum and ionomer in the porous cathode to maximize the performance of a PEM fuel cell." *AICHE Journal* 63 (11):4895–4910. doi:10.1002/aic.15826.

Xing, L., Y. Wang, P. K. Das, K. Scott, and W. D. Shi. 2018. "Homogenization of current density of PEM fuel cells by in-plane graded distributions of platinum loading and GDL porosity." *Chemical Engineering Science* 192:699–713. doi:10.1016/j.ces.2018.08.029.

Zamel, N., X. Li, J. Shen, J. Becker, and A. Wiegmann. 2010. "Estimating effective thermal conductivity in carbon paper diffusion media." *Chemical Engineering Science* 65 (13):3994–4006. doi:10.1016/j.ces.2010.03.047.

Zenyuk, I. V., P. K. Das, and A. Z. Weber. 2016. "Understanding impacts of catalyst-layer thickness on fuel-cell performance via mathematical modeling." *Journal of the Electrochemical Society* 163 (7):F691–F703. doi:10.1149/2.1161607jes.

17

Combustion in Porous Media for Porous Burner Application

Muhammad Abdul Mujeebu

CONTENTS

17.1 Introduction 361
17.2 Fundamentals of PMC 362
    17.2.1 PMC History and Principle 362
    17.2.2 Flame Stabilization in PMC 363
    17.2.3 PMC with Liquid Fuel 363
    17.2.4 PMC Modeling 364
    17.2.5 Staged Combustion 364
    17.2.6 Reciprocating Flow Combustion 365
    17.2.7 Reverse Combustion 365
17.3 Development of PMC Burners 365
    17.3.1 Household PM Burners 365
    17.3.2 Microscale Applications 366
    17.3.3 PMB for Electricity Generation 367
    17.3.4 PMB for Combustion of Low CV Fuels 367
    17.3.5 PMB for Hydrogen and Syngas Production 367
    17.3.6 Catalytic PMB 368
    17.3.7 PMB for Industrial Applications 368
    17.3.8 PMB with Heat Exchanger 368
    17.3.9 Flexible PMB 368
    17.3.10 PMBs for Miscellaneous Applications 369
17.4 Conclusion 369
Nomenclature 369
References 370

17.1 Introduction

Fast depletion of fossil fuel reserves and the environmental threat associated with the excessive use of fossil fuels have prompted all nations to encourage the use of renewable and clean energy technologies on one hand, and to impose adoption of energy efficient and eco-friendly combustion techniques on the other hand. In the context of the latter solution, which is the focus of this chapter, a substantial amount of research has been carried out on proposing novel combustion techniques. Among them, porous media combustion (PMC) has demonstrated its capability as one of the excellent options. In PMC, the combustion occurs inside the voids of a solid porous matrix, and the heat exchange between burning gas mixture and the porous medium (PM) permits the heat energy being stored within the solid matrix, causing the flame temperatures to be much higher than the adiabatic temperature. It offers high-power densities, high turn-down ratios, good flame stability, and the capability of burning lean mixtures and low-calorific fuels. Due to its promising advantages, there has been significant increase of research in this area in the past decades. The history and fundamentals of PMC, and its diverse applications and research progress, are well established in the literature (Abdul Mujeebu 2016; Abdul Mujeebu et al. 2009a, b, c, d; Abdul Mujeebu et al. 2010; Howell et al. 1996; Kamal and Mohamad 2006; Mujeebu et al. 2009; Pantangi and Mishra 2006; Wood and Harris 2008). However, this chapter provides a brief outline of the fundamentals of PMC and development of PMC-based burners for various applications.

17.2 Fundamentals of PMC

17.2.1 PMC History and Principle

The concepts of heat recuperation and excess enthalpy combustion were first investigated by Weinberg (1971), who stated that by heat recirculation, the flame temperatures could be higher than the theoretical maximum adiabatic flame temperatures. In 1973, Hardesty and Weinberg (1973) demonstrated theoretically that the heat transfer from the combustion products to the reactants extends the flammability limits of a given fuel and increases the flame temperatures above the adiabatic flame temperatures. Later on, as a means to realize an excess enthalpy flame, PMC was introduced by Takeno and Sato (1979), who proposed an internal heat recirculation mechanism by the insertion of a semi-infinite, high-conductivity porous solid into a one-dimensional flame zone. This mechanism was analytically investigated by Deshies and Joulin (1980) who ascertained the existence of an increased flame temperature. This effort was pursued further by Buckmaster and Takeno (1981), Kotani and Takeno (1982), Kotani et al. (1985); Takeno and Hase (1983) and Takeno et al. (1981) who reported replacing a semi-finite solid by finite solid Takeno et al. (1981), studying the flashback and blow-off mechanisms of excess enthalpy flames (Buckmaster and Takeno 1981), and the influence of the solid length and heat loss on PMC performance (Takeno and Hase 1983), as well as experimental studies on PMC (Kotani and Takeno 1982; Kotani et al. 1985).

Unlike in free flames, PMC takes place within a solid PM, which has interconnected voids that are big enough to allow combustion and with a relative high permeability. When the gaseous reactants enter the porous matrix, the matrix is hotter due to the intense heat transfer across the solid by radiation and conduction from the combustion and products zone to the reactants zone. The higher solid temperatures result in an additional heat transfer to the reactants that would not occur in a free flame, and consequently to higher reaction rates and flame temperatures. Further downstream, heat is transferred by convection as well, from the gas to the solid upstream, as illustrated in Figure 17.1. The capacity of the reactants being preheated, and therefore the extent of the internal energy recirculation, depends mainly on the porous media characteristics. The above features qualify PMC to be excellent for burning lean fuel–air mixtures

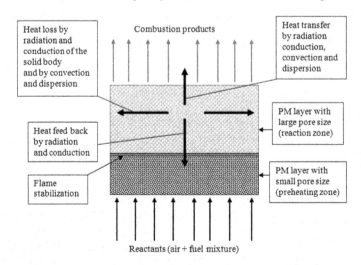

FIGURE 17.1 Heat transfer phenomena in PMC.

Combustion in Porous Media for Porous Burner Application

and low heat content fuels. Another advantage of the internal energy recirculation is that the turn-down ratio is increased, which is a benefit in developing burners for a wide range of applications.

Based on the relative motion of flame and the PM, PMC can be classified as transient and stationary; in the former, the flame propagates freely through the PM (filtration combustion), while in the latter the flame is stabilized in a specific region in the PM. PMC is further categorized according to the flame stabilization mechanism, as surface-stabilized and submerged. In the surface-stabilized mode, flame is stabilized at the surface of the PM, while in the submerged combustion mode, the flame is stabilized within the PM. PMC burners are also classified as premixed and non-premixed, based on whether the fuel and air (or oxygen) are mixed or not before being fed to the burner. All these categories have potential for different applications.

17.2.2 Flame Stabilization in PMC

The flame velocities in PMC are higher compared to laminar free flames for a given mixture (Babkin et al. 1991; Kotani and Takeno 1982; Sathe et al. 1991; Takeno and Sato 1979). One way of stabilizing the flame inside the matrix is through a sudden or gradual change in the burner cross-sectional area (Zhdanok et al. 1998; Malico et al. 2000). However, this strategy suffers from low turn-down ratio and flashback at small firing rates. Another way of stabilizing the flame is through an active control of the flow rates of air and fuel and flame velocity through temperature measurements. Once again, the possible turn-down ratios are small, which may be a disadvantage for certain applications. A third flame stabilization mechanism is obtained if one cools the reaction zone. This cooling can be achieved through radiation or actively using, for instance, water-cooling tubes. A fourth flame stabilization mechanism is achieved if one "plays" with the characteristics of the porous matrices. Burners that have two different porous media zones in series, the so-called two-layer porous burners, are common (Hsu et al. 1993; Rumminger et al. 1996; Trimis and Durst 1996). In such burners, the premixed reactants enter the burner through a small-pore PM, the preheating zone, and then pass to a larger pore medium, the combustion zone, where the reaction takes place (Figure 17.1). A third PM zone may be present to enhance the heat transfer from the hot combustion product to a heat exchanger. The stabilization of the flame through multi-layered porous burners allows high turn-down ratios (Abdul Mujeebu and Malico 2015).

The possibility of sustaining combustion inside a porous matrix is determined by two opposite effects: at one side, the solid walls decelerate the reaction, and on the other side, the heat recirculation promoted by the matrix accelerates the reaction (Abdul Mujeebu and Malico 2015). For combustion to occur inside a solid PM, the modified Péclet number (P_e) should be higher than 65 (Trimis and Durst 1996). In the context of PMC, this non-dimensional number is defined as:

$$P_e = \frac{S_L d_m C_p \rho}{k} \tag{17.1}$$

where S_L is the laminar flame speed, d_m is the equivalent diameter of the PM, C_p is the specific heat of the gas mixture, ρ is its density, and k is its thermal conductivity. Whenever the Péclet number is below 65, the flame is quenched inside the porous matrix, and no flame can be sustained. In the two-layer porous burner-type described above, the preheating layer is made of a PM where the Péclet number is below 65 for safety reasons. In this way, no flame can enter this zone, and flashback is prevented. On the other side, the combustion zone is made of a PM where the Péclet number is above or equal to 65, and the flame is not quenched. The flame stabilization in PMC has been a topic of research (Bubnovich et al. 2010; Catapan et al. 2011).

17.2.3 PMC with Liquid Fuel

There are two mechanisms for liquid-fuel-fired PMC burners: one is fuel spraying and the other is fuel vaporizing. The former is normally employed for industrial and civil applications, such as boilers and furnaces. The main drawback of this mechanism is the lack of homogeneity in combustion, especially in large load ranges, because the flame is relatively large and the stability of the flame is affected by aerodynamics among the air and the droplets. The fuel-vaporizing technique ensures nearly homogeneous combustion and smaller flame and lesser soot emission, compared to the spraying type. However, the electrical power requirement for fuel vaporization is not attractive, which reduces the overall thermal-to-electric conversion efficiency in power generation. This technology has promising industrial and domestic applications, such as incineration of liquid hazardous waste, which is difficult to incinerate in conventional burners owing to its low energy content and the existence of chlorinated species. Mujeebu et al. (2009c) provided a comprehensive review of the fundamentals, modeling, and applications of liquid fuel combustion in PM.

17.2.4 PMC Modeling

The extremely complex PMC phenomenon is yet to be understood properly. Hence, it has been a challenging task for experts in combustion modeling to formulate a realistic PMC model by incorporating factors, such as multi-step reaction kinetics, temperature non-equilibrium between the solid and gaseous phases, and radiation. Due to the presence of both solid and fluid (gas or liquid) media, governing equations must be developed for both the phases. The effects of conduction and radiation, as well as convection between the solid and the gas, are incorporated in the solid-phase equation. Conduction, chemical energy release due to combustion, and convection with the solid phase are included in the gas-phase equation. The recent studies show that researchers have gained significant command in modeling and simulation of various PMC problems. A comprehensive review of all the various modeling trends was provided by Mujeebu et al. (2010). However, in the later progress, many researchers have come up with sufficiently accurate models for various PMC applications, such as hydrogen production from biomass (Toledo et al. 2016), household burners (Panigrahy et al. 2016), microcombustors (Bani et al. 2018), electricity generation (Henriquez-Vargas et al. 2015), and internal combustion (IC) engines (Zhou et al. 2014).

17.2.5 Staged Combustion

Figure 17.2 illustrates the staged combustion in porous media, wherein the combustion takes place in two successive zones of porous matrix. The premixing zone facilitates premixing as well as flow straightening. The premixed fuel-rich mixture enters the first stage where the rich combustion occurs, and the products of combustion are added with suitable quantities of air and fuel in the intermediate mixing zone, to form a lean mixture to be fed into the second stage. The temperature in the second stage is maintained at a level

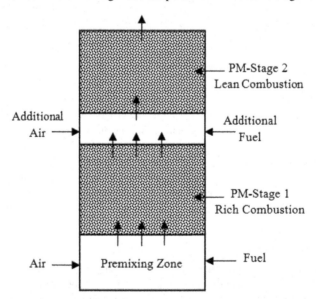

FIGURE 17.2 Two-stage combustion in porous media.

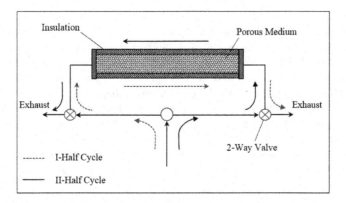

FIGURE 17.3 Schematic of reciprocating flow mechanism in PMC.

that is low enough to prevent NOx formation either by thermal or prompt mechanisms. This arrangement could be extended to a three-stage combustion where a lean mixture is burned in the first stage, a rich mixture in the second one, and a lean mixture in the third one (Abdul Mujeebu et al. 2009a).

17.2.6 Reciprocating Flow Combustion

In reciprocating flow combustion (Figure 17.3), the fresh mixture flows in one direction so that the gas and solid temperatures reach a maximum at the exit side. Then the flow direction is reversed by means of valves. On the reverse flow half-cycle, the fresh mixture is exposed to higher solid temperatures (resulted by the combustion products in the previous half-cycle) while entering. This process enhances the magnitude of heat recycling compared to the unidirectional flow combustion, thereby increasing the degree of excess enthalpy (Hanamura et al. 1993; Hoffmann et al. 1997). This technique has been widely explored for developing burners for various applications, such as hydrogen and syngas production, electricity generation, and steam boilers.

17.2.7 Reverse Combustion

Reverse combustion is facilitating the oxidant flow in a direction opposite to the flame propagation in a combustible PM. This technique has applications in in situ fossil fuel recovery. Detailed insight into this interesting technique was provided by Dindi et al. (1990). De Soete (1967) investigated reverse combustion of methane-oxygen-nitrogen mixtures propagating in sand of calibrated grain sizes, while Britten and Krantz (1985) analyzed the dynamics of reverse combustion by the method of activation energy asymptotics; further developments in this method have not been reported in the open literature.

17.3 Development of PMC Burners

17.3.1 Household PM Burners

There is a great potential for PMC for household cooking applications, and a lot of research and development works have been done in this direction (Malico and Mujeebu 2015). Kakati et al. (2007) experimented incorporating PM medium in a household pressurized kerosene stove. They found that the use of PM could achieve a 34% savings in fuel and 10%–11% increase in thermal efficiency with appreciable control over CO, unburned hydrocarbon (UHC), and NOx emissions. Pantangi et al. (2007) compared the efficiencies, emissions, and energy cost for the conventional LPG cooking stoves with and without the use of PM. The PM tested included metal balls, pebbles, and metal chips. The use of PM could yield a thermal efficiency of 73% whereas it was maximum 69% for the conventional stoves. The CO was found to decrease from 225 to 118, and the energy savings was estimated to be 10%. In a subsequent study,

Muthukumar et al. (2011) proposed a porous radiant burner (PRB) for domestic cooking by using LPG fuel. Compared to the conventional burner, the PRB was claimed to achieve about 6% improvement in thermal efficiency at an equivalence ratio of 0.68, 1.24 kW power intensity, and 310°C ambient temperatures. The CO and NOx emissions were 9–16 ppm and 0–0.2 ppm, respectively, while for the conventional burner, the corresponding values were 50–225 ppm and 2–7 ppm. Further studies were conducted on the proposed PRB, focusing on detailed characterization, heat transfer analysis, and usability and effectiveness of incorporating PRB in the conventional kerosene pressure stove (Mishra et al. 2015a, b; Mishra and Muthukumar 2018; Muthukumar and Shyamkumar 2013; Pantangi et al. 2011, 2016; Pradhan et al. 2016; Sharma et al. 2016a, b. Another study by this team (Panigrahy and Mishra 2018) has assessed the compatibility of dimethyl ether (DME) with the household PM burner (PMB), as an alternative to LPG fuel. The total heat generation rate, the gas- and solid-phase temperatures, and radiant efficiencies with DME were observed to be higher than those with LPG, for similar input conditions.

Yoksenakul and Jugjai (2011) built a self-aspirating PM burner (SPMB) of submerged flame type, as a replacement to conventional gaseous fuel free-flame burners. The SPMB was made by a packed bed of alumina spheres. An output radiation efficiency of 23% could be achieved at a turn-down ratio of 2.65 while the firing rate ranged from 23 to 61 kW. The CO emission was less than 200 ppm, and NOx emission was less than 98 ppm. Mujeebu et al. (2011a, b, c) have developed prototypes of two-layer household LPG burners based on submerged and surface combustion modes. For the submerged mode, the preheating zone was made of porcelain foam of pore density 25 ppcm (pores per centimeter), and the reaction zone was a discrete structure made up of alumina spheres of 30 mm size. The corresponding zones in the surface-stabilized mode were built by alumina foams of 26 ppcm and 8 ppcm. Thermal efficiencies of 59% and 71%, respectively, were reported for a thermal load of 0.62 kW, with an NOx emissions reduction by about 75% compared to the conventional stove.

Ali (2014) studied the performance of conventional LPG cooking stove with different PM and found that the thermal efficiency could be improved by 10.71% with ball bearing as PM. In a similar study, Wu et al. (2014) demonstrated that, by using metallic PM in a conventional Bunsen flame burner, a flat flame could be stabilized while improving the operating range, turn-down ratio, and pollution emissions. Herrera et al. (2015) developed and tested a domestic LPG-fueled PM burner made with a combination of a bed of Al_2O_3 particles obtained from grinding residues and SiSiC (silicon-infiltrated silicon carbide) ceramic foam. Lapirattanakun and Charoensuk (2017) designed a PMB operating on wasted vegetable oil (WVO) for cooking application. The PM was made of a 2-cm diameter of spherical ceramic balls. In the proposed burner, steam was used to atomize the WVO droplets, to entrain air into the combustion zone, and to reduce soot and CO emission. The maximum achievable thermal efficiency was 42%, and the combustion efficiency was around 99.5%. Similar efforts on developing PMBs for household applications include those of Iral and Amell (2015), Laphirattanakul et al. (2016), and Pradhan et al. (2018).

17.3.2 Microscale Applications

The high-energy density of hydrocarbon fuels creates a great opportunity to develop combustion-based micropower generation systems to meet increasing demands for portable power devices, microunmanned aerial vehicles, microsatellite thrusters, and microchemical reactors and sensors (Ju and Maruta 2011). Top batteries currently available (lithium) have an energy density of 1.2 MJ/kg (0.6 MJ/kg for an alkaline battery). Thus, a miniature combustion device with a mere 3% system efficiency would compete with top batteries simply from the fact that the fuel is easily replaceable. Although higher efficiencies are needed for combustion systems to displace batteries, the high efficiencies obtained in large-scale power systems encourage the development of miniaturized power generation devices using combustion, with the expectation that devices with competitive efficiencies can be developed (Fernandez-pello 2002). PMC has demonstrated its excellence as one of the feasible combustion techniques for microscale applications, and this area has attracted a great deal of research attention.

Li et al. (2008) employed PMC as a fuel-holding mechanism, in order to improve the performance of miniaturized liquid-fuel burners. Li et al. (2010) and Chou et al. (2010) investigated a planar microcombustor developed for the combustion of premixed H_2–air, which was employed as emitter in a microthermophotovoltaic system. The effects of flow conditions and position of porous media on the wall temperature distribution were studied. A mini-scale porous media combustor with heat recuperation was set up by Xu et al. (2011) to study the stability of lean combustion and its emission. The porous media was of 20 mm diameter and 140 mm in length. It was established that, for a mixture mass flow rate of 0.163 g/s, the extinction limit was extended to an equivalence ratio of 0.40 in the methane combustion and 0.39 in the propane combustion. Several other studies have recently been reported on the development and characterization of microcombustors (Pan et al. 2015; Li et al. 2016a, b, c; Janvekar et al. 2017; 2018a, b).

17.3.3 PMB for Electricity Generation

There has been a growing interest in the recent past, in developing PMBs integrated with thermoelectric (TE) and thermophotovoltaic (TPV) modules (Mustafa et al. 2017). Following the works of Hanamura et al. (2005), and Qiu and Hayden (2007), Bubnovich et al. (2013) have developed and tested a prototype burner for TE generation. The heat produced in the burner was converted to electricity by an external TE module. The maximum voltage and current yielded were 503 mV and 150 mA, respectively. Henríquez-Vargas et al. (2013) performed numerical analysis on direct thermal to electric energy conversion in a reciprocal flow porous media burner embedded with two layers of thermoelements. The aim of the study was to find out the lean combustibility limit in order to maximize conversion efficiency with minimum fuel consumption. A two-temperature-resistance model for finite-time thermodynamics was developed for the TE element energy fluxes. Maximum values for current and system efficiency obtained were 44.3 mA and 2.5%, respectively. This study was also performed on a continuous flow PMB (Henriquez-Vargas et al. 2013), and later on, they focused on detailed numerical studies on the PMB-TE units Donoso-García and Henríquez-Vargas 2015; Henriquez-Vargas et al. 2015). Ismail et al. (2013) developed a mini-cogeneration system which, while functioning as a domestic cooking stove, could generate a voltage of 9.3V to power a mobile phone charger, with the help of thermoelectric cells attached on the outer surfaces of the burner body. Similarly, Mustafa et al. (2015a, b, c, 2016) developed and tested PMB-TE and PMB-TPV systems operating on gaseous and liquid fuels. Few other developments include those reported by Mueller et al. (2013), Wang et al. (2016, 2017) Zeng et al. (2017), Bani et al. (2018), and Wu et al. (2018).

17.3.4 PMB for Combustion of Low CV Fuels

Combustion of low calorific value (CV) gaseous fuels poses issues with respect to flame stability as well as combustion efficiency. However, the internal heat recuperation feature of PMC has been exploited for the combustion of these fuels. This is often accomplished by inserting a PM in the flame region, which enhances the heat transfer by conduction and radiation toward the upstream region, thereby increasing the combustion temperature and widening the lean combustibility range. This eventually reduces CO, UHC, and NOx emissions. Focusing on these kinds of burners, Francisco Jr et al. (2010, 2012, 2013) have studied the effects of fuel composition on the flame stability, flame temperature, and pollutant emissions. In a subsequent study, Al-attab et al. (2015) investigated a PMB for the combustion of producer gas generated by biomass gasification. The heat recovered from the PMB was used for producing hot air that could be used for the drying process in small industries. The lowest CO and NOx emissions were 6 ppm and 230 ppm, respectively, while heat recovery effectiveness was up to 93% with overall system efficiency of 54%. Song et al. (2017) experimented with axial and radial gradually varied premixed PMBs with annular heat recirculation for combustion of ultralow CV gases. The effect of the firing rate on temperature profiles, flame stability, and CO emission was studied for different CVs. Development and testing of a reciprocating-flow PMB for the combustion of biogas was reported by de Araújo et al. (2013). This burner showed a wide equivalence ratio range ($0.10 < \phi < 1$), energy extraction efficiencies above 90%, and ultralow CO and NOx emissions (below 1 ppm). Similar works were also reported by Gao et al. (2011), Barajas et al. (2012), and Keramiotis and Founti (2013).

17.3.5 PMB for Hydrogen and Syngas Production

A promising breakthrough in the application of PMC is production of hydrogen and syngas from various hydrocarbon fuels, biomass, and solid wastes. A substantial amount of research was done, and is currently in good progress, on the development and characterization of PMBs for this purpose. The author has presented a comprehensive review on this topic (Abdul Mujeebu 2016). Anger et al. (2011) employed PMB as the heat source for reforming in the Supercritical Water Reforming (SCWR) process for the production of syngas from glycerine. As the researchers claim, the proposed burner technology allows effective and low-emission combustion of different fuels in a high-performance modulation.

17.3.6 Catalytic PMB

The lean flammability limit of PMBs can be extended by coating catalysts on the surface of PM, and many researchers were interested in developing burners based on catalytic combustion in PM. Robayo et al. (2014) studied the effect of La-Sr-Fe-Cr-Ru-based perovskite catalysts on matrix-stabilized PMC. By coating the SiC PM with perovskite catalysts, it was possible to lower the minimum stable ϕ and improve the combustion efficiency. In similar studies (Feng and Qu 2016; Feng et al. 2016; Qu and Feng 2015), the perovskite catalyst was coated on alumina pellets and zirconia foam to develop PMBs for the premixed combustion of methane, while a catalytically active $Ce_{0.8}Gd_{0.2}O_{1.9}$ coating was applied on MgO-stabilized ZrO_2 porous ceramics for the heterogeneous combustion of methane (Terracciano et al. 2017). Another catalyst employed was manganese-based submicroparticles stabilized with oleic acid for combustion of heavy oil in a quartz PM (Galukhin et al. 2017).

17.3.7 PMB for Industrial Applications

PRBs are developed to utilize the benefits of infrared heating. Infrared heating is widely used by industry for manufacturing processes, which include paper and wood drying, powder coating, annealing, food browning and baking, plastics curing and forming, destruction of volatile organic compounds (VOC), and boilers (Baukal 2003). The heat transfer performance of infrared heaters depends on the temperature of the source and the absorption characteristics of the product being heated. Since the radiation output is proportional to the fourth power of the temperature, the temperature of the source dominates the performance of the device. This temperature depends on the emissivity of the radiating matrix, the rate of energy release of the reactants, and the heat transfer rates between the hot combustion products and the porous media. Infrared heating can provide significantly faster heating times compared to the convective heat transfer. These advantages applied to industrial applications can allow reduced oven lengths, increased conveyor speeds, and improved surface finish of products, such as coated paper that might be sensitive to air currents associated with convective drying. While the radiative heat transfer rates can be high, the combustion temperatures of radiant burners are relatively low and the combustion is very efficient, allowing the emissions of NOx and CO to be low (Abdul Mujeebu et al. 2009a). The sideway-faced porous radiant burner (SFPRB) (Devi and Sahoo 2017) and self-aspiring radiant tube burner (SRTB) (Chuenchit and Jugjai 2012) are excellent examples of the recent progress in developing PMBs for industrial heating applications.

17.3.8 PMB with Heat Exchanger

Banerjee and Saveliev (2018) numerically studied the heat capture in a PMC-based counterflow heat exchanger. The discrete PM employed was made of alumina (Al_2O_3) balls, and the fuel used was methane. The maximum heat extraction efficiency obtained was 60% at 1300 K. Chuenchit and Jugjai (2012) studied the effect of tube length on combustion characteristics of a SRTB combined with PM heat exchanger. Makmool et al. (2016) developed PMB integrated with a condensing porous heat exchanger (CPHE) in order to improve the durability and reliability of the PMB.

17.3.9 Flexible PMB

The idea of flexible PMB was explored at the Combustion and Engine Research Laboratory of King Mongkut's University of Technology in Thailand (Sompu et al. 2015). The proposed flexible PM burner (FPMB) could burn both liquid and gaseous fuels and be operated in either premixed or non-premixed combustion mode. The stable operating range of the FPMB was found to be at firing rates from 8 to 11 kW and equivalence ratios (ϕ) from 0.4 to 0.95, with appreciably low CO and NOx emissions.

17.3.10 PMBs for Miscellaneous Applications

Apart from the aforementioned applications, plenty of other applications have also been reported, and most of these applications updated until the year 2009 were mentioned in the author's previous review (Mujeebu et al. 2009). However, to exemplify few of them and the recent progress in the development of PMBs, the following are worth noting: automobile engines and gas turbines (Cheng and Pan 2018; Kornilov et al. 2012), cylindrical PMB (Fursenko et al. 2016; Khosravy-el-hosseini et al. 2013), oil recovery (Monmont et al. 2012), and PMBs for the combustion of hydrogen (Su et al. 2014; 2016), ammonia (Nozari et al. 2017), natural gas-syngas blend (Arrieta et al. 2017), and low-concentration coal mine methane (Dai et al. 2015a, b, c, 2018).

17.4 Conclusion

A brief account of PMC and the development of PM burners for various applications are presented in this chapter. Substantial works were carried out since the advent of this technology, and in the recent past, there has been a growing interest in exploiting this technique for various practical applications. A comprehensive review of all those works was beyond the limit of this chapter. However, maximum effort was made to outline the various attempts for developing porous media burners for different applications. The works cited are only a few from the wide literature, and this information would hopefully be useful for deciding future directions of research in this area. The laboratory-level works on PMC are plenty, and hence the future works should focus on real applications.

NOMENCLATURE

Abbreviations

CV	calorific value
DME	dimethyl ether
FPMB	flexible porous medium burner
IC	internal combustion
LPG	liquefied petroleum gas
PM	porous medium
PMC	porous medium/media combustion
PRB	porous radiant burner
SCWR	supercritical water reforming
SFPRB	sideway-faced porous radiant burner
SPMB	self-aspirating porous medium burner
TE	thermoelectric
TPV	thermophotovoltaic
VOC	volatile organic compound/s

Greek Letters

Al_2O_3	alumina
Ce	cerium
CO	carbon monoxide
CO_2	carbon dioxide
C_p	specific heat at constant pressure
Cr	chromium
d_m	equivalent diameter of PM

Fe	iron
Gd	gadolinium
H_2	hydrogen
k	thermal conductivity
K	Kelvin
kg	kilogram
kW	kilowatts
La	lanthanum
MgO	magnesium oxide
MJ	megajoules
mm	millimeter
NO	nitrogen monoxide
NOx	oxides of nitrogen
P_e	Péclet number
ppcm	pores per centimeter
ppm	parts per million
Ru	ruthenium
SiC	silicon carbide
SiSiC	silicon-infiltrated silicon carbide
S_L	laminar flame speed
Sr	strontium
UHC	unburned hydrocarbon
V	volts
W	watts
WVO	wasted vegetable oil
ZrO_2	zirconium oxide
ρ	density
φ	equivalence ratio

REFERENCES

Abdul Mujeebu M., "Hydrogen and syngas production by superadiabatic combustion: A review," *Applied Energy*, vol. 173, pp. 210–224, 2016.

Abdul Mujeebu M., and I. Malico, "Porous media combustion technology," in *Energy Science and Technology Vol. 12: Energy Management*, J. N. Govil, U. C. Sharma, R. Prasad, and S. Sivakumar, Eds. Studium Press LLC, Boca Raton, FL, 2015, pp. 293–315.

Abdul Mujeebu M., M. Z. Abdullah, A. A. Mohamad, and M. Z. Abu Bakar, "Trends in modeling of porous media combustion," *Progress in Energy and Combustion Science*, vol. 36, pp. 627–650, 2010.

Abdul Mujeebu M., M. Z. Abdullah, M. Z. Abu Bakar, A. A. Mohamad, R. M. N. Muhad, and M. K. Abdullah, "Combustion in porous media and its applications: A comprehensive survey.," *Journal of Environmental Management*, vol. 90, pp. 2287–2312, 2009a.

Abdul Mujeebu M., M. Z. Abdullah, M. Z. Abu Bakar, A. A. Mohamad, R. M. N. Muhad, and M. K. Abdullah, "Corrigendum to 'Combustion in porous media and its applications: A comprehensive survey" [Journal of Environmental Management 90 (2009) 2287–2312]," *Journal of Environmental Management*, vol. 91, p. 550, 2009b.

Abdul Mujeebu M., M. Z. Abdullah, M. Z. Abu Bakar, A. A. Mohamad, and M. K. Abdullah, "A review of investigations on liquid fuel combustion in porous inert media," *Progress in Energy and Combustion Science*, vol. 35, no. 2, pp. 216–230, 2009c.

Abdul Mujeebu M., M. Z. Abdullah, M. Z. A. Bakar, A. A. Mohamad, and M. K. Abdullah, "Applications of porous media combustion technology: A review," *Applied Energy*, vol. 86, no. 9, pp. 1365–1375, 2009d.

Al-attab K. A., J. C. Ho, and Z. A. Zainal, "Experimental investigation of submerged flame in packed bed porous media burner fueled by low heating value producer gas," *Experimental Thermal and Fluid Science*, vol. 62, pp. 1–8, 2015.

Ali S. M. M., "Performance of domestic LPG cooking stove with porous media," *Open Access Library Journal*, vol. 1, no. e864, pp. 1–5, 2014.

Anger S., D. Trimis, B. Stelzner, Y. Makhynya, and S. Peil, "Development of a porous burner unit for glycerine utilization from biodiesel production by supercritical water reforming," *International Journal of Hydrogen Energy*, vol. 36, no. 13, pp. 7877–7883, 2011.

Arrieta C. E., A. M. García, and A. A. Amell, "Experimental study of the combustion of natural gas and high-hydrogen content syngases in a radiant porous media burner," *International Journal of Hydrogen Energy*, vol. 42, no. 17, pp. 12669–12680, 2017.

Babkin V. S., A. A. Korzhavin, and V. A. Bunev, "Propagation of premixed gaseous explosion flames in porous media," *Combustion and Flame*, vol. 87, no. 2, pp. 182–190, 1991.

Banerjee A., and A. V. Saveliev, "High temperature heat extraction from counterflow porous burner," *International Journal of Heat and Mass Transfer*, vol. 127, pp. 436–443, 2018.

Bani S., J. Pan, A. Tang, Q. Lu, and Y. Zhang, "Micro combustion in a porous media for thermophotovoltaic power generation," *Applied Thermal Engineering*, vol. 129, 2018.

Barajas P. E., R. N. Parthasarathy, and S. R. Gollahalli, "Combustion characteristics of biofuels in porous-media burners at an equivalence ratio of 0.8," *Journal of Energy Resources Technology*, vol. 134, no. 2, pp. 021004, 2012.

Baukal C. E. J., *Industrial Burners*. CRC Press LLC, Boca Raton, FL, 2003.

Britten J. A., and W. B. Krantz, "Linear stability of planar reverse combustion in porous media," *Combustion and Flame*, vol. 60, pp. 125–140, 1985.

Bubnovich V. I., N. Orlovskaya, L. A. Henríquez-Vargas, and F. E. Ibacache, "Experimental thermoelectric generation in a porous media burner," *International Journal of Chemical Engineering and Applications*, vol. 4, no. 5, pp. 301–304, 2013.

Bubnovich V., M. Toledo, L. Henríquez, C. Rosas, and J. Romero, "Flame stabilization between two beds of alumina balls in a porous burner," *Applied Thermal Engineering*, vol. 30, no. 2–3, pp. 92–95, 2010.

Buckmaster J., and T. Takeno, "Blow-off and flashback of an excess enthalpy flame," *Combustion Science and Technology*, vol. 25, no. 3–4, pp. 153–158, 1981.

Catapan R. C., A. a. M. Oliveira, and M. Costa, "Non-uniform velocity profile mechanism for flame stabilization in a porous radiant burner," *Experimental Thermal and Fluid Science*, vol. 35, no. 1, pp. 172–179, 2011.

Cheng Y.-C., and K.-L. Pan, "Simulation and design of a scramjet combustor enhanced by a porous cylinder burner," *2018 AIAA Aerospace Sciences Meeting*, January 8–12, Kissimmee, FL, 2018.

Chou S. K., W. M. Yang, J. Li, and Z. W. Li, "Porous media combustion for micro thermophotovoltaic system applications," *Applied Energy*, vol. 87, no. 9, pp. 2862–2867, 2010.

Chuenchit C., and S. Jugjai, "Effect of tube length on combustion characteristics of a self-aspirating Radiant Tube Burner (SRTB)," *Journal of Research and Applications in Mechanical Engineering*, vol. 1, no. 2, pp. 19–23, 2012.

Dai H., and B. Lin, "Scale effect of ceramic foam burner on the combustion characteristics of lowconcentration coal mine methane," *Energy & Fuels*, vol. 28, no. 10, pp. 6644–6654, 2014.

Dai H., B. Lin, K. Ji, and Y. Hong, "Two-dimensional experimental study of superadiabatic combustion in a packed bed burner," *Energy & Fuels*, vol. 29, p. 5311–5321, 2015a.

Dai H., B. Lin, K. Ji, C. Wang, Q. Li, Y. Zheng, and K. Wang, "Combustion characteristics of lowconcentration coal mine methane in ceramic foam burner with embedded alumina pellets," *Applied Thermal Engineering*, vol. 90, pp. 489–498, 2015b.

Dai H., B. Lin, C. Zhai, Y. Hong, and Q. Li, "Subadiabatic combustion of premixed gas in ceramic foam burner," *International Journal of Heat and Mass Transfer*, vol. 91, pp. 318–329, 2015c.

Dai H., Q. Zhao, B. Lin, S. He, X. Chen, Y. Zhang, Y. Niu, and S. Yin, "Premixed combustion of lowconcentration coal mine methane with water vapor addition in a two-section porous media burner," *Fuel*, vol. 213, no. May 2017, pp. 72–82, 2018.

de Araújo W. C., W. M. Barcellos, P. G. Ferreira, F. N. A. Freire, A. R. S. Camelo, and R. V. M. Araújo, "Biogas combustion on reciprocal flow porous burner with energy extraction," in *22nd International Congress of Mechanical Engineering (COBEM 2013)*, November 3–7, Ribeirão Preto, Brazil, 2013, no. Cobem, pp. 5509–5520.

De Soete G., "Stability and propagation of combustion waves in inert porous media," *Symposium (International) on Combustion*, vol. 11, no. 1, pp. 959–966, 1967.

Deshaies B., and G. Joulin, "Asymptotic study of an excess-enthalpy flame," *Combustion Science and Technology*, vol. 22, no. 5–6, pp. 281–285, 1980.

Devi S., and N. Sahoo, "Impact of geometric parameters of fuel-air distribution system on the temperature variation and emission of a sideway faced porous radiant burner (SFPRB)," in *ASME 2017 Gas Turbine India Conference*, Bangalore, India, December 7–8, 2017.

Dindi H., J. A. Britten, and W. B. Krantz, "Combustion and dielectric breakdown instabilities in porous media," *Earth Science Reviews*, vol. 29, no. 1–4, pp. 401–417, 1990.

Donoso-García P., and L. Henríquez-Vargas, "Numerical study of turbulent porous media combustion coupled with thermoelectric generation in a recuperative reactor," *Energy*, vol. 93, pp. 1189–1198, 2015.

Feng X. B., and Z. G. Qu, "Lean methane premixed combustion over a catalytically stabilized zirconia foam burner," *International Journal of Green Energy*, vol. 13, no. 14, pp. 1451–1459, 2016.

Feng X. B., Z. G. Qu, and H. B. Gao, "Premixed lean methane/air combustion in a catalytic porous foam burner supported with perovskite LaMn0.4Co0.6O3catalyst with different support materials and pore densities," *Fuel Processing Technology*, vol. 150, pp. 117–125, 2016.

Fernandez-pello A. C., "Micro power generation using microcombustion," *Proceedings of the Combustion Institute*, vol. 29, pp. 883–899, 2002.

Francisco J. W., M. Costa, R. C. Catapan, and A. A. M. Oliveira, "Combustion of hydrogen rich gaseous fuels with low calorific value in a porous burner placed in a confined heated environment," *Experimental Thermal and Fluid Science*, vol. 45, pp. 102–109, 2013.

Francisco Jr R. W., M. Costa, R. C. Catapan, and A. A. M. Oliveira, "Combustion of hydrogen rich gaseous fuels with low calorific value in a porous burner placed in a confined heated environment," *Experimental Thermal and Fluid Science*, vol. 45, pp. 102–109, 2012.

Francisco R. W., F. Rua, M. Costa, R. C. Catapan, and A. A. M. Oliveira, "On the combustion of hydrogen-rich gaseous fuels with low calorific value in a porous burner," *Energy and Fuels*, vol. 24, no. 2, pp. 880–887, 2010.

Fursenko R., A. Maznoy, E. Odintsov, A. Kirdyashkin, S. Minaev, and K. Sudarshan, "Temperature and radiative characteristics of cylindrical porous Ni–Al burners," *International Journal of Heat and Mass Transfer*, vol. 98, pp. 277–284, 2016.

Galukhin A. V, M. A. Khelkhal, A. V Eskin, and Y. N. Osin, "Catalytic combustion of heavy oil in the presence of manganese-based submicroparticles in a quartz porous medium," *Energy & Fuels*, vol. 31, pp. 11253–11257, 2017.

Gao H., Z. Qu, W. Tao, Y. He, and J. Zhou, "Experimental study of biogas combustion in a two-layer packed bed burner," *Energy and Fuels*, vol. 25, no. 7, pp. 2887–2895, 2011.

Hanamura K., R. Echigo, and S. A. Zhdanok, "Superadiabatic combustion in a porous medium," *International Journal of Heat and Mass Transfer*, vol. 36, no. 13, pp. 3201–3209, 1993.

Hanamura K., T. Kumano, and Y. Iida, "Electric power generation by super-adiabatic combustion in thermoelectric porous element," *Energy*, vol. 30, no. 2–4 SPEC. ISS, pp. 347–357, 2005.

Hardesty D. R., and F. J. Weinberg, "Burners producing large excess enthalpies," *Combustion Science and Technology*, vol. 8, no. 5–6, pp. 201–2014, 1973.

Henriquez-Vargas L., J. Loyola, D. Sanhueza, and P. Donoso, "Numerical study of reciprocal flow porous media burners coupled with thermoelectric generation," *Journal of Porous Media*, vol. 18, no. 3, pp. 257–267, 2015.

Henriquez-Vargas L., M. Maiza, and P. Donoso, "Numerical study of thermoelectric generation within a continuous flow porous media burner," *Journal of Porous Media*, vol. 16, no. 10, pp. 933–944, 2013.

Herrera B., K. Cacua, and L. Olmos-Villalba, "Combustion stability and thermal efficiency in a porous media burner for LPG cooking in the food industry using Al2O3 particles coming from grinding wastes," *Applied Thermal Engineering*, vol. 91, pp. 1127–1133, 2015.

Hoffmann J. G., R. Echigo, H. Yoshida, and S. Tada, "Experimental study on combustion in porous media with a reciprocating flow system," *Combustion and Flame*, vol. 111, no. 1–2, pp. 32–46, 1997.

Howell J. R., M. J. Hall, and J. L. Ellzey, "Combustion of hydrocarbon fuels within porous inert media," *Progress in Energy and Combustion Science*, vol. 22, pp. 121–145, 1996.

Hsu P. F., W. D. Evans, and J. R. Howell, "Experimental and numerical study of premixed combustion within nonhomogeneous porous ceramics," *Combustion Science and Technology*, vol. 90, no. 1–4, pp. 149–172, 1993.

Iral, L., and A. Amell, "Performance study of an induced air porous radiant burner for household applications at high altitude," *Applied Thermal Engineering*, vol. 83, pp. 31–39, 2015.

Ismail A. K., M. Z. Abdullah, M. Zubair, Z. A. Ahmad, A. R. Jamaludin, K. F. Mustafa, and M. N. Abdullah, "Application of porous medium burner with micro cogeneration system," *Energy*, vol. 50, no. 1, pp. 131–142, 2013.

Janvekar A. A., M. A. Miskam, A. Abas, Z. A. Ahmad, T. Juntakan, and M. Z. Abdullah, "Effects of the preheat layer thickness on surface/submerged flame during porous media combustion of micro burner," *Energy*, vol. 122, 2017.

Janvekar A. A., M. Z. Abdullah, Z. A. Ahmad, A. Abas, A. A. Hussien, P. S. Kataraki, M. Mohamed, A. Husin, and K. Fadzli, "Investigation of micro burner performance during porous media combustion for surface and submerged flames Investigation of micro burner performance during porous media combustion for surface and submerged flames," *Materials Science and Engineering*, vol. 370, p. 012049, 2018a.

Janvekar A. A., M. Z. Abdullah, Z. A. Ahmad, A. Abas, A. K. Ismail, A. A. Hussien, P. S. Kataraki, M. H. H. Ishak, M. Mazlan, and A. F. Zubair, "Experiential study on temperature and emission performance of micro burner during porous media combustion Experiential study on temperature and emission performance of micro burner during porous media combustion," *OP Conference Series: Materials Science and Engineering*, Vol. 370, No. 1, IOP Publishing, 2018b.

Ju Y., and K. Maruta, "Microscale combustion: Technology development and fundamental research," *Progress in Energy and Combustion Science*, vol. 37, no. 6, pp. 669–715, 2011.

Kakati S., P. Mahanta, and S. K. Kakoty, "Performance analysis of pressurized kerosene stove with porous medium inserts," *Journal of Scientific and Industrial Research*, vol. 66, no. 7, pp. 565–569, 2007.

Kamal M. M., and A. A. Mohamad, "Combustion in porous media," *Proceedings of the Institution of Mechanical Engineers Part A: Journal of Power and Energy*, vol. 220, pp. 487–508, 2006.

Keramiotis C., and M. a. Founti, "An experimental investigation of stability and operation of a biogas fueled porous burner," *Fuel*, vol. 103, pp. 278–284, 2013.

Khosravy-el-hosseini M., D. R. Heris, Q. Dorostihassankiadeh, and H. Biglarian, "Investigation of cylindrical porous burner behavior under various equivalence ratios," *Journal of Basic and Applied Scientific Research*, vol. 3, no. 10, pp. 59–67, 2013.

Kornilov, V. N., S. Shakariyants, and L. P. H. de Goey, "Novel burner concept for premixed surfacestabilized combustion," in *ASME Turbo Expo 2012 GT2012*, June 11–15, Copenhagen, Denmark, 2012, pp. 1–7.

Kotani Y., and T. Takeno, "An experimental study on stability and combustion characteristics of an excess enthalpy flame," *Symposium (International) on Combustion*, vol. 19, no. 1, pp. 1503–1509, 1982.

Kotani Y., H. F. Behbahani, and T. Takeno, "An excess enthalpy flame combustor for extended flow ranges," *Symposium (International) on Combustion*, vol. 20, no. 1, pp. 2025–2033, 1985.

Laphirattanakul P., A. Laphirattanakul, and J. Charoensuk, "Effect of self-entrainment and porous geometry on stability of premixed LPG porous burner," *Applied Thermal Engineering*, vol. 103, pp. 583–591, 2016.

Lapirattanakun A., and J. Charoensuk, "Developement of porous media burner operating on waste vegetable oil," *Applied Thermal Engineering*, vol. 110, pp. 190–201, 2017.

Li J., S. K. Chou, Z. W. Li, and W. M. Yang, "Experimental investigation of porous media combustion in a planar micro-combustor," *Fuel*, vol. 89, no. 3, pp. 708–715, 2010.

Li J., Y. Wang, J. Chen, J. Shi, and X. Liu, "Experimental study on standing wave regimes of premixed H2-air combustion in planar micro-combustors partially filled with porous medium," *Fuel*, vol. 167, pp. 98–105, 2016a.

Li J., Q. Li, J. Shi, X. Liu, and Z. Guo, "Numerical study on heat recirculation in a porous micro-combustor," *Combustion and Flame*, vol. 171, pp. 152–161, 2016b.

Li J., Q. Li, Y. Wang, Z. Guo, and X. Liu, "Fundamental flame characteristics of premixed H2-air combustion in a planar porous micro-combustor," *Chemical Engineering Journal*, vol. 283, pp. 1187–1196, 2016c.

Li Y., C. Yei-chin, N. S. Amade, and D. Dunn-rankin, "Progress in miniature liquid film combustors: Double chamber and central porous fuel inlet designs," *Experimental Thermal and Fluid Science*, vol. 32, pp. 1118–1131, 2008.

Makmool U., N. Pinta, P. Homhuan, J. Kittichaiyanan, A. Kaewpradap, and S. Jugjai, "Reliability improvement and optimization of condensing porous heat exchanger (CPHE) integrated with porous medium urner (PMB)," in *The 7th TSME International Conference on Mechanical Engineering*, December 13–16, Chiang Mai, Thailand, 2016.

Malico I., and M. A. Mujeebu, "Potential of porous media combustion technology for household applications," *International Journal of Advanced Thermofluid Research*, vol. 1, no. 1, pp. 50–69, 2015.

Malico I., X. Y. Zhou, and J. C. F. Pereira, "Two-dimensional numerical study of combustion and pollutants formation in porous burners," *Combustion Science and Technology*, vol. 152, no. 1, pp. 57–79, 2000.

Mishra N. K., and P. Muthukumar, "Development and testing of energy efficient and environment friendly porous radiant burner operating on liquefied petroleum gas," *Applied Thermal Engineering*, vol. 129, pp. 482–489, 2018.

Mishra N. K., S. C. Mishra, and P. Muthukumar, "Performance characterization of a medium-scale liquefied petroleum gas cooking stove with a two-layer porous radiant burner," *Applied Thermal Engineering*, vol. 89, pp. 44–50, 2015a.

Mishra V. K., S. C. Mishra, and D. N. Basu, "Combined mode conduction and radiation heat transfer in a porous medium and estimation of the optical properties of the porous matrix," *Numerical Heat Transfer, Part A: Applications*, vol. 67, no. 10, pp. 1119–1135, 2015b.

Monmont F. B. J., D. E. A. Van-Odyck, and N. Nikiforakis, "Experimental and theoretical study of the combustion of n-triacontane in porous media," *Fuel*, vol. 93, pp. 28–36, 2012.

Mueller K. T., O. Waters, V. Bubnovich, N. Orlovskaya, and R. H. Chen, "Super-adiabatic combustion in Al2O3 and SiC coated porous media for thermoelectric power conversion," *Energy*, vol. 56, pp. 108–116, 2013.

Mujeebu M. A., M. Z. Abdullah, and A. A. Mohamad, "Development of energy efficient porous medium burners on surface and submerged combustion modes," *Energy*, vol. 36, no. 8, pp. 5132–5139, 2011c.

Mujeebu M. A., M. Z. Abdullah, M. Z. A. Bakar, A. A. Mohamad, and M. K. Abdullah, "Applications of porous media combustion technology: A review," *Applied Energy*, vol. 86, no. 9, 1365–1375, 2009.

Mujeebu M. A., M. Z. Abdullah, M. Z. A. Bakar, and A. A. Mohamad, "A mesoscale premixed LPG burner with surface combustion in porous ceramic foam," *Energy Sources, Part A: Recovery, Utilization and Environmental Effects*, vol. 34, no. 1, 2011b.

Mujeebu M. A., M. Z. Abdullah, M. Z. Abu Bakar, and A. A. Mohamad, "Development of premixed burner based on stabilized combustion within discrete porous medium," *Journal of Porous Media*, vol. 14, no. 10, pp. 909–917, 2011a.

Mustafa K. F., S. Abdullah, M. Z. Abdullah, and K. Sopian, "A review of combustion-driven thermoelectric (TE) and thermophotovoltaic (TPV) power systems," *Renewable and Sustainable Energy Reviews*, vol. 71, pp. 572–584. 2017.

Mustafa K. F., S. Abdullah, M. Z. Abdullah, and K. Sopian, "Combustion characteristics of butane porous burner for thermoelectric power generation," *Journal of Combustion*, p. ID 121487, 2015a.

Mustafa K. F., S. Abdullah, M. Z. Abdullah, and K. Sopian, "Comparative assessment of a porous burner using vegetable cooking oil-kerosene fuel blends for thermoelectric and thermophotovoltaic power generation," *Fuel*, vol. 180, no. pp. 137–147, 2016.

Mustafa K. F., S. Abdullah, M. Z. Abdullah, and K. Sopian, "Experimental analysis of a porous burner operating on kerosene-vegetable cooking oil blends for thermophotovoltaic power generation," *Energy Conversion and Management*, vol. 96, pp. 544–560, 2015c.

Mustafa K. F., S. Abdullah, M. Z. Abdullah, K. Sopian, and A. K. Ismail, "Experimental investigation of the performance of a liquid fuel-fired porous burner operating on kerosene-vegetable cooking oil (VCO) blends for micro-cogeneration of thermoelectric power," *Renewable Energy*, vol. 74, pp. 505–516, 2015b.

Muthukumar P., P. Anand, and P. Sachdeva, "Performance analysis of porous radiant burners used in LPG cooking stove," *International Journal of Energy and Environment*, vol. 2, no. 2, pp. 367–374, 2011.

Muthukumar, P., and P. I. Shyamkumar, "Development of novel porous radiant burners for LPG cooking applications," *Fuel*, vol. 112, pp. 562–566, 2013.

Nozari H., O. Tuncer, and A. Karabeyoglu, "Evaluation of ammonia-hydrogen-air combustion in SiC porous medium based burner," *Energy Procedia*, vol. 142, pp. 674–679, 2017.

Pan J. F., D. Wu, Y. X. Liu, H. F. Zhang, A. K. Tang, and H. Xue, "Hydrogen/oxygen premixed combustion characteristics in micro porous media combustor," *Applied Energy*, vol. 160, pp. 802–807, 2015.

Panigrahy S., and S. C. Mishra, "The combustion characteristics and performance evaluation of DME (dimethyl ether) as an alternative fuel in a two-section porous burner for domestic cooking application," *Energy*, vol. 150, pp. 176–189, 2018.

Panigrahy S., N. K. Mishra, S. C. Mishra, and P. Muthukumar, "Numerical and experimental analyses of LPG (liquefied petroleum gas) combustion in a domestic cooking stove with a porous radiant burner," *Energy*, vol. 95, pp. 404–414, 2016.

Pantangi V. K., A. S. S. R. K. Kumar, S. C. Mishra, and N. Sahoo, "Performance analysis of domestic LPG cooking stoves with porous media," *International Energy Journal*, vol. 8, pp. 139–144, 2007.

Pantangi V. K., and S. C. Mishra, "Combustion of gaseous hydrocarbon fuels within porous media: A review," *Proceedings of 1st National Conference on Advances in Energy Research*, IIT Bombay, India, pp. 455–461, 2006.

Pantangi V. K., S. C. Mishra, P. Muthukumar, and R. Reddy, "Studies on porous radiant burners for LPG (liquefied petroleum gas) cooking applications," *Energy*, vol. 36, no. 10, pp. 6074–6080, 2011.

Pradhan P., P. C. Mishra, and B. B. Samantaray, "Performance and emission analysis of a novel porous radiant burner for domestic cooking application performance and emission analysis of a novel porous radiant burner for domestic," *Heat Transfer Engineering*, vol. 39, no. 9, pp. 784–793, 2018.

Pradhan P., P. C. Mishra, and K. B. Sahu, "Thermal performance and emission characteristics of newly developed porous radiant burner for cooking applications," *Journal of Porous Media*, vol. 19, no. 5, pp. 441–452, 2016.

Qiu K., and A. C. S. Hayden, "Thermophotovoltaic power generation systems using natural gas-fired radiant burners," *Solar Energy Materials and Solar Cells*, vol. 91, no. 7, pp. 588–596, 2007.

Qu Z. G., and X. B. Feng, "Catalytic combustion of premixed methane/air in a two-zone perovskitebased alumina pileup-pellets burner with different pellet diameters," *Fuel*, vol. 159, pp. 128–140, 2015.

Robayo M. D., B. Beaman, B. Hughes, B. Delose, N. Orlovskaya, and R.-H. Chen, "Perovskite catalysts enhanced combustion on porous media," *Energy*, vol. 76, pp. 477–486, 2014.

Rumminger M. D., R. W. Dibble, N. H. Heberle, and D. R. Crosley, "Gas temperature above a porous radiant burner: Comparison of measurements and model predictions," *Symposium (International) on Combustion*, vol. 26, no. 1, pp. 1755–1762, 1996.

Sathe S. B., M. R. Kulkarni, R. E. Peck, and T. W. Tong, "An experimental and theoretical study of porous radiant burner performance," *Symposium (International) on Combustion*, vol. 23, no. 1, pp. 1011–1018, 1991.

Sharma M., P. Mahanta, and S. C. Mishra, "Usability of porous burner in kerosene pressure stove: An experimental investigation aided by energy and exergy analyses," *Energy*, vol. 103, no. x, pp. 251–260, 2016a.

Sharma M., S. C. Mishra, and P. Mahanta, "Effect of burner configuration and operating parameters on the performance of kerosene pressure stove with submerged porous medium combustion," *Applied Thermal Engineering*, vol. 107, pp. 516–523, 2016b.

Sompu P., A. Kaewpradap, and S. Jugjai, "Performance and stability of flexible porous medium burner (FPMB)," in *The 6th TSME International Conference on Mechanical Engineering*, December 16–18, Cha-Am, Hua-Hin, Petchburi, Thailand, 2015.

Song F., Z. Wen, Z. Dong, E. Wang, and X. Liu, "Ultra-low calorific gas combustion in a graduallyvaried porous burner with annular heat recirculation," *Energy*, vol. 119, pp. 497–503, 2017.

Su S. S., S. J. Hwang, and W. H. Lai, "On a porous medium combustor for hydrogen flame stabilization and operation," *International Journal of Hydrogen Energy*, vol. 39, no. 36, pp. 21307–21316, 2014.

Su S. S., W. H. Lai, and S. J. Hwang, "Experimental study of the heat recovery rate in a porous medium combustor under different hydrogen combustion modes," *International Journal of Hydrogen Energy*, vol. 41, no. 33, pp. 15043–15055, 2016.

Takeno T., and K. Hase, "Effects of solid length and heat loss on an excess enthalpy flame," *Combustion Science and Technology*, vol. 31, no. 3–4, pp. 207–215, 1983.

Takeno T., and K. Sato, "An excess enthalpy flame theory," *Combustion Science and Technology*, vol. 20, no. 1–2, pp. 73–84, 1979.

Takeno T., K. Sato, and K. Hase, "A theoretical study on an excess enthalpy flame," *18th Symposium (International) on Combustion, The Combustion Institute*, vol. 18, no. 1, pp. 465–472, 1981.

Terracciano A. C., S. De Oliveira, D. Vazquez-Molina, F. J. Uribe-Romo, S. S. Vasu, and N. Orlovskaya, "Effect of catalytically active Ce0.8Gd0.2O1.9coating on the heterogeneous combustion of methane within MgO stabilized ZrO2porous ceramics," *Combustion and Flame*, vol. 180, pp. 32–39, 2017.

Toledo M., C. Rosales, and C. Silvestre, "Numerical simulation of the hybrid filtration combustion of biomass," *International Journal of Hydrogen Energy*, vol. 41, no. 46, pp. 21131–21139, 2016.

Trimis D., and F. Durst, "Combustion in a porous medium-advances and applications," *Combustion Science and Technology*, vol. 121, no. 1–6, pp. 153–168, 1996.

Wang Y., H. Zeng, Y. Shi, T. Cao, N. Cai, X. Ye, and S. Wang, "Power and heat co-generation by microtubular flame fuel cell on a porous media burner," *Energy*, vol. 109, pp. 117–123, 2016.

Wang Y., Y. Shi, T. Cao, H. Zeng, N. Cai, X. Ye, and S. Wang, "A flame fuel cell stack powered by a porous media combustor," *International Journal of Hydrogen Energy*, vol. 43, no. 50, pp. 22595–22603, 2017.

Weinberg F. J., "Combustion temperatures: The future?" *Nature*, vol. 233, no. 5317, pp. 239–241, 1971.

Wood S., and A. T. Harris, "Porous burners for lean-burn applications," *Progress in Energy and Combustion Science*, vol. 34, pp. 667–684, 2008.

Wu C.-Y., K.-H. Chen, and S. Y. Yang, "Experimental study of porous metal burners for domestic stove applications," *Energy Conversion and Management*, vol. 77, pp. 380–388, 2014.

Wu H., M. Kaviany, and O. C. Kwon, "Thermophotovoltaic power conversion using a superadiabatic radiant burner," *Applied Energy*, vol. 209, no. August 2017, pp. 392–399, 2018.

Xu K., M. Liu, and P. Zhao, "Stability of lean combustion in mini-scale porous media combustor with heat recuperation," *Chemical Engineering & Processing: Process Intensification*, vol. 50, no. 7, pp. 608–613, 2011.

Yoksenakul W., and S. Jugjai, "Design and development of a SPMB (self-aspirating, porous medium burner) with a submerged flame," *Energy*, vol. 36, no. 5, pp. 3092–3100, 2011.

Zeng H., Y. Wang, Y. Shi, and N. Cai, "Biogas-fueled flame fuel cell for micro-combined heat and power system," *Energy Conversion and Management*, vol. 148, pp. 701–707, 2017.

Zhdanok S. A., K. V. Dobrego, and S. I. Futko, "Flame localization inside axis-symmetric cylindrical and spherical porous media burners," *International Journal of Heat and Mass Transfer*, vol. 41, no. 22, pp. 3647–3655, 1998.

Zhou L., M. Xie, and K. Hong, "Numerical study of heat transfer and combustion in IC engine with a porous media piston region," *Applied Thermal Engineering*, vol. 65, no. 1–2, pp. 597–604, 2014.

Index

A

ablation, 265–266
ablative systems, 265–266
absorbent, 138
acoustic, 271, 273, 275, 283–284
active cooling, 265–266
Adaptive Mesh Refinement, 269
adiabatic, 265, 267, 270, 273, 279
aerospace, 287
agriculture, 87
air conditioning, 310, 314
air-cooled condensers, 317–318, 322
aluminum foam, 158, 164, 167–169
analytical approach, 235–238
anisotropic, 166, 169
anisotropically, 14
anisotropic pore network model, 90
annealing method, 291
anode, 342–343, 345, 354
aquifers, 138
area goodness factor, 322, 324, 333
Arrhenius, 23
asphalt, 130
automotive, 287

B

bi-directional, 233, 235
bidisperse, 184, 186
bi-disperse porous medium (BDPM), 137–148
bifurcation phenomenon, 14
blowing ratio, 267–268, 277–278, 282–283
blunt bodies, 263
boundary layer, 263–271, 273–281
brazing process, 314, 322
brine, 88–89
Bruggeman approximation, 348–349
buoyancy, 20, 22, 28
burner, 363–368

C

calorific value (CV), 367
capillary imbibition, 125–127, 132–133
capillary pressure, 125–127
capillary tube, 125–126
carbon fibers, 346
casting method, 289–290
catalyst, 368
catalyst layer (CL), 342, 345–347
catalytic PMB, 368

cathode, 342–343, 345, 354
central scheme, 269
Chapman–Enskog, 174, 176, 179, 189–190, 192
chemical reactions, 264
choking, 258
Colburn factor, 312–313
combustion, 361–364, 366, 368–369
    reciprocating flow combustion, 365
compact heat exchangers, 311–312, 333
composite materials, 265
compression work, 176, 178, 193
conduction, 349
conjugate heat transfer, 173–175, 181, 183–186
constant wall heat flux, 13–14
contact angle, 126–127
continuum approach, 3
convection, 349
cooling tower, 31–33
copper foams, 296–297, 299–304
corrugated, 318–319
Couette flow, 189
cryogenic systems, 104
crystalline rock, 204
CT scanning method, 291

D

Darcy regime, 8
Darcy's equation, 7
Darcy's law, 349, 355
Darcy velocity, 6, 8
deposition method, 289–290
diffusion approximation, 242–243, 248–249
diffusion-driven drying, 105–106
diffusion media, 345
directional-hemispherical, 234–235
direct numerical simulation (DNS), 269
discrete velocity, 176, 178–179
displacement thickness, 268–270, 276, 281
double-diffusive, 193
double distribution function (DDF) models, 176
drainage process, 89
drying, 87–89, 94
dry porous medium, 21–23
dual porosity model, 205
ducts partially filled, 37, 46–47
Dupuit–Forchheimer relationship, 6

E

effective thermal conductivity, 37, 40–44, 46–47
effusion cooling, 266

electrolyte membrane, 344, 349–350
electromagnetic, 233
electronic cooling, 55
electrons, 342–343, 345–346, 349–350
electron transport, 350–351
energy, 176–179, 189–193
energy performance ratio (EPR), 68
energy transport, 356
enhanced geothermal system (EGS), 203
entropy, 66, 69, 72
evaporation process, 87–89, 97, 104–107, 109, 116–117
excess enthalpy, 362, 365
exhaust gas recirculation (EGR) systems, 318
experimental
    approach, 234–235
    reconstruction methods, 291
    research progress, 296–304
extinction coefficient, 233, 235–237, 252–253
extrapolation, 182–183

F

Faraday's constant, 350
film cooling technique, 266–268, 270, 276, 278
fin, 312–319, 321, 323, 327
finite difference (FD), 174
finite volume (FV) method, 174, 244–246
fissure, 137, 139, 145
flame arrestors, 310
flame stabilization, 363
flat-plate solar collector, 300
flexible PMB, 368–369
flow instabilities, 256–259
fluid-to-solid heat transfer coefficient, 4
foamed wrapped tube banks, 321
foaming method, 289–290
foam ligaments, 327
foam-paraffin composite, 302
food, 287
forced convection, 37, 44–47, 138–139, 148
Forchheimer coefficient, 7
Forchheimer regime, 8
form drag coefficient, 7
fouling, 318–319, 328–331
Fourier's law, 349
f-phase, 137, 139, 143–145
fractal, 127
fractured oil reservoirs, 132
fractures network, 204
free convection, 20, 23–27, 29–32
friction factor, 312–314, 323
fuel cells, 321, 342
fully developed flow, 40, 44–45, 47

G

gas diffusion layer (GDL), 342, 345–347
gas phase transport, 354–355

gas-saturated rocks, 127
Gaussian field method, 291–292
geothermal energy, 203
geothermal reservoir, 205, 209–211, 213, 215
Gibbs free energy, 344
Graetz problem, 46
graphite foams, 318
Grashof number, 29
gravitational acceleration, 126
greenhouse gas (GHG), 341

H

Hagen–Poiseuille equation, 126
Hartmann number, 60–61
heat dissipation, 264
heat exchangers, 3, 10
heat extraction process, 205–211, 213–215, 220, 223
heat flux splitting, 14
heat pipes, 103, 108–109
heat sinks, 67–68, 72
heat storage, 266
heat transfer enhancement, 55, 73
heavy oil, 125, 128
heterogeneity, 204
homogeneous, 40, 43
honeycomb material, 288–289
hot spots, 256–259
hurst exponent, 90, 92–93
HVAC, 312–317
hydrodynamic model, 177, 180
hydrodynamics, 175–176, 187–189
hydrodynamic stability, 72
hydrogen fuel cell, 342
hydrophobic poly-tetra-fluoro-ethylene, 347
hydrostatics, 109
hypersonic, 263–266, 270–284

I

imbibition, 103, 106–107
impermeable wall, 14
incompressible flow, 73
inertia, 106, 112
inertial permeability, 256
infiltration rate, 125
inhomogeneities, 268
injection well, 210, 212–213, 220, 222–223
insulation, 289
interface tension, 125, 132–133
interfacial heat transfer, 40–41, 47
internal energy, 176–179, 183
internal heat transfer coefficient, 11
interpolation, 182–183
invasion percolation (IP), 89
isoflux, 40, 46
isolated pore, 154
isothermal, 40, 46
isotropic porous media, 93–94

Index

K

kinematic viscosity, 176, 179, 188–189
Kozeny constant, 268

L

large eddy simulation (LES), 269
lattice BGK (LBGK)
    energy, 176–179
    hydrodynamics, 175–176
lattice Boltzmann method (LBM), 174–194
lattice gas automata (LGA), 174
Lewis number, 60
ligament, 77
linear stability analysis, 144, 148
liquid phase transport, 355
liquid-vapor phase, 106
lithium ion batteries, 301
localized slots, 276–280
local thermal equilibrium (LTE), 11, 189–191
local thermal non-equilibrium (LTNE), 11, 39, 47–48,
    192–193, 230
louvered fin, 312–315, 317–318, 327

M

Mach number, 263–264, 266, 268, 270, 276
macroscopic, 4
magnetic field, 60–61, 66, 68
mass diffusion, 348–349
melting point, 290
membrane, 342, 344–345, 347–348, 350–351
metal foam heat exchanger, 309–333
metallic foams, 3
metallurgy, 287
microchannel, 66, 68
microcombustor, 364, 366–367
micro-encapsulated, 67
microorganisms, 72
microporous layers, 346–347
microscale applications, 366–367
microscopic, 3–4, 7
    velocity, 175, 180
micro-tomography, 158, 167–169
midspan, 279–280
migration, 350
mixed convection, 55, 69–73, 80
mixed forced convection, 142–143
mixture model, 74–77
moisture, 21, 24
molecular dynamics (MD), 175
monodisperse porous medium (MDPM), 138, 140, 143,
    145, 147–148
Monte Carlo (MC) method, 246–247
morphology, 310
multiphase flow, 88
multi-relaxation time LBM (MRT-LBM), 179–180
multi-speed (MS), 176

N

nanofluids in porous media
    forced convection, 61–69
    mixed convection, 69–73
    natural convection, 55–61
nanoparticles, 55, 60–61, 66–69, 72–73
natural convection
    BDPM, 144–147
    in partly porous box, 28–30
    in porous-saturated box, 24–27
    TDPM, 148–149
Navier–Stokes equation, 7–8
Newton's law of cooling, 205
noise attenuation, 310
non-Darcy flow, 205–206, 208, 211, 218–223
non-dimensional, 139, 141, 146
non-premixed, 363, 368–369
nonstoichiometric, 228
no-slip, 173–174, 181–182, 184
numerical instability, 176
Nusselt number, 12

O

Oberbeck–Boussinesq approximation, 20
Ohmic loss, 343
oil recovery, 125, 128, 132–133
oil viscosity, 128, 132–133
open-celled foams, 37–44, 47–48
open pore, 154
oxidation, 21

P

packed bed, 3, 8, 12
parabolic collector, 67
particle image velocimetry, 20
passive cooling, 265–266
passive scalar (PS) models, 176
Péclet number, 363
pellets, 138
periodic porous structure, 154, 160–161
permeability, 4, 6–7
pharmaceutical, 287
phase change materials, 301
pitch, 322–323
pixel, 164
plenum chamber, 267, 276–277
PM burner (PMB), 366–369
Poiseuille flow, 189–190
polarization, 343–344
polymer electrolyte, 345
polymer-electrolyte fuel cells (PEFCs), 342–351
pool boiling, 28
pore density, 312, 314, 317, 325, 331–332
pore-network simulation, 91–92
pore-scale approach, 3
pore-size distribution (PSD), 88–89

pore space, 88–89, 94, 97
porosity, 4, 6–7, 10, 14
porous burner, 363
porous-fluid interface, 10
porous material, 3, 288–289
  characteristics, 290–291
  preparation methods, 289–290
  reconstruction methods, 291–292
porous media, radiative properties, 233
  analytical approach, 235–238
  experimental approach, 234–235
  statistical approach, 238–242
porous media combustion (PMC), 361–369
porous medium, 3, 11, 13–14
  continuity equation in, 6
  heat transfer in, 11–15
  momentum equation in, 6–10
porous radiant burner (PRB), 366, 368
post-collision, 179–181
power plants, 319, 322
p-phase, 137, 139, 143–145
PPI, 164–165, 167–168
premixed, 363–364, 366–369
pressure drop, 312, 314–315, 317–318, 320–325, 327, 331–332
prism voxel, 164–165
probability, 238–241
production wells, 204, 206–207, 210–211, 213, 219–220
propellant management devices (PMD), 104–105, 109–111
protons, 342–343, 349–350, 356
pumping power, 68–69, 72
pyrolysis, 265–266

R

radiation, 228–234, 244, 248, 250–253
radiative conductivity, 243
radiative transfer equation (RTE), 14, 242–249
Rayleigh–Darcy number, 27
Rayleigh number, 26–27
reactants flow rates, 344–345
reciprocating flow combustion, 365
reflectivity, 233–234, 237, 255
refraction index, 233
refrigeration, 310, 314–318
relative humidity, 344–345
representative elementary volume, 4, 10
reservoir, 203–206, 208, 210–215, 220–223
reverse combustion, 365
reversible efficiency, 345
Reynolds number, 29
rock-fracturing, 203
Rosseland, 42–43
  approximation, 242

S

saturation distribution, 96–97
saturation temperature, 318
scanner, 164

scattering, 232, 235, 243
  albedo, 15, 232–234, 241, 243, 251
  coefficient, 232, 236, 238
Schiller and Naumann relations, 75
second order extrapolation, 183
sedimentary rocks, 204
seepage flow, 207, 209, 211, 213, 221
seepage velocity, 6
sequence slice grouping method, 291
sheet drying, 105–106, 111–113
shell-and-tube, 68–69
single phase model, 73–74, 207
sintering method, 289
soil, 87
solar thermal absorbers, 231
solid-fluid interface, 173–174, 181, 183, 186
solid matrix, 19
specific heat capacity, 209–210, 217–218
specific surface area, 4, 11
spherical harmonics, 243–244
spontaneous imbibition, 125–126, 128, 132–133
staged combustion, 364–365
stagnant thermal conductivity, 157–158
stagnation, 263–264, 266, 277
standard LBM, 175–183
statistical approach, 238–242
steady-state, 22–23, 25, 27
Stefan tube, 92
stiffness, 310
stochastic porous medium, 155
strain isolation, 310
stratified porous media, 88, 95
streamtube, 127
streamwise, 264, 267, 270–271, 274, 276, 278–279, 283–284
subsurface, 87
superconductors, 287
supercritical CO_2 ($SCCO_2$), 209
superficial velocity, 6
surface tension, 104–105, 110
syngas production, 365, 367–368

T

thermal contact resistance, 314, 320, 327–328
thermal control, 287
thermal diffusivity, 179, 190–191
thermal dispersion, 154, 157, 159, 166
thermal efficiency, 255
thermal energy storages, 310–311
thermal epoxy resin, 314
thermal expansion, 20
thermally developing flow, 45–46
thermal management, 287–290, 294, 301–304
thermal protection system (TPS), 265–266
thermal radiation, 41–43
thermal resistance, 25–27, 30
thermochemical, 227–228
thermodynamic efficiency, 345
thermodynamic equilibrium, 112
thermoelectric, 319–320

Index 381

thermophotovoltaic (TPV), 367
three-dimensional, 205
tomography, 154–155, 164, 166
topology, 183
tortuosity, 126–127
transient, 183, 192–194
transmissivity, 233–235, 237–238
transpiration cooling technique, 266–268,
    280–281, 283
tri-disperse porous media (TDPM), 137
   forced convection in, 148
   natural convection in, 148–149
two-dimensional (2D), 264, 275
two-equation thermal model, 205
two-phase mixture model, 74

U

ultra high temperature ceramics (UHTCs), 266
unit cell, 159–161, 170
unit layer, 159–161

V

vapor, 88–89, 91–93, 95, 97
viscous dissipation, 60, 66, 69, 73
viscous drag, 22, 28–30
viscous losses, 109, 111
void, 19–20, 25
volatile hydrocarbons, 87

volume-averaged method, 153–154
   governing equations, 156–157
   transport properties, 157–159, 166
volume fraction, 140–144, 148
volumetric heat transfer coefficient, 192
volumetric receiver, 227
volumetric solar absorbers, 252–259
volumetric solar receiver, 227–228, 230, 248–249, 252
vortex shedding, 39
vortical, 266, 278

W

wall-cooling effect, 273
waste energy recovery, 319–320
waste heat, 319–320
WENO-centered-difference (WENO-CD), 269, 274
wet porous medium, 23–24
wettability, 88
wetting, 125–127, 130–131
wicking process, 103–119

X

X-ray, 154, 158, 164–165, 167–169

Z

zeotropic, 332
zero gravity conditions, 110–111